9th Asia Conference on Mechanical and Aerospace Engineering (ACMAE 2018)

Journal of Physics: Conference Series
Volume 1215

Singapore
29 - 31 December 2018

ISBN: 978-1-5108-8786-2
ISSN: 1742-6588

Printed from e-media with permission by:

Curran Associates, Inc.
57 Morehouse Lane
Red Hook, NY 12571

proceedings.com

Some format issues inherent in the e-media version may also appear in this print version.

Institute of Physics
Dirac House, Temple Back
Bristol BS1 6BE UK

Phone: 44 1 17 929 7481
Fax: 44 1 17 920 0979

techtracking@iop.org

Additional copies of this publication are available from:

Curran Associates, Inc.
57 Morehouse Lane
Red Hook, NY 12571 USA
Phone: 845-758-0400
Fax: 845-758-2633
Email: curran@proceedings.com
Web: www.proceedings.com

Table of contents

Volume 1215

9th Asia Conference on Mechanical and Aerospace Engineering 29–31 December 2018, Singapore

Accepted papers received: 27 March 2019
Published online: 22 May 2019

Preface

Papers

Aircraft design and aerodynamic performance simulation

Aviation Control Engineering and Applications

Space communication and key technologies

Materials and Mechanical Engineering

Preface

It is our great pleasure to welcome you to 2018 The 9th Asia Conference on Mechanical and Aerospace Engineering (ACMAE 2018) which will be held in Singapore during December 29-31, 2018. ACMAE 2018 is dedicated to issues related to Mechanical and Aerospace Engineering topics.

Following the successful conference in Yokohama, Japan last year, the conference focus on Asia's research, achivements, markets, industy, etc. The technological innovation advances at an ever-dramatic pace, we believe that face-to-face communication is essential. With a solid reputation, ACMAE is where you will meet and network with engineering leaders, business innovators, researchers, academics and policy makers from across the globe. Don't miss the chance, hope to see you at ACMAE 2018.

The proceedings present a selection from papers submitted to the conference from universities, research institutes and industries. The papers selected depended on their quality and their relevancy to the conference. The volume tends to present to the readers the recent advances in the field of Aerospace Mechatronics and Avionics Systems, Mechanical Engineering in Aerospace and various related areas, such as Aerospace Communications, Aerospace Engineering and Management and Electronic Systems, etc..

We would like to thank all the authors who have contributed to this volume and also to reviewers, speakers, chairpersons, sponsors and all the conference participants for their support to ACMAE 2018.

ACMAE 2018 Organizing Committee
December 14th, 2018

IOP Conf. Series: Journal of Physics: Conf. Series **1215** (2019) 011002 doi:10.1088/1742-6596/1215/1/011002

Conference Committees

Conference Chairs

Prof. Ian McAndrew, Capitol Technology University, USA

Conference Advisory Chair

Prof. Wen Chih-Yung, The Hong Kong Polytechnic University, Hong Kong

Conference Program Co-Chairs

Prof. Kenji Uchiyama, Nihon University, Japan

Prof. Nam Seo Goo, Konkuk University, South Korea

Prof. Yoshifumi Yokoi, National Defense Academy of Japan, Japan

Steering Co-chairs

Prof. Simon Barrans, University of Huddersfield, UK

Prof. Ibrahim Ozkol, Istanbul Technical University, Turkey

Publication Chair

Prof. Renfu Li, Huazhong University of Science and Technology, China

International Publicity Co-chairs

Prof. Elena Vishnevskaya, Embry Riddle Aeronautical University, Germany

Prof. Peter Monka, TU Kosice, Faculty of Manufacturing Technologies, Slovakia

Conference Technical Committees

Prof. Weihua Zhang, National University of Defense Technology, China

Prof. Venkatraman Ramamoorthi, Sastra Univeristy, Thanjavur, India

Assoc. Prof. Irina V. Menshova, Russian Academy of Sciences, Russia

Prof. Ming-Liang Liao, Air Force Institute of Technology, Taiwan

IOP Conf. Series: Journal of Physics: Conf. Series **1215** (2019) 011002 doi:10.1088/1742-6596/1215/1/011002

Peer review statement

All papers published in this volume of *Journal of Physics: Conference Series* have been peer reviewed through processes administered by the proceedings Editors. Reviews were conducted by expert referees to the professional and scientific standards expected of a proceedings journal published by IOP Publishing.

Simulation of the Store Separation from Internal Weapons Bay Using M_SST DES Model

Junfeng Gong[1] *, Fang Wang[2]

[1] UAV Research Institute, Northwestern Polytechnical University, Xi'an, China
[2] Xi'an ASN Technology Group Co., Ltd., Xi'an, China

gong.junfeng@163.com

Abstract. The internal carriage of stores on an aircraft's merit is especially importance for attack aircraft. For the safe carriage and separation of stores, it is important to know the flow field environment the store experiences during flight. A CFD investigation was conducted to determine cavity flow characteristics. M-SST DES Method are used as numerical simulate method. Compared with the wind tunnel test results, the open cavity flows are studied. The numerical simulation results are in good agreement with the wind tunnel test results, thus verifying the credibleness of the M-SST DES Method. As a typical research model, AEDC's 'Wing-Pylon-Finned Store' wind tunnel test model is computed in this paper. Last, this paper investigates the store's separation characteristic, which is internal carriage on and release from aircraft. Those results are also indicated that the DES method of numerical simulation flow-filed can embody the unsteady aerodynamic random factors of the store separation from internal weapons bay.

1. Introduction

For the modern and high performance aircraft, stealth has become an important concern and basic precondition. And other request includes the maneuverability and cruise performance. Internal carriage of stores on an aircraft's merit is especially importance for high performance attack aircraft.

Weapon bay flow is a typical cavity flow aerodynamics that is complicated and oscillations. Oscillations are driven by the instabilities in the shear layer and vortex shedding, with a strong coupling between the shear layer and flow inside the cavity. Most Computational Fluid Dynamics (CFD) research conducted on cavity flows indicated that the turbulence is key role to play in driving the flow mechanisms. DES (Detached-Eddy Simulation) approach is theoretically a good way of taking advantage of the merits of both RANS (Reynolds Averaged Navier-Stokes) and LES (Large Eddy Simulation) approaches. Simulate results, computer capability and time cost are integrate balance, DES is selected for its unique characteristic [1-3].

The main objective of this paper is to simulate of the store separation from an aircraft's internal weapons bay using M_SST DES model. First open cavity flow simulation result are compare with wind tunnel test result. Then store separation simulation is validated. Last internal store separation from an aircraft simulation are research and analyzed.

2. M_SST DES Method

The notable modification to the k–w model came from Menter in 1994. For DES with the two-equation k–w model, the only modification is in the dissipation term

$$-\beta^{*}\rho\omega k \tag{1}$$

The M_SST turbulent length scale is defined by $l_{k-\omega}$

$$l_{k-\omega} = k^{1/2} / \beta^* \omega \tag{2}$$

In M_SST DES Method, l is given by

$$\tilde{l} = \min(l_{k-\omega}, C_{DES}\Delta) \tag{3}$$

In equation (3), $C_{DES}\Delta = 0.65$. Δ is the metric of the grid size, which essentially computes the size of the maximum cell length.

$$\Delta = \max(\Delta_x, \Delta_y, \Delta_z) \tag{4}$$

For unstructured grid, Δ is the length that local cell's center point to neighborhood cell's center point.

When space between the cell (i.e. $C_{DES}\Delta$) is great than the M_SST turbulent length scale ($l_{k-\omega}$), RANS is triggered. LES is activated when the converse occurs. This boundary between LES and RANS is therefore completely dependent on the geometry and on the density of the computational domain. For near the wall mesh, the ω is larger and k is a limited value. So $l_{k-\omega}$ is likely to be always smaller than the actual space between the cells near the wall so RANS will remain active there. When the cells are far to the wall the $l_{k-\omega}$ is rapidly increase, so $C_{DES}\Delta$ is used and LES is activated.

3. Test cases and result for cavity flow

Cavity flows can be described in many ways. Generally speaking, its can be divided by the free-stream Mach number. Now the typical way of categorising cavity flows is however directly based on the flow characteristics that occur in the cavity. Which flow is triggered in the cavity is strongly dependent on the cavity length-to-depth ratio (L/D) ratio. The categories include open cavity flow, closed cavity flow and transitional cavity flow[4,5].

The experimental data used to validate the numerical results in this paper were provided by Plentovich, Robert et al.[6]. A rectangular, three-dimensional cavity was mounted in a flat plate. The cavity length was 14.4 in. The cavity width was 9.6 in. the depth of the cavity was 2.4 in. The free-stream Mach number of the wind tunnel test was 0.6. The static-pressure was 144694Pa and the static-temperature was 340.15 K. The numerical model states are same to the states of the wind tunnel test.

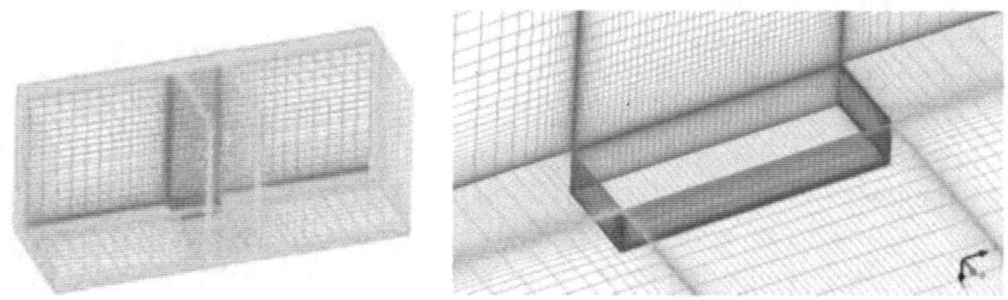

Figure 1. CFD grid of the open cavity model (L/D=6)

Fig.1 gives the grid of the numerical model. The mesh is y-symmetry and "H" type. The pressure-far-field boundary conditions is apply to flow field. In the bottom solid surface, the no-slip condition is applied.400000 nodes are gridded within the cavity. In general, the first node close to the wall has y+ between 30~300.Initially, flow-field of the cavity is computed by RANS method, and given the result of the Cavity flows. At 0 sec, unsteady flow-field is initialized by this steady flow-field result. unsteady flow-field is computed by M_SST DES Model method. Enough flow time passed, initial steady flow-field influence can be ignored.

IOP Conf. Series: Journal of Physics: Conf. Series **1215** (2019) 012001 doi:10.1088/1742-6596/1215/1/012001

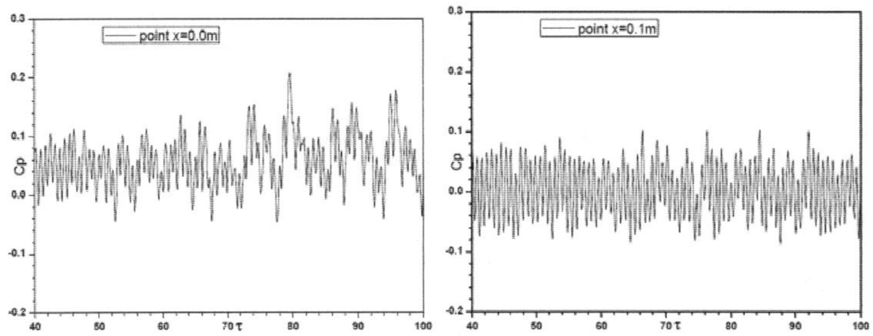

Figure 2. Pulsatile pressure of the measure points in symmetry line of the cavity's bottom

Fig.2 gives the pulsatile pressure of the four measure points in symmetry line of the cavity's bottom. The pulsatile pressures are looked like high frequency and random. It is foreseen that it is also related to the sampling frequency of the pressure value. However, on the whole, the pressure pulsation range is relatively stable, and the average value is stable.

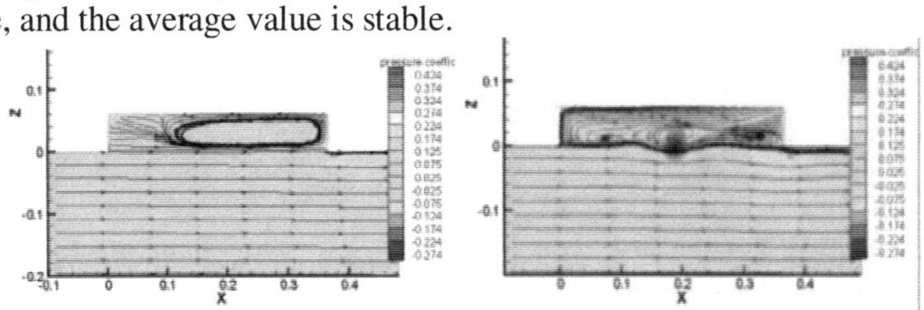

Figure 3. Symmetry plane calculation results for the open cavity(L/D=6)

Fig.3 shows the symmetry plane calculation results for the open cavity based on the RANS and DES methods, respectively. It can be seen that the calculation result based on the RANS method is stable. The results based on the DES method are variable and fluctuating, indicating the unsteady nature of the open cavity flow.

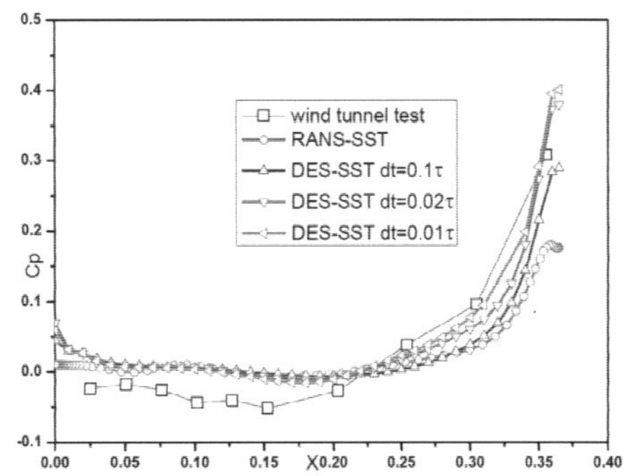

Figure 4. Symmetrical line pressure distribution in bottom of the open cavity (L/D=6)

Fig.4 shows the comparison of the CFD calculation results with the experimental results for open cavity bottom pressure. It can be seen that the calculation result based on the SST-based DES method is closer to the experimental value than the SST-based RANS method. Since the DES method is essentially an unsteady method, the amount of computation is much higher than that of the RANS method, so the time step of unsteady flow field calculation cannot be too small. Therefore, the time steps given in this paper are $\Delta t = 0.1\tau$, $\Delta t = 0.05\tau$, and $\Delta t = 0.01\tau$. Calculate 10 steps, 20 steps, and 100 steps respectively in the flow through the cavity.

The results show that as the calculated number of steps decreases, the calculation results are approximated to the RANS method. As the number of calculation steps increases, the calculation result approaches the experimental value. The related paper also confirmed that the time step has an effect on the simulation results of the DES method. Simone Crippa studied the Half-Span Delta Wing using the RANS method and the DES method. It was found that as the time step increases, the DES simulation result approaches the RANS method simulation result. In general, the reason for this phenomenon is that as the time step is reduced, the DES method can discriminate and capture smaller time scale vortices. Especially for the pressure pulsation, as the time step decreases, the pressure pulsation of the higher frequency flow field, ie noise, can be resolved. For practical engineering calculations, a balance needs to be struck between the accuracy of the calculations and the consumption. When the time step is small to a certain extent, the calculation accuracy of the flow field brought by the smaller time step is limited, and the computational resource consumption increases rapidly. Referring to the calculation result of Fig. 4, the time step of the cavity flow DES calculation in this paper is determined to be $\Delta t = 0.01\tau$.

4. Test Cases and Result for store separation

4.1 Technology of Coupling solve

Flow chart of Coupling solve is laid out as follows:

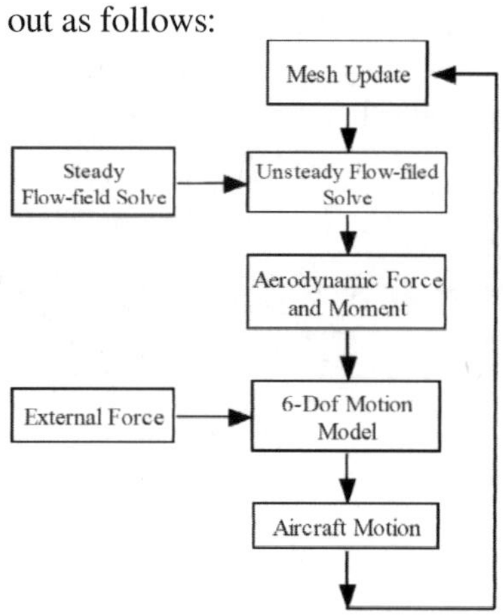

Figure 5. Flow chart of Coupling solve

Initially, the aircraft doesn't separate from the parent aircraft, steady flow-field is computed. At 0 sec, unsteady flow-field is initialized by the steady flow-field result.

At each time step, unsteady flow-field is integrated to give the aerodynamic forces and moments of the separating stores. External forces including gravity, ejector and other external forces, the aerodynamic forces and moments are passed to six-degree-of-freedom motion of separation stores model. Through this model, the new positions and attitudes are computed for the next time step. Relying on the information, the unstructured dynamic mesh model updates the grid of the unsteady flow-field. The next time step unsteady flow-field is solved. As this sequence, iterative time advancement is executed.

4.2 The Simulation Result Validated

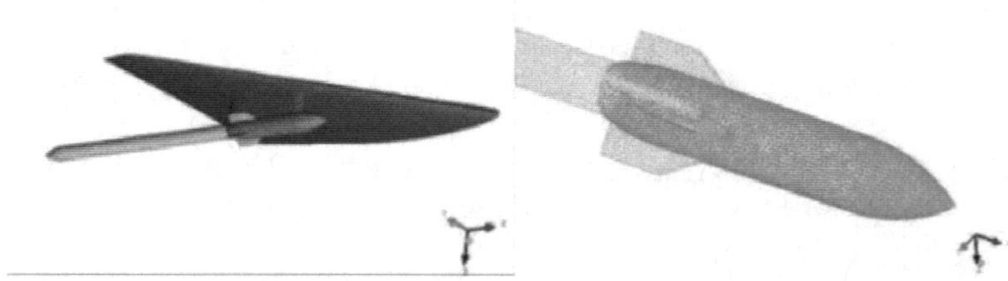

Figure 6. Geometry and surface grid of AEDC wind tunnel test model

AEDC [7] wind tunnel test model is showed in Fig.6. The detailed geometry definitions and experimental data for wind tunnel test can be found in Ref.(13). The state of test and CFD is Ma=1.2, H=12000 m. Angle of attack of the wing model is 0 deg. The quantity of the volume mesh is about 800000. Unsteady flow-field time step is taken as 0.0005 second.

IOP Conf. Series: Journal of Physics: Conf. Series **1215** (2019) 012001 doi:10.1088/1742-6596/1215/1/012001

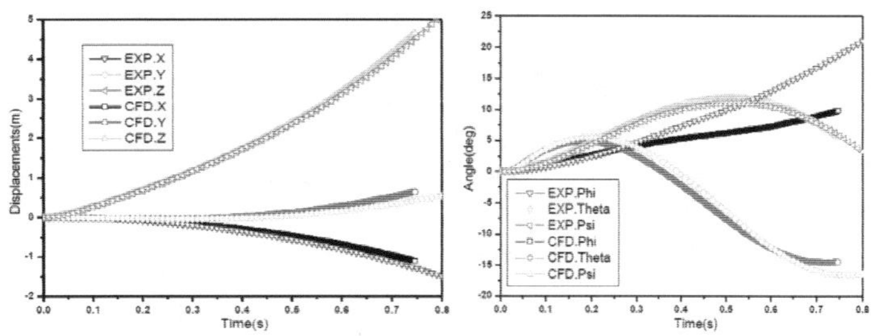

a) Linear Displacements b) Angular Displacements

Figure 7. Comparison displacement results of the experiment and CFD

Linear displacements graphs are shown in Fig.7a. Angular displacements graph are given in Fig.7b. CFD simulation linear and angular displacements are in very good agreement with experimental data in short time range. Along with time, the difference of CFD and experimental data became bigger. Especially, after t=0.3 second, the roll angular displacement error is greater than others. This difference mainly comes from that aerodynamic roll moment error is sensitive to roll inertia. Roll inertia is far smaller than pitch inertia and yaw inertia.

5. Results and Discussion

The cavity flow is complex and there is a constant unsteady oscillation. Therefore, when performing unsteady weapon separation simulations, whether it is the RANS calculation flow field or the DES calculation flow field, the separation of the flow field at the initial time is uncertain. This paper attempts to study the influence of the flow field oscillation with missile on the separation characteristics of the missile under different initial flow fields. As in the previous section, the RANS model was first used to simulate the flow field a cavity with weapons. After sufficient flow field time iterations, the effect of the initial flow field can be ignored.

The calculation of cavity flow does not use the URANS method, and any flow field after full iteration using the RANS method is used as the initial flow field of the unsteady DES simulation flow. After the DES method simulates the flow field for an unfixed period of time, three different moments of flow are taken as the initial moments for the missile to separate from the bay. Thereafter, the initial flow field at this moment was used to perform missile separation simulation. In this paper, the three initial moments of the simulation of the flow field by the DES method are 0.2s, 0.22s, and 0.24s, respectively.

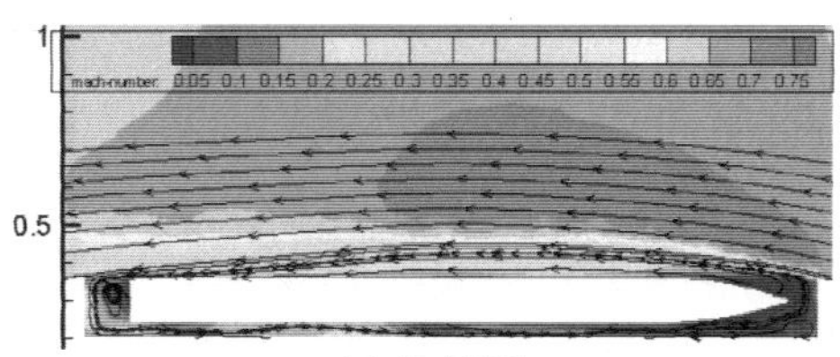

(a) RANS

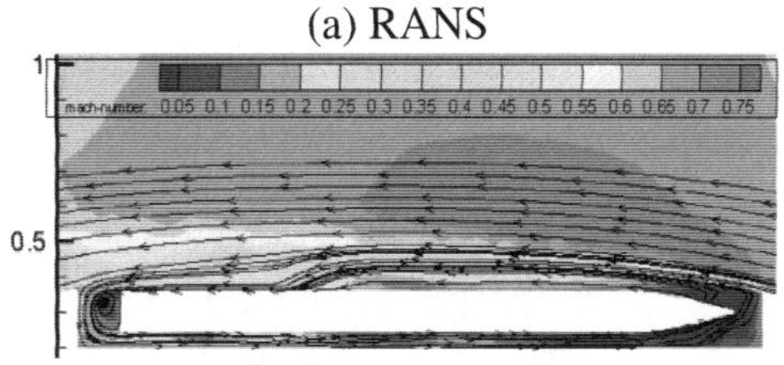

(b) DES-SST (t=0.20s)

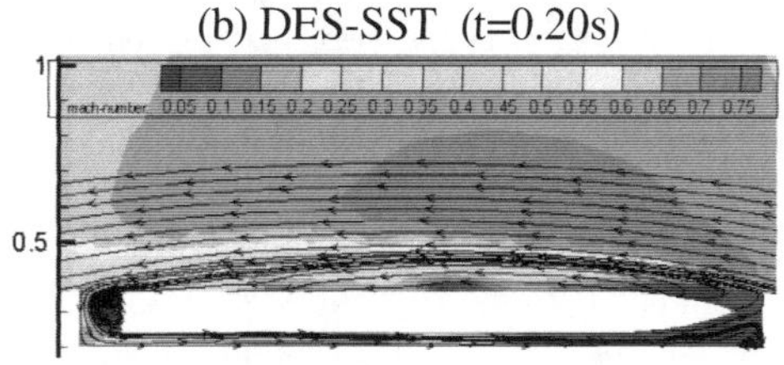

(c) DES-SST (t=0.22s)

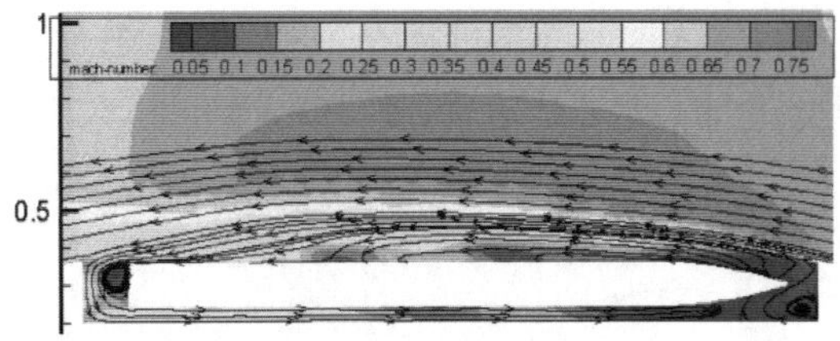

(d) DES-SST (t=0.24s)

Figure 8. Flow line of the symmetry plane of the weapon bay at the initial moment of separation

Fig.8 shows the flow line of the symmetry plane of the weapon bay at the initial moment of separation. The flow at several initial moments is different. However, on the whole, there is a very strong flow shear layer in the upper part of the weapon bay. The shear layer interferes with the upper part of the weapon bay, causing the shear layer to fluctuate continuously. Under the action of the shear layer, the cavity forms a circular flow. The air flow above the missile flows backwards and the flow below the missile flows forward. Due to the complex flow of the cavity, there is a very strong three-dimensional effect.

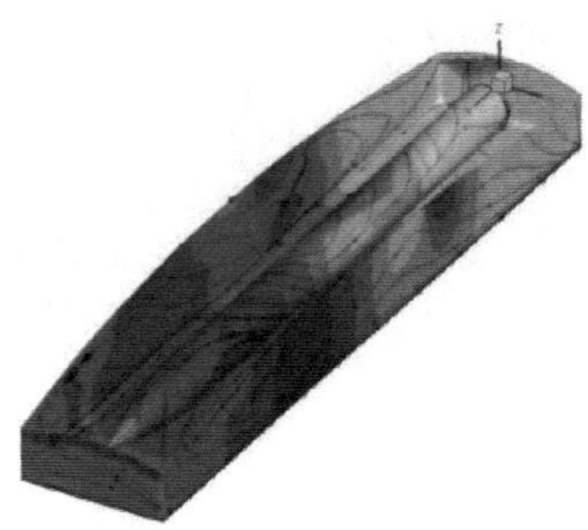

Figure 9. Flow field state of the cavity containing the missile

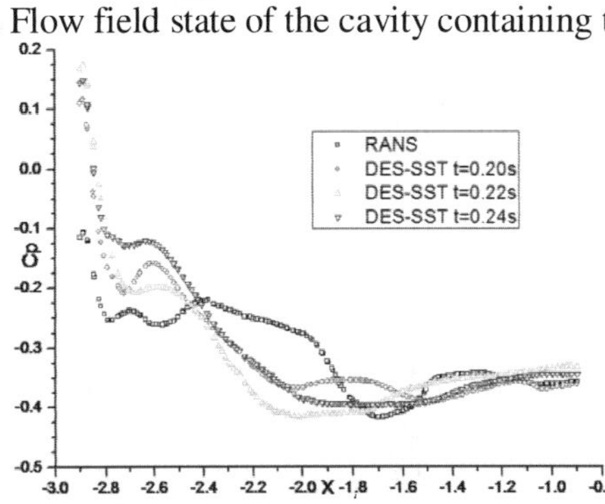

Figure 10. Pressure distributed on the symmetry line at the bottom of the cavity

Although the inflow and boundary conditions are completely symmetric, it can be seen from Fig.9 that there is no symmetry in the flow, and the left and right flows also seem to fluctuate randomly.

Fig.10 shows the pressure distribution on the symmetry line at the bottom of the cavity, showing that the surface pressure calculated by the DES-SST method is also pulsating. The following will study the changes in the field separation characteristics of a missile under the action of this pressure pulsation flow.

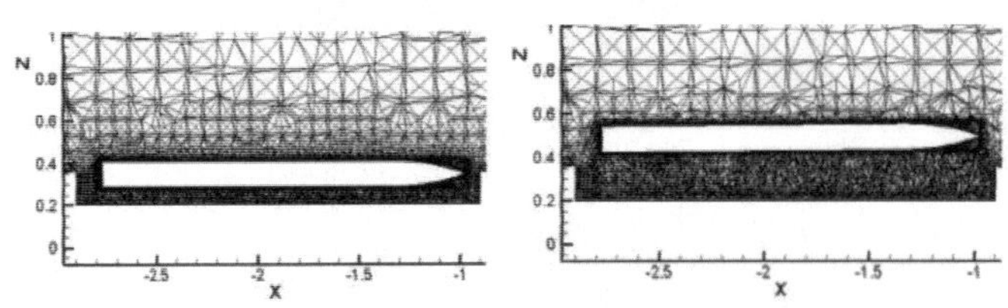

(a) t=0.10s (b) t=0.2s

Figure 11. Mesh of the symmetry plane of the flow field during the separation process

Fig.11 shows the meshing of the moving mesh during weapon separation. It can be seen from the figure that the viscous mesh of the boundary layer of the missile adopts the subdomain control method that follows the movement of the missile body and does not deform as a rigid body. There is no decline in the quality of this part of the grid. This is very insignificant for viscous flow calculations.

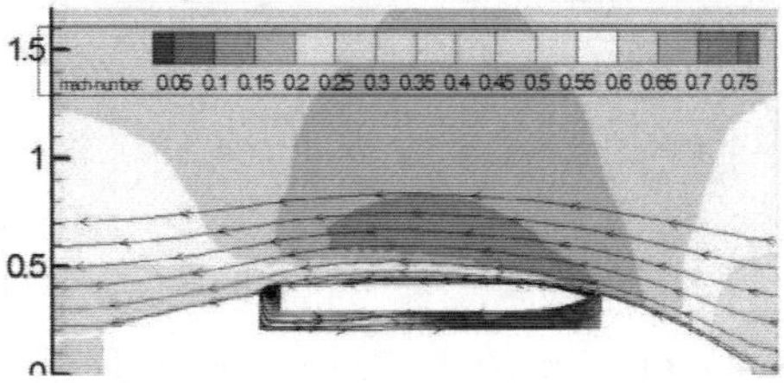

(a) t=0.10s

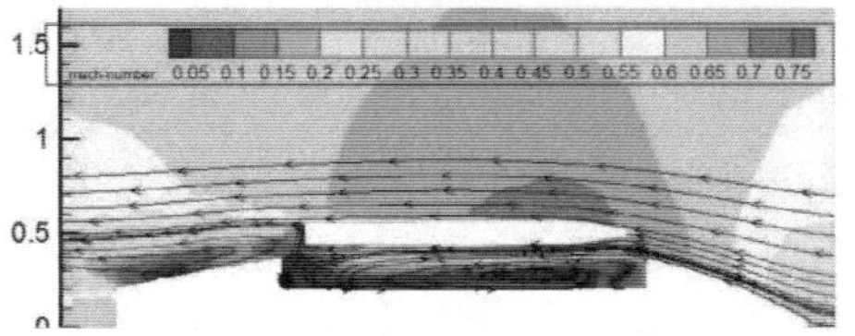

(b) t=0.20s

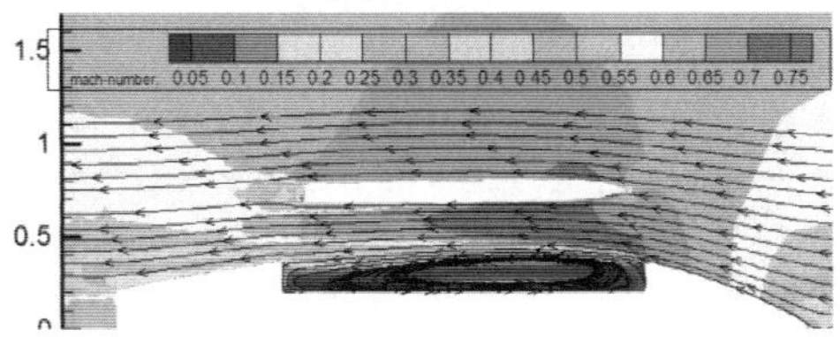

(c) t=0.30s

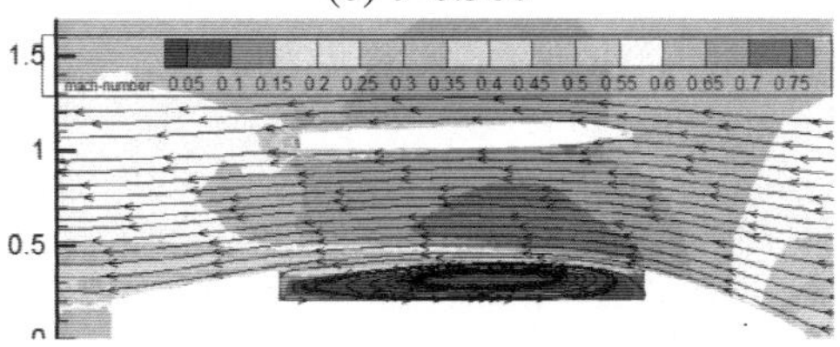

(d) t=0.40s

Figure 12. Symmetrical surface flow lines at different moments in missile separation (Example 2)

Fig.12 shows the symmetry plane streamline at different moments for missile separation in Example 2 (initial separation time T=0.22s). As the missile passes through the shear layer above the cavity, the complexity of the cavity flow greatly increases. At this point, the high-speed shear layer outside the cavity, under the blocking and guidance of the projectile, rushes into the lower part of the cavity, opposite to the local forward flow direction of the cavity. The impact of the two makes the flow of the cavity more complicated.

Since then, with the gradual upward movement of the projectile, the shear layer completely enters the lower part of the projectile, and relatively smoothly reaches the back of the cavity, and the flow of the cavity is gradually stabilized. Before this, the annular flow of the ballistic body was re-established and gradually increased, lengthened, and strengthened. The flow evolved into a relatively simple cavity flow.

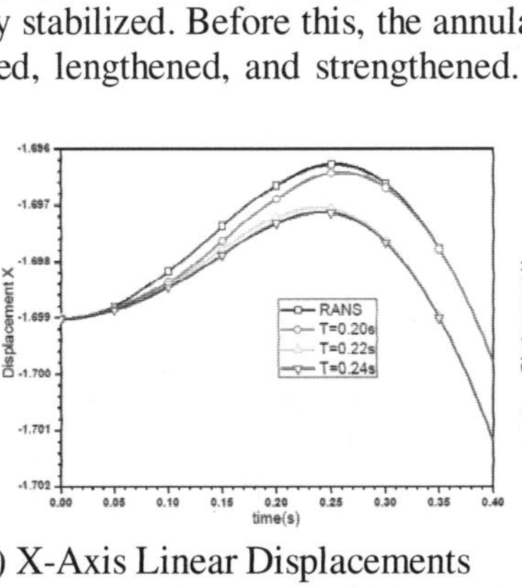

(a) X-Axis Linear Displacements

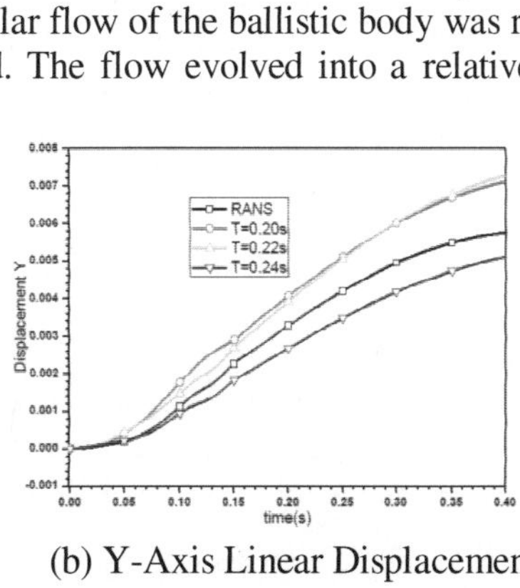

(b) Y-Axis Linear Displacements

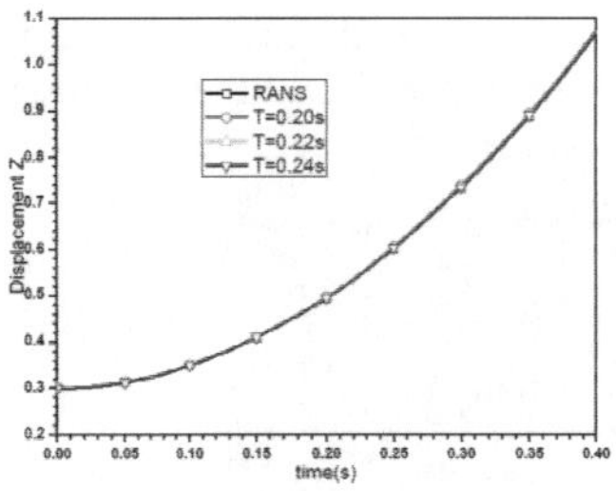

(c) Z-Axis Linear Displacements

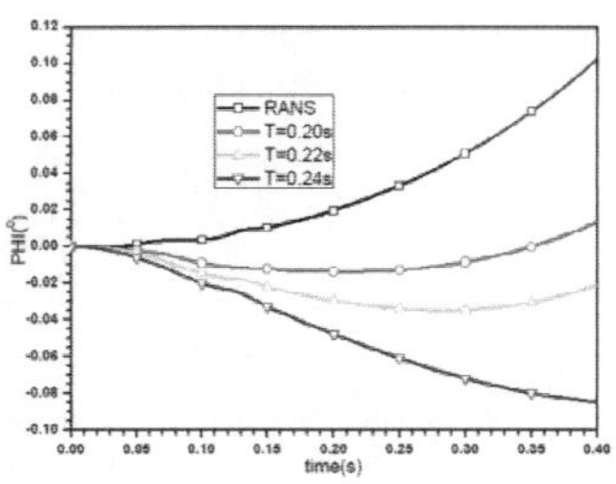

(d) Roll Angular Displacements

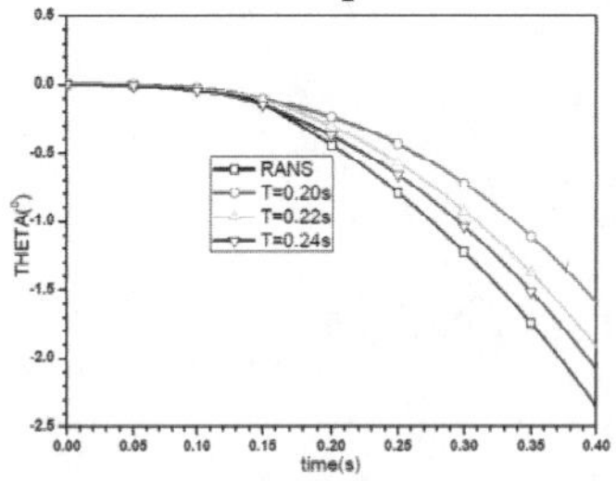

(e) Pitch Angular Displacements

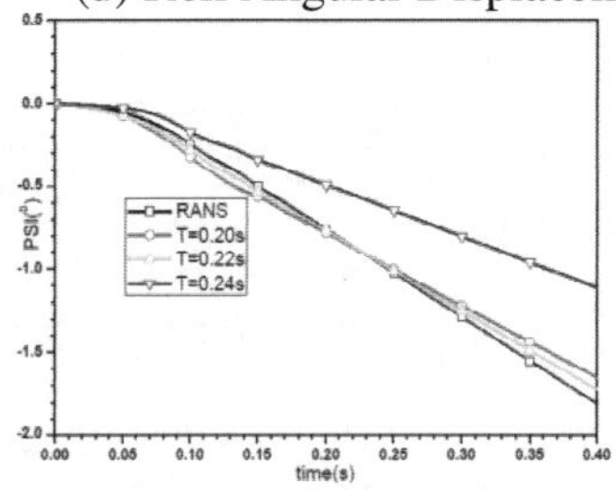

(f) Yaw Angular Displacements

Figure 13. Effect of different separation initial conditions on missile separation characteristics

Fig. 13 shows the weapon separation characteristics of the four different initial conditions in the upper section. It can be seen that the different separation initial states due to the complexity of the cavity flow affect the weapon separation characteristics of the buried weapon bay. First of all, calculating the symmetry of the model does not cause the missile's lateral motion to not appear. The calculation results are in accordance with actual experience. There is no complete symmetry in practice. From the above we can also see that even with the URANS method, this phenomenon can be foreseen.

Because of the complexity of the cavity flow, the DES method inevitably has the difference in the initial state of the flow field, but from the calculation results, it can be seen that the separation characteristics are similar. With the increase of the dynamic pressure of the flow field, when the aerodynamic interference force and the gravity of the weapon reach the same magnitude, the separation characteristics of the weapon will be more sensitive to the initial state. Using this method, the dispersion of the separation characteristics can be given through multiple calculations.

6. Conclusion

Due to the broad application prospects of buried weapons in aircraft, this paper studies the flow of buried weapon cavities. Due to the complexity of the cavity flow, this paper decided to use the DES method based on the SST model. Firstly, the principle and characteristics of the DES method are analyzed in detail and the DES method based on the M-SST model is given.

Then the DES method based on the SST model was used to simulate the cavity flow, the open cavity flow was calculated, and the flow characteristics of such cavity were analyzed. By comparing the numerical simulation of the cavity flow with the experimental results, a numerical simulation method for the flow of the bullet chamber in the DES method based on the SST model was established.

At the end of this paper, we attempted to establish a non-structural dynamic grid, based on the DES method of the SST model, and used this method to simulate the separation of weapons from the buried weapon bay.

Finally, a preliminary attempt was made to establish a DES method based on the SST model using an unstructured dynamic grid to simulate the separation characteristics of weapons from the buried weapon bay. This method was successfully used to simulate the process of weapon's separation from the aircraft's buried weapon bay.

This method was successfully used to simulate the process of weapon's separating from buried weapon bay of the aircraft. This paper considers the unsteady nature of the cavity flow itself, using the DES method to reflect the aerodynamic random factors that are not unique to the separation path of the actual weapon separation process.

7. References

[1] Werner Haase, Marianna Braza, Alistair Revell. DESider-A European Effort on Hybrid RANS-LES Modelling. Results of the European-Union Funded Project, 2004–2007

[2] Shia-Hui Peng, Werner Haase. Advances in Hybrid RANS-LES Modelling. Papers Contributed to the 2007 Symposium on Hybrid RANS-LES Methods, Corfu, Greece, 17–18 June 2007

[3] Shia-Hui Peng, Piotr Doerffer, Werner Haase. Progress in Hybrid RANS-LES Modelling. Papers Contributed to the 3rd Symposium on Hybrid RANS-LES Methods, Gdansk, Poland, June 2009

[4] Stallings, Robert L., Jr.; and Wilcox, Floyd J., Jr,; Experimental Cavity Pressure Distributions at Supersonic Speeds. NASA TP-2683,1987.

[5] Tracy,M.B.; Plentovich, E. b.; and Chu, Julio; Measurements of Fluctuating Pressure in a Rectangular Cavity in Transonic Flow at High Reynolds Numbers, NASA TM-4363, 1992

[6] E.B.Plentovich, Robert L. Stallings, M. B. Tracy. Experimental Cavity Pressure Measurements at Subsonic and Transonic Speeds. NASA-TP-3358, 1993

[7] Fox, John H., "23. Generic Wing, Pylon, and Moving Finned Store" RTO-TR-26, Verification and Validation Data for Computational Unsteady Aerodynamics, October 2000

Topology Optimization Design of Compliant Mechanism of Composite Wing Leading Edge

Lijie Zhao[1], Kai Li[1], Yingying Chang[1] and Jingkui Li[2]

[1]Laboratory of General Aviation, Shenyang Aerospace University, Liaoning Shenyang 110136, China
[2]Civil Aviation Institute, Shenyang Aerospace University, Shenyang 110136, China

Abstract. This paper introduces a topology optimization design method for flexible mechanism of composite wing leading edge based on discrete material optimization (DMO) with composite material .The DMO method is combined with the topology optimization to select the deviation of the actual displacement and the target displacement of the 10 coordinate points on the initial curve of the leading edge as the objective function, and the volume fraction of each phase material is used as the constraint to establish the composite material optimization model, which is solved by the OC optimization criterion method. The MATLAB codes are compiled to calculate the optimization problem and a distinct result image is obtained. The CATIA software is used to model the topology image in three dimensions, and the model is imported into Hyperworks software for simulation analysis. The results show that this complaint mechanism can achieve a maximum deflection of 9.1 degrees, which verifies the feasibility of the method.

1. Introduction

The research of compliant mechanism has become a hotspot of structural optimization design [1]. Compared to the traditional wings, compliant wings can automatically change the front and rear edge structures shape as the flight conditions change, thereby improving the flight performance [2]. The idea of the wing with the flexible mechanism was originally proposed by KOTA.S [3]. G. Matthew [4] used the basic structure method to optimize flexible mechanism of the wing trailing edge. Huang Jie [5] obtained a clear wing leading edge topology structural by using SIMP stiffness-density interpolation. Santer M et al [6] uses load path method to design wing leading edge structure. Ge Wenjie [7] used a combination of SIMP interpolation and genetic algorithm to obtain a stable topology result.

Compared with isotropic materials, fiber-reinforced composite materials are widely used in aerospace, automotive and many other industries due to their superior material properties [8]. The fiber angles of the composite material directly determines the properties of the material and the orientation optimization has become the research focus of scholars. Callahan and Weeks [9] first used genetic algorithms to optimize the design of constant stiffness laminates. Xiu Yingjun [10] combined the neural network with genetic algorithm to optimize orientation. J.Stegmann and E.Lund [11,12] proposed a composite optimization method based on multi-phase material topology optimiza tion principle—discrete material optimization (DMO).

2. Discrete Material Optimization (DMO) Principle

The basic idea in the Discrete Material Optimization (DMO) principle is essentially an extension of the ideas used in multiphase material structural topology optimization introduced by Sigmund [13]. This methodology can be stated as: for all elements in the structure find one distinct material from a set of pre-defined candidate materials such that the objective function is minimized. This was first introduced as multiphase topology optimization by Sigmund who used it for designing materials with extreme thermal expansion [12]. Common to the parametrization used in [13] is that materials are assumed to be isotropic, the maximum number of phases involved is three—two distinct materials and void. The DMO formulation extends the scope of application by considering multiple phases and furthermore by assuming to be orthotropic materials.

According to the theory of composite material discrete material optimization (DMO) method, carbon fiber composite materials with 30° and 45° common fiber angles are labelled as three-phase materials with void materials together, and the interpolation scheme can be expressed as[11,12]:

$$D_\beta^e = \sum_\alpha^m (x_{\alpha,\beta})^p D_{\alpha,\beta} \qquad 0 \le x_{\alpha,\beta} \le 1 \tag{1}$$

Where D_β^e denotes the elastic constitutive matrix of the β element, $D_{\alpha,\beta}$ denotes the elastic constitutive matrix of the α th candidate material of the β th element, and m is the number of species of the candidate material , $x_{\alpha,\beta}$ represents the artificial density of the candidate material of the β th elements, and p is the penalty index.

3. Objective function of wing leading edge

The deviation between the actual displacement of the 10 discrete points on the leading edge of the selected wing and the target displacement is the objective function, and the mathematical model is established by taking the volume fraction of each candidate material as the constraint. The expression is:

$$\begin{aligned}
&find: x = \{x_{\alpha,\beta}\} \quad (\alpha = 1,2,\cdots,\mathrm{m}; \beta = 1,2,\cdots,N) \\
&min: f(x) = \sum_i^n \omega_i \sqrt{\left(u_{x,i} - u'_{x,i}\right)^2 + \left(u_{y,i} - u'_{y,i}\right)^2} \\
&s.t. \begin{cases} KU = F \\ K\tilde{U} = L \\ \sum_{\beta=1}^N x_{\alpha,\beta} v_\alpha \le V_\alpha^* \\ 0 < x_{\alpha,\beta} \le 1 \end{cases}
\end{aligned} \tag{2}$$

The sensitivity of objective function is expressed as:

$$\begin{aligned}
\frac{\partial f(x)}{\partial x_{a,\beta}} = \sum_i^n \omega_i \Bigg\{ \left[\left(u_{x,i} - u'_{x,i}\right)^2 + \left(u_{y,i} - u'_{y,i}\right)^2 \right]^{-1/2} * \\
\left[\left(u_{x,i} - u'_{x,i}\right) \frac{\partial u_{x,i}}{\partial \rho_{a,\beta}} + \left(u_{y,i} - u'_{y,i}\right)^2 \frac{\partial u_{y,r}}{\partial \rho_{a,\beta}} \right] \Bigg\}
\end{aligned} \tag{3}$$

Where: n is the number of selected displacement output points, $u_{x,i}$, $u'_{x,i}$, $u_{y,i}$, $u'_{y,i}$ are the x and y direction of the actual displacement and target displacement of i -th output point, respectively; and ω_i is the weighting factor of i -th output point The value ranges from $0 \le \omega_i \le 1$, and the sum of the weight factors is 1, that is, $\sum_i^n \omega_i = 1$. x is the design variable vector, $x_{\alpha,\beta}$ is the density of the material phase α in element β, N is the total number of discrete elements, K is the total stiffness matrix, F is the external force vector, L is the adjoint load vector, U is the displacement vector, $\tilde{U}$ is the accompanying displacement vector, V_α^* is the target volume of material phase α . In any element of the design domain, the sum of the density and volume of the alternative phase material is 1, respectively.

According to the finite element theory, the stiffness matrix of the i -th unit is expressed as:

$$k_e = \int_{\Omega_e} B^T D B d\Omega_e \tag{4}$$

Where B is strain matrix, D is material property matrix. Based on (1) the stiffness matrix can be computed as:

$$k_e = \int_{\Omega_e} B^T \left(\sum_{\alpha}^{m} (x_{\alpha,\beta})^p D_{\alpha,\beta} \right) B d\Omega_e \tag{5}$$

The overall stiffness matrix of the structure can be expressed as:

$$K = \sum_{\beta}^{N} \int_{\Omega_e} B^T \left(\sum_{\alpha}^{m} (x_{\alpha,\beta})^p D_{\alpha,\beta} \right) B d\Omega_e \tag{6}$$

Finally, the displacement u_e can be written as:

$$u_e = \sum_{\beta}^{N} \tilde{u}_{\beta}^T k_e u_\beta \tag{7}$$

The sensitivity of displacement is obtained as:

$$\frac{\partial u_e}{\partial x_{a,\beta}} = -\tilde{u}_{\beta}^T \frac{\partial k_e}{\partial x_{a,\beta}} u_e \tag{8}$$

The objective function and sensitivity expression of the composite wing leading edge flexible mechanism model based on discrete material optimization (DMO) method can be obtained by combining (1) to (7).

The OC (Optimization Criteria) method is used to solve the topology optimization problem [14]. The expression is shown as :

$$x_{a,\beta}^{new} = \begin{cases} max(x_{min}, \rho_{a,\beta} - m_\beta) \\ if \;\; x_{a,\beta} B_\beta^\eta \le max(x_{min}, \rho_{a,\beta} - m_\beta) \\ x_a B_\beta^\eta \\ if \;\; max(x_{min}, x_{a,\beta} - m_\beta) < x_{a,\beta} B_\beta^\eta < min(1, x_{a,\beta} + m_\beta) \\ min(1, x_{a,\beta} + m_\beta) \\ if \;\; min(1, x_{a,\beta} + m_\beta) \le x_{a,e} B_\beta^\eta \end{cases} \tag{9}$$

s.t.

$$B_\beta = -\frac{\partial f(x) / \partial x_{a,\beta}}{\lambda \partial \left(\sum_{i=1}^{N} x_{a,\beta} v_\beta \right) \Big/ \partial x_{a,\beta}} \tag{10}$$

m_β is the moving limit constant,which is taken as 0.2, η is the damping factor which is 0.4. λ is the lagrangian operator.

In order to avoid the formation of checkerboard patterns, mesh-independency filter works described in [15] is introduced. The mesh-independency filter function is expressed as:

$$\overline{\frac{\partial f(x)}{\partial x_{a,\beta}}} = \frac{1}{(\gamma, x_{a,\beta}) \sum_{d \in N_e} H_d} \sum_{d \in N_e} H_d x_{a,d} \frac{\partial f(x)}{\partial x_{a,\beta}} \tag{11}$$

Where: N_e is the number of all units with the center distanceΔ (e , d) of the unit e being smaller than the filter radius r_{min} . γ takes 10^{-3}. H_d is the weight coefficient, and the expression is :

$$H_d = max(0, r_{min} - \Delta(e,d)) \tag{12}$$

The flow chart of topology optimization design is shown in Figure 1.

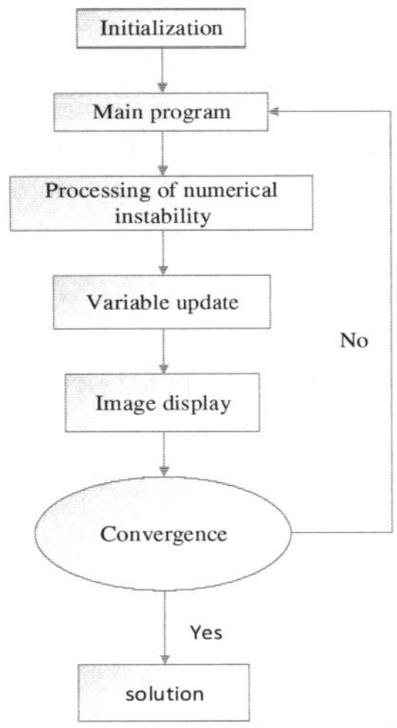

Figure 1. The process of topology optimization.

4. Optimization settings

4.1. Topology optimization design of flexible wing leading edge structure

The NACA2418 airfoil leading edge is chosen as the research object. The initial design domain is shown in Figure 2, the gray area is the design field and the white area is the non-design field. Coordinates, target displacement and weight factors of the output points are listed in table 1.

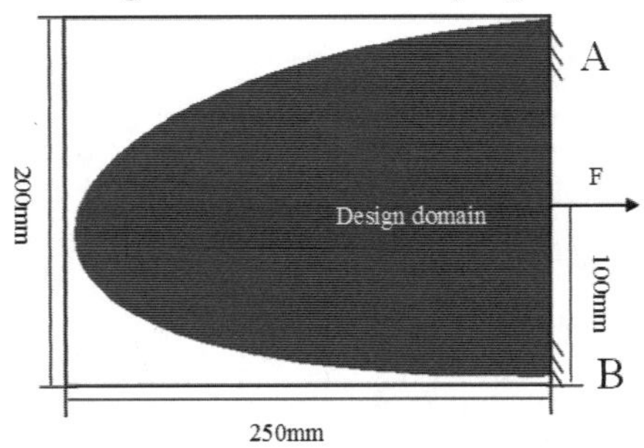

Figure 2. Initial design area.

Table 1. Output Point Coordinates, Target Displacement and Weight Factor.

Number	Coordinates		Target Displacement		Weight Factor
	X/mm	Y/mm	ΔX/mm	ΔY/mm	
1	220	16	-15.23	9.95	0.04
2	180	33	-11.36	23.38	0.08
3	120	55	-7.31	30.78	0.11
4	60	90	-0..53	38.17	0.15
5	10	140	9.82	43.69	0.2
6	15	180	15.80	38.17	0.16
7	75	207	20.38	29.70	0.12
8	150	211	21.45	21.41	0.09
9	210	219	18.07	5.74	0.05

The design domains are all meshed by 250×200 rectangular finite element. The design domain boundary is defined by the airfoil function. The candidate materials used are Carbon fiber reinforced composites with the orthotropic properties E_L =54GPa, E_T =8.6GPa, v_{LT} =0.3, G_{LT} =9GPa oriented at 30°, 45°. The volume fraction constraints is chosen of 0.16:0.14:0.7. F=150N, p=3 are employed.

4.2. Optimization results and simulation analysis

The optimization results of the wing leading edge structure with multiphase by MATLAB are shown in Figure 3 As shown in Figure 4, the black area composite fiber has a 30° angle and the blue area composite fiber has a 45° angle.

According to the structure diagram of Figure 4, the topology optimization structure is modeled by CATIA software, as shown in Figure. 4.

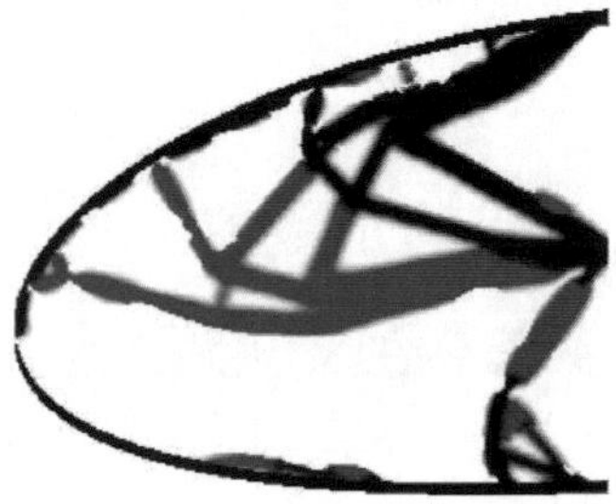

Figure 3. Results of optimization by MATLAB.

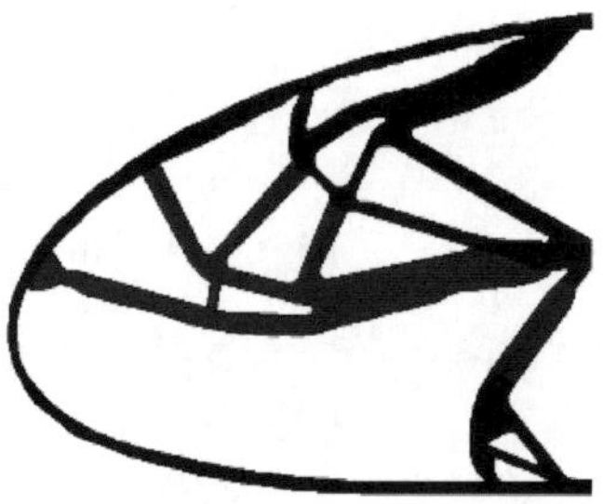

Figure 4. Geometric reconstruction by CATIA.

The 3D model is imported into hyperworks software for simulation analysis. The results are shown in Figure 5-6.

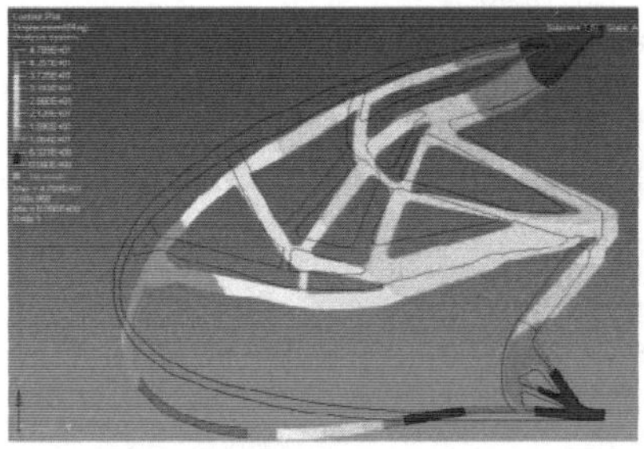

Figure 5. Displacement distributions.

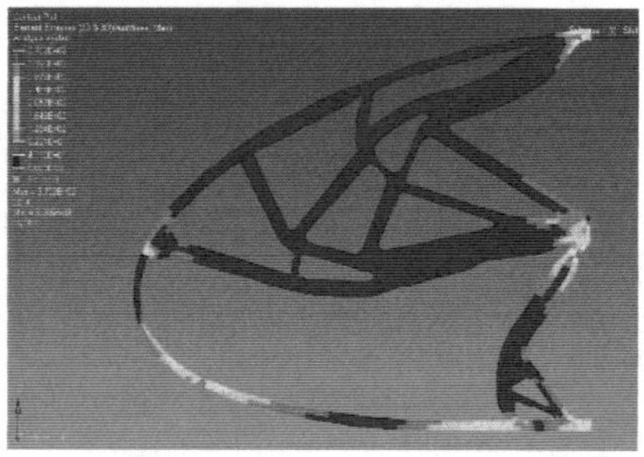

Figure 6. Stress distributions.

As is shown above, the simulation results shows that topology optimization structure of the leading edge can achieve continuous deformation with a maximum deflection angle of 9.1 degrees. The maximum stress of the unit is 78.45MPa, which is smaller than the yield strength of the carbon fiber composite, which meets the strength requirements.

5. Conclusion

This paper proposed a topology optimization method for compliant airfoil leading edge with multiphase composite by using discrete material optimization (DMO). The topology optimization model is obtained to minimize the least square error (LSE) between deformed curve and desired aerodynamics shape by using DMO material interpolation scheme. In the present paper DMO is introduced The DMO method is derived from multiphase material optimization and the element stiffness is computed from a weighted sum of candidate materials. The aim of optimization is to select the material that minimizes the target the most from the set of candidate materials. The candidate materials can be either isotropic or orthotropic with a given fiber angle. So far, the major drawback of DMO method is the large number of design variables, which increases the computational cost. In this paper, the OC method is used to solve the topology optimization problem because of the less quantity of finite element meshes and design variables. However, the optimization algorithm MMA is the best choice when solving topology optimization problems with a large number of design variables.

A distinct result of two angles composite materials is obtained by using OC methods and sensitivity filtering approaches to solve the topology optimization problem and ensure the existence of solutions. The simulation result obtained by Hyperworks shows that the leading wing compliant mechanism can achieve a maximum deflection of 9.1°, which verifies the feasibility of the method.

6. References

[1] YU Jing-jun, HAO Guang-bo, CHEN Gui-min. Research progress on flexible mechanism and its application[J]. Journal of Mechanical Engineering, 2015,51(13): 53-68.

[2] Ge Wen-jie. Research on the Development Status and Key Technologies of Adapting to Wings[J]. Aviation Weapons,2010,4(2):8-12.Luigi T.De Luca, *Propulsion physics* (EDP Sciences, Les Ulis, 2009)

[3] Kota S. Shape control of adaptive structures using com pliant mechanisms[R]. Stand From 298(Rev.2-89), Prescribed by ANSI std. Z39. 18, 2000: 1-7.

[4] Matthew G. Development of a variable camber compliant aircraft tail using structural optimization [D]. Blacksburg: Virginia Polytechnic Institute and State University, July 2003.

[5] HUANG Jie, GE Wen-jie, YANG Fang. Topology optimization of flexible mechanism for continuous change of shape of wing leading edge[J]. Acta Aeronautica et Astronautica Sinica, 2007, 28(4): 988-992.

[6] M. Santer, S. Pellegrino. Topological Optimization of compliant adaptive wing structure. AIAA Journal, vol.47, March 2009, pp.523-534.

[7] GE Wen-Jie, ZHU Peng-Gang, LIU Shi-Li, et al. Multi-objective Optimization of Wing Front Deformation Based on Flexible Mechanism[J]. Journal of Northwestern Polytechnical University, 2010, 28(2): 211-217.

[8] Li Xiao-ping, Zhang Xiao-ping, Wang Hong-wei. The development of carbon fiber and its application status[J]. High-Tech Fibers and Applications, 2005, 30(5):28-34.

[9] Callahan KJ,Weeks GE Optimum design of composite laminates using genetic algorithms[J]. Compos Eng,1992,2(3):149-160.

[10] Xiu Ying-ying, Cui De-gang. Composite layer laminate stability optimization design [J]. Engineering Mechanics, 2005, 22(6): 212-216.

[11] Stegmann J,Lund E. Discrete optimization of gen-eral composite shell structures. International Journal for Numerical Methods in Engineering 2005,62(14):2009-2027.

[12] Stegmann J,Lund E.Discrete Fiber Angle Optimization of General Shell Structures using a Multi-Phase Material Analogy.In:Proc.of the 5th world Congress on Structural and Multidisciplinary Optimization,WCSMO5,Venice,Italy,2003.

[13] Sigmund O, Torquato S. Design of materials with ex treme thermal expansion using a three-phase topology optimization method[J]. Smart Structures and Materials' 97. International Society for Optics and Photonics, 1997: 52-60.

[14] Sigmund O.A 99 line topology optimization code written in Matlab[J]. Structural & Multidisciplinary Optimization, 2001, 21(2): 120-127.

[15] E. Andreassen, A. Clausen, M. Schevenels, B. S. Lazarov, O.Sigmund. Efficient topology optimization in MATLAB using 88 lines of code. Structural & Multidisciplinary Optimization, vol. 43, January 2011, pp. 1-16.

IOP Conf. Series: Journal of Physics: Conf. Series **1215** (2019) 012003 doi:10.1088/1742-6596/1215/1/012003

Aircraft Surface Trajectory Prediction Method Based on LSTM with Attenuated Memory Window

Zhiming Ma, Mengfei Yao, Tao Hong and Bo Li *

School of Aeronautics and Astronautics, University of Electronic Science and Technology of China, 611731, Chengdu, China

libo@uestc.edu.cn

Abstract. With the rapid development of the global air transport industry, airport surface traffic is increasingly busy. The safety hazards of taxiing conflict still exist during the airport operation for the initially planned taxiing path about aircraft, which directly affects the operational safety and efficiency of the airport surface. Using the correlation and dependence between the position sequences of aircraft glide motion, a method based on Long Short Term Memory networks (LSTM) is used. Combining the change of the motion state of the surface aircraft, the attenuation memory window is introduced to improve the hidden layer structure to further enhance the prediction accuracy in the LSTM model, and it is compared and verified under different parameters. The method realizes the target of surface position prediction for the aircraft in future period, lays the foundation for aircraft surface taxiing conflict detecting, and avoids taxiing conflict.

1. Introduction

The calculation of the surface trajectory for a certain period of time becomes a key factor affecting the rationality and safety of path planning [1]. Common approaches to trajectory prediction are point mass models, probability models and machine learning models. In terms of the point mass models, reference [2] established the dynamic model of aircraft sliding by multi dimension stress analysis of the aircraft to realize the prediction of aircraft runway slide trajectory. Reference [3] introduced a nominal velocity profile on the basis of reference [2], combined with the BADA data, and established the aircraft surface taxiing trajectory prediction model. In terms of the probability models, using Monte Carlo method, track prediction is realized indirectly by estimating wind speed and airspeed [4]. Reference [5] achieved trajectory prediction by using extended Kalman Filter method to estimate UAV motion status. In terms of the machine learning models, neural networks was applied to predict an aircraft trajectory in the vertical plane, which is trained by a set of real historical trajectory [6]. reference [7] train model to realize trajectory prediction by historic data and GLM (Generalized Linear Models) with a stepwise regression.

Because of the strong correlation and dependence of the position sequence of the aircraft's gliding motion, the time series method can be used to analyze and interpret the implicit relationship between the sequence values. The circulatory neural network is an important method in the field of artificial intelligence in dealing with time series problems [8]. Because of this, based on historical data, this paper uses a method of recurrent neural networks, combined with changes in the movement state of aircraft scenes, introduces attenuation memory mechanisms to improve the hidden layer structure in LSTM(Long Short Term Memory networks), and further improve the accuracy of trajectory prediction models.

2. Trajectory data preprocessing

According to the aircraft historical data set provided by Chongqing Jiangbei Airport, select historical trajectory data on a straight taxiway. Use the following formula:

$$T_k = k \cdot T, \quad k \in N^* \tag{1}$$

Where k is the sampling factor and is a positive integer. T is the period of the scene tracking radar acquisition trajectory data. The equidistant sampling obtains n track point sequences expressed as:

$$\left[(x_1, y_1, v_1), (x_2, y_2, v_2), \cdots, (x_n, y_n, v_n)\right] \tag{2}$$

(x_t, y_t, v_t) respectively indicate the latitude, longitude and speed of the track point at time t.

Differences between t and t+1 entries in the trajectory sequence, ie, first-order differential processing:

$$x_t^{'} = x_{t+1} - x_t, \; y_t^{'} = y_{t+1} - y_t, v_t^{'} = v_{t+1} - v_t \tag{3}$$

Among them, x_t 、 y_t and v_t respectively correspond to the basic items of the difference processing. Get a new n-1 track point sequence represented by:

$$\left[(x_1^{'}, y_1^{'}, v_1^{'}), (x_2^{'}, y_2^{'}, v_2^{'}), \cdots, (x_{n-1}^{'}, y_{n-1}^{'}, v_{n-1}^{'})\right] \tag{4}$$

Use the sklearn tool in the machine learning library to scale the first-order difference track point sequence to the range [-1,1] with the MinMaxScaler function and save the output parameter scaler of the function, then use the normalized data with the machine The pandas tool in the learning library is converted to a supervised learning sequence. The operation flow is shown in Figure 1.

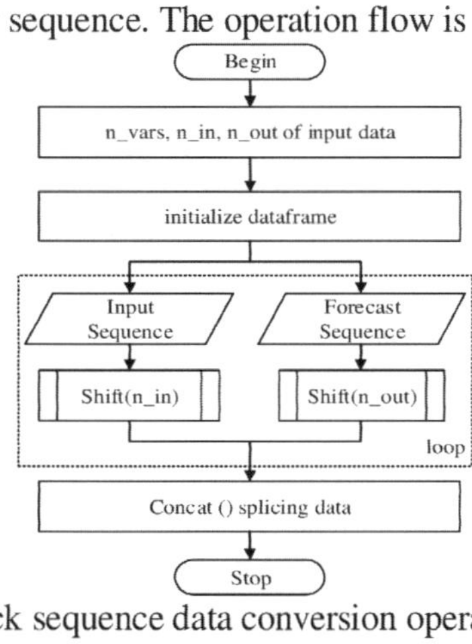

Figure 1. Track sequence data conversion operation flow chart.

n_vars indicates the number of input data variables, n_in indicates the number of input sequences, and n_out indicates the number of predicted sequences. It is constructed using the DataFrame() function. The data object, Shift() function, respectively, constructs the input sequence data and the predictive sequence tag data, and concatenates them with the Concat() function to form the final supervised learning track point sequence data. The processed data sample format is as shown in Table 1.

Table 1. Supervised learning sequence data format.

x_{t-3}	y_{t-3}	v_{t-3}	x_{t-2}	y_{t-2}	v_{t-2}	x_{t-1}	y_{t-1}	v_{t-1}	x_t	y_t	v_t	x_{t+1}	y_{t+1}	v_{t+1}
x_1	y_1	v_1	x_2	y_2	v_2	x_3	y_3	v_3	x_4	y_4	v_4	x_5	y_5	v_5
x_2	y_2	v_2	x_3	y_3	v_3	x_4	y_4	v_4	x_5	y_5	v_5	x_6	y_6	v_6
x_3	y_3	v_3	x_4	y_4	v_4	x_5	y_5	v_5	x_6	y_6	v_6	x_7	y_7	v_7
x_4	y_4	v_4	x_5	y_5	v_5	x_6	y_6	v_6	x_7	y_7	v_7	x_8	y_8	v_8

3. Construction of trajectory prediction model based on LSTM with attenuated memory window

According to the characteristics of the time series problem, the trajectory prediction at the current moment is related to the scene aircraft movement situation at the past moment. Analyze the scene movement state of the aircraft over a long period of time, according to the change of speed, can roughly be reduced to three types of movement: acceleration, deceleration and uniform movement, as shown in Figure 2. During the time period $t_1 \sim t_2$, the aircraft did speed up; during the time period $t_2 \sim t_3$, the aircraft did a uniform movement; during the time period $t_3 \sim t_4$, the aircraft did a deceleration movement. According to the memory property of the LSTM unit, when the position prediction is performed within the time period $t_2 \sim t_3$, the data stream generated during the time period $t_1 \sim t_2$ will cause an illusion of acceleration motion state disturbance at the current moment of the movement, resulting in an increase in the prediction error.

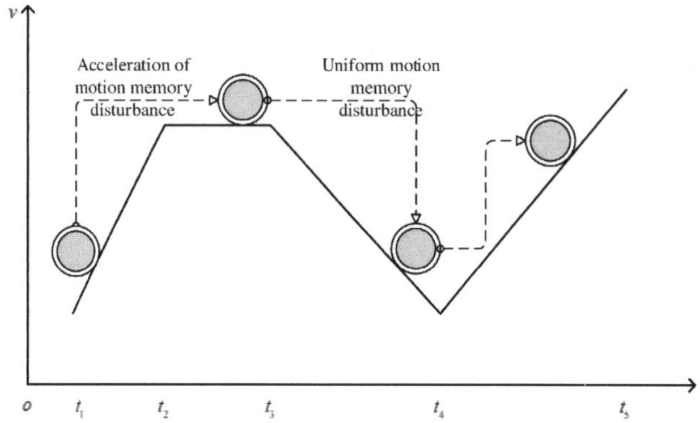

Figure 2. Scene movement status changes.

Aiming at the problem of historical memory perturbation of the current state of movement, a mechanism is needed to downplay the correlation between different state-of-motion data over time. To this end, attenuation memory mechanisms are introduced to weaken the influence of this long-period subtle influence, which helps build more accurate trajectory prediction models. Therefore, based on the construction of the LSTM-based trajectory prediction model, combined with the fluidity of the data in the model training process, the data flow in the hidden layer is weakened by the exponential decay of the memory window sliding with the training set to the current moment. The effect of movement status. Specifically, the attenuation coefficient is introduced into the hidden layer, which can be adjusted according to the characteristics of the change in the trajectory data. The value in this paper is 0.9, which is applied to the cell state element. The LSTM framework with improved hidden layer is established as shown in the Figure 3.

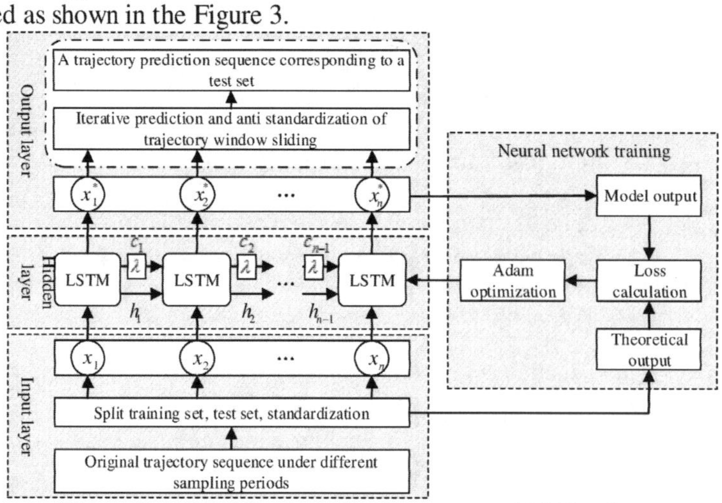

Figure 3. LSTM framework with improved hidden layer.

The input layer data of the predictive model framework is derived from the original trajectory sequence under different sampling periods, and is processed by first-order difference, maximum value standardization, and supervised learning sequence conversion. The dimension of the input data is M*3, including latitude, longitude, and speed. In the hidden layer, the flow of information delivered to the next moment by each LSTM unit includes two states, cell state and hidden state. In the output layer, the tag data is the position sequence and velocity normalized at the next time. Network training, and reverse propagation update weights until the loss value converges.
The expression of mathematical relations is as follows:

$$c_t = \lambda^W \prod_{i=1}^{W} c_{t-i} \tag{5}$$

Where t represents the most recent data moment within the sliding window and W represents the sliding window size.

4. Model validation and result analysis

4.1. Comparison of prediction results under different activation function

Figure 4, Figure 5, and Figure 6 show the prediction models constructed using LSTM for the three activation functions, respectively, and perform better trajectory predictions for 10 trainings.

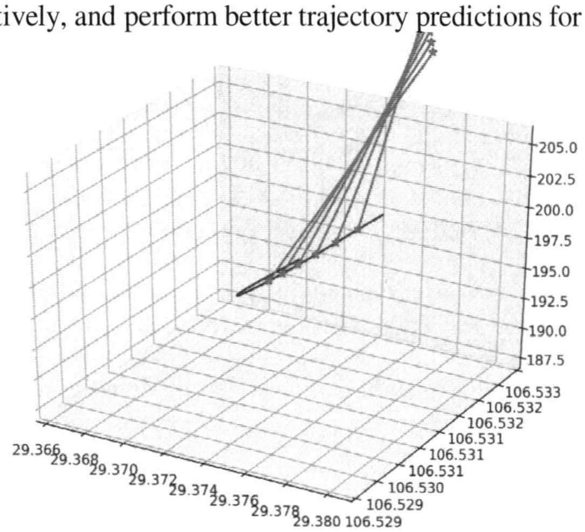

Figure 4. trajectory prediction effect with sigmoid activation function.

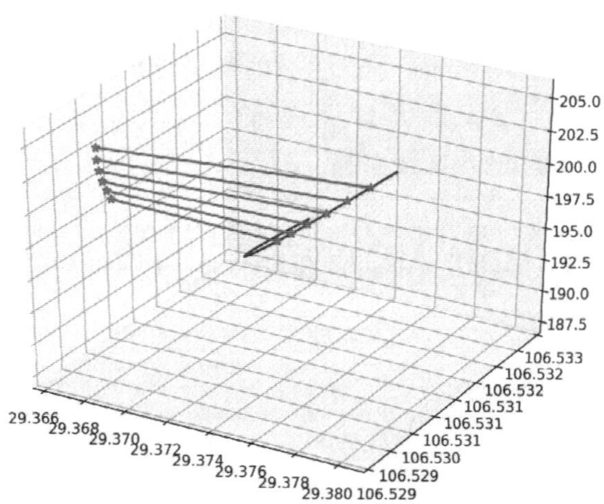

Figure 5. trajectory prediction effect with tanh activation function.

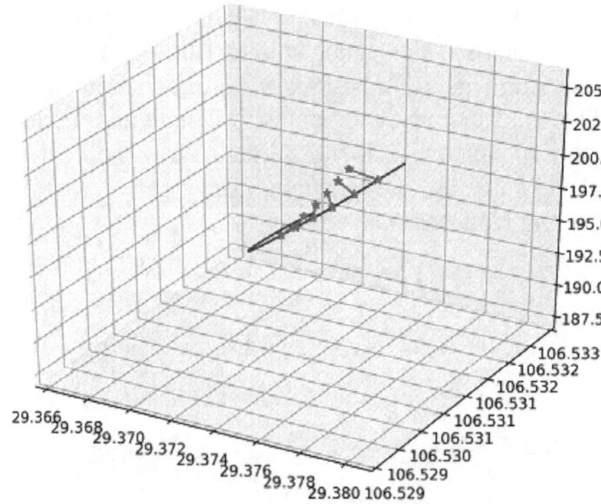

Figure 6. trajectory prediction effect with relu activation function.

In the three figures, the line with a star indicates the degree of deviation of the trajectory prediction. From the comparison of the three graphs, the prediction deviation under the Relu activation function is the smallest and the prediction performance is the best.

4.2. Comparison of prediction results with attenuated memory window

In order to illustrate the impact of the introduction of the fading memory window on the prediction effect, the paper sets the attenuation memory window to 0, 20, and 40 under the condition of selecting the Relu activation function. The configuration of other model parameters is consistent with Table 2.

Table 2. Verification environment configuration used by the LSTM framework

Parameter item	Experimental parameters
GPU	GTX1080
Deep Learning Framework	Keras/TensorFlow
Activation function	Sigmoid/Tanh/Relu
loss function	MSE
Optimization method	Adam
Number of neurons	10/ 20/ 30/ 50/ 80/ 100
Epochs	50
Training Set/ Test Set	400/ 6

Among them, Fig 6 can represent the prediction result that the attenuation memory window is 0, Fig 7 and Fig 8 show the prediction result that the attenuation memory window is 20 and 40 respectively.

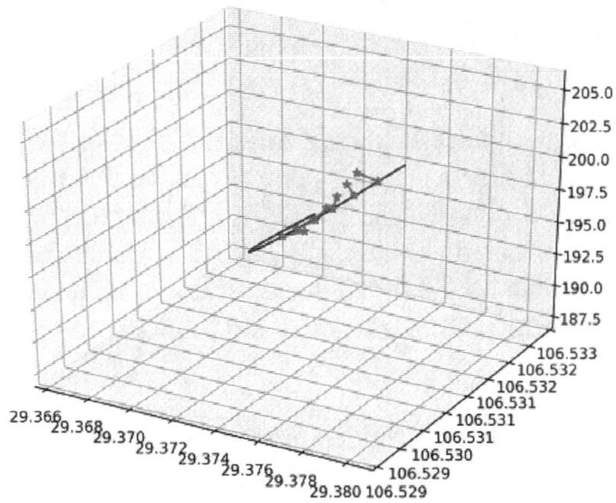

Figure 7. Window size of 20.

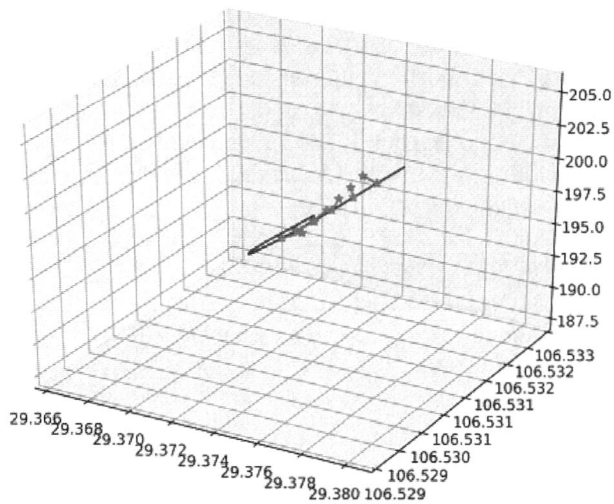

Figure 8. Window size of 40.

From the comparison of the three graphs, it can be seen that the accuracy of the trajectory prediction model with improved attenuation memory window has been further improved, and with the expansion of the attenuation memory window, the prediction effect of the model is relatively better. In order to more clearly compare the degree of optimization of the model, the paper separately calculates the root mean square error (RMSE) and training time of the three, as shown in Table 3.

Table 3. Comparison of prediction errors with fading memory window.

Model	Windows number	Prediction error（RMSE）	Training time(min)
LSTM	0	0.001589	6min
Improved LSTM	20	0.000934	10min
	40	0.000317	13min

It can be seen from the table that with the increase of the number of windows, the improved LSTM prediction model with an attenuation memory window can further enhance the accuracy of the prediction model, thus verifying the improved performance of the improved model. At the same time, however, the computational complexity of the predictive model also increases in multiples, and the training duration increases significantly.

5. Conclusion

In the paper, firstly, the historical gliding data are sampled, first-order differential, standardized, and other preprocessing. Secondly, the surface movement state of aircraft is analyzed, and the attenuation memory window is introduced to improve the hidden layer structure, so that the trajectory prediction model based on LSTM is constructed. Finally, the model is compared and verified under different parameters. The results show that the introduction of the attenuated memory window can greatly improve the prediction effect of the prediction model and reduce the prediction error.

6. References

[1] Ayhan S and Samet H 2016 Aircraft Trajectory Prediction Made Easy with Predictive Analytics *Proc. Int. Conf. on the 22nd ACM SIGKDD (California)* 173–214

[2] Liao C W 2015 Research on Forecasting Method of Aircraft Track Glide Track *Naval Ship Electronic Engineering* **8** 127-129

[3] Zhou L, Tang X M and Tang J 2015 Aircraft Scene Glide 4D Trajectory Prediction Based on Velocity Profile Fitting. *Aeronautical Computing Technology* **1** 35-39

[4] Lymperopoulos I and Lygeros J 2010 Sequential Monte Carlo methods for multi - aircraft trajectory prediction in air traffic management[J]. *International Journal of Adaptive Control & Signal Processing* **24** 830-849

[5] Prevost C G, Desbiens A and Gagnon E 2007 Extended Kalman Filter for State Estimation and Trajectory Prediction of a Moving Object Detected by an Unmanned Aerial Vehicle[C] *American Control Conference* 1805-1810

[6] Fablec Y L and Alliot J M 2007 Using Neural Networks to predict aircraft trajectories *International Conference on Artificial Intelligence*

[7] Leege A D, Paassen M V and Mulder M 2006 A Machine Learning Approach to Trajectory Prediction *AIAA Guidance, Navigation, and Control (GNC) Conference*

[8] Z. C. Lipton, J. Berkowitz and C. Elkan 2015 A Critical Review of Recurrent Neural Networks for Sequence Learning *Computer Science* **4**

Acknowledgments

This work is supported by Science & Technology Department of Sichuan Province under Grant 2017GZ0047 & 2018GZ0077 & 2018GZ0069.

Aerodynamic Investigations on SC-0414 Airfoil with Small High Lift Devices

T D Nguyen[1*], M Kashitani[1] and M Taguchi[1]

[1] Department of Aerospace Engineering, National Defense Academy of Japan
1-10-20 Hashirimizu Yokosuka, Kanagawa, 239-8686, Japan

em57051@nda.ac.jp

Abstract. Gurney flap (GF) is well-known that one of the most attractive plain flaps because of the simple configuration and actually effective in improving lift of the airfoil. Many studies were conducted but the effects of GF on the various airfoil types need to be further investigated. This study aimed to clarify the effect of GF in case of the Supercritical airfoil, SC-0414 airfoil. The Ansys Fluent software is used to perform the flow simulations. The result in airfoil baseline is compared with the experiment result obtaining from wind tunnel testing. The flow simulations focus on analyzing the lift coefficient when increasing the height of Gurney flaps. The velocity of the simulations, as well as the experiment is U = 12 m/s and the Reynolds number Re is 1.6×10^5. In the smoke wind tunnel, the flow field are clearly observed by smoke lines. The lift coefficient results of the flow simulations show the similar qualitative trend as the conventional airfoil model with Gurney flap.

1. Introduction

Gurney flap (GF) is well known that one of the most attractive plain flaps because of the simple configuration and actually effective in improving lift of the airfoil. GF was first used in race cars, but it is also expected to improve aerodynamics performance for the airplane wings. Many previous experimental studies about effects of GF [1 - 3] were performed and obtained the impressive results on enhancement life coefficient. The results highlight that the GF significantly increased the pre-stall lift and lift-to-drag ratio at small angles of attack (AOAs) on low-speed flow. In the specific case, the GF also decreases the drag of the airfoil. So, the GF is expected to actually used in the planes and UAV (unmanned aerial vehicle) to improve the aerodynamic performance.

The Supercritical Airfoil (SC airfoil) was focused when NASA effort to developing practical two-dimensional airfoils with good transonic behaviors while retaining acceptable low-speed characteristics [4]. The SC airfoils have to development of the Supercritical wings (SCW) now used worldwide on business jets, airliners and transports, and numerous military aircraft [5]. To improve the aerodynamic performance, the combination of SC airfoil and high-lift device are considered. However, investigations about the effects of GF in case of the transonic airfoils is insufficient and the aerodynamic characteristics of the transonic airfoil with GF need to be investigated more.

This study aims to clarify the aerodynamic characteristics, specially on lift coefficient and flow fields of Supercritical airfoil, SC-0414 airfoil installed GF by flow simulation and smoke wind tunnel experiment.

2. Numerical and Experimental method

2.1. Numerical simulation

In the flow simulation, the chord length of SC-0414 airfoil mode is 0.2 m, the flow velocity is 12 m/s, and Reynolds numbers based on chord length and velocity is $Re = 1.6 \times 10^5$, as well as the condition of the wind tunnel experiment. Figure 1 shows the grid distribution near the GF on this simulation. The Grid-generation by ICEM software constructs the C-type structured grid, with 340 points distributing on the surface of the airfoil and 400 points in case of the airfoil with GF. The y+ value of the grid is about 1, it means that the height of first grid nearest the airfoil is about 2.4 x 10-5 m. The Spalart - Allmaras (S-A) one-equation model was used to simulated in all condition [6]. In previous calculation for high-lift devices flow, the S-A model performed as well as high-order turbulence models [7].

The PISO (pressure-implicit with splitting of operators) algorithm was used to steady simulation. The pressure term of the second-order upwind scheme, the moment terms of third-order central MUSCL scheme, and viscosity of first-order upwind scheme were used to perform simulation [1]. For all of the simulations, converge criteria is required less than 10-6 for the continuity and 10-5 for other terms.

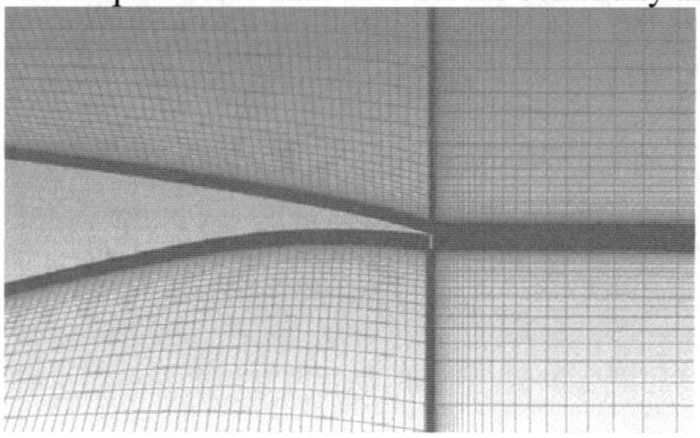

Figure 1 Grid generation.

2.2. Smoke wind tunnel experiment setup

Figure 2 shows the low-speed smoke wind tunnel is used to conduct the experiment. The width of test section is 150 mm, the height is 1500 mm and the length is 2000 mm, respectively. The airfoil is fixed on a 600 mm diameter plywood turntable. The experiments are performed from -5 degrees to 8 degrees of angle of attack. The tunnel flow velocity U_∞ is 12 m/s and Reynolds number Re is 1.6×10^5. Smoke is generated by smoke generator Kanto-Kogaku SC-3. Nikon D50 camera with shutter speed 1/15 s was selected to record flow pattern and G3 Data was used to measure the position or displacement of the smoke line.

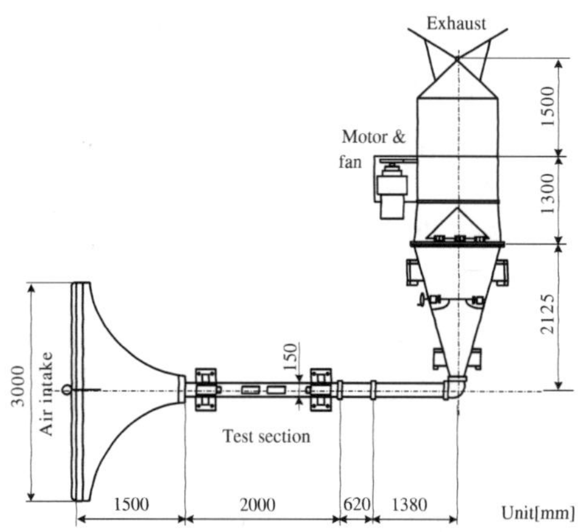

Figure 2 Smoke wind tunnel [8].

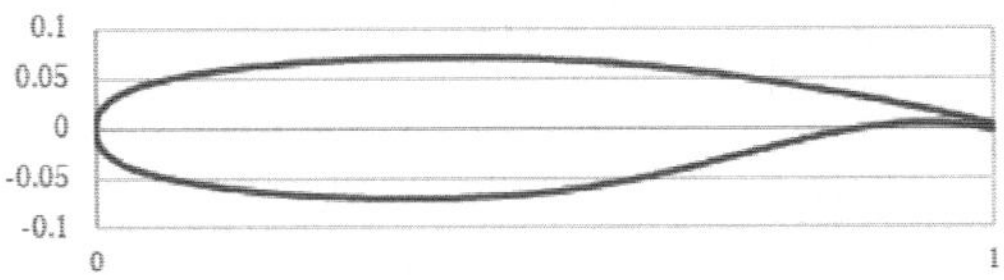

Figure 3 SC-0414 airfoil model.

Table 1 Detail information (SC-0414).

	Value	Position (from the front)
Maximum thickness	14 % *c*	36 % *c*
Maximum camber	1.5 % *c*	83 % *c*

Figure 3 and Table 1 show the airfoil model used in the experiment. The chord length of the airfoil model *c* is 200 mm and the width is 150 mm, the same as the width of the wind tunnel, to reduce the three dimensional effects. The GFs are set up on the airfoils with scotch tape. The hight of GF (*h*) is 2 mm, 4 mm, 6 mm in proportion to 1%, 2%, 3% of airfoil's chord length. The flaps thickness is 0.05 mm.

In the smoke tunnel, the streamline can be visualized by smoke line. By Yamana's method [9 - 10], the lift coefficient can be estimated from flow visualization results, that is obtained by measuring the characteristics of specific points in the smoke line near the airfoil. Moreover, Yamaguchi et al. [11] modified the method able to apply with the present condition. The theoretical approach to the modified Yamana's method was discussed with the effects of the alteration of the center pressure caused by flaps [11]. Several studies were conducted by modified Yamana's method and get the accurate results [8] [11].

3. Results and discussion

3.1. Baseline

Figure 4 shows the lift coefficient of the flow simulation and the smoke wind tunnel experiment. The results are compared with reference data [12] taken from Xfoil software prediction (Reynolds number *Re* of reference data is 2×10^5). The lift slope of the experimental data is about 0.1128, reference data is about 0.1179 and simulation data is about 0.0965. The lift coefficient of the simulation results has slight difference at any angle of attack with the reference data and the experiment results but the lift slope shows the qualitatively same trend. The experiment results also show good agreement with the reference data.

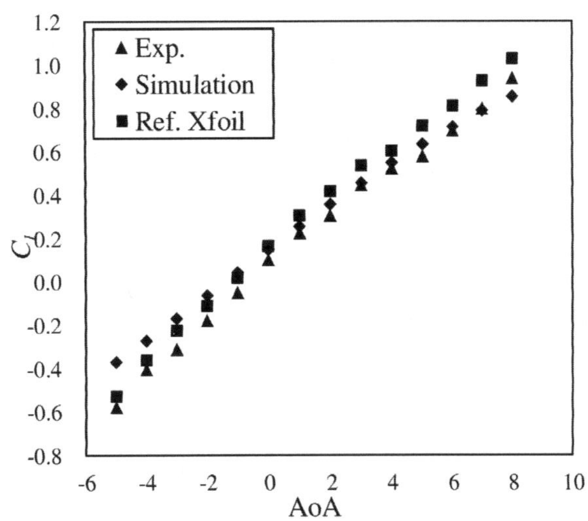

Figure 4 Lift Coefficient of SC-0414 in clean condition.

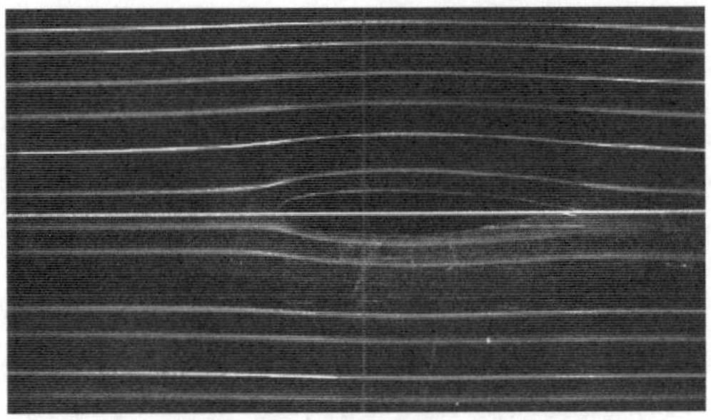

(a) SC airfoil clean condition

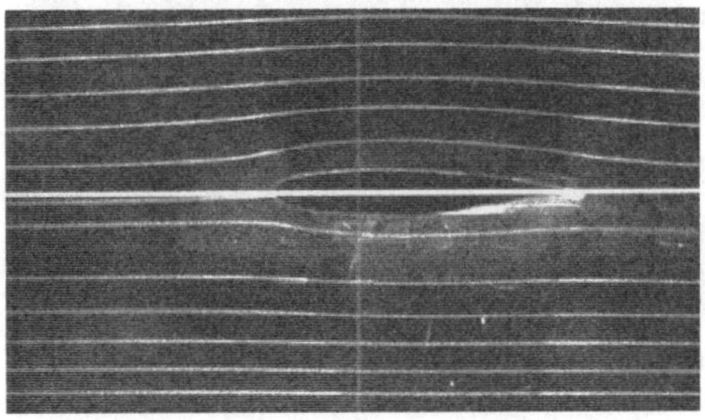

(b) SC airfoil with 0.01 *c* GF

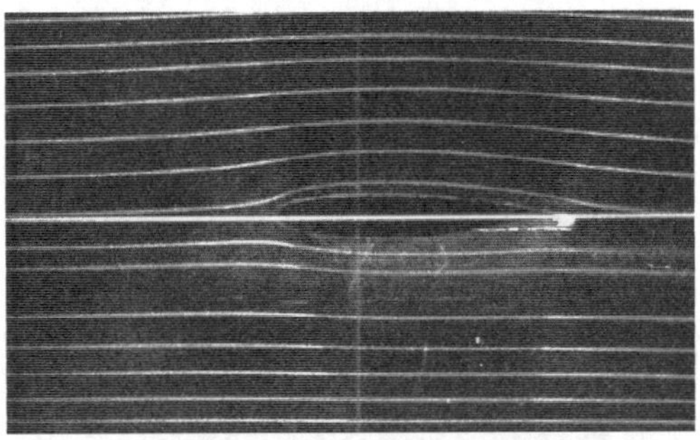

(c) SC airfoil with 0.02 *c* GF.

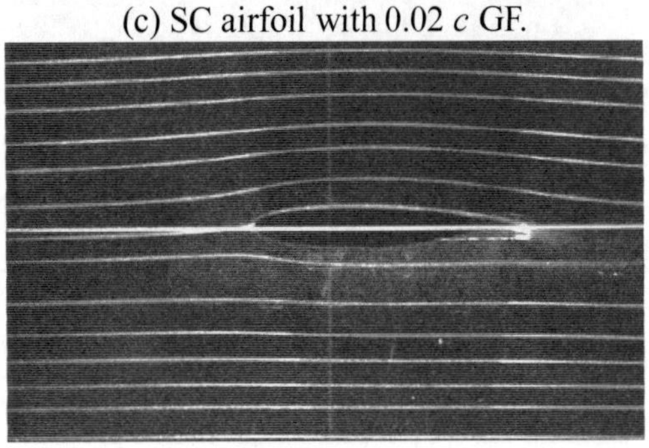

(d) SC airfoil with 0.03 *c* GF.

Figure 5 Flow visualization around airfoil model. (U_∞=12m/s, Re=1.6×10^5, AoA=0 deg.).

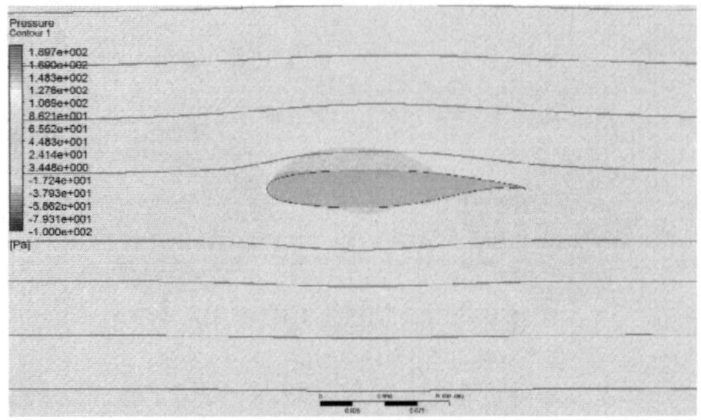

(a) SC airfoil clean condition.

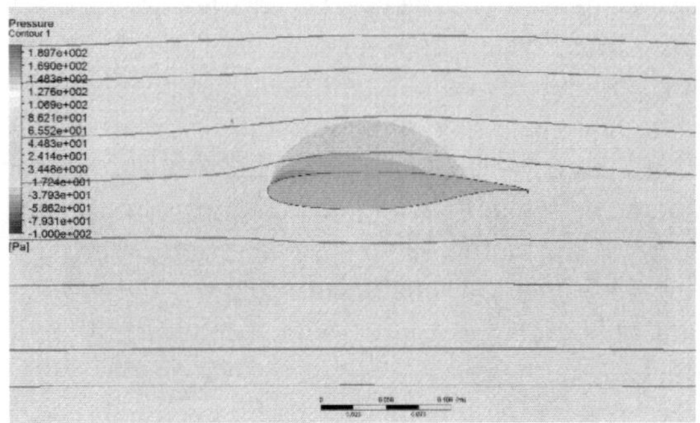

(b) SC airfoil with 0.01 *c* GF

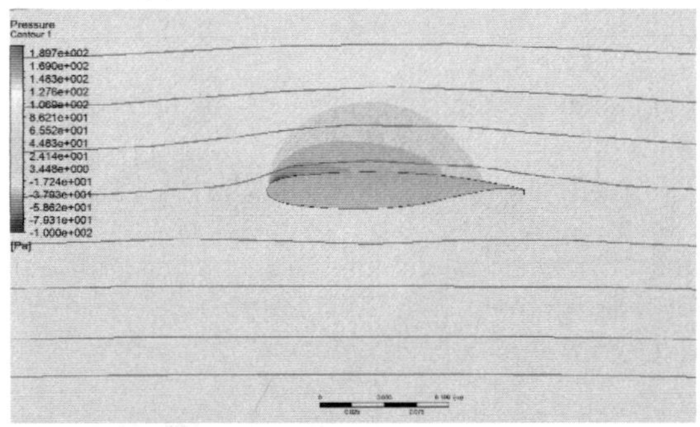

(c) SC airfoil with 0.02 *c* GF.

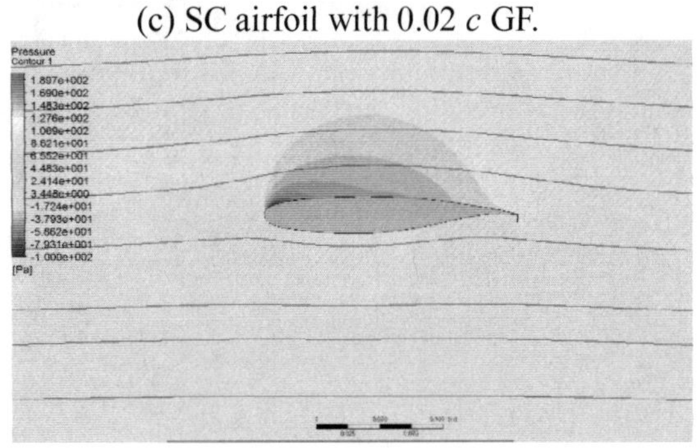

(d) SC airfoil with 0.03 *c* GF.

Figure 6 Pressure contour by CFD Fluent. (U_∞=12m/s, Re=1.6×10^5, AoA=0 deg.).

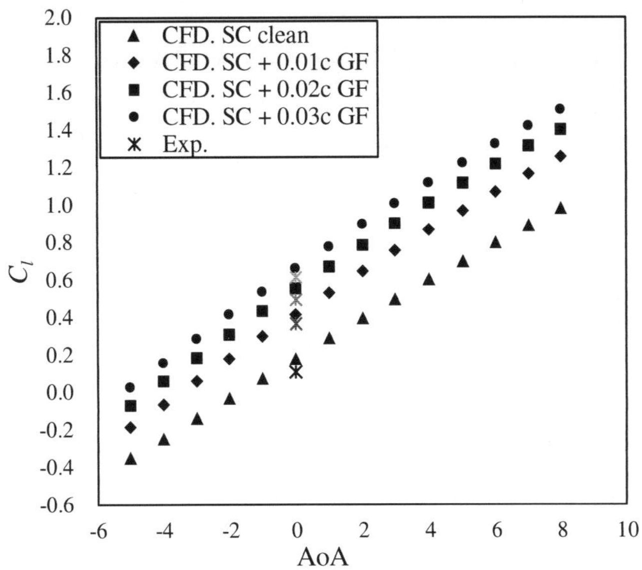

Figure 7 Lift Coefficient of SC-0414 airfoil with GF.

3.2. Effects of Gurney flap

Figure 5 shows the flow fields around the SC-0414 airfoil models. The angle of attack AoA is 0 degrees. The figure (a) shows the clean condition, (b) h/c=0.01, (c) h/c=0.02, (d) h/c=0.03, respectively (h is the height of GF, c is the airfoil model's chord length). The steady smoke line was observed. From figure 5 (a) to (d), the picture does not show the clear difference but the curves of smoke lines near the airfoil are changed when increasing the height of GF. By Yamana's method which is mentioned above, the lift coefficient can be estimated through measure the distance between the smoke line located near the airfoil model and reference line (the yellow line in the pictures of figure 5). In this study, we only focus on simulation results.

Figure 6 shows the pressure contour calculated by CFD. The red lines represent the streamlines. The angle of attack AoA is 0 deg. The flap heights of the pictures are the same as the figures 5. Because of the original camber of the SC airfoil (difference with symmetry airfoils), the high pressure at the lower surface of trailing edge can be observed in figure (a).

Figure 6 shows that the pressure at the upper and lower surface are changed clearly when GF height increases. At figure (d), the wide high-pressure area is observed at the lower surface. The GF make the pressure area around the airfoil are the difference in (a), (b), (c) and (d).

Figure 7 shows the lift coefficient calculated by CFD simulation and experiment results at 0 deg. The experiment results are calculated by modified Yamana's method, as well as in Figure 4. The experiment results show the same trend with the simulations. When the height of the flap increases, the lift coefficient increases too. The lift slope is about 0.118, nearly the same in all condition. The nearly same lift slope showed the qualitatively same trend of lift coefficient value. The case of the airfoil installed the 1% of the airfoil's chord length of GF shows the impression of improving the lift. The results show that in case of the SC-0414 airfoil, GF has the similar effects like other simple plain flaps, as well as Ref.13.

4. Conclusion

In this study, the flow visualization experiment in smoke wind tunnel and simulation were performed. The following results were obtained;

From -5 to 8 deg, the steady smoke lines are clearly observed. The visualization experiment results show the qualitatively same with CFD results.

In case of the SC-0414 airfoil, GF has the similar effects like other simple plain flaps. The lift slopes are nearly the same in clean condition even when GF is installed. When the height of the flap increases, the lift coefficient increases too.

Reference

[1] T. Yu, J. J. Wang and P. F. Zhang 2011 Numerical Simulation of Gurney flap on RAE-2822 Supercritical Airfoil, *Journal of Aircraft* **48** 5, pp 1565–1575.

[2] Mark D. Maughmer and Gotz Bramesfeld 2008 Experimental Investigation of Gurney Flaps, *Journal of Aircraft* **45** 6, pp 2062–2067.

[3] Libin Daniel and Lance W. Traub 2013 Effect of Aspect Ratio on Gurney-Flap Performance, *Journal of aircraft* **50** 4.

[4] Charles D. Harris 1990 NASA Supercritical Airfoils, A Matrix of Family-Related Airfoils, *NASA Technical Paper* 2969.

[5] The Supercritical Airfoil 2004 (*NASA Technology Facts*, TF–2004–13 DFRC).

[6] Spalart, P. R., and Allmaras 1992 S. R., A One-Equation Turbulence Model for Aerodynamic Flows, *AIAA Paper* pp 1992–0439.

[7] Spalart, P. R., and Rumsey, C. L. 2007 Effective inflow conditions for turbulence Models in Aerodynamics Calculations, *AIAA Paper*, pp 2007–2544.

[8] M. Kashitani, Y. Yamaguchi, Y. Kai and K. Kusunose 2009 Study on Busemann Biplane Airfoil in Low Speed Smoke Wind Tunnel, *Transactions of the Japan Society for Aeronautical and Space Sciences 52*, pp 213–219.

[9] M. Yamana 1965 Measurement of Section Lift Coefficient in a Two-Dimensional Smoke Wind-Tunnel, *J. J SASS* 13 141, pp. 321–325.

[10] M. Yamana and H. Nakaguchi 1979 Theory of Airplane Design, *Yokendo Press*, pp. 645-648, (in Japanese).

[11] Y. Yamaguchi, M. Kashitani, and K. Sato 2002 An Experiment on Aerodynamic Characteristics of Airfoil Constructed with Flat Plates in a Low Reynolds Number Flow, *Bulletin of National Defense Academy*, 40-9, pp 117–126.

[12] http://airfoiltools.com/airfoil/details?airfoil=sc20414-il#polars.

[13] D. P. Raymer 2012 Aircraft Design: *A conceptual Approach Fifth edition*, AIAA Educational Series, pp 419–428.

Analysis on the Aerodynamic Characteristics of a Continuous Whole Variable Camber Airfoil

Anmin Zhao[2], Zou Hui[1], Haichuan Jin[2], Dongsheng Wen[2*]

[1]School of Aeronautical Science and Engineering, Beihang University, Beijing, China
[2]National key Laboratory of Human Machine and Environment Engineering, Beihang University, Beijing, China

d.wen@buaa.edu.cn

Abstract. In this paper, based on the genetic algorithm, a 4-section optimization airfoil structure is designed, which can change entire camber of the airfoil. For this new type of integral changeable camber airfoil, its first structure cam be deflected downward by 30 degrees, the second section is fixed, the third stage be capable of deflect downward by 30 degrees, and the fourth section be able to bend up and down by 30 degrees. In addition, the computational fluid dynamics method is adopted to numerically simulate the aerodynamic performance of variable camber airfoils. The numerical results observe that the used of varying camber airfoil has better been stalling characteristics and larger lift coefficient comparison with the Clark Y airfoil and NACA114 airfoil.

1. Introduction

In terms of basic aerodynamics and practical flight application, variable-camber UAV (Unmanned Aerial Vehicle) has great potential [1-3]. In the field of aerodynamics, variable camber wings can improve the flight performance of UAV and adapt to a variety of missions; in the field of bionic science, it can help understand the complex mechanisms of animal flight. Besides, it can effectively delay the flow separation and improve stall characteristics for UAV with low Reynolds number [4-6]. For low-velocity vehicles, the control methods of the leading edge and trailing edge flaps are not only difficult to satisfy the multi-task requirements of the aircraft, but also the complex hinges on the system greatly increase the structural weight, which easily leads to configurational noise, vibration and fatigue damage [7]. Therefore, it is particularly important to design a wing that can maintain an optimal aerodynamic shape under various missions and environments.

Since the 1980s, the flexible technology of variable camber wing had been studied [8], which was listed as one of the key technologies of future aircraft. Based on research and analysis results, it was shown that in the implementation of adaptive deformation of the wing, from the traditional rigid structure to the compliant mechanism, there is a certain amount of research from the conventional materials and structures to the new smart materials and structures [2, 9-14]. Among them, the wing camber change was reached by using of compliant mechanisms, new intelligent material and structures. However, due to the fact that the study of functional materials, structures, and compliant mechanisms is still in an immature stage, and the drive system based functional materials have a relatively low efficiency in practical applications, which are difficult to meet the needs of current feasible engineering.

In order to overcome the problem, a novel entire variable camber wing was achieved by authors. In this paper, it is worth noting that the airfoil is optimized designed by genetic algorithm. The new elastic membrane method is adopted to realize the continues smooth surface of the wing. The research results shown that optimize design airfoil has better aerodynamic characteristics and larger stalled angle of attack.

2. Optimize airfoil design

In order to be available to define the deflection angle of the variable camber airfoil, a symmetrical airfoil, NACA 0012, is selected. Figure 1 shows the maximum bending angle range of the airfoil.

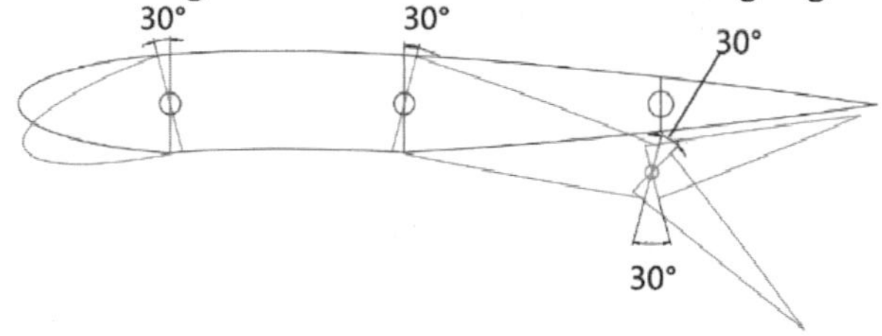

Figure 1. The maximum range of the variable camber airfoil

Based on the XFOIL software for aerodynamics calculation, genetic algorithm and the corresponding objective function, a set of aerodynamic shape optimization method for the overall change camber airfoil is established. Shown in Figure 2-a, in the process of the change of airfoil camber, the four-typical state of airfoil was chosen to research. To verify the aerodynamic performance advantages of variable camber airfoil profile of 2D, Clark Y airfoil and NACA114 airfoil was selected, which were used as analysis of the results, as shown in figure 2-b.

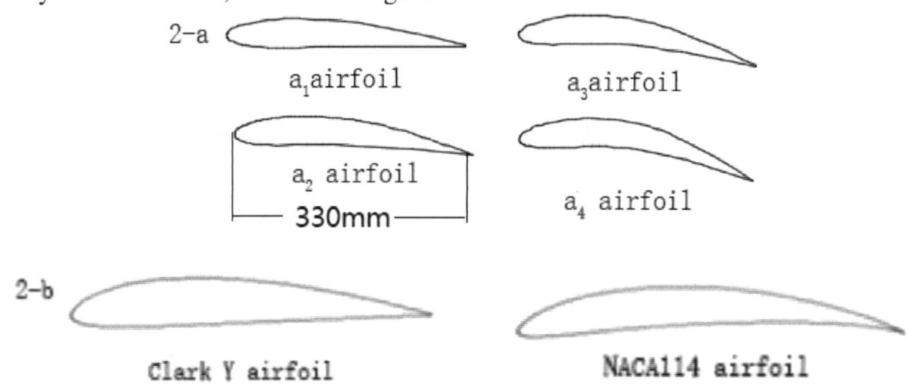

Figure 2. Schematic diagram of 2D airfoil: (2-a) 4-variable camber airfoil profiles, (2-b) the Clark Yairfoil and NACA114 airfoil

3. Numerical simulation

3.1 CFD validation

A two-dimensional airfoil, NACA4412, is selected to validate the accuracy of the computational simulation. In order to obtain the precise simulation of the flow around variable camber airfoil, a C-type structured grid is adopted in the ICEM CFD 15.0[15], as shown schematically in Figure 3. The total number of mesh cells of computation field around the airfoil profile is 108500, the flow field is 20L away from the far-field boundary condition, and the quality of grid is 0.86~0.93. The reference length L is equal to 330mm, which corresponds to the airfoil chord length.

The boundary layer of normal velocity is very large in the adjacent airfoil region, which decreases from a relatively high value to the wall velocity. Therefore, it is necessary for boundary grid Y spacing value set to calculate the flow field of the near wall region. In this study, NACA Y+ wall distance estimation online is adopted, and the height of the first layer grid is calculated to be 0.001mm. The Spalart-Allmaras (SA) solver is adopted in this study.

IOP Conf. Series: Journal of Physics: Conf. Series **1215** (2019) 012005 doi:10.1088/1742-6596/1215/1/012005

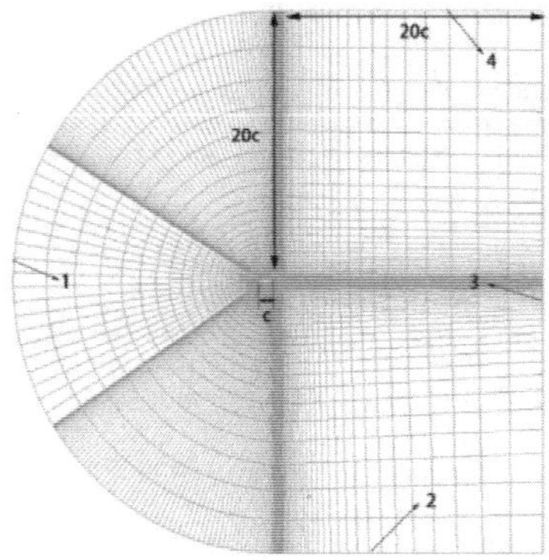

Figure 3. Schematic of computation mesh.

The far-field boundary conditions are described as follows: Mach number is equal to 0.215, Reynold number is equal to 3e+6, and the angle of attack is range from 0° ~ 10°, with the far-field constant velocity. The simulation results are displayed in Figure 4.

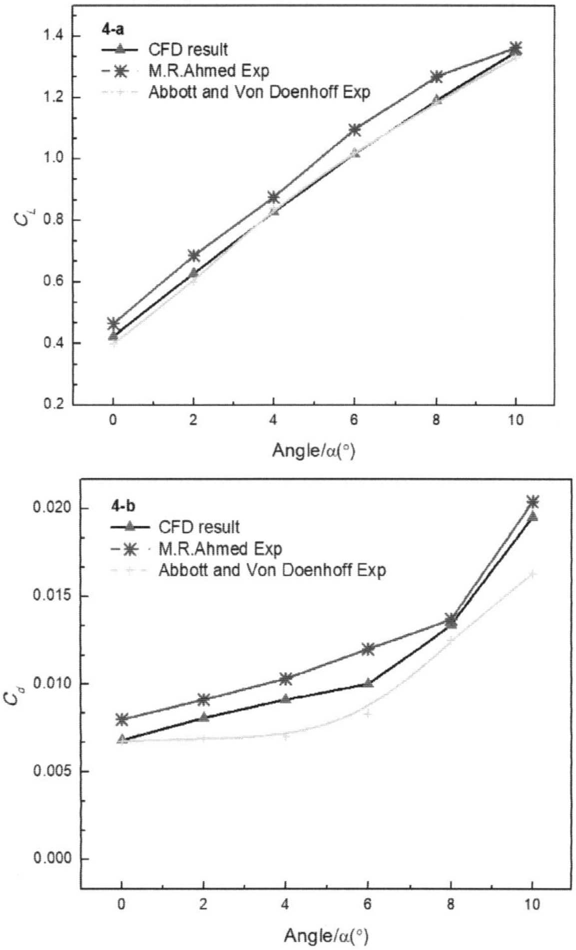

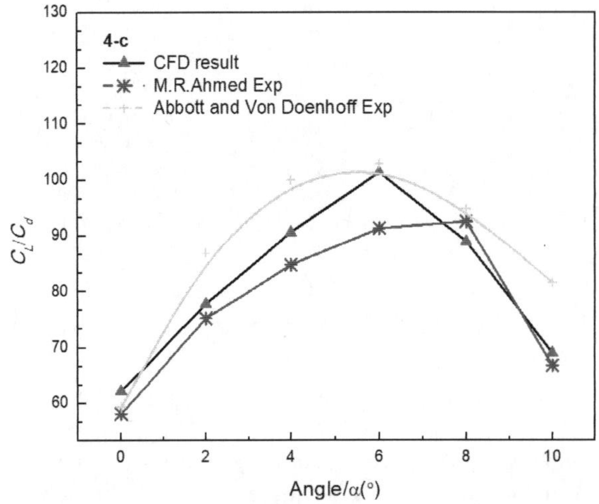

Figure 4. Calculation and experimental characteristic coefficient versus AOA (Angle of attack) for NACA4412.

Figure 4-a show the effect of angle of attack on the lift coefficient distribution. The lift curve agrees well with experimental data within the range of the angle of attack. Figure 4-b and 4-c illustrates the airfoil drag coefficients and lift-to-drag ratios. It can be seen that the results of numerical computations are within the measured data range of the M.R.Ahmed[16] and Abbott et al[17] wind tunnels test. In general, the simulation results before stalling are considered to be believable.

3.2 Aerodynamic characteristics of the optimize variable camber airfoil (VCA)

3.2.1 Grid independence. In order to verify grid independence, three types of mesh with the increase grid density are employed to investigate lift and drag coefficients based on Clark Y airfoil in Figure 2 and α= 0 °～14 °. The far-field flow is 18m/s and other parameters are the same as in section 3.1. The calculation results in Table 1 and Figure 5 show that the deviation of calculation of differences grid density are lower than 1.5% before lower than 8°. When the angle of attack become larger than 8°, the compute results deviate, which might have due to stalling of airfoil or unsteady vortex cause the flow separation. To reduce the costs and obtain a relatively high accuracy, the normal grid density is chosen in present research.

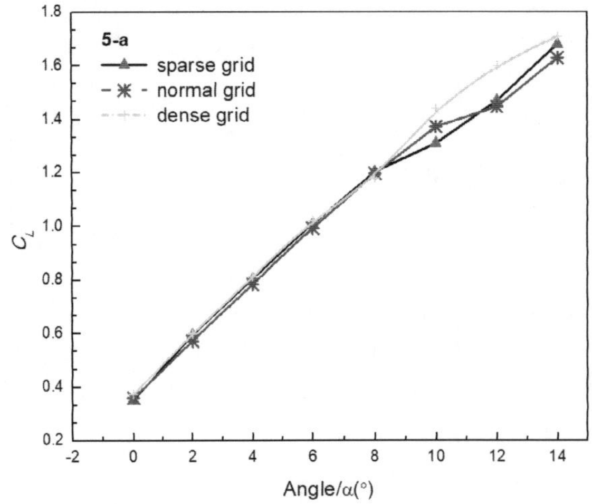

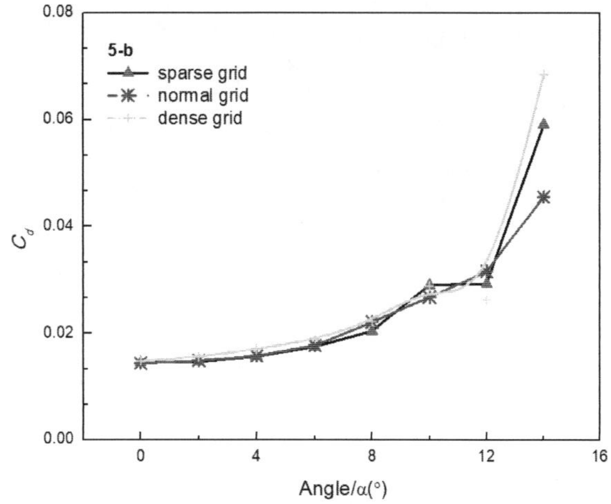

Figure 5. Lift and drag coefficient distribution of Clark Y airfoil with different grid density.

Table 1 Mesh independence study for variable camber airfoil at Re=406508 and angle of attack of 0 ° to 10 °.

Type	Grid cells	α	C_l	C_d
Sparse grid	45600	0	0.353	0.0145
		4	0.805	0.0157
		8	1.202	0.0204
Normal grid	123400	0	0.359	0.0144
		4	0.786	0.0158
		8	1.196	0.0221
Dense grid	384000	0	0.373	0.0148
		4	0.813	0.0170
		8	1.178	0.0222

3.2.2 2D characteristics of the optimization VCA. A calculation numerical simulation is performed to evaluate the aerodynamic characteristics of the optimization VCA. The normal grid is selected, and the boundary conditions are set adopting the same as conditions in section 3.2.

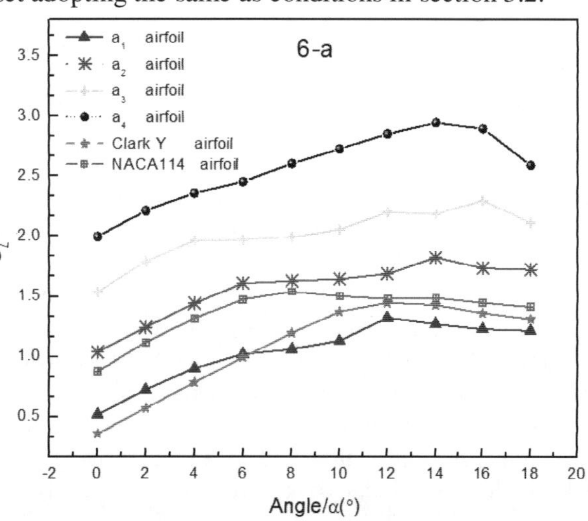

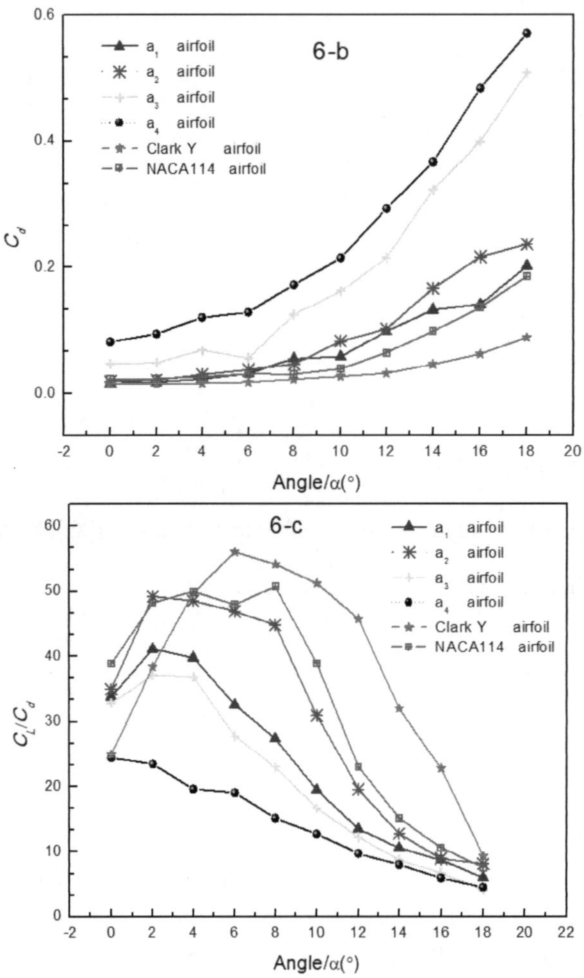

Figure 6. Aerodynamic coefficient versus AOA airfoil: 6-a lift coefficient versus angle of attack, 6-b drag coefficient versus angle of attack, and 6-c lift-to drag versus angle of attack.

Figure 6-a shows that the lift coefficient increases at first and then turns to descending with the AOA increasing. The numerical results indicate that the variable camber airfoil has a larger stalling angle and a greater maximum lift coefficient than the Clark Y airfoil and NACA114 airfoil. The lift coefficient increases with the increase of airfoil camber with the same angle of attack, and it show an obvious enhances trend toward the camber increase. In addition, the lift coefficient of the Clark Y airfoil and NACA114 airfoil is between a_1 airfoil and a_2 airfoil, which demonstrate that the optimized airfoil in this paper has high lift and large angle of stall characteristics.

The drag coefficient distribution with different airfoil camber is shown in Figure 6-b. According to this result, the resistance coefficient increases when the AOA of the far-field air increase. As it is is shown in Figure 6-b, the drag coefficient increases obviously when the airfoil camber is greater than a_1 airfoil. Besides, the drag coefficients of the a_1 airfoil and NACA114 airfoil are almost identical, which also indicates that the optimally designed airfoil has a similar resistance characteristic of the conventional airfoil. It is worth noting that when the angle of attack is in the range of 12° to 14°, the drag coefficient increases significantly and it as well corresponds to the airfoil stall characteristics.

Figure 6-c shows that for variable camber airfoil, decambering can increase the lift-to-drag ratio; for the camber increases even further, the situation reversed. In addition, the lift-to-drag ratio of a_2 airfoil is approximately identical with NACA114 airfoil. The above shows that the optimized design of variable camber airfoil has excellent aerodynamic performance.

IOP Conf. Series: Journal of Physics: Conf. Series **1215** (2019) 012005 doi:10.1088/1742-6596/1215/1/012005

4. Conclusions

In this paper, an optimal design variable camber airfoil based on genetic algorithm is proposed. Besides, two-dimensional numerical simulation methods are applied to variable camber airfoil to investigate the effect of varying camber airfoil on the aerodynamic performance of UAV. The following conclusions can be drawn as

1) As compared with Clark Y airfoil and NACA114 airfoil, adjustable camber airfoils have better aerodynamic performance and stalled characteristics.

2) For variable camber airfoil, decambering can increase the lift-to-drag ratio; for the camber increases even further, the situation reversed.

Further study is necessary to explore the appropriate number of wing rib sections and investigates the effects of the relative rotation angle between different sections on aerodynamic performance of the whole AVCW UAV. The developed complete AVCW UAV is a potential future candidate for UAV field.

5. Acknowledgement

The work was supported by the Innovation and Entrepreneurship Foundation No. YCSJ-01-2018-01 to the Beihang university, and the Beihang Aviation Innovation Practice Base. We also would like to express appreciation for the financial support by Wing Flying Technologies Co., Ltd No LYFH-01-2018.

6. Reference

[1] F. Gandhi, P. Anusonti-Inthra, Skin design studies for variable camber morphing airfoils, Smart Materials & Structures, 17 (2008).

[2] H. Zhu, W.D. Liu, C.S. Zhao, Morphing Aircraft and Its Morph-driving Techniques, Machine Building & Automation, DOI (2010).

[3] M. Bashir, P. Rajendran, Morphing Wing Technologies, 2017.

[4] O.M. Curet, A. Carrere, R. Waldman, K.S. Breuer, Aerodynamic Characterization of a Wing Membrane with Variable Compliance, Aiaa J., 52 (2014) 1749-1756.

[5] L. Liu, W. Yin, F. Dai, Y. Liu, J. Leng, Static aeroelastic deformation of flexible skin for continuous variable trailing-edge camber wing, in: M.N. GhasemiNejhad (Ed.) Active and Passive Smart Structures and Integrated Systems 20112011.

[6] B.K. Stanford, Static and Dynamic Aeroelastic Tailoring with Variable-Camber Control, Journal of Guidance Control and Dynamics, 39 (2016) 2522-2534.

[7] H. Takahashi, T. Yokozeki, Y. Hirano, Development of variable camber wing with morphing leading and trailing sections using corrugated structures, Journal of Intelligent Material Systems and Structures, 27 (2016) 2827-2836.

[8] S. POWERS, L. WEBB, E. FRIEND, W. LOKOS, Flight test results from a supercritical mission adaptive wing with smooth variable camber, 28th National Heat Transfer Conference, 1992, pp. 4101.

[9] E.J. Abdullah, C. Bil, S. Watkins, Adaptive airfoil control system using shape memory alloy actuator for unmanned aerial vehicle, Crc Press-Taylor & Francis Group, Boca Raton, 2011.

[10] J.D. Bartley-Cho, D.P. Wang, C.A. Martin, J.N. Kudva, M.N. West, Development of high-rate, adaptive trailing edge control surface for the Smart Wing Phase 2 wind tunnel model, Journal of Intelligent Material Systems and Structures, 15 (2004) 279-291.

[11] H.D. Li, H.S. Ang, Preliminary airfoil design of an innovative adaptive variable camber compliant wing, J. Vibroeng., 18 (2016) 1861-1873.

[12] S. Kota, J.A. Hetrick, R. Osborn, D. Paul, E. Pendleton, P. Flick, C. Tilmann, Design and application of compliant mechanisms for morphing aircraft structures, Smart Structures and Materials, 2003.

[13] S. Barbarino, R. Pecora, L. Lecce, A. Concilio, S. Ameduri, L. De Rosa, Airfoil Structural Morphing Based on SMA Actuator Series: Numerical and Experimental Studies, Journal of Intelligent Material Systems and Structures, 22 (2011) 987-1004.

[14] W. Yin, T. Fu, J. Leng, Structural shape sensing for variable camber wing using FBG sensors, Proceedings of SPIE - The International Society for Optical Engineering 7292 (2009) 72921H-72921H-72910.
[15] A. Fluent, ANSYS FLUENT 14.5 Theory Guide, DOI (2012).
[16] M.R. Ahmed, T. Takasaki, Y. Kohama, Aerodynamics of a NACA4412 Airfoil in Ground Effect, Aiaa J., 45 (2007) 37-47.
[17] I.H. Abbott, A.E. Von Doenhoff, Theory of Wing Sections, General Information, DOI (1959).

IOP Conf. Series: Journal of Physics: Conf. Series **1215** (2019) 012006 doi:10.1088/1742-6596/1215/1/012006

Numerical Experiment of Flow Around a Compound Airfoil

Yoshifumi Yokoi*

Department of Mechanical Engineering, National Defense Academy of Japan, 1-10-20 Hashirimizu, Yokosuka 239-8686, Japan

yokoi@nda.ac.jp

Abstract. A numerical simulation around a compound airfoil was performed using the vortex method. The compound airfoil consists of a front airfoil, a fixed wing, and a rear airfoil, and it assumes the rudder of a submarine. As for the shape of the front airfoil and the rear airfoil, NACA0012 was used. Reynolds number was $Re = 3.8\times10^5$. In the numerical simulation, the angle of attack of a front airfoil and a rear airfoil was changed, and the magnitude of the gap between these movable airfoils and the fixed wing was changed. The time history of the fluid force of acting on an airfoil, and the flow pattern were investigated. The total lift was greatly produced, when the front airfoil and rear airfoil attached the same attack angle as the same direction. It was found that the total lift with the large one where the gap between these movable airfoils and the fixed wing is larger can be obtained. It was found that the rear airfoil governs the flow of the compound airfoil.

1. Introduction

The airfoil is a device for generating lift and is an important component in mechanical engineering. Therefore, the performance of airfoils and that of the machine equipped with airfoils have been well researched, and many characteristic tests have been performed [1]. As a method to obtain a large lift, giving an attack angle to an airfoil, speeding up, and making area of an airfoil wide are known. Here, when giving an attack angle to the airfoil, it is cautious of a direction of movement changing. It is the situation which needs a large lift at the time when velocity is late. For example, it is at the time of landing of the airplane and entry into port of the ship. Since the rudder is prepared before the propeller as for the submarine, effectiveness of a rudder worsens rather than the time of the cruise at the time of entry into port. Therefore, the measures which advance effectiveness of the rudder are required.

Yokoi [2] investigated the aspect of the interference flow wheal simulation. Furthermore, Yokoi et. al. [3] performed the numerical simulation of an interference flow about two symmetrical airfoils by which close setting was carried out, and they reported the flow characteristic around those airfoil. Yokoi [4] performed the numerical simulation of flow around an airfoil of tandem arrangement from the layout of the fin of a "tuna", and showed the fluid force characteristics and flow patterns. In tandem arrangement, if the interval of two airfoils is narrowed, they will contact and will become one body. Accordingly, the two airfoils will constitute one total airfoil (compound airfoil). Such the "combined airfoil" is already used also in the airplane or the ship. Since the compound airfoil is promising as an apparatus which obtains a high lift, there are many examples of study [5-9].

In this study, the numerical simulation which used the vortex method about the compound airfoil which arranged the movable airfoil in front and in rear of one fixed wing was performed. As for the movable airfoil, the NACA0012 airfoil type was used. Reynolds number *Re* based on the airfoil chord length was $Re = 3.8\times10^5$. The flow pattern and the fluid force characteristic were investigated.

2. Numerical calculations

2.1. Calculation apparatus
The numerical experiment apparatus was a consisted of simulation software and a notebook type computer (NEC; Lavie) as calculation hardware which are on the market. The soft ware which named "UzuCrise 2D ver.1.1.3 rev.H (College Master Hands Inc., 2006)" is used.

2.2. Calculation method
The method of analyzing the numerical simulation of this study is a vortex method based on the Lagrange analysis. The vortex method is a direct viscid-inviscid interaction scheme, and the emanation of velocity shear layers due to boundary layer separation is represented by introduction of discrete vortices with viscous core step by step of time. The separating shear layers were represented the discreet vortices which were introduced at the separation points. The details of calculation technique of vortex method and accuracy of calculation are shown in references [10] and [11].

2.3. Shape of compound airfoil
The compound airfoil consists of a front airfoil, a fixed wing (stay), and a rear airfoil, and it assumes the rudder of a submarine. The shape of the compound airfoil is shown in figure 1. As for the shape of the front airfoil and the rear airfoil, NACA0012 was used. The full length of the compound airfoil is 3 m, and the length of the movable airfoil is 1 m, respectively. The full length of the compound airfoil is 3 m, and the length of the movable airfoil is 1m, respectively. The gap is set up between the movable airfoil and fixed wing. The movable airfoil (front airfoil and rear airfoil) was divided into 80 panels, over which the vortices were distributed. The fixed wing was divided into 58 panels, over which the vortices were distributed.

2.4. Calculation conditions
Two-dimensional calculations were performed for incompressible and viscous flows. Here, the target object is assumed as the rudder of the submarine, and the velocity at the time of coming alongside the quay is assumed. So, the fluid is water, and the chord length of the airfoil is set as $C = 1.0$ m. Although the cruising speed of the submarine is 20 knots, at the time of coming alongside the quay, the velocity reduces to about 1 knot. So, the main flow velocity U used in the calculation was set as 0.5 m/s. The Reynolds number was set as $Re = 3.8\times10^5$. Every calculation continued to more than non-dimensional time $T = N\Delta t\ U/C = 5$ at intervals of time step $\Delta t = 0.01$ s (N is the number of times of calculation (= 1000 times)). Here, the non-dimension time $T = 5$ is equivalent to 10 seconds in real time. The calculation area was from -2 m to 15 m in x direction, and ±10 m in y direction. The origin was set at lift center which is 1/4 chord length of airfoil from the leading edge. Here for convenience, the front side airfoil was called the 1st airfoil, and the rear side airfoil was called the 2nd airfoil.

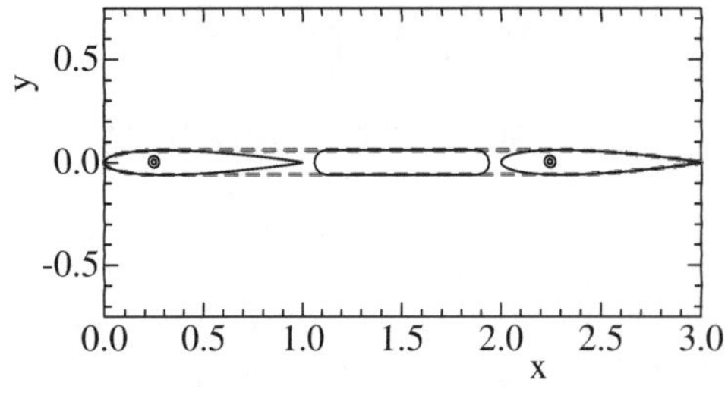

Figure 1. Shape of the compound airfoil; tandem arrangement of a movable airfoil, a fixed wing, and a movable airfoil.

2.5. Calculation parameters

The main parameter in the numerical experiment was the gap between moving airfoil and fixed wing and the angles of attack of moving airfoils. The gap size is two kinds and is 60 mm and 15 mm. Since it is provided before and after the fixed wing, the length of the fixed wing is changed. The length of those fixed wings is 880 mm, and 970 mm, respectively. There are two kinds of situations of the angle of attack of the movable airfoil. One of which is the case that the angle of attack of both wings is the same. This case is 3 kinds. The combination of the angle of attacks is the following. (α1°, α2°) = (5°,5°), (10°,10°) and (15°,15°). Another is the case where an angle of attack is attached to airfoil of one of the two. This case is 6 kinds. The combination of the angle of attacks is the following. (α1°, α2°) = (0°,5°), (0°,10°), (0°,15°), (5°,0°), (10°,0°) and (15°,0°).

3. Results and discussions

3.1. Performance of fixed wing

The flow pattern and the fluid force characteristic of the fixed wing were investigated. Since the fixed wing was used without the angle of attack, it was carried out about the single fixed wing in the case of having no angle of attack. The flow pattern of the fixed wing and the time history of fluid force are shown in figure 2. In the figure, the length of a fixed wing is the case of 880 mm and 970mm. Although the length of those fixed wings differed, since the shape was the same, it was found that both of flow patterns and fluid force characteristics become the same. The drag coefficient of NACA0012 airfoil in calculation of this technique was $C_D = 0.0072$, the drag coefficient of this fixed wing is $C_D = 0.0035$, and it was shown that the fixed wing of this proposal is smaller than NACA0012 type airfoil. It proved that the shape of the fixed wing of the proposal of this is a good shape.

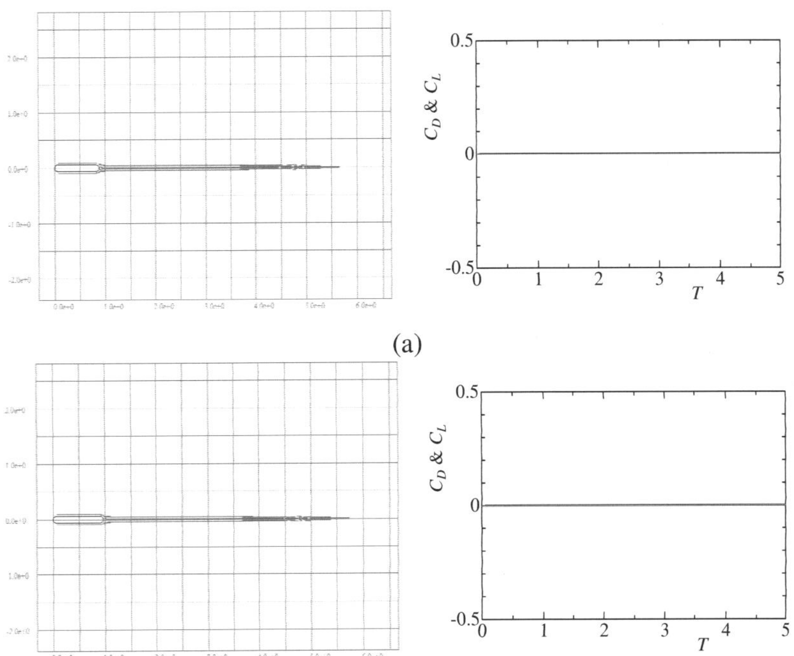

(a)

(b)

Figure 2. The flow pattern of a fixed wing, and the time history of the fluid force characteristic, (a) the wing length is 880 mm, $C_D = 0.0035$, $C_L = 0.0$, (b) the wing length is 970 mm, $C_D = 0.0035$, $C_L = 0.0000$.

3.2. Performance of compound airfoil

3.2.1. In the case of same angle of attack. The flow pattern of the compound airfoil and the time history of fluid force are shown in figures 3 and 4. There is no turbulence in the 1st airfoil and the lift is increasing. On the other hand, since there is inflow from the 1st airfoil and the fixed wing in the 2nd airfoil, turbulence has occurred. In the case of 880 mm fixed wing, it is shown that the lift of the 2nd airfoil is smaller than the lift of the 1st airfoil. In the fixed wing, the negative lift has occurred temporarily. When the angle of attack is 15 degrees, the separation bubble is formed by the suction side of the 1st airfoil. In the case of 970 mm fixed wing, although the lift has occurred temporarily in the fixed wing, on the whole, it becomes a negative lift. The lift coefficients in each condition are shown in table 1. Here, the NACA0012 type lift coefficient obtained by this calculation technique was 0.5400 in 5 degrees of angle of attacks, was 1.1831 in 10 degrees of angle of attacks, and was 1.6403 in 15 degrees of angle of attacks. In any case, a total lift is larger than the case of a single NACA0012 type airfoil. When the case of 880 mm fixed wing is compared with the case of 970 mm fixed wing, the one where the gap between the movable airfoil and the fixed wing is larger seems to be convenient for the use.

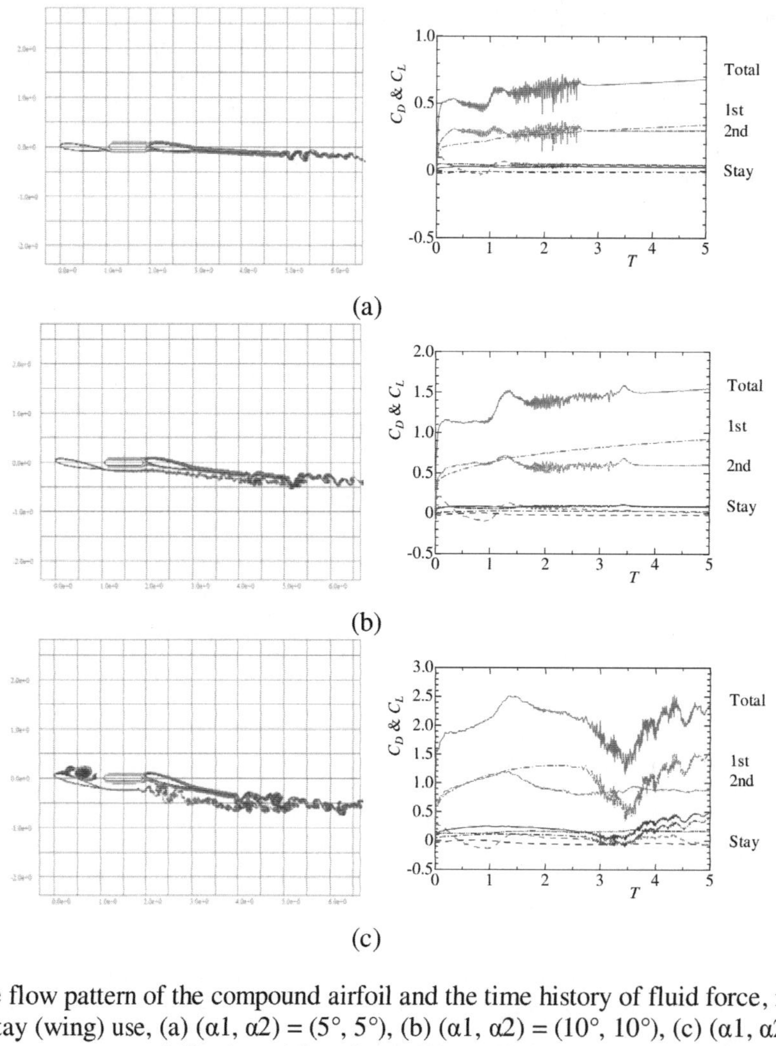

Figure 3. The flow pattern of the compound airfoil and the time history of fluid force, in the case of 880 mm stay (wing) use, (a) (α1, α2) = (5°, 5°), (b) (α1, α2) = (10°, 10°), (c) (α1, α2) = (15°, 15°), the red line shows lift coefficient.

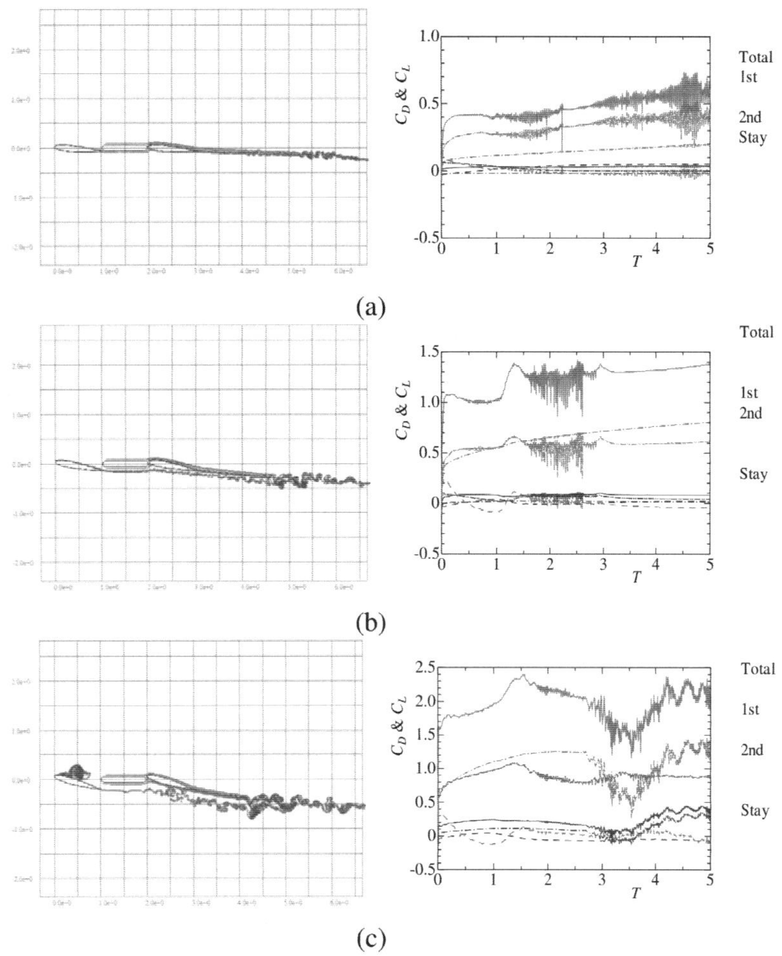

Figure 4. The flow pattern of the compound airfoil and the time history of fluid force , in the case of 970 mm stay (wing) use, (a) (α1, α2) = (5°, 5°), (b) (α1, α2) = (10°, 10°), (c) (α1, α2) = (15°, 15°), the red line shows lift coefficient.

Table 1. The value of the lift coefficient in each condition in case an angle of attack is in both airfoils, at the time $T = 5$.

Wing type	S880				S970			
(α1,α2)	1st	2nd	stay	total	1st	2n	stay	total
(5, 5)	0.3416	0.2994	0.0382	0.6792	0.1986	0.4280	-0.0097	0.6169
(10, 10)	0.9236	0.6050	0.0154	1.5440	0.8054	0.6163	-0.0420	1.3797
(15, 15)	1.5383	0.8900	-0.0707	2.3576	1.3528	0.8753	-0.0726	2.1555

3.2.2. The angle of attack is attached to one airfoil. This situation has two kinds of cases. The one is the case where an angle of attack is attached to the 1st airfoil. Another is the case where an angle of attack is attached to the 2nd airfoil. The flow pattern of each case and the time history of fluid force are shown in figures 5 and 6. The lift coefficients in each condition are shown in tables 2 and 3. Here, the underlined values in the table are larger than the values for the single airfoil. In the case where an angle of attack is attached to the 1st airfoil, no separation bubble can be seen there be on all airfoils. A negative lift can be seen have occurred in the fixed wing from the time history. It is found that it is smaller than the case where the total lift of the compound airfoil has a single NACA0012 type airfoil. Even when an angle of attack is attached to the 2nd airfoil, no separation bubble can be seen there be on all airfoils. The separation produced in both of the cases at the time of 15 degrees of angle of attacks is not produced. Although there is no angle of attack in the 1st airfoil, the lift has occurred on the 1st airfoils. The lift of the 1st airfoil at the time is smaller than the lift of the 2nd airfoil. Direction of the 2nd airfoil is found seeing synthetically and governing the overall direction of the flow. Although the total lift of the compound airfoil is larger than the case of the NACA0012 type airfoil, the value does not exist more than twice.

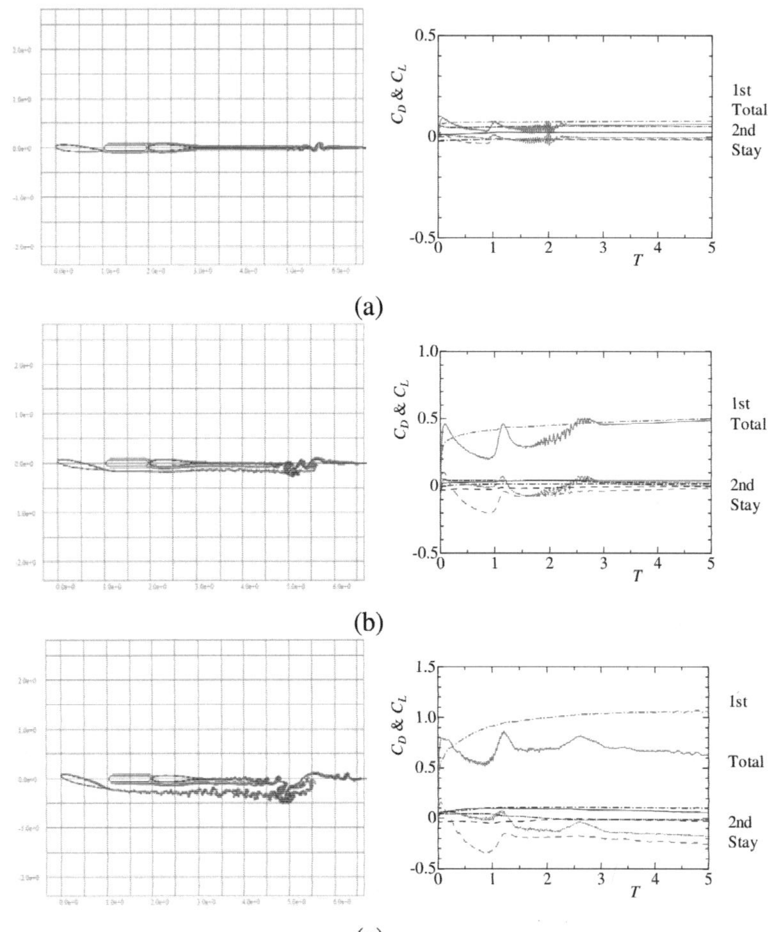

Figure 5. The flow pattern of the compound airfoil and the time history of fluid force, in the case of 880 mm stay (wing) use, in case an angle of attack in the 1st airfoil, (a) α1 = 5°, (b) α1 = 10°, (c) α1 = 15°, the red line shows lift coefficient.

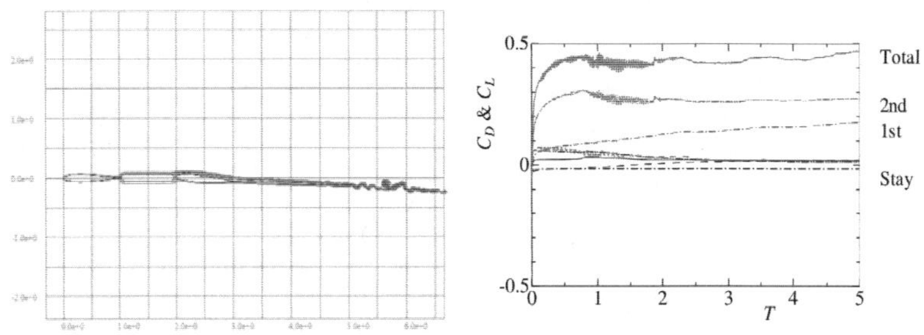

(a)

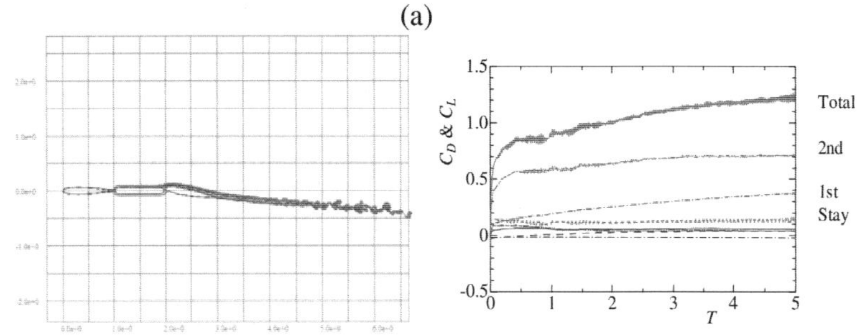

(b)

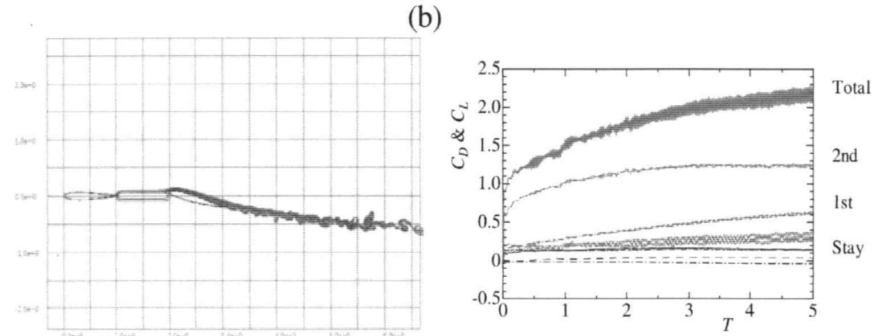

(c)

Figure 6. The flow pattern of the compound airfoil and the time history of fluid force, in the case of 880 mm stay (wing) use, in case an angle of attack in the 2nd airfoil, (a) α2 = 5°, (b) α2 = 10°, (c) α2 = 15°, the red line shows lift coefficient.

Table 2. The value of the lift coefficient in each condition in case an angle of attack is in the 1st airfoil, at the time $T = 5$.

Wing type	S880				S970			
(α1,α2)	1st	2nd	stay	total	1st	2n	stay	total
(5, 0)	0.0781	-0.0046	-0.0127	0.0608	-0.0026	0.0127	-0.1332	-0.1231
(10, 0)	0.4992	0.0033	-0.0171	0.4854	0.4722	-0.1319	0.0479	0.3882
(15, 0)	1.0562	-0.1752	-0.2556	0.6254	1.0082	0.1269	-0.3707	0.7644

Table 3. The value of the lift coefficient in each condition in case an angle of attack is in the 2nd airfoil, at the time $T = 5$.

Wing type	S880				S970			
(α1,α2)	1st	2nd	stay	total	1st	2n	stay	total
(0, 5)	0.1754	0.2726	0.0192	0.4672	0.1552	0.3558	-0.0271	0.4839
(0, 10)	0.3802	0.7217	0.1579	<u>1.2598</u>	0.1830	0.5759	0.1292	0.8881
(0, 15)	0.6358	1.2461	0.3531	<u>2.2350</u>	0.4768	1.1642	0.2779	<u>1.9189</u>

4. Conclusions

Numerical simulations of initial flow around the compound airfoil with angle of attack were performed by use of the vortex method. The following conclusions were obtained.

(1) In the single test, the fixed wing obtained the drag coefficient lower than the movable airfoil.

(2) Although the total lift of the compound airfoil is larger than the case of the NACA0012 type airfoil, the value does not exist more than twice.

(3) The performance of the compound airfoil of the case where the gap between the movable airfoil and the fixed wing is large is good.

(4) Direction of the 2nd airfoil is governing the overall direction of the flow.

Acknowledgment

The greater part of this study was performed during the author's stay in TU-Berlin (Berlin, Germany). Gratitude is expressed to Professor Dr. -Ing. Christian Oliver Paschereit and Dr. -Ing. Christian Navid Nayeri for the help with the calculation enviroment. Thanks are due to the staff of the National Defense Academy of Japan (NDA) for the opportunity to perform research in TU-Berlin. Furthermore, the author is thankful to Mr. Kouhei Ishii and Mr. Teppei Makita, who helped in the numerical experiment and who are the student staff of the author's laboratory at the NDA.

References

[1] Abbott I H, Doenhoff A E and Stivers L S 1945 *National Advisory Committee for Aeronautics* 824
[2] Yokoi Y 2015 *Applied Mechanics and Materials* **798** 609
[3] Yokoi Y, Nayeri C N and Paschereit C O 2016 *Proc. 7th Int. Conf. on Vortex Models (Rostock)* paper no. Yokoi et.al.
[4] Yokoi Y 2017 *Proc. 8th Int. Conf. on Mechanical and Aerospace Engineering (Prague)* 447
[5] Inumaru N and Kitamura N 1968 *Technical Memorandum of National Aerospace Lab.* TM-140
[6] Tachi K 1988 *The Journal of Japan Institute of Navigation* **98** 25
[7] Jia W, Nakamura Y, Yasuhara M and Kuwabara K 1989 *Trans. of the Japan Scoety for Aeronautical and Space Science* **37-428** 430
[8] Yoshimura Y, Sasaki N and Takekawa M 1997 *Journal of Marine Science and Technology* **181** 191
[9] Hiraoka T and Kajitani H 2003 *Bulletin of Sojo University* **28-1** 141
[10] Kamemoto K 1993 *Japan Society of Computational Fluid Dynamics* **2-1** 20
[11] Kamemoto K 1994 *Japan Society of Computational Fluid Dynamics* **2-2** 28

IOP Conf. Series: Journal of Physics: Conf. Series **1215** (2019) 012007 doi:10.1088/1742-6596/1215/1/012007

Research on Automatic Tool Delivery for CNC Workshop of Aircraft Equipment Manufacturing

Ping Zeng[1], Fei Wu[1], Tongxing Zhi[1], Le Xiao[1] and Shaowei Zhu[2.]

[1]Dalian Start Hi-tech Developing Co.,Ltd. Dalian 116023, P.R.China
[2]Avic Chengdou Aircraft Indutrial（Group）Co.,Ltd. Chengdu 610091, P.R.China

Abstract. Aiming at the inefficiency of tool distribution in CNC workshop of aircraft equipment manufacturing, and adapting to the particularity of distribution without large network environment, The solution of intelligent tool delivery using AGV based on machine vision and QR Code navigation is proposed. Image processing of AGV path ribbon, path navigation error analysis and QR Code position calibration are studied in the planned path to achieve machine vision plus QR Code redundant navigation positioning. The AGV can complete the accurate loading and unloading tool of the robot library through the QR Code interaction mode with the machine tool. To realize automatic distribution of tool in CNC workshop of aircraft equipment manufacturing.

1. Introduction

Tools are an important part of the machining system. Its variety in great quantities, usually wait before and after the processing, distribution and storage aspects consumes a lot of time, become the bottleneck of further shorten the production cycle, seriously impedes the development of NC machining efficiency and the realization of the mode of production on time.

Therefore, the development of professional transport equipment to achieve tool distribution and production operations of close coordination and collaborative integration, To improve production efficiency and reduce production cost has important engineering value and practical significance.

AGV (Automated Guided Vehicle) has many advantages of good flexibility, high degree of automation and high intelligent level. It is the main equipment for distributing tool in the CNC workshop at present. However, due to the special requirements of the aircraft equipment manufacturing industry, the workshop has no large network environment, which makes AGV unable to communicate with the tool distribution center in real time, It is difficult to realize effective scheduling and management. Therefore, the existing AGV can not directly meet the requirements of too distribution in aircraft equipment manufacturing industry.

Based on the production practice of aircraft manufacturing enterprises, AGV navigation [1,2] and QR Code redundant positioning technology in the absence of large network environment are studied on the basis of existing laboratory environment and resources. An automatic tool delivery and transportation system based on AGV scheduling was established [3], so as to realize the automatic delivery of tools in the CNC workshop of aircraft assembly manufacturing [4].

2. Research on AGV machine vision navigation technology

2.1. Realization principle

Visual navigation and guidance method is to brush the ribbon on the AGV driving path, and the ribbon should have obvious color difference with the driving path. A photographic sensor is installed on the

AGV to compare the pictures taken continuously with the stored pictures, and the offset signal is output to the driving control system. The control system corrects the direction of AGV by calculation, realizes the navigation of AGV[5], precise control and accurate position positioning. The process is shown in Figure 1.

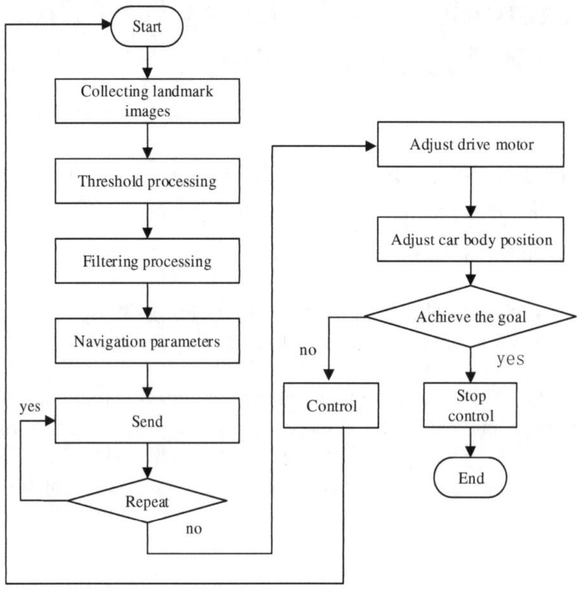

Figure 1. Visual navigation AGV control flow chart

2.2 Image preprocessing

There are noise points in the collected images, we need to process the images.

2.2.1 Grayscale processing of color map. In order to reduce the complexity of image recognition, we need to grayscale the image. It determined on the specific color values of the palette, and the three components of the grayscale are equal. When the image is 8 bits, there are 256 color values in the palette, each of which is exactly from 0 to 255, and all three components are equal. When the image is 4-bit, there are 16 color values in the palette, divided equally into 255 color values, three components are equal. When converted to two, the palette has four colors, equally spaced 255 colors, and three components are equal. When the image is 1 bits, there are two color values in the palette, which are 0 and 255, respectively, indicating black and white. With the formula Y=Y/ (1<< (8- conversion digit)), the color image can be transformed into a grayscale image.

*2.2.2 Image enhancement.*Image enhancement is to transform the information of the image, to highlight the image features of interest, to suppress the unnecessary image features, so that the information presented by the image can be highlighted. In the whole process of image enhancement, without considering the impact of image quality degradation, the image after processing will be different from the original image.

*2.2.3 Two value processing of grayscale image.*Binary image processing is the processing of the image into black or white. The threshold value of the 256 brightness levels of the gray image is selected, and the contour features of the image can still be seen after binarization. In image processing, image binarization is very important, especially in the actual application of image processing, to binary image processing system has a lot of, want to undertake the greyscale image processing and analysis, the gray image binarization processing, image binarization, image characteristics exist in 0 or 255 pixels, and the data processing small, make processing is very simple.In order to get a relatively clear binary image, interconnecting regions are usually used.When the pixel value of the grayscale map is greater than the threshold value and is selected as the area of interest, its gray value is represented by

255, otherwise these pixel points are excluded from the area of interest, and the gray value is represented by 0.The threshold method can be used to distinguish certain objects from other objects if the gray value of a particular object is uniform and the same level as that of other objects.If the difference between an object and the image background cannot be reflected in the gray value, the feature difference can be transformed into the gray difference, and then the threshold segmentation can be used to process the image.The threshold can be adjusted dynamically, and the binarization of images can also be observed dynamically .

2.3 Edge detection of path labeling line image based on Laplacian operator

There are many kinds of operators for edge detection, such as Candy operator, Sobel operator, Laplacian operator, etc. Sobel operator uses the first derivative information, the extracted image edge is fuzzy; Candy operator edge detection is more accurate, but the processing of more data, poor real-time. Laplacian operator is a directionless operator. It uses second-order derivative information. Compared with Candy operator, Laplacian operator has less computation and strong real-time performance, and the edge information extracted is clear. Therefore, the Laplacian operator is used in this study.

Laplacian is used to detect the marker image by the second derivative of gray value.The identification line image has the characteristic of impulse edge, and the range of pulse edge is determined by detecting the upper and lower peak of the first derivative, and two edge points of the identification line are obtained by detecting the two second derivatives of the impulse profile.

In the process of image processing, the Laplacian operator function can be realized by template convolution.In the template, the coefficient of the corresponding center pixel is positive, and the coefficient of other neighboring pixels is negative, and the sum of all coefficients is zero, so that there is no gray deviation [4].

Several operators, including Lpalacain, are applied to perform edge detection on the same identification line image. The results of edge detection are shown in figure 2.By comparing the detection results, the edge extraction of Laplacian operator is more accurate, and the extraction results of sobel and Canny operator are more clear, which provides convenience for the extraction of parameters of relative position of car body.

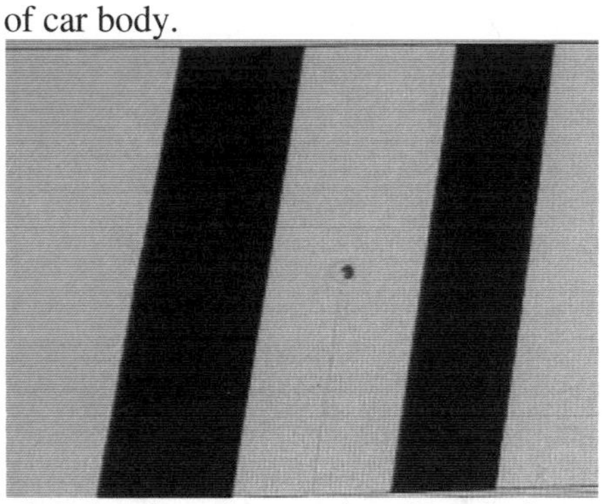

Figure 2. Edge detection results of Laplacian operator

2.4 Use parallelogram to find the deflection Angle

The parallelogram is extracted from the collected image, and the result is shown in Figure 3. Where O (x, y) is the intersection of two diagonal lines of a parallelogram, F (x1, y1) is the midpoint on the edge of cd. The abscissa of O (x, y) and the ordinate of F (x1, y1) are taken to mark y1, and the line segment OE obtained is perpendicular to the edge cd. The algorithm of AForge.NET is used to evaluate line segments OF, OE and EF, and cosine theorem is used to calculate the degree of angleα. As shown in Figure 3.

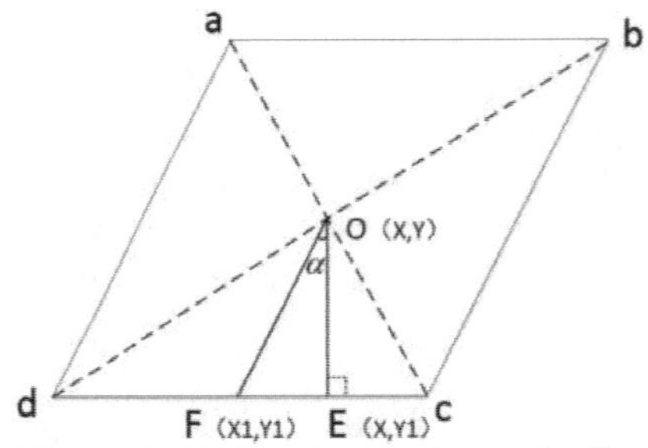

Figure 3. Deflection angle result diagram

Therefore, the image acquired by the camera is processed, the path and background information are distinguished by threshold images, the path is obtained by binarization images, the parallelogram in the field of vision is judged, the central position of the car body is obtained according to the geometric principle, and the position information of AGV in the coordinate system is obtained[7].

3 Realization of precise navigation based on machine vision plus QR Code redundant location

QR Code is easy to control communication, has no interference with acoustooptic,can be reliably distinguished in oblique angle or dim light, and can also read code in the case of high-speed movement of the product. Therefore, the precise navigation of tool distribution is achieved by using ground color belt machine vision and QR Code redundancy positioning.

3.1 QR Code position calibration on path

For the distribution route layout of the workshop, it is necessary to mark the QR Code position at the intersection, turn and straight path.The QR Code information of path calibration includes attitude, control, position and other information[9] . As shown in Figure 4-5.

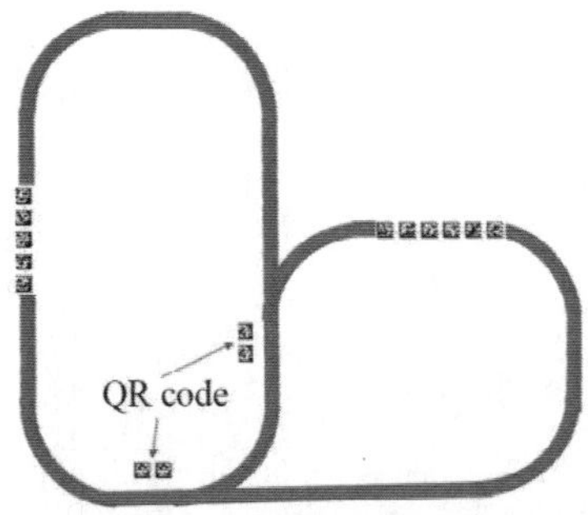

Figure 4. QR Code position

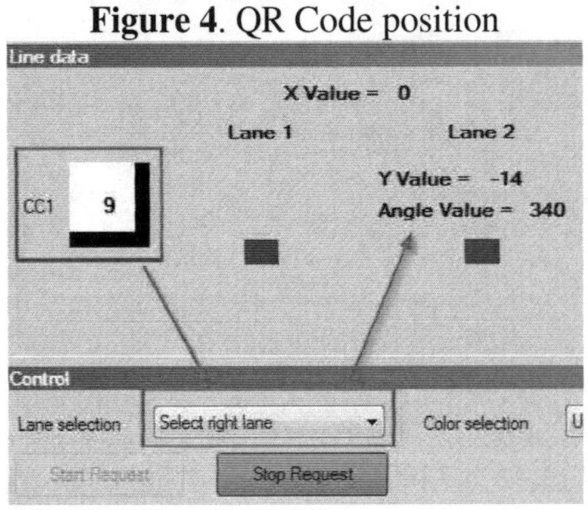

Figure 5. Debugging of visual positioning system

3.2 Realization of precise navigation and location

According to the experimental environmental conditions and application scenarios, the communication between the code reader on AGV and the AGV on-board control system is carried out in a small environment using protocols[6].

When AGV goes to the ground QR Code position, the vehicle's Code reader will feedback the ground QR Code position X-axis offset, Y-axis offset and Angle offset to the vehicle's control system. The control system will adjust the vehicle's position successively until it reaches the allowable error range [7].The flow of its walking algorithm is shown in figure 6:

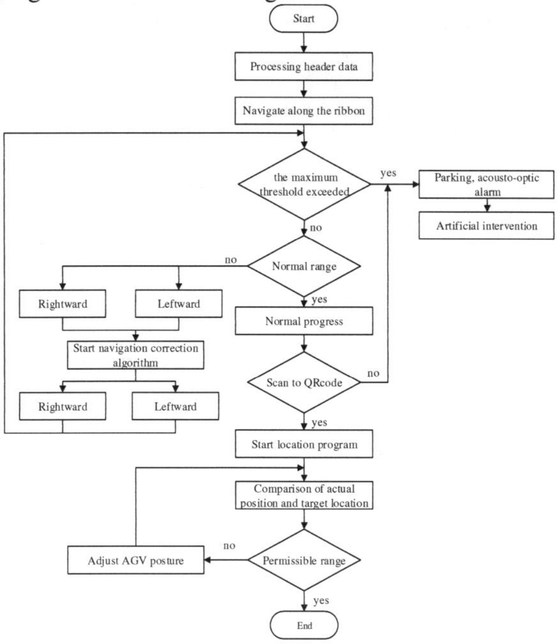

Figure.6. AGV navigation and positioning process

The following table is a ground QR Code calibration point (target position is 32 mm X axis, 5 mm Y axis, angle -10°), AGV travel speed of 1 km / h, nine times into the QR Code measured data. As shown in Table 1.

Table 1. AGV entering QR Code calibration precision test

Target / Time	X(0.1mm) 32	Y(0.1mm) 5	A(0.1°) -10	result
1	39	41	7	OK
2	41	15	7	OK
3	43	8	7	OK
4	42	-21	7	OK
5	39	18	5	OK
6	31	7	7	OK
7	42	36	4	OK
8	33	25	7	OK
9	43	0	6	OK

According to the accuracy test of the actual operation result of the workshop, the control precision of the running track is (+4 mm), which can maintain the precision control of the running track and ensure the running speed and stability. Meet the actual requirements of workshop tool operation. Compared with the traditional AGV navigation, this scheme is easy to install, low in cost, easy to maintain in the later stage and high in accuracy [8].

4 Tool delivery and transportation process

The workshop sets up the intelligent tool delivery system, hereinafter referred to as the dispatching system, which is mainly responsible for AGV scheduling and information acquisition.

Set the QR Code reader at the key node of the machine tool, and the reader is connected to the machine tool.The dispatching system can monitor and control AGV through communication with the machine control system and the QR Code reader[8]. As shown in figure 7.

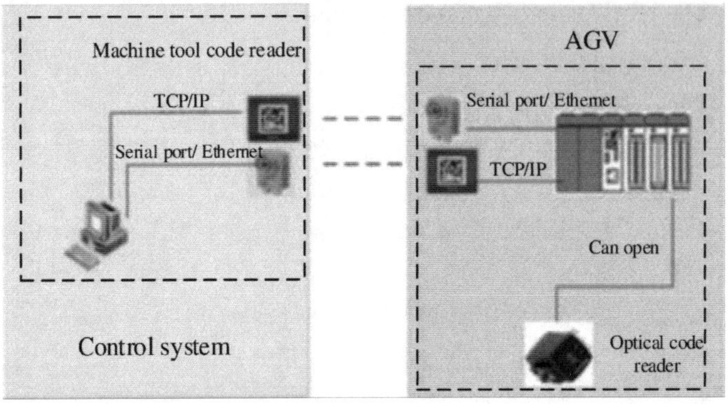

Figure 7. Information interaction between machine tool and AGV

4.1 Tool delivery process

The dispatching system puts the machine task into the queue according to the priority. When the machine task is assigned to AGV, the dispatching system sends the QR code written with the logistics task information to the AGV display screen. At the same time, it generates the logistics route according to the task type, resource type and the target machine tool information, and writes it into the AGV navigation system. AGV receives task instructions, goes to the tool handling station to pick up the tool in accordance with the prescribed route, interacts with the QR code reading and writing terminal of the tool loading and unloading station, holds up the tool tray (loaded with the tool), and carries out transportation according to the path planning route [6], as shown in Figure 8. When the AGV reaches the target machine tool, it interacts with the QR code reading and writing terminal of the machine tool [10]. The robot next to the machine tool loads the tools on the AGV onto the tool holder. After completion, AGV returns to the tool loading and unloading station according to the original way, places the tool tray in the loading and unloading station, returns to the waiting area, the dispatching system reads the relevant information, and closes the task. As shown in Figure 8.

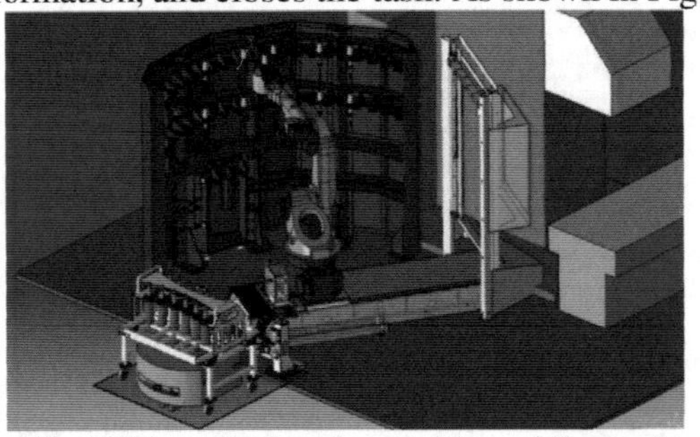

Figure 8. AGV running to the position of the tool holder

4.2 Tool recovery process

The dispatching system receives the task of recovering the tool sent by the machine tool and puts it into the task queue according to the task priority.When the task is assigned to AGV, the dispatch system sends the QR code of the logistics task to the display screen of the car. At the same time, the logistics route is generated according to the task type, resource type and target machine tool, and written into the AGV navigation system [7].When the AGV receives the task, first go to the tool loading and unloading station according to the navigation route, and lift the tool tray, and then go to the position of the target machine tool according to the predetermined route.Adjust the position to the buffer area of the tool bank. After the machine tool QR code reader recognizes the task information of the car, unload the tool that needs to be unloaded to the tool tray of AGV.After unloading, AGV goes

IOP Conf. Series: Journal of Physics: Conf. Series **1215** (2019) 012007 doi:10.1088/1742-6596/1215/1/012007

to the tool loading and unloading station to unload the tool.The AGV returns to the waiting area, and the system reads the information and closes the task.

5 Conclusion

Redundant navigation positioning technology of machine vision and QR code not only enables AGV to maintain the accuracy control of running track along the planned path, but also ensures the running speed and stability of AGV. Moreover, it is easy to install and maintain.On the basis of accurate navigation, information interaction between AGV and machine tool via QR code.Realize accurate tool loading and unloading of the tool bank robot.In the process of tool distribution, recycling and loading and unloading, humanized operation is realized to avoid mistakes, cost is reduced, production efficiency is improved, and good implementation experience is provided for other CNC workshops without wireless network environment.

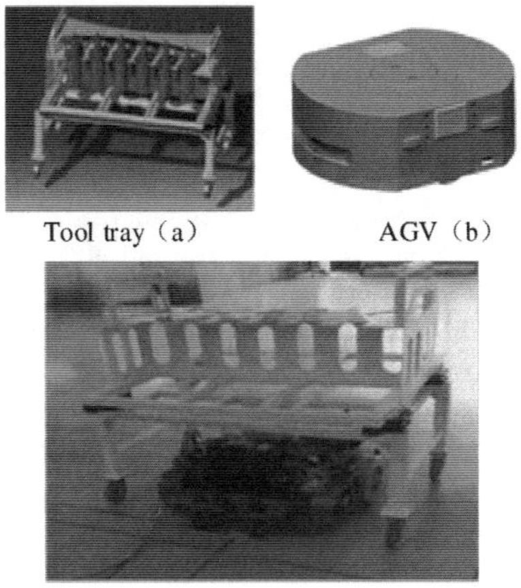
Tool tray（a） AGV（b）

Figure 9. Tool tray and AGV whole

6 Reference

[1] Z.M.Wang, J.Y.Liu and H.Yue, *Summary of research on mobile robot autonomous navigation technology,* Journal of Tianjin vocational and Technical Teachers College, 2004, **6** (4): p.32-34.

[2] P.Zhou, Y.M. Wang, Y Zhao and Y.F.Xu, *Two classes of segmentation methods in autonomous navigation of unstructured natural images,*Journal of Agricultural Machinery, 2004, **4** (2): p. 12-14.

[3] H.G.Zhen, X.H.Qi. *An intelligent vehicle path recognition algorithm based on non uniform line ,* Electronic products world, 2010, **11** (8):P 40-42.

[4] W.Wei,*Intelligentcontroltechnology,*Machinery Industry Press, 2007.

[5] W.Chen ,B.C.Li and H.T Sun, *Research on AGV fuzzy optimal control based on visual navigation* ,China Mechanical Engineering, 2006.

[6] H. Qian. *Design of mobile robot navigation system based on fuzzy control*, Harbin Institute of Technology, 2013.

[7] Z.S.Zhang, *Designand research of autonomous mobile robot based on visual navigation*, Harbin Institute of Technology, 2008.

[8] Q.D.Liu, *Research on AGV system guided by machine vision*, Harbin Institute of Technology, 2008.

IOP Conf. Series: Journal of Physics: Conf. Series **1215** (2019) 012008 doi:10.1088/1742-6596/1215/1/012008

CFD Simulation of Unsteady Interaction Between Rotor and Fuselage for Canard Rotor Wing in Hovering Using Overset Grids

Pang Chao[1*], Gao Zhenghong[1]

1 Northwestern Polytechnical University, School of Aeronautics, 710072 Xi`an, China

pangs_c@163.com

Abstract. Different from the traditional helicopter, there would be a severe interaction between rotor and fuselage with canard and horizontal tail during hovering for canard rotor wing (CRW) aircraft. The un-steady RANS (URANS) equations discretized on overset grid are solved in this paper to simulate this kind of aerodynamic interaction. A UH60-A rotor hovering case is used to validate the numerical method. Subsequently, the numerical method is employed on two configurations, i.e., Isolated rotor configuration and Full CRW configuration under numerous collective pitch angle. The force and moment coefficients for different parts with respect to different collective pitch angles are obtained. The result shows that downwash of rotor has a stronger impact on fuselage than canard wing and horizontal tail. Compared with isolated rotor, fuselage with canard wing and horizontal tail has a slight impact on the trust and torque of the rotor in full CRW aircraft configuration.

1. Introduction

The concept of canard rotor wing (CRW) aircraft is firstly proposed by Rutherford etc. al [1], which can be regarded as a combination conception of the fixed wing and the helicopter. The advantages in vertical landing and taking off of helicopter as well as the high-speed cruise of fixed wing are gathered in CRW. The most famous demonstrator vehicle is Dragon fly X-50A [2], which was assisted by NASA Ames and so on.

Helicopter mode, conversion mode and fixed wing mode are three typical modes during the whole fight. Helicopter mode is used to vertically take off or land. Fixed wing mode is used to cruise with high speed. Aerodynamic lifting forces are transferred from rotor to canard wing and horizontal tail with the increasing free stream velocity during conversion mode. Lifting force for CRW aircraft in cruise state mainly comes from canard wing and horizontal tail, which means that area of those lifting surface must be big enough to provide sufficient lifting force. Meanwhile, interaction between those fixed wings and rotor will be quite severe due to the big area of canard wing and horizontal tail, which causes the aircraft hard to control.

Researchers have done many experiments [3-6] including wind tunnel experiment and flying test to explore the aerodynamic interaction between different parts of CRW. As for numerical simulation, Saeid [7] etc. al. has used momentum disk source method to simulate the hovering configuration of CRW aircraft near the ground, where the ground may have a “ground effect” on the rotor. Due to plenty of assumptions, the accuracy of momentum source method is less than that of URANS method. Li etc. al. [8] has solved URANS equations to simulate downwash effect on canard wing and horizontal tail during hovering. But they didn’t consider the fuselage of CRW aircraft as well as their effects on the rotor. Sun etc. al. [9] has solved the URANS equations to study the interactions during the forward fight. But they consider only one case, which was not sufficient for designers to use.

In this paper, URANS equations combined with SA turbulence model are solved on the overset grid to simulate a full CRW aircraft configuration as well as an isolated rotor configuration. CRW aircraft model (Fig. 1) in this paper is self-designed by NPU (Northwestern Polytechnical University) Smart Loong work team. The rotor is trapezoid without twist. Due to special aerodynamic characters, airfoil of rotor must be symmetric in flow direction (x direction), which is now an elliptic airfoil. The canard wing and horizontal tail were assigned far away from the rotor. A T type tail is designed for this CRW aircraft. All of these shape characters were designed to avoid the severe unsteady aerodynamic interaction of CRW aircraft.

The rest of this paper is outlined as follows. Section 2 gives a brief introduction to numerical schemes including governing equations of CFD simulation and grid assemble method for overset grids. A hovering case for UH60-A rotor will be simulated using CFD simulation based on overset grids to validate the method in section 3. The result shows good agreement between CFD simulation and experiment. In section 4, the aerodynamic interactions between rotor and fuselage are analyzed respectively, which shows a severe effect on total thrust as well as torque for rotor. Some conclusions are summarized in section 5.

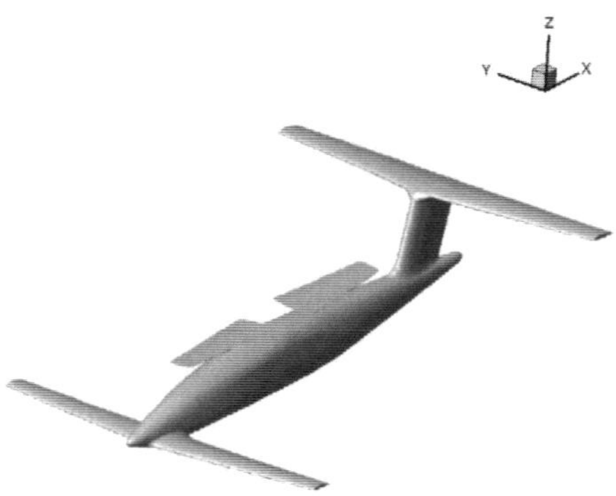

Figure 1. The geometry of a CRW aircraft.

2. Numerical Methods

2.1. Governing Equations

Unsteady RANS equations are used in this paper, which can be derived from the Navier-Stokes (NS) equations. Unsteady compressible NS equations in Cartesian coordinate can be written as:

$$\frac{\partial}{\partial t}\iiint_V \mathbf{Q}dV + \iint_S \mathbf{F}_i \cdot \mathbf{n}dS = \iint_S \mathbf{F}_v \cdot \mathbf{n}dS \tag{1}$$

where S is the surface of control volume; n denotes surface normal vector; V represents volume of control volume; **Q** is solution vector, $\mathbf{Q}=[\rho, \rho u, \rho v, \rho w, \rho E]$, ρ is density, u, v and w are three components of velocity, E is total energy; $\mathbf{F}_i$ denotes inviscid flux; $\mathbf{F}_v$ represents viscous flux. The 1-eq SA turbulence model is employed in this paper.

A finite volume method is used to discretize these equations. Roe upwind scheme is employed to get inviscid flux. Central difference scheme is used to obtain viscous flux. The implicit dual time step time marching scheme is adopted to get time accurate solutions. Multigrid algorithm is applied in pseudo time marching to accelerate convergent rate in one physical time step.

Characteristic boundary conditions based on Riemann invariant is set for far filed boundary condition and the adiabatic no-slip boundary condition is set for all viscous walls

2.2. Geometry and Grids

Geometry of CRW aircraft in this paper is shown in Fig. 1. The radius of the rotor is 1.397m, whose angular speed is 870 rpm. Collective pitch angle of rotor varies from 0°to 13°. The area of canard wing and horizontal tail is 1.2 m^2 and 1.5 m^2 respectively.

The overset grids for the full CRW configuration can be divided into two parts, i.e., near-body grids and off-body grids (Fig. 2). 3-D hyperbolic equations are solved to get near-body grids, and off-body grid is a series of Cartesian grids. The outer grid boundary is far enough away from the aircraft in all directions (nearly 20 times of rotor radii).

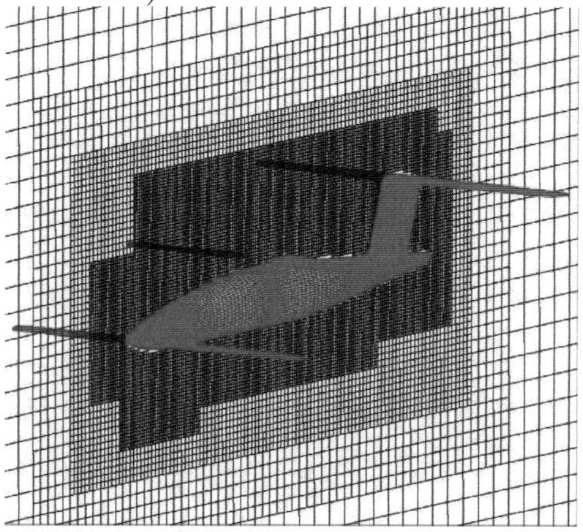

Figure 2. Overset grids for a CRW aircraft. (total vertex number is 13 million; red lines mean surface grid line; black lines mean a cutoff plane of off-body grids)

2.3. Grid Assemble

Grid assemble is an important step in overset grid method, which includes hole cutting and interpolation.

Chiu and Meakin [10] proposed hole-map method for hole cutting. This method contains 4 major steps.

1. Make a Cartesian box around all of the closed surfaces.
2. Set the x-y plane of the Cartesian box as a "map". A serial of rays in z direction can be obtained from the "map", then get the intersection point between these rays and all closed surfaces (water tight surfaces).
3. Use Eq. (2) to verify the position in the "map" for an arbitrary point P.

$$i_p = \mathrm{int}[(x_p - x_{\min})\frac{i_{\max} - 1}{x_{\max} - x_{\min}}] + 1$$

$$j_p = \mathrm{int}[(y_p - y_{\min})\frac{j_{\max} - 1}{y_{\max} - y_{\min}}] + 1 \qquad (2)$$

Compare the z value of point P with it of those intersection points to verify the P position roughly.

4. Use bilinear interpolation equation defined in Eq. (3) to determine whether point P is a hole point or not.

$$B_P = C_1 B_{ip,jp} + C_2 B_{ip+1,jp} + C_3 B_{ip,jp+1} + C_4 B_{ip+1,jp+1}$$

$$C_1 = \frac{(x_{ip+1} - x_p)(y_{ip+1} - y_p)}{\Delta x \Delta y}$$

$$C_2 = \frac{(x_p - x_{ip})(y_{ip+1} - y_p)}{\Delta x \Delta y}$$

$$C_3 = \frac{(x_{ip+1} - x_p)(y_{p+1} - y_{ip+1})}{\Delta x \Delta y}$$

$$C_4 = \frac{(x_p - x_{ip})(y_p - y_{ip})}{\Delta x \Delta y} \qquad (3)$$

$B_{ip,jp}$, $B_{ip+1,jp}$, $B_{ip,jp+1}$, $B_{ip+1,jp+1}$ are obtained from step 3. When B_p is less than 0.5, point P is considered as hole point.

The interpolation procedure in grid assembly aims at finding donor cells for those boundary cells without boundary condition (like hole boundary). Inverse map method [11] can find the nearest donor cells for boundary cells, as well as locate them in the donor cells. If one boundary cell cannot find a donor cell, this boundary cell will become an orphan cell. The solution variables would be obtained by trilinear interpolation using Eq. (4).

$$\mathbf{Q_{bp}} = \mathbf{Q}_1 + \mathbf{Q}_2\xi + \mathbf{Q}_3\eta + \mathbf{Q}_4\zeta + \mathbf{Q}_5\xi\eta + \mathbf{Q}_6\eta\zeta + \mathbf{Q}_7\xi\zeta + \mathbf{Q}_8\xi\eta\zeta \tag{4}$$

where $\mathbf{Q}_{bp}$ represents the solution vector for one of the eight vertices in boundary cells; $\mathbf{Q}_{1\sim8}$ denotes the solution vector for eight vertices in donor cells respectively; ξ, η, ζ represent relative position in donor cells for boundary cells vertices, which are between 0 and 1.

3. UH-60A Validation Case

3.1. UH60-A Rotor and Flow Conditions

The blade of UH60-A rotor has a nonlinear twist, and the tip is swept. Fig. 3 shows the platform of blade. The rotor contains 4 blades whose radius is 8.17m. Section geometry of rotor contains two airfoil i.e., SC1094-R8 and SC1095. Rotor solidity is 0.0826.

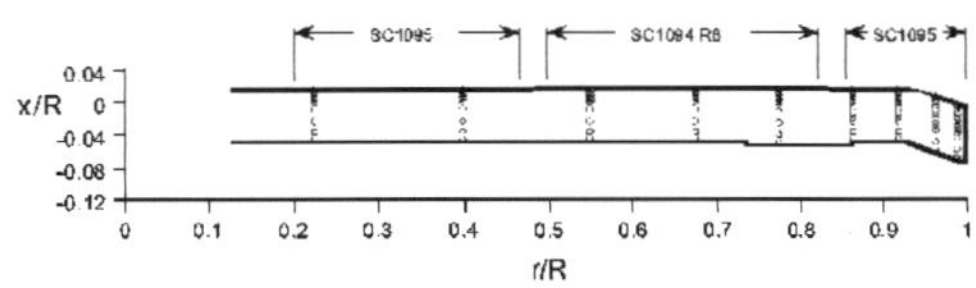

Figure 3. UH60-A blade platform geometry.

The validation case used in this section is hovering status. Tip Mach number of rotor is 0.628, and the collective pitch angle is 9°. Reynold number is 2.75×106.

3.2. Validation Result

Total number of vertices in grid is 9 million with 12 near-body grids as well as 27 off-body grids. Fig. 4 shows the near-body grids and the off-body grids. Boundary of off-body grids is located 3 blade radii [12] far from the rotation center of rotor.

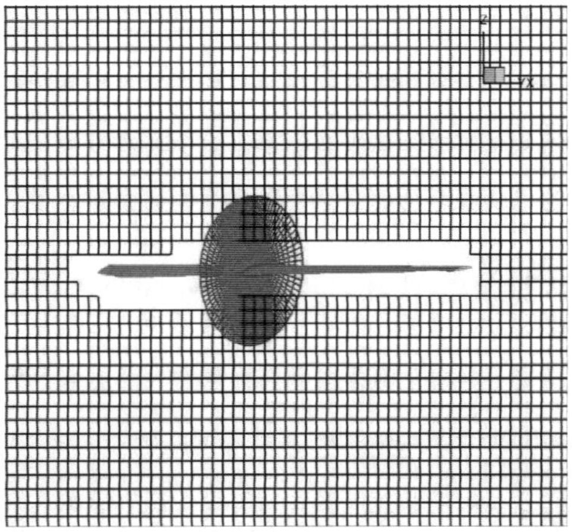

Figure 4. Overset grids for UH60-A blade. (red lines mean cut-off plane of near-body grids; black lines mean cut off of off-body grids)

IOP Conf. Series: Journal of Physics: Conf. Series **1215** (2019) 012008 doi:10.1088/1742-6596/1215/1/012008

The experiment section pressure coefficient can be found in Ref [13], where 6 sectional stations are chosen to compare. Note that reference velocity used to calculate the pressure coefficient is local speed instead of blade tip velocity. The calculation results compared with experiment data are shown in Fig. 5.

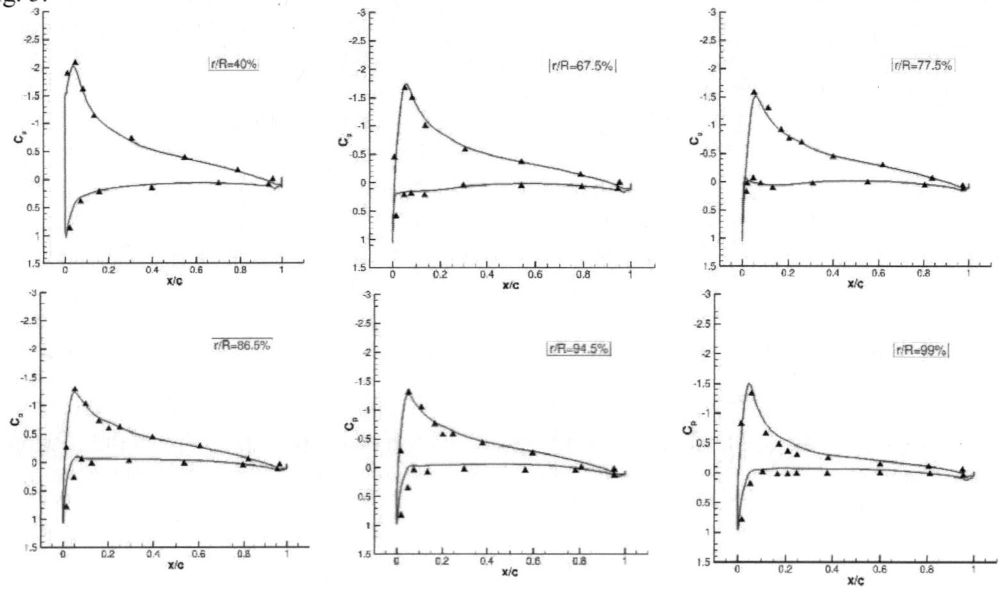

Figure 5. Comparisons of the CFD result (red line) about sectional pressure distribution with experimental data. (black delta symbol ▲)

Results show good agreement between CFD and experiment in blade root and tip region. In the area which is very close to blade tip, CFD simulation cannot correctly predict the pressure distribution due to a strong 3-D effect.

Table 1 gives the thrust and torque coefficient comparison. Note that reference velocity used to calculate those coefficients is blade tip speed, and reference area is rotor disk area. Reference length is the radius of rotor.

Table 1. Comparisons of the CFD result about performance with experimental data.

	CFD	EXP	EROOR
C_T/σ	0.083	0.085	2.3%
C_Q/σ	0.0067	0.0069	2.8%
FM	0.72	0.73	1.3%

Errors for all coefficients are within 5%. Thus, CFD solver used in this paper has great accuracy to predict rotor aircraft in hovering status.

4. Results and Discussions

To better understand the interaction between rotor and other parts of CRW aircraft, two different configurations are simulated using CFD method in this section. One of these two configurations is an isolated rotor without any other parts, another is the full CRW configuration which includes all parts of this aircraft.

There are five revolutions in each calculation case. The physical time step is set as 1 degree per step based on the 30 sub-iterations.

4.1. Aerodynamic Effect on Rotor Performance

Fig. 6 shows the difference in thrust and torque coefficient of rotors in those two configurations. Reference velocity is blade tip velocity which is nearly 0.374 Ma. Reference area and length are the area and radius of rotor respectively. Note that force and moment coefficient shown in Fig. 6 is an average of instant value in the final revolution. Torque coefficient is not zero when thrust coefficient is nearly zero due to skin friction and a separation region near the trailing edge of this elliptic airfoil [14]. Trailing edge separation region will always exist, which will cause an additional pressure drag. Meanwhile, this additional pressure drag will make the FM value lower than normal rotors with sharp trailing edge.

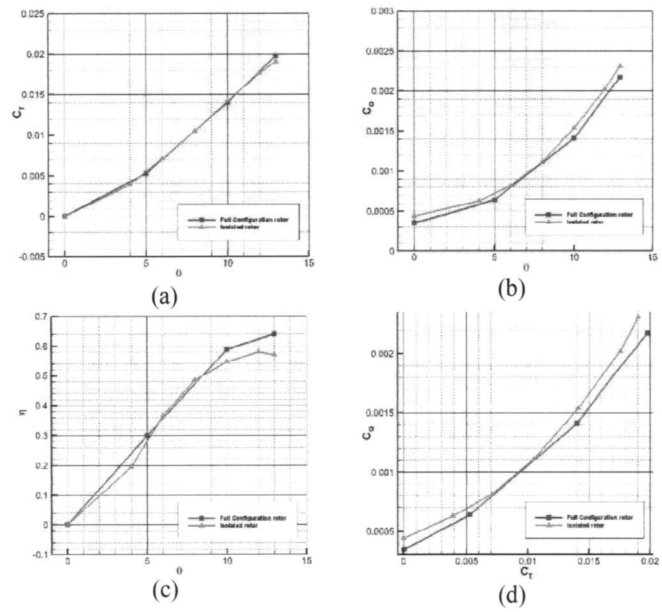

Figure 6. Performance comparison for the rotor between two configuration: rotor of full CRW (red line) configuration and isolated rotor(green line)

According to Fig. 6 (a), it can be observed that other parts of CRW aircraft have little effects on thrust coefficient of rotor unless collective pitch angle is big enough. Torque coefficient of rotor in full CRW configuration is obviously less than that of isolated rotor. It means that less energy will be needed to produce the same magnitude of thrust (Fig. 6 d).

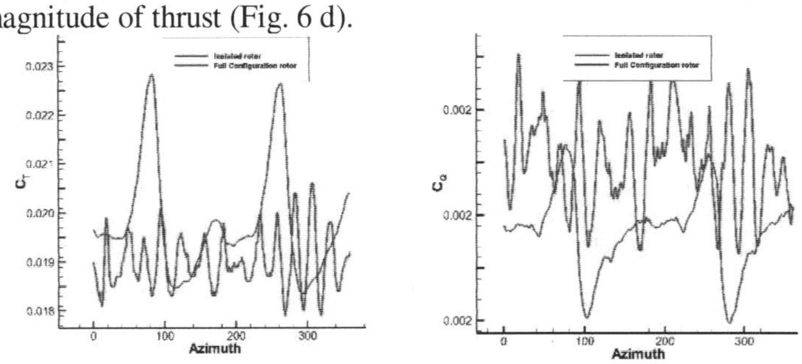

Figure 7. Instant coefficients of rotor in last revolution. (red line full configuration rotor; blue line isolated rotor)

Fig. 7 separately gives instant thrust and torque coefficient with respect to azimuth angle in the last revolution of rotor with 13°collective pitch angle. In this case, the azimuth angle equals 0 when rotor blade lie in y direction. The fluctuation of these coefficient is very small due to hovering status. There is an evident difference between these two configurations. In full CRW configuration, the coefficients have an obvious variation compared to isolated configuration when the blade of rotor is approaching the fuselage. The reason is that fuselage will resist the downwash flow induced by rotor while they are

IOP Conf. Series: Journal of Physics: Conf. Series **1215** (2019) 012008 doi:10.1088/1742-6596/1215/1/012008

closed to each other. This kind of resistance can produce an additional local pitch angle for the blades. Thereof coefficient for both thrust and torque is raised up rapidly.

4.2. Aerodynamic Effect on Fuselage

Fig 8 shows the downwash flow of rotor. The canard wing and horizontal tail of this CRW aircraft is designed as far as possible from the rotor disk. Velocity above the rotor disk is much smaller than it be-low the rotor disk form momentum theory. T type tail and the long distance mentioned above will help those fixed wings to avoid effect of rotor downwash flow. In this full CRW configuration, main aero-dynamic interaction will appear between fuselage and rotor.

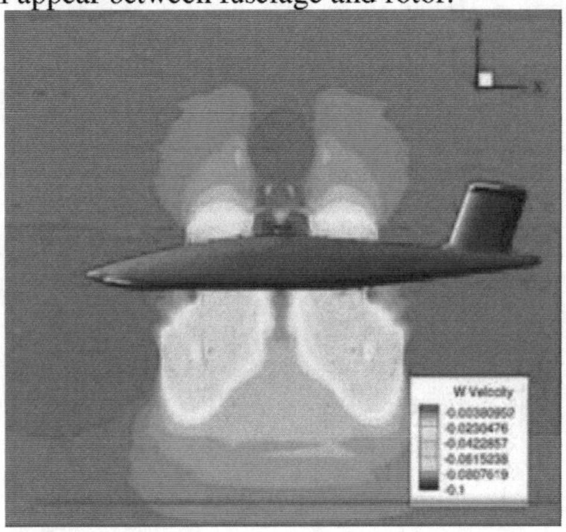

Figure 8. The downwash flow velocity of rotor.

Fuselage under the rotor disk can be considered as a cylinder, which would produce a loss of thrust for full CRW configuration due to the large separation area. With collective pitch angle increasing, this kind of thrust loss becomes more obviously. Thrust loss is getting smaller because of the additional thrust of the rotor mentioned in previous section when the collective pitch angle is 13 degree.

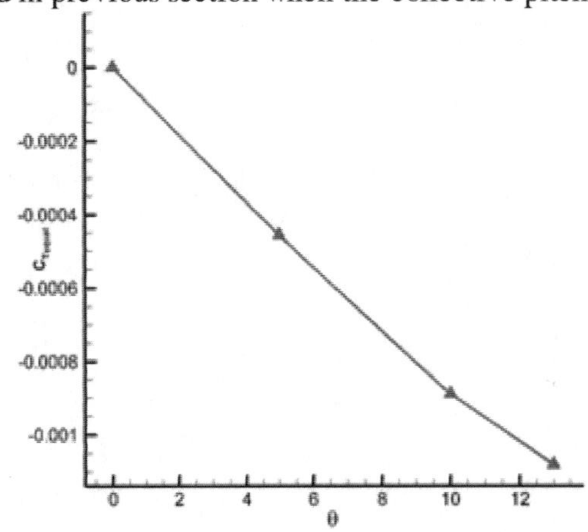

Figure 9. Thrust loss due to existence of fuselage.

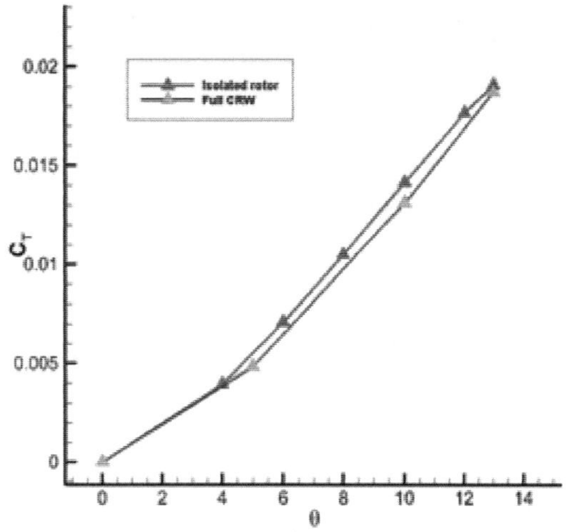

Figure 10. Total thrust coefficient for full CRW (blue line) compared with isolated rotor (red line). Thrust loss of full CRW configuration has an effect on the fight performance during helicopter mode, like hover ceiling, rate of climb and so on.

Conclusion and Discussion

1. The CFD method based on overset grids has well accuracy in predicating the rotor aircraft in hovering status.
2. Other parts of CRW aircraft have little effect on the thrust of rotor unless collective pitch angle is big enough. But these parts will drop the torque coefficient, which will improve the figure of merit.
3. The downwash of rotor only has obvious effect on a finite region under rotor disk for hovering status. Make fixed wings far away from or above the rotor disk is a sufficient method to avoid the downwash flow from the rotor disk.
4. Fuselage under the rotor disk will suffer from the downwash and it will cause an obvious thrust loss for the total aircraft. This point should be considered while designing.

References

[1] Rutherford J W, Bass S M and Larsen S D 1993 Canard rotor/wing- A revolutionary high-speed rotorcraft concept. In: *1993 AIAA/AHS/ASEE Aerospace Design Conference,*

[2] Mitchell C A and Vogel B J 2003 The canard rotor wing (CRW) aircraft - A new way to fly. In: *AIAA\ICAS International Air and Space Symposium and Exposition: The Next 100 Years, 2003, July 14, 2003 - July 17, 2003,* (Dayton, OH, United states: American Institute of Aeronautics and Astronautics Inc.)

[3] Ao H E, Gao Z H, Deng Y P, Liang-Ming L I, Tian L and Huang J T 2013 A wind tunnel study on aerodynamic characteristics of CRW plane by the rotor/wing *Journal of Experiments in Fluid Mechanics* **27** 13-6

[4] BASS S, THOMPSON T, RUTHERFORD J and Swanson S 1993 Low-speed wind tunnel test results of the Canard Rotor/Wing concept. In: *11th Applied Aerodynamics Conference,* p 3412

[5] Bass S M, Thompson T L and Rutherford J W 1995 Fixed-wing performance predictions of the canard rotowwing concept based on wind tunnel test results. In: *13th Applied Aerodynamics Conference, 1995, June 19, 1995 - June 22, 1995,* (San Diego, CA, United states: American Institute of Aeronautics and Astronautics Inc, AIAA) pp 1003-12

[6] Deng Y P, Gao Z H and Hao Z 2009 Experimental Investigation on Aerodynamic Interactions of Canard Rotor/Wing Aircraft in Hover and Low Speed Forward Flight *Journal of Experimental Mechanics*

[7] Niazi S, Rajagopalan R G, Thompson T L and Janakiram R D 2005 Simulation of unsteady aerodynamics of the unmanned CRW dragonfly aircraft hovering near the ground. In: *AHS International Specialists' Meeting - Unmanned Rotorcraft: Design, Control and Testing,*

January 18, 2005 - January 20, 2005, (Chandler, AZ, United states: American Helicopter Society) pp 439-59

[8] Yibo L and Dongli M 2012 Numerical Simulation of Rotor-aerodynamic Surface Interaction in Hover Using Moving Chimera Grid *Chinese Journal of Aeronautics* **25** 342-8

[9] Sun W. Z. Gao J J C 2016 Interactive aerodynamic characteristics of canard rotor wing aircraft in helicopter forward flight. *Acta Aeronautica et Astronautica Sinica* **8** 2498-506

[10] Chiu I-T and Meakin R 1995 On automating domain connectivity for overset grids. In: *33rd Aerospace Sciences Meeting and Exhibit,* p 854

[11] MEAKIN R 1991 A new method for establishing intergrid communication among systems of overset grids. In: *10th Computational Fluid Dynamics Conference,* p 1586

[12] Strawn R and Ahmad J 2000 Computational modeling of hovering rotors and wakes. In: *38th Aerospace Sciences Meeting and Exhibit,* p 110

[13] LORBER P, STAUTER R and LANDGREBE A 1989 A comprehensive hover test of the airloads and airflow of an extensively instrumented model helicopter rotor. In: *AHS, Annual Forum, 45 th, Boston, MA,*

[14] Sun W, Gao Z, Du Y and Xu F 2015 Mechanism of unconventional aerodynamic characteristics of an elliptic airfoil *Chinese Journal of Aeronautics* **28** 687-94

Nonlinear Aeroelastic Modeling of a Folding Wing Structure

Peicheng Li[1], Yingge Ni[2], Chi Hou1, Xiaopeng Wan[1,*] and Meiying Zhao[1]

[1]School of Aeronautics, Northwestern Polytechnical University, Xi'an 710072, Shaanxi Province, China
[2]Department of Smart Structures/SHM, Aircraft Strength Research Institute of China, Xi'an 710065, Shaanxi Province, China

wanxp@nwpu.edu.cn

Abstract. The nonlinear aeroelastic response of a folding wing with the hinge free-play nonlinearity is investigated. A nonlinear structural dynamic model is established based on the modal synthesis of free-substructures by modification of the connected relationships, and is unnecessary to re-establish it if the stiffness changes. The nonlinear aeroelastic equation is obtained by rational function approximation of the unsteady aerodynamic force, and solved by Runge-Kutta method to predict the aeroelastic response of the folding wing. The results show that the free-play at hinges changes the aeroelastic characteristic of the folding wing, leading to the occurrence of the limit cycle oscillation within a certain velocity range. Moreover, the sensitivity to the free-play is different for the inboard and outboard hinges. If the free-play is controlled within a certain value, the motion of nonlinear aeroelastic system of the folding wing can be considered linear approximately.

1. Introduction

With the development of aircraft design technologies and for the purpose of improving the performance of the aircraft in different flight conditions, some aircraft designers turned their attention to the morphing aircraft from traditional fixed-wing aircraft [1], because morphing aircraft can change shape and size in-flight to enable a single vehicle to perform multiple mission roles. The majority of shape change occurs in the wing, resulting in a wide variety of changes in aerodynamic and structural features [2].

Many factors, such as large deformation, free-play nonlinearity, etc., can contribute to the nonlinear aeroelastic response of a folding wing [3]. Among them, the hinge free-play is often the main source of nonlinearity. Lee et al [4] analyzed the nonlinear aeroelastic response of a folding wing considering hinge free-play nonlinearity using ZAERO and MSC.NASTRAN and obtained the condition for the limit cycle oscillation (LCO), in their research works, fictitious mass method [5] was employed to improve the accuracy and efficiency of modal-based structural analysis. Bae et al [6,7] and Kim et al [8,9] studied the nonlinear aeroelastic response of an aircraft wing with control surface using the fictitious mass modal approach and the description function approach. Their numerical results showed the nonlinearities changed the flutter characteristics of the wing and LCO occurred within different velocities ranges. Yang et al [10] developed a free-interface component mode synthesis method which could be used to establish governing equations and to analyze the characteristics of nonlinear aeroelastic systems, and the results of the numerical simulations and the wind tunnel tests indicated the same trends and critical velocities.

Two methods have been developed for structural nonlinear aeroelastic analysis. The first one is the fictitious mass mode synthesis method. In this method, a fictitious mass is added to the mass matrix corresponding to the degree of freedom where structural changes will occur to improve local structural characteristics. However there is not a definite standard to quantify the fictitious mass. It is possible that different values could be used for different structures, which limits to the applicability of this method in nonlinear aeroelastic modeling. Furthermore, it may cause numerical difficulties if the value of fictitious mass becomes too large. Another kind of method is Craig-Bampton component mode synthesis. In this method, the dynamics of a structure is described by selected sets of normal modes of individual component structures, plus a set of static vectors that account for the coupling at each interface where component structures are connected. Since the degrees of freedom at the interfaces still remain, which results to the equation with high order and the computational cost of the general aerodynamic influence coefficient matrix.

In the present work, the aeroelastic equation for a folding wing with structural nonlinearities is established and the nonlinear aeroelastic analysis is carried out. The free-play structural dynamic equation is derived by the free interface component mode synthesis, in which the continuity at interfaces is modified and the nonlinear internal force is taken into consideration. The unsteady aerodynamic force is introduced into the structural dynamic equation by the virtual work principle. Finally, the aeroelastic equation for a folding wing with multi-degree of freedom structural nonlinearities is achieved and the aeroelastic responses are obtained.

2. Nonlinear structural dynamic model of a folding wing based on the free interface component mode synthesis

The folding wing consists of three components: the fuselage, the inboard wing and the outboard wing, denoted as A, B, C, respectively. The material has a Young's modulus of $7.1\times10^{10}\,\mathrm{Pa}$, Poisson's ratio of 0.33 and density of $2.7\times10^{3}\,\mathrm{kg/m^{3}}$, representing a typical aluminum alloy. All components are modeled by plates and discretized using the CQUAD4 elements with the thickness of 2 mm in MSC.NASTRAN. Each node of this element has six degrees of freedom: three translations, u_x, u_y, u_z, and three rotations, $\theta_x, \theta_y, \theta_z$. A hinge between component A and component B is model by a set of torsional spring and the multi-point constraint (MPC). The rotational degree of freedom θ_x is coupled using torsional springs with the stiffness of K_A while the other five degrees of freedom are defined via MPC. Similarly, hinge between components B and C is modeled by a set of torsional spring with the stiffness of K_B and the MPC. Figure 1 shows the finite element model for the folding wing.

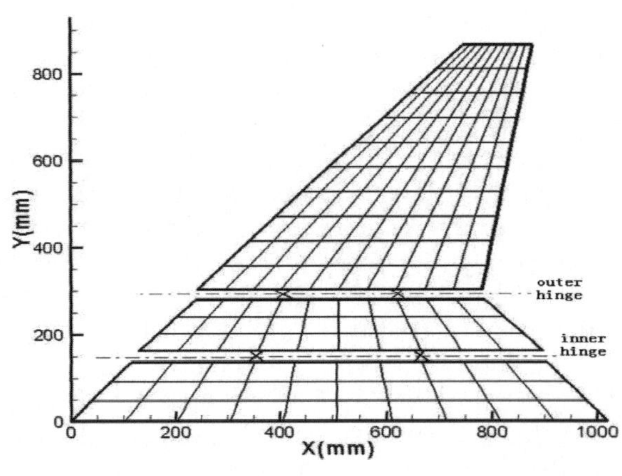

Figure 1. Finite element model for a folding wing (X represent hinges position).

2.1. Structural Dynamic Equation

The detailed finite element model of a folding wing typically has hundreds of thousands degrees of freedom, which makes the component mode synthesis method a natural choice for structural dynamic analysis. It enables the dynamic analysis to be performed on highly reduced matrices to save computational time. It also makes modifications possible to reduce the number of design iterations [11]. Hou first introduced the free interface component mode synthesis method. But this method was found to result in unacceptable errors in certain cases. Later Rubin [12] and Craig and Chang [13] introduced modifications to overcome this problem and their modified free interface component mode synthesis method has been widely used since. However, the elastic connection and the nonlinear internal force are not taken into consideration in the traditional free interface component mode synthesis. In the present work, modifications are made in the continuity at the interface so that this method can be used to establish the nonlinear structural dynamic equation. Figure 2 illustrates the procedure for obtaining the nonlinear structural dynamic equation.

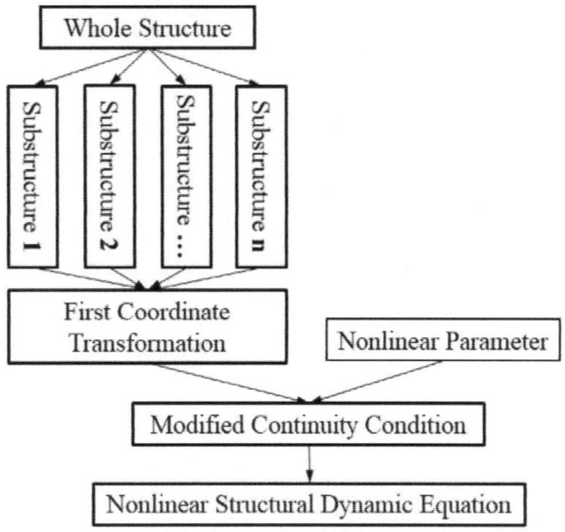

Figure 2. Procedure of nonlinear structural dynamic equation.

First, the structural dynamic equation of each undamped substructure can be written as:

$$\mathbf{M}^{\lambda}\ddot{\mathbf{u}}^{\lambda} + \mathbf{K}^{\lambda}\mathbf{u}^{\lambda} = \mathbf{f}^{\lambda} = \mathbf{B}^{\lambda T}\mathbf{f}_J^{\lambda} \quad (\lambda = \mathrm{A,B,C}) \tag{1}$$

where $\mathbf{M}$ is the mass matrix, $\mathbf{K}$ is the stiffness matrix, $\mathbf{u}$ is the displacement vector, $\mathbf{f}$ is the external force, $\mathbf{f}_J$ is the internal force applied to the degrees of freedom at interfaces, and $\mathbf{B}$ is the Boolean matrix.

The nodal displacement vector for each substructure includes displacements of the internal nodes and the displacements of the interfacial nodes. In preparation for the component mode synthesis, one can write the structural dynamic equation of each substructure in a rearranged and partitioned form. In present model, the hinges are modeled using the torsional spring and the MPC. Therefore the degrees of freedom at the interfaces are divided into two parts, displacements given by the MPC and displacements due to the torsional springs. Here, substructure A is taken as an example and its dynamic equation is as follows:

$$\begin{bmatrix} \mathbf{M}_{ii}^{A} & \mathbf{M}_{im}^{A} & \mathbf{M}_{is}^{A} \\ & \mathbf{M}_{mm}^{A} & \mathbf{M}_{ms}^{A} \\ sym & & \mathbf{M}_{ss}^{A} \end{bmatrix} \begin{bmatrix} \ddot{\mathbf{u}}_{i}^{A} \\ \ddot{\mathbf{u}}_{m}^{A} \\ \ddot{\mathbf{u}}_{s}^{A} \end{bmatrix} + \begin{bmatrix} \mathbf{K}_{ii}^{A} & \mathbf{K}_{im}^{A} & \mathbf{K}_{is}^{A} \\ & \mathbf{K}_{mm}^{A} & \mathbf{K}_{ms}^{A} \\ sym & & \mathbf{K}_{ss}^{A} \end{bmatrix} \begin{bmatrix} \mathbf{u}_{i}^{A} \\ \mathbf{u}_{m}^{A} \\ \mathbf{u}_{s}^{A} \end{bmatrix} = \begin{bmatrix} \mathbf{0} \\ \mathbf{f}_{m}^{A} \\ \mathbf{f}_{s}^{A} \end{bmatrix} \tag{2}$$

In the above equation, subscript i represents the terms refers to the internal nodal, subscript m represents the terms refers to the MPC, and subscript s represents the terms refers to the torsional spring. When the degrees of freedom at the interfaces are set free, the modal analysis of the substructure A is carried out as follows:

$$\mathbf{u}^{A} = \begin{bmatrix} \mathbf{u}_{i}^{A} \\ \mathbf{u}_{m}^{A} \\ \mathbf{u}_{s}^{A} \end{bmatrix} = \mathbf{\Phi}_{k}^{A}\mathbf{p}_{k}^{A} + \mathbf{\psi}_{d}^{A}\mathbf{f}_{J}^{A} = [\mathbf{\Phi}_{k}^{A} \quad \mathbf{\psi}_{d}^{A}] \begin{bmatrix} \mathbf{p}_{k}^{A} \\ \mathbf{f}_{J}^{A} \end{bmatrix} \tag{3}$$

IOP Conf. Series: Journal of Physics: Conf. Series **1215** (2019) 012009 doi:10.1088/1742-6596/1215/1/012009

where $\mathbf{p}_k^A$ is the modal coordinates, $\mathbf{\Phi}_k^A$ is the kept modes, $\boldsymbol{\psi}_d^A = (\mathbf{K}^A)^{-1}\mathbf{B}^{AT} - \mathbf{\Phi}_k^A(\mathbf{\Lambda}_k^A)^{-1}\mathbf{\Phi}_k^{AT}\mathbf{B}^{AT}$ is the residual modes.

Substituting Eq. (3) into Eq. (2) leads to:

$$\bar{\mathbf{M}}^A\ddot{\bar{\mathbf{p}}}^A + \bar{\mathbf{K}}^A\bar{\mathbf{p}}^A = \bar{\mathbf{f}}^A \tag{4}$$

where $\bar{\mathbf{M}}^A = [\mathbf{\Phi}_k^A \quad \boldsymbol{\psi}_d^A]^{\mathrm{T}}\tilde{\mathbf{M}}^A[\mathbf{\Phi}_k^A \quad \boldsymbol{\psi}_d^A] = \begin{bmatrix} \bar{\mathbf{M}}_k^A & \mathbf{0} \\ \mathbf{0} & \bar{\mathbf{M}}_d^A \end{bmatrix}$, $\bar{\mathbf{K}}^A = [\mathbf{\Phi}_k^A \quad \boldsymbol{\psi}_d^A]^{\mathrm{T}}\tilde{\mathbf{K}}^A[\mathbf{\Phi}_k^A \quad \boldsymbol{\psi}_d^A] = \begin{bmatrix} \bar{\mathbf{K}}_k^A & \mathbf{0} \\ \mathbf{0} & \bar{\mathbf{K}}_d^A \end{bmatrix}$,

$\bar{\mathbf{p}}_k^A = [\mathbf{p}_k^{A\mathrm{T}} \quad \mathbf{f}_J^{A\mathrm{T}}]$, $\bar{\mathbf{f}}^A = [\mathbf{\Phi}_k^A \quad \boldsymbol{\psi}_d^A]^{\mathrm{T}}\mathbf{B}^{A\mathrm{T}}\mathbf{f}_J^A$.

Similarly, the above procedures can be applied to substructures B and C. Therefore the uncoupling structural dynamic equation can be expressed as:

$$\bar{\mathbf{M}}\ddot{\bar{\mathbf{p}}} + \bar{\mathbf{K}}\bar{\mathbf{p}} = \bar{\mathbf{f}} \tag{5}$$

where $\bar{\mathbf{M}} = diag(\bar{\mathbf{M}}^A, \bar{\mathbf{M}}^B, \bar{\mathbf{M}}^C)$, $\bar{\mathbf{K}} = diag(\bar{\mathbf{K}}^A, \bar{\mathbf{K}}^B, \bar{\mathbf{K}}^C)$, $\bar{\mathbf{p}} = \begin{bmatrix} \bar{\mathbf{p}}^{AT} & \bar{\mathbf{p}}^{BT} & \bar{\mathbf{p}}^{CT} \end{bmatrix}^T = \begin{bmatrix} \mathbf{p}_k^{AT} & \mathbf{f}_m^{AT} & \mathbf{f}_s^{AT} & \mathbf{p}_k^{BT} & \mathbf{f}_m^{BaT} & \mathbf{f}_s^{BaT} & \mathbf{f}_m^{BbT} & \mathbf{f}_s^{BbT} & \mathbf{p}_n^{CT} & \mathbf{f}_m^{CT} & \mathbf{f}_s^{CT} \end{bmatrix}$,

$$\bar{\mathbf{f}} = \begin{bmatrix} \bar{\mathbf{f}}^{AT} \\ \bar{\mathbf{f}}^{BT} \\ \bar{\mathbf{f}}^{CT} \end{bmatrix} = \begin{bmatrix} \begin{bmatrix} \mathbf{\Phi}_k^A \\ \mathbf{\Psi}_d^A \end{bmatrix} \mathbf{B}^{AT} \begin{bmatrix} \mathbf{f}_m^{AT} \\ \mathbf{f}_s^{AT} \end{bmatrix}^{\mathrm{T}} & \begin{bmatrix} \mathbf{\Phi}_k^B \\ \mathbf{\Psi}_d^B \end{bmatrix} \mathbf{B}^{BT} \begin{bmatrix} \mathbf{f}_m^{BaT} \\ \mathbf{f}_s^{BaT} \\ \mathbf{f}_m^{BbT} \\ \mathbf{f}_s^{BbT} \end{bmatrix}^{\mathrm{T}} & \begin{bmatrix} \mathbf{\Phi}_k^C \\ \mathbf{\Psi}_d^C \end{bmatrix} \mathbf{B}^{CT} \begin{bmatrix} \mathbf{f}_m^{CT} \\ \mathbf{f}_s^{CT} \end{bmatrix}^{\mathrm{T}} \end{bmatrix}^{\mathrm{T}}.$$

The continuity condition at the interface includes the displacement continuity and the force/moment continuity. The displacement continuity can be expressed as:

$$\begin{cases} \mathbf{u}_m^A = \mathbf{u}_m^{Ba}, \mathbf{u}_m^{Bb} = \mathbf{u}_m^C \\ \mathbf{u}_s^{Ba} = \mathbf{u}_s^A + \boldsymbol{\delta}_1, \mathbf{u}_s^C = \mathbf{u}_s^{Bb} + \boldsymbol{\delta}_2 \end{cases} \tag{6}$$

The force/moment continuity can be expressed as:

$$\begin{cases} \mathbf{f}_m^A + \mathbf{f}_m^{Ba} = \mathbf{0}, \mathbf{f}_m^{Bb} + \mathbf{f}_m^C = \mathbf{0} \\ \mathbf{f}_s^A = K_A\boldsymbol{\delta}_1,\ \mathbf{f}_s^{Ba} = -K_A\boldsymbol{\delta}_1,\ \mathbf{f}_s^A + \mathbf{f}_s^{Ba} = \mathbf{0} \\ \mathbf{f}_s^{Bb} = K_B\boldsymbol{\delta}_2,\ \mathbf{f}_s^C = -K_B\boldsymbol{\delta}_2,\ \mathbf{f}_s^{Bb} + \mathbf{f}_s^C = \mathbf{0} \end{cases} \tag{7}$$

Because of the torsional springs at interfaces, the force and moment are equal at both sides but the corresponding displacements are not equal, there should be a relative shift $\boldsymbol{\delta}$. By modal coordinate transformation for each substructure, the displacement condition at interfaces can be expressed as:

$$\begin{cases} \mathbf{B}_m^A(\mathbf{\Phi}_k^A\mathbf{p}_k^A + \boldsymbol{\psi}_{dm}^A\mathbf{f}_m^A) = \mathbf{B}_m^{Ba}(\mathbf{\Phi}_k^B\mathbf{p}_k^B + \boldsymbol{\psi}_{dm}^{Ba}\mathbf{f}_m^{Ba}) \\ \mathbf{B}_m^{Bb}(\mathbf{\Phi}_k^B\mathbf{p}_k^B + \boldsymbol{\psi}_{dm}^{Bb}\mathbf{f}_m^{Bb}) = \mathbf{B}_m^C(\mathbf{\Phi}_k^C\mathbf{p}_k^C + \boldsymbol{\psi}_{dm}^C\mathbf{f}_m^C) \\ \mathbf{B}_s^{Ba}(\mathbf{\Phi}_k^B\mathbf{p}_k^B + \boldsymbol{\psi}_{ds}^{Ba}\mathbf{f}_s^{Ba}) = \mathbf{B}_s^A(\mathbf{\Phi}_k^A\mathbf{p}_k^A + \boldsymbol{\psi}_{ds}^A\mathbf{f}_s^A) + \boldsymbol{\delta}_1 \\ \mathbf{B}_s^C(\mathbf{\Phi}_k^C\mathbf{p}_k^C + \boldsymbol{\psi}_{ds}^C\mathbf{f}_s^C) = \mathbf{B}_s^{Bb}(\mathbf{\Phi}_k^B\mathbf{p}_k^B + \boldsymbol{\psi}_{ds}^{Bb}\mathbf{f}_s^{Bb}) + \boldsymbol{\delta}_2 \end{cases} \tag{8}$$

where $\mathbf{B}_m^\lambda$ and $\mathbf{B}_s^\lambda$ represent the projection matrix related to MPC and torsional spring respectively, and $\boldsymbol{\psi}_{dm}^\lambda$ and $\boldsymbol{\psi}_{ds}^\lambda$ represent the residual modes related to MPC and torsional spring respectively. By considering the force/moment continuity condition, one can obtain:

$$\begin{cases} \mathbf{f}_m^A = -\mathbf{f}_m^{Ba} = (\mathbf{B}_m^A\boldsymbol{\psi}_{dm}^A + \mathbf{B}_m^{Ba}\boldsymbol{\psi}_{dm}^{Ba})^{-1}(-\mathbf{B}_m^A\mathbf{\Phi}_k^A\mathbf{p}_k^A + \mathbf{B}_m^{Ba}\mathbf{\Phi}_k^B\mathbf{p}_k^B) \\ \mathbf{f}_s^A = -\mathbf{f}_s^{Ba} = (\mathbf{B}_s^A\boldsymbol{\psi}_{ds}^A + \mathbf{B}_s^{Ba}\boldsymbol{\psi}_{ds}^{Ba})^{-1}(-\mathbf{B}_s^A\mathbf{\Phi}_k^A + \mathbf{B}_s^{Ba}\mathbf{\Phi}_k^B - \boldsymbol{\delta}_1) \\ \mathbf{f}_m^C = -\mathbf{f}_m^{Bb} = (\mathbf{B}_m^{Bb}\boldsymbol{\psi}_{dm}^{Bb} + \mathbf{B}_m^C\boldsymbol{\psi}_{dm}^C)^{-1}(\mathbf{B}_m^{Bb}\mathbf{\Phi}_k^C - \mathbf{B}_m^C\mathbf{\Phi}_k^C) \\ \mathbf{f}_s^C = -\mathbf{f}_s^{Bb} = (\mathbf{B}_s^C\boldsymbol{\psi}_{ds}^C + \mathbf{B}_s^{Bb}\boldsymbol{\psi}_{ds}^{Bb})^{-1}(\mathbf{B}_s^C\mathbf{\Phi}_k^C + \mathbf{B}_s^{Bb}\mathbf{\Phi}_k^B + \boldsymbol{\delta}_2) \end{cases} \tag{9}$$

where $\mathbf{A}_{11}=\mathbf{B}_m^A\boldsymbol{\psi}_{dm}^A+\mathbf{B}_m^{Ba}\boldsymbol{\psi}_{dm}^{Ba}$, $\mathbf{G}_{11}=\mathbf{B}_m^A\boldsymbol{\Phi}_k^A$, $\mathbf{G}_{12}=\mathbf{B}_m^{Ba}\boldsymbol{\Phi}_k^B$, $\mathbf{A}_{22}=\mathbf{B}_s^A\boldsymbol{\psi}_{ds}^A+\mathbf{B}_s^{Ba}\boldsymbol{\psi}_{ds}^{Ba}$, $\mathbf{G}_{21}=\mathbf{B}_s^A\boldsymbol{\Phi}_k^A$, $\mathbf{G}_{22}=\mathbf{B}_s^{Ba}\boldsymbol{\Phi}_k^B$, $\mathbf{A}_{33}=\mathbf{B}_m^{Bb}\boldsymbol{\psi}_{dm}^{Bb}+\mathbf{B}_m^C\boldsymbol{\psi}_{dm}^C$, $\mathbf{G}_{31}=\mathbf{B}_m^{Bb}\boldsymbol{\Phi}_k^C$, $\mathbf{G}_{32}=\mathbf{B}_m^C\boldsymbol{\Phi}_k^C$, $\mathbf{A}_{44}=\mathbf{B}_s^C\boldsymbol{\psi}_{ds}^C+\mathbf{B}_s^{Bb}\boldsymbol{\psi}_{ds}^{Bb}$, $\mathbf{G}_{41}=\mathbf{B}_s^{Bb}\boldsymbol{\Phi}_k^B$, $\mathbf{G}_{42}=\mathbf{B}_s^C\boldsymbol{\Phi}_k^C$, the above equation can be expressed:

$$\mathbf{p}=\begin{bmatrix}\mathbf{p}_k^A\\ \mathbf{f}_m^A\\ \mathbf{f}_s^A\\ \mathbf{p}_k^B\\ \mathbf{f}_m^{Ba}\\ \mathbf{f}_s^{Ba}\\ \mathbf{f}_m^{Bb}\\ \mathbf{f}_s^{Bb}\\ \mathbf{p}_k^C\\ \mathbf{f}_m^C\\ \mathbf{f}_s^C\end{bmatrix}=\mathbf{Sq}=\begin{bmatrix}\mathbf{I} & \mathbf{0} & \mathbf{0} & \mathbf{0} & \mathbf{0}\\ -\mathbf{A}_{11}^{-1}\mathbf{G}_{11} & \mathbf{A}_{11}^{-1}\mathbf{G}_{12} & \mathbf{0} & \mathbf{0} & \mathbf{0}\\ -\mathbf{A}_{22}^{-1}\mathbf{G}_{21} & \mathbf{A}_{22}^{-1}\mathbf{G}_{22} & \mathbf{0} & -\mathbf{A}_{22}^{-1} & \mathbf{0}\\ \mathbf{0} & \mathbf{I} & \mathbf{0} & \mathbf{0} & \mathbf{0}\\ \mathbf{A}_{11}^{-1}\mathbf{G}_{11} & -\mathbf{A}_{11}^{-1}\mathbf{G}_{12} & \mathbf{0} & \mathbf{0} & \mathbf{0}\\ \mathbf{A}_{22}^{-1}\mathbf{G}_{21} & -\mathbf{A}_{22}^{-1}\mathbf{G}_{22} & \mathbf{0} & \mathbf{A}_{22}^{-1} & \mathbf{0}\\ \mathbf{0} & -\mathbf{A}_{33}^{-1}\mathbf{G}_{31} & \mathbf{A}_{33}^{-1}\mathbf{G}_{32} & \mathbf{0} & \mathbf{0}\\ \mathbf{0} & -\mathbf{A}_{44}^{-1}\mathbf{G}_{41} & \mathbf{A}_{44}^{-1}\mathbf{G}_{42} & \mathbf{0} & -\mathbf{A}_{44}^{-1}\\ \mathbf{0} & \mathbf{0} & \mathbf{I} & \mathbf{0} & \mathbf{0}\\ \mathbf{0} & \mathbf{A}_{33}^{-1}\mathbf{G}_{31} & -\mathbf{A}_{33}^{-1}\mathbf{G}_{32} & \mathbf{0} & \mathbf{0}\\ \mathbf{0} & \mathbf{A}_{44}^{-1}\mathbf{G}_{41} & -\mathbf{A}_{44}^{-1}\mathbf{G}_{42} & \mathbf{0} & \mathbf{A}_{44}^{-1}\end{bmatrix}\begin{bmatrix}\mathbf{p}_k^A\\ \mathbf{p}_k^B\\ \mathbf{p}_k^C\\ \boldsymbol{\delta}_1\\ \boldsymbol{\delta}_2\end{bmatrix} \tag{10}$$

Substituting Eq. (10) into Eq. (5) and pre-multiplying $\mathbf{S}^T$, one can obtain:

$$\tilde{\mathbf{M}}\ddot{\mathbf{q}}+\tilde{\mathbf{K}}\mathbf{q}=\tilde{\mathbf{f}} \tag{11}$$

where $\tilde{\mathbf{M}}=\mathbf{S}^T\bar{\mathbf{M}}\mathbf{S}$, $\tilde{\mathbf{K}}=\mathbf{S}^T\bar{\mathbf{K}}\mathbf{S}$, $\tilde{\mathbf{K}}=\mathbf{S}^T\bar{\mathbf{K}}\mathbf{S}$. In general, the mass and stiffness matrices are coupled. But since the size of the matrix is reduced, the computational efficiency is improved.

Now consider the right hand term $\mathbf{S}^T\bar{\mathbf{f}}$. The general forces corresponding to the modal coordinates $\mathbf{p}_k^A$, $\mathbf{p}_k^B$, $\mathbf{p}_k^C$ are zero, which is consistent with the result of the traditional free interface component mode synthesis. The following discusses the general forces corresponding to $\boldsymbol{\delta}_1$, $\boldsymbol{\delta}_2$.

$$\begin{aligned}&-(\mathbf{A}_2^{-1})^{\mathrm{T}}\boldsymbol{\psi}_{ds}^{A\mathrm{T}}\mathbf{B}_s^{A\mathrm{T}}\mathbf{f}_s^A+(\mathbf{A}_2^{-1})^{\mathrm{T}}\boldsymbol{\psi}_{ds}^{Ba\mathrm{T}}\mathbf{B}_s^{Ba\mathrm{T}}\mathbf{f}_s^{Ba}\\ &=-(\mathbf{A}_2^T)^{-1}(\boldsymbol{\psi}_{ds}^{A\mathrm{T}}\mathbf{B}_s^{A\mathrm{T}}+\boldsymbol{\psi}_{ds}^{Ba\mathrm{T}}\mathbf{B}_s^{Ba\mathrm{T}})\mathbf{f}_s^{Ba}\\ &=-(\boldsymbol{\psi}_{ds}^{A\mathrm{T}}\mathbf{B}_s^{A\mathrm{T}}+\boldsymbol{\psi}_{ds}^{Ba\mathrm{T}}\mathbf{B}_s^{Ba\mathrm{T}})^{-1}(\boldsymbol{\psi}_{ds}^{A\mathrm{T}}\mathbf{B}_s^{A\mathrm{T}}+\boldsymbol{\psi}_{ds}^{Ba\mathrm{T}}\mathbf{B}_s^{Ba\mathrm{T}})\mathbf{f}_s^{Ba}\\ &=-\mathbf{f}_s^{Ba}\end{aligned} \tag{12}$$

$$\begin{aligned}&-(\mathbf{A}_4^{-1})^{\mathrm{T}}\boldsymbol{\psi}_{ds}^{Bb\mathrm{T}}\mathbf{B}_s^{Bb\mathrm{T}}\mathbf{f}_s^{Bb}+(\mathbf{A}_4^{-1})^{\mathrm{T}}\boldsymbol{\psi}_{ds}^{C\mathrm{T}}\mathbf{B}_s^{C\mathrm{T}}\mathbf{f}_s^{C}\\ &=-(\mathbf{A}_4^T)^{-1}(\boldsymbol{\psi}_{ds}^{Bb\mathrm{T}}\mathbf{B}_s^{Bb\mathrm{T}}+\boldsymbol{\psi}_{ds}^{C\mathrm{T}}\mathbf{B}_s^{C\mathrm{T}})\mathbf{f}_s^{Bb}\\ &=-(\boldsymbol{\psi}_{ds}^{C\mathrm{T}}\mathbf{B}_s^{C\mathrm{T}}+\boldsymbol{\psi}_{ds}^{Bb\mathrm{T}}\mathbf{B}_s^{Bb\mathrm{T}})^{-1}(\boldsymbol{\psi}_{ds}^{C\mathrm{T}}\mathbf{B}_s^{C\mathrm{T}}+\boldsymbol{\psi}_{ds}^{Bb\mathrm{T}}\mathbf{B}_s^{Bb\mathrm{T}})\mathbf{f}_s^{Bb}\\ &=-\mathbf{f}_s^{Bb}\end{aligned} \tag{13}$$

As can be seen from (12) and (13), the generalized force of the whole folding wing structure contains the torque by the torsional spring at the interfaces. The final generalized force can be expressed as:

$$\tilde{\mathbf{f}}=\begin{bmatrix}\mathbf{0}^{\mathrm{T}} & \mathbf{0}^{\mathrm{T}} & \mathbf{0}^{\mathrm{T}} & K_A\boldsymbol{\delta}_1{}^{\mathrm{T}} & K_B\boldsymbol{\delta}_2{}^{\mathrm{T}}\end{bmatrix}^{\mathrm{T}} \tag{14}$$

In this case, the generalized force depends on K_A , K_B and $\boldsymbol{\delta}_1$, $\boldsymbol{\delta}_2$. Obviously, the nonlinear internal force can also be explicitly expressed in the structural dynamics equation.

2.2. Aeroelastic equation in state space

In this section, the aerodynamic force for a folding wing is obtained by the doublet lattice method. By modal coordinate transformation of the aerodynamic force, the aeroelastic equation for the folding wing can be written as

$$\tilde{\mathbf{M}}\ddot{\mathbf{q}}+\tilde{\mathbf{K}}\mathbf{q}=\tilde{\mathbf{f}}+q_d\mathbf{S}^{\mathrm{T}}\mathbf{QSq} \tag{15}$$

Note that the modal matrix of each substructure is not affected as $\tilde{\mathbf{f}}$ changes, and it is not necessary to recalculate $\mathbf{Q}$. The nonzero elements in $\tilde{\mathbf{f}}$ due to the nonlinear internal forces caused by the torsional spring can be expressed as nonlinear functions of the generalized coordinates. The free-play nonlinearity is generally modeled by expressing the internal force as a piecewise nonlinear function of the generalized displacement. Let θ be the generalized displacement at the hinge, figure 3 shows the free-play nonlinearity given in reference 3. In this model the torque is expressed as a piecewise nonlinear function of θ.

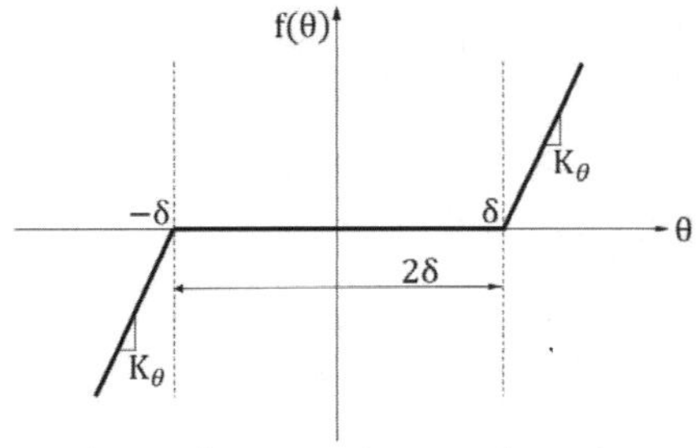

Figure 3. Free-play nonlinearity.

$$f(\theta)=\begin{cases}K_\theta(\theta-\delta),(\theta>\delta)\\0,(-\delta\le\theta\le\delta)\\K_\theta(\theta+\delta),(\theta<-\delta)\end{cases} \tag{16}$$

where δ is the amount of clearance. When δ is set to zero, the moment is a linear function of θ. Generally, the unsteady aerodynamic influence coefficient matrix is calculated for discrete reduced frequency rather than as a continuous function of the circular frequency. Thus, the aerodynamic influence coefficient matrices can be approximated by rational functions. There are many methods of rational function approximation. The minimum state method is used here for simplicity and fast computation time. Introducing the aerodynamic states $\mathbf{q}_a$, one can obtain the aeroelastic equation in the state space:

$$\begin{Bmatrix}\dot{\mathbf{q}}\\\ddot{\mathbf{q}}\\\dot{\mathbf{q}}_a\end{Bmatrix}=\begin{bmatrix}\mathbf{0}&\mathbf{I}&\mathbf{0}\\-\bar{\mathbf{M}}^{-1}\bar{\mathbf{K}}&-\bar{\mathbf{M}}^{-1}\bar{\mathbf{D}}&q_d\bar{\mathbf{M}}^{-1}\mathbf{D}_s\\\mathbf{0}&\mathbf{E}_s&(V/b)\mathbf{R}_s\end{bmatrix}\begin{Bmatrix}\mathbf{q}\\\dot{\mathbf{q}}\\\mathbf{q}_a\end{Bmatrix}+\bar{\mathbf{M}}^{-1}\begin{Bmatrix}\mathbf{0}\\\tilde{\mathbf{f}}\\\mathbf{0}\end{Bmatrix} \tag{17}$$

where $\bar{\mathbf{M}}=\tilde{\mathbf{M}}-q_d(b/V)^2\mathbf{S}^{\mathrm{T}}\mathbf{A}_2\mathbf{S}$, $\bar{\mathbf{D}}=-q_d(b/V)\mathbf{S}^{\mathrm{T}}\mathbf{A}_1\mathbf{S}$, $\bar{\mathbf{K}}=\tilde{\mathbf{K}}-q_d\mathbf{S}^{\mathrm{T}}\mathbf{A}_0\mathbf{S}$. Here, $\mathbf{A}_0$, $\mathbf{A}_1$, $\mathbf{A}_2$, $\mathbf{D}_s$ and $\mathbf{E}_s$ are unknown matrices in the minimum state approximation, which can be calculated from the least square fit, $\mathbf{R}_s$ is the aerodynamic lag matrix, and b is the reference length. Integrating Eq. (17) gives the time response for a nonlinear system. Here, the Runge-Kutta algorithm is used to solve the response for the folding wing.

3. Analysis and discussion

The DMAP language is used to extract the mass and stiffness matrices of each component in MSC.NASTRAN. The unsteady aerodynamic forces are achieved using Doublet Lattice Method (DLM). Finally, the aeroelastic equation of motion in state space form is expressed as the minimum state approximation. The six kept modes of each substructure are used to establish the nonlinear structural dynamic equation for the folding wing based on the free interface component mode synthesis. At the interface, ten degrees of freedom are coupled by the MPC and torsional springs. Adding the four aerodynamic states to the dynamic equation, the order of the final aeroelastic equation in state space is 48. A MATLAB program is written based on the Runge-Kutta algorithm to solve the response for the folding wing.

3.1. Model validation

In order to verify the present model, which introduces generalized displacements at the inner and outer hinges, the natural frequency and the flutter boundary for the linear system are compared with those by MSC.NASTRAN, here $K_A=K_B=1\mathrm{Nm/rad}$. The corresponding natural frequency of the folding wing is given in table 1. Since the substructures are connected by a hinge and the remaining nodes at the interface are free, a lower first-order frequency occurs. The free interface mode synthesis introduces the residual mode, which makes up the error caused by the truncation of the high order modes. The results show a good agreement with MSC.NASTRAN. The corresponding mode shapes are showed in figure 4.

Table 1. Comparative result of natural frequencies.

Mode order	Present (Hz)	MSC.NASTRAN (Hz)	Deviation
1	0.3545	0.3915	-9.45%
2	3.5245	3.6686	-3.93%
3	10.5703	10.5182	0.50%
4	20.2140	19.0352	6.19%
5	26.6760	25.3796	5.11%
6	28.3108	27.0206	4.77%

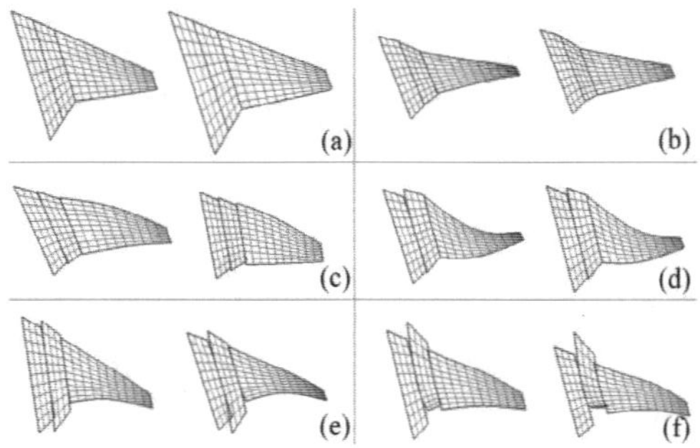

Figure 4. Mode shapes comparison (in each group, left: present; right: MSC.NASTRAN) (a) The first mode shape; (b) The second mode shape; (c) The third mode shape; (d) The fourth mode shape; (e) The fifth mode shape; (f) The sixth mode shape.

The first mode is dominated by the first bending mode of component C. The second mode is dominated by the second bending mode of component C. The third mode is dominated by the first bending mode of component B and the first torsion of component C. As for the last three modes, it is not easy to tell the dominated shape.

Table 2 compares the flutter boundary and shows that the result of the present method is in good agreement with MSC.NASTRAN.

Table 2. Comparison of flutter boundary.

Present(m/s)	MSC.NASTRAN(m/s)	Deviation
31.25	33.03	-5.39%

3.2. Free play nonlinear aeroelastic analysis

In order to analyze the effect of the free play on the aeroelastic responses for the folding wing, three different clearances ($\delta = 0.02°, 0.2°, 1.0°$) are taken into consideration. The initial condition is a vertical displacement of 0.01mm. For convenience, the tip is selected for motion characteristics. ($V_f = 31.25 m/s$ is the linear flutter velocity by the present method).

3.2.1 Inner hinge with free play nonlinearity. Some simulations show that there are three different kinds of motion, the convergent motion, limit cycle oscillation and divergent oscillation, when the inner hinge has the free-play nonlinearity. The motion map is shown in figure 5.

When the clearance δ equals to $0.02°$, the limit cycle oscillation occurs within a much smaller range of the velocity. The response is a divergent oscillation occurs when the velocity is greater than the linear flutter velocity. If the velocity is less than the flutter velocity, the response is damped out as time proceeds because the aerodynamic damping exists in the nonlinear aeroelastic system. The response becomes complex as the velocity increases and the limit cycle oscillation and divergent motion take place. The range of the velocity corresponding to the damped motion decreases as the clearance increases.

3.2.2 Outer hinge with free play nonlinearity. Similarly, three kinds of motion occur when the outer hinge has free-play nonlinearity.

The motion map of folding wing with free play at the outboard hinge is shown in figure 6. As the clearance increases, the range of the velocity corresponding to the damped motion greatly decreases when the free-play exists in the outer hinge. As for $\delta = 0.2°$ and $\delta = 1.0°$, the ranges of the velocity corresponding to the limit cycle oscillation and divergent motion both increase. When the clearance δ equals to $0.02°$, a harmonic oscillation occurs at about the linear flutter velocity. The response is divergent when the velocity is greater than a critical value. Obviously, there exists this characteristic in the linear aeroelastic system. A typical limit cycle oscillation occurs near the linear flutter velocity in the case of $\delta = 0.02°$. At the same time, the response is divergent at the velocity greater than the linear flutter velocity, which is the characteristic of an approximately linear or weakly nonlinear aeroelastic system. As the clearance increases, the velocity at which the response becomes divergent decreases.

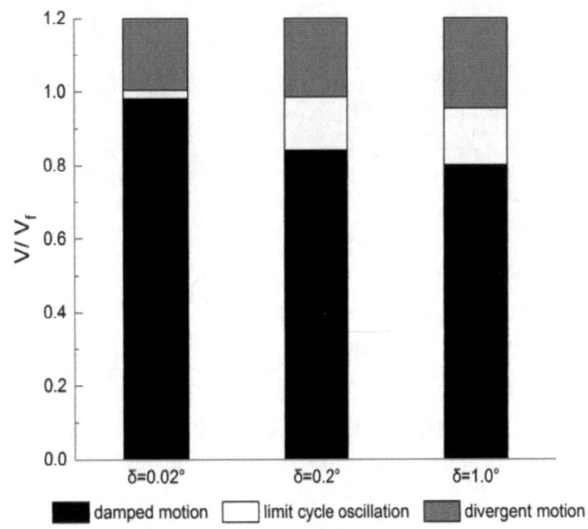

Figure 5. Motion map of folding wing with free play at the inboard hinge.

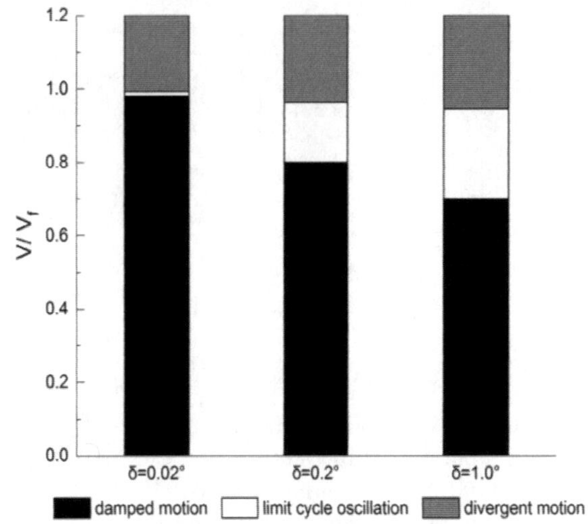

Figure 6. Motion map of folding wing with free play at the outboard hinge.

4. Conclusions

In this study, the aeroelastic equation for a folding wing with structural nonlinearities is established and the dynamic response is obtained. The following conclusions can be drawn from the present study:

1)The nonlinear structural dynamic equation is established based on the free interface component mode synthesis, in which the nonlinear internal force can be explicitly expressed in the structural dynamics equation. When the stiffness changes, it is not necessary to reestablish the structural dynamic model. Moreover, the order of the final aeroelastic equation is equal to the summation of kept modal degree of freedom and the generalized coordinates, which is relatively low. While comparing the frequencies of each mode order and flutter boundary obtained by the aeroelastic equation and MSC.NASTRAN, all the relative errors are no more than 10%, hence this method attractive for engineering applications.

2)The aeroelastic equation is derived from the integration of the nonlinear structural equation and the unsteady aerodynamic force. The aeroelastic response analysis is simplified, which does not require to recalculate the generalized aerodynamic influence matrix when the structural stiffness changes.

3)When the free-play exists in a folding wing and the clearance is within a certain value, the system can be regarded as an approximately linear system. The limit cycle oscillation will occur at some clearance values.

References

[1] Sofla A Y N, Meguid S A, Tan K T, Yeo W K. Shape Morphing of Aircraft Wing: Status and Challenges. Mater. Design, **31** (2010), 1284-1292.

[2] Lee D H, Weisshaar T A. Aeroelastic Studies on a Folding Wing Configuration, 46th AIAA/ASME/ ASCE/AHS/ASC Structures, Structural Dynamics & Materials Conference, Austin, Texas, USA, AIAA 2005-1996, 2005.

[3] Lee B H K, Price S J, Wong Y S. Nonlinear Aeroelastic Analysis of Airfoil: Bifurcation and Chaos. Prog. Aerosp. SCI., **35** (1999), 205~334.

[4] Lee D H, Chen P C. Nonlinear Aeroelastic Studies on a Folding Wing Configuration with Free-play Hinge Nonlinearity. 47th AIAA/ASME/ASCE/AHS/ ASC Structures, Structural Dynamics & Materials Conference, Newport, Rhode Island, USA, AIAA 2006-1734, 2006.

[5] Karpel M, Raveh D. Fictitious Mass Element in Structural Dynamics. AIAA J., **34** (1996), 607-613.

[6] Bae J S, Inman D J, Lee I. Effects of Structural Nonlinearity on Subsonic Aeroelastic Characteristcs of an Aircraft Wing with Control Surface. J. Fluid. Struct., **19** (2004), 747-763.

[7] Bae J S, Yang S M, Lee I. Linear and Nonlinear Aeroelastic Analysis of Fighter-type Wing with Control Surface. J. Aircraft, **39** (2002), 697-708.

[8] Lee I, Kim S H. Aeroelastic Analysis of a Flexible Control Surface with Structural Nonlinearity. J. Aircraft, **32** (1995), 868-874.

[9] Kim S H, Lee I. Aeroelastic Analysis of a Flexible Airfoil with a Freeplay Non-linearity. J. Sound Vib., **193** (1996), 823-846.

[10] Yang N, Wang N, Zhang X, Liu W. Nonlinear Flutter Wind Tunnel Test and Numerical Analysis of Folding Fins with Freeplay Nonlinearities. Chinese J. Aeronaut., **29** (2016), 144-159.

[11] Shanmugam A, Padmanabhan C. A Fixed-free Interface Compoent Mode Synthesis Method for Rotordynamic Analysis. J. Sound Vib., **297** (2006), 664-679.

[12] Rubin S. Improved Component-Mode Representation for Structural Dynamic Analysis. AIAA J., **13** (1975), 995-1006.

[13] Craig R R, Chang C J. On the Use of Attachment Modes in Substructure Coupling for Dynamic Analysis. AIAA/ASME 18th Structural Dynamics & Material Conference, San Diego, USA, AIAA 77-405, 1977.

IOP Conf. Series: Journal of Physics: Conf. Series **1215** (2019) 012010 doi:10.1088/1742-6596/1215/1/012010

Calibration of Aircraft Airdata Systems Using Adaptive Filtering Algorithm

Majeed Mohamed[1], Nidhin Joy[2] and Asok Kumar[2]

[1]Principal Scientist, National Aerospace Laboratories, Bangalore, India
[2]Department of Electrical Engineering, College of Engineering, Trivandrum, India

Abstract. This paper focuses on the calibration of aircraft airdata system to obtain online accurate Angle of Attack(AOA) and Angle of Sideslip (AOSS) from measured quantities contain error. For this purpose, an adaptive version of the extended Kalman filter (EKF) is employed to estimate accurately flight path trajectories of an aircraft. The investigations are made with flight simulated data with turbulence effects. Results are analyzed and found that adaptive EKF provides more accurate results over EKF.

1 Introduction

Calibration of aircraft airdata system is referred to the reconstruction of measured trajectories accurately from measured air data contain error. Typical airdata system of aircraft contains the sensors of pitot tubes with vanes and pressure probes. Precise knowledge of Angle of Attack (AOA) and Angle of Sideslip (AOSS) is essential for many flight applications such as in-flight simulation, aircraft modeling and model following control [1][2]. In the flight control architecture, AOA and AOSS are used as feedback control signals and hence accurate information of flow angles are crucial for unstable fighter aircraft. Moreover, transport aircraft also requires the accurate value of AOA for the stall warning system (SWS) for the safe operation of flight.

The airdata measurement is solely dependent on the data from the sensors installed on the aircraft and such sensor data are prone to errors during the time of measurement due to the flow distortion at aircraft body and the sensors misalignment. Multiple sensors are placed at different points of an aircraft based on the minimum variation of static pressure across the aircraft geometry for the accurate measurements of the same quantity of airdata. The errors caused by the pressure field and by flow angularity are called position errors and it will introduce variation in readings of the measurand. For accounting those errors, flow distortion effect is taken into the consideration by wind tunnel testing with use of the scaled version of the aircraft, and resulting correction part to the flight measurement is called as position error correction [3]. Nevertheless, there may be errors in such a correction process due to the scaled version of the aircraft and calm testing environment equivalent to ideal condition of an atmosphere which is not the case in actual flight. Moreover, these airdata measurements are influenced by several flight variables that may vary over a very wide range; thus, airdata sensors calibration must be treated as multidimensional and nonlinear problem [4]. To this end, online aircraft modeling method has attracted more and more research attention for the design of adaptive controller to maintain flight stability and high performance in uncertain environment where, both process and measurement noise occur [5], [6].

Measured flight data can contain a considerable amount of noise. Moreover, there might be biases and unobserved states in the system model which must be estimated by using filtering techniques. The most popular nonlinear filtering technique is the extended Kalman filter (EKF) [7]. Although EKF is widely used, proper knowledge of its noise characteristics is necessary, failing of this might lead to divergence of the filter. To avoid these problems, an adaptive extended Kalman filter might be applied, which automatically tunes the filter noise covariance matrices Q and R to adapt insufficiently known prior values. There have been many investigations in the area of adaptive filter and parameter estimation algorithm [8-12].

The calibration of aircraft airdata system can be viewed as state and estimation problem of flight path reconstruction (FPR) [2]. The main goal of the investigation is to obtain online accurate estimates of AOA and AOSS from the flight measurements. For this, we have proposed an adaptive filtering algorithm to obtain accurate estimates of Angle-of-Attack (AOA) and Angle-of-Sideslip (AOSS) in real time. The special attention for the presence of turbulence during the flight maneuvers is considered through the flight simulation and estimated its flight path trajectories as an aircraft gets affected by the adverse effect of atmospheric disturbance. The approach used for the adaptive version is based on estimating the covariance of the process (system dynamics) and measurement noises for this estimation problem by using the covariance matching technique [2]. Therefore, the proposed adaptive filter to estimate aircraft flow angles is needed to be derived from the covariance matching technique, by employing filter residuals to adjust the noise statistical properties. The proposed adaptive filter composed of an extended Kalman filter [12] as the main filter and two parallel Kalman filters for estimating main filter process noise statistics. By estimating the process and measurement noise covariance, the proposed algorithm will be able to compensate the estimation errors resulting from the insufficient knowledge of noise statistics in the estimation problem. The main contributions of this paper are

1. The investigation of calibration of aircraft airdata system has been carried out with the simulated data of a fighter aircraft in the presence of turbulence. Adaptive version of extended Kalman filter is used for this purpose.
2. The estimated results are compared with the estimates obtained from the extended Kalman filter and true value to demonstrate the efficacy of the proposed approach of Adaptive EKF.

The rest of the paper is organized as follows. Section 2 presents the problem formulation for the calibration of aircraft airdata system, and adaptive version of extended Kalman filter is described in section 3 in detail. The estimation results from simulated data are given in section 4 with discussion. Section 5 describes concluding remarks of the paper.

2 Problem formulation

For the reconstruction of measured trajectories from noisy airdata, aircraft dynamics is represented by the first order differential equation as follows [13]:

$$\begin{aligned}
\dot{u} &= -(q_m - \Delta q)w + (r_m - \Delta r)v - g\sin\theta\, A_x;\; u(t_0) = u_0 \\
\dot{v} &= -(r_m - \Delta r)u + (p_m - \Delta p)w - g\cos\theta\sin\phi + A_y;\; v(t_0) = v_0 \\
\dot{w} &= -(p_m - \Delta p)v + (q_m - \Delta q)u - g\cos\phi\sin\theta + A_z;\; w(t_0) = w_0 \\
\dot{\phi} &= -(p_m - \Delta p) + (q_m - \Delta q)\sin\phi\tan\theta + (r_m - \Delta r)\cos\phi\tan\theta;\; \phi(t_0) = \phi_0 \\
\dot{\theta} &= -(q_m - \Delta q)\cos\phi + (r_m - \Delta r)\sin\phi;\; \theta(t_0) = \theta_0 \\
\dot{h} &= u\sin\theta - v\sin\phi\cos\theta - w\cos\phi\cos\theta;\; h(t_0) = h_0
\end{aligned} \tag{1}$$

where p, q and r are the projection of the angular rate vector along the aircraft body axis and their biases in measurements are defined by $(\Delta p, \Delta q, \Delta r)$; θ and ϕ are the pitch and the roll angles respectively; u, v, w are inertial speed projections along the aircraft body axis. The linear accelerations (A_x, A_y, A_z) at the centre of gravity are computed from the accelerations measured by the accelerometer at a point away from the centre of gravity of an aircraft [1]. The state trajectories u, v, w, ϕ, θ, h of aircraft dynamics in (1) can be reconstructed from the following measurement equation:

$$\begin{aligned}
V_m &= \sqrt{u^2+v^2+w^2}+\eta_V \\
\bar{q}_m &= 0.5\rho\left(u^2+v^2+w^2\right)+\Delta\bar{q}+\eta_{\bar{q}} \\
\phi_m &= \phi+\eta_\phi \\
\theta_m &= \theta+\eta_\theta \\
h_m &= h+\eta_h \\
\alpha_m &= K_\alpha \tan^{-1}\left(\frac{w}{u}\right)+\Delta\alpha+\eta_\alpha \\
\beta_m &= K_\beta \sin^{-1}\left(\frac{v}{\sqrt{u^2+v^2+w^2}}\right)+\Delta\beta+\eta_\beta
\end{aligned} \tag{2}$$

where η is associated additive measurement noise. The density of air ρ can be computed from the actual measurement of static pressure P_s using universal gas law $\rho = P_s / RT_s$. R is the gas constant and T_s the static temperature K_α and K_β are the scale factors, and $\Delta\alpha$, $\Delta\beta$ are the biases used to model the measurement errors in (2). These are the sensor calibration parameters obtained by estimation. The sensor offset corrections for the flow angles measurements can be done through computation of the velocity components along the three body-fixed axes at an off centre of gravity location from u, v and w. We can compute the AOA and AOS from the reconstructed trajectories of u, v, w as follows [12]:

$$\alpha = \tan^{-1}\left(\frac{w}{u}\right) \text{ and } \beta = \sin^{-1}\left(\frac{v}{\sqrt{u^2+v^2+w^2}}\right) \tag{3}$$

By taking the state $x = [u, v, w, \phi, \theta, h]^T \in R^6$, input $u_m = \left[A_x, A_y, A_z, p, q, r\right]^T \in R^6$, measurement $z = [V_m, \bar{q}_m, \phi_m, \theta_m, h_m, \alpha_m, \beta_m]^T \in R^7$ and parameter $\Theta = \begin{bmatrix} \Delta p, \Delta q, \Delta r, \Delta A_x, \Delta A_y, \\ \Delta A_z, \Delta\bar{q}, K_\alpha, \Delta\alpha, K_\beta, \Delta\beta \end{bmatrix} \in R^{11}$, the postulated aircraft system dynamics in (2) and (3) can be described by the following continuous-discrete state space model:

$$\begin{aligned}
\dot{x}(t) &= f\left[x(t), u_m(t), \Theta, w(t)\right]; x(t_0) = x_0 \\
y(t) &= h[x(t), \Theta] \\
z(k) &= y(k)+\eta(k), k = 1, 2, \ldots N
\end{aligned} \tag{4}$$

where f and h are the general non linear functions. The measurement vector z is sampled at N discrete time steps with a fixed sampling time Δt and k is the discrete time index. The measurement noise vector $\eta(k)$ in (4) is assumed to be a sequence of independent zero mean white Gaussian noise with covariance matrix R, and $w(t)$ represents the noise in measured inputs u_m with covariance matrix Q. As a result, the estimation of aircraft airdata system can be viewed as state and parameter estimation problem for the reconstruction of flight path trajectories. The constant sensor parameters Θ can be considered as output of an auxiliary dynamic system presented in (5)

$$\dot{\Theta} = 0 \tag{5}$$

This dual estimation problem can be further reduced into a state estimation problem by defining a augmented state vector as:

$x_a = \begin{bmatrix} x \\ \Theta \end{bmatrix}$ (6) and thus final extended system becomes

$$\begin{aligned} x_a &= f_a\left[x_a(t), u_m(t), w(t)\right] \\ y(t) &= h_a\left[x_a(t)\right] \\ z(k) &= y(k) + v(k) \end{aligned} \tag{7}$$

where the augmented variables are denoted by a subscript 'a'. The most widely used nonlinear estimation algorithm is extended Kalman filter [12], but it is hard to tune the values of noise covariances Q and R when dealing with significant noise variations in sensor measurements. To avoid this problem, an adaptive filtering algorithm is proposed to estimate the flight path trajectories.

3 Adaptive filtering algorithm

Introduced adaptive extended Kalman filter (AEKF) consists of main filter as an EKF and two parallel Kalman filters for estimating process and measurement noise covariance matrices for main filter. The block diagram of AEKF is given in Figure 1 and it explains the working of the estimation algorithm without knowing the information of noise covariances. The first parallel filter handles state estimation by the process noise. The residual is estimated from the predicted and corrected states of the main filter, which is given as input to the first parallel filter [14]. For an EKF as a main filter, the linearized system is calculated based on (7) through the following derivatives with the Jacobians defined by:

$$\begin{aligned} \mathrm{A}(k) &= \frac{\partial \mathrm{fa}(\mathrm{x}(t), \mathrm{u}(t), 0)}{\partial \mathrm{x}} \,|\, \hat{x}(k) \\ \mathrm{B}(k) &= \frac{\partial \mathrm{fa}(\mathrm{x}(t), \mathrm{u}(t), 0)}{\partial \mathrm{u}} \,|\, \hat{x}(k) \\ \mathrm{C}(k) &= \frac{\partial \mathrm{ha}(\mathrm{x}(t))}{\partial \mathrm{x}} \,|\, \hat{x}(k) \end{aligned} \tag{8}$$

The state transition matrix ϕ and its integral ψ at each discrete point are given by,

$$\phi(k+1) = I + A(k)\Delta t + A^2(k)\frac{\Delta t^2}{2!} + \ldots \tag{9}$$

$$\psi(k+1) = I\Delta t + A(k)\frac{\Delta t^2}{2!} + A^2(k)\frac{\Delta t^3}{3!} + \ldots \tag{10}$$

By calculating the transition matrix ϕ and its integral ψ, the following standard Kalman filter equations of time propagation and measurement correction are used with assumption of initial state and covariance matrix at first [1][15].

$$\begin{aligned} \hat{x}(k+1) &= \hat{x}(k) + \int_{t_k}^{t_{k+1}} f\left(x(t), \bar{u}_m(t), 0\right) dt \\ \tilde{P}(k+1) &\approx \Phi(k+1)\hat{P}(k)\Phi^T(k+1) \\ &\quad + \psi(k+1)B(k)Q(k)B^T(k)\psi^T(k+1) \\ K(k) &= \tilde{P}C^T(k)\left[C(k)\tilde{P}(k)C^T(k) + R(k)\right]^{-1} \end{aligned} \tag{11}$$

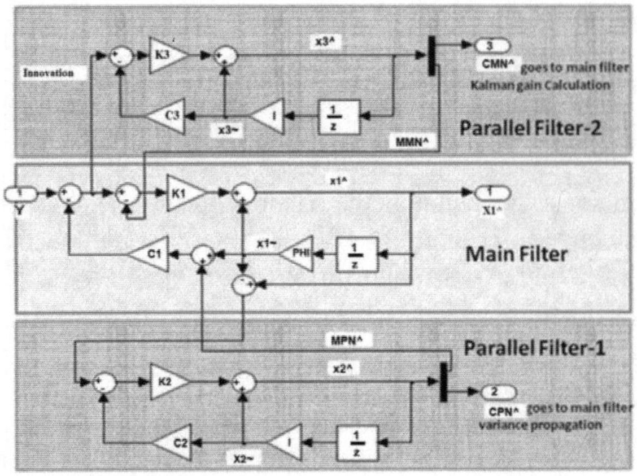

Figure 1. Block diagram of adaptive filtering

$$\hat{x}(k) = \tilde{x}(k) + K(k)[z(k) - h(\tilde{x}(k))]$$

$$\hat{P}(k) = [I - K(k)C(k)]\tilde{P}(k)$$

3.1 Estimation of process noise covariance

The parallel filter determines the mean (MPN) and covariance (CPN) and is fed back to the main filter [16]. Noise dynamics is modeled as a random walk to handle its stochastic nature. The residual constitute part of the parallel filter measurement vector as

$$MPN_m(j+1) = \hat{x}(j+1) - \hat{x}(j) - \int_{t_j}^{t_{j+1}} f((x(t), 0, dt) \tag{12}$$

which is equivalent to,

$$MPN_m(j+1) = \hat{x}(j+1) - \tilde{x}(j+1) \tag{13}$$

The associated covariances can be calculated as

$$C_{MPN}(j) = \left(MPN_m(j) - \widehat{MPN}(j)\right)\left(MPN_m(j) - \widehat{MPN}\right)^T \tag{14}$$

MPN_m is the main filter process noise mean measurement vector. C_{MPN} is the main filter process noise mean covariance matrix. The main filter process noise covariance measurement matrix MPN_m could be obtained by expanding the residual defined in (13) and relating it to its variance calculated in (14). The steps are well described by Myers and Leathrum, (1976, 1981), and the result is:

$$CPN_m(j) = C_{MPN}(j) - \left\{\Phi(j)\hat{P}(j-1)\Phi^T(j) - \hat{P}(j)\right\} \tag{15}$$

Only diagonal terms from the covariance measurement matrix are considered. A correction is desired to keep covariance matrix positive semi definite, changing negative values into zero. The Eq. (13) and (15) define the measurement model for the first parallel filter. The associated dynamic model to describe process noise is as a random walk, just as it was proposed for wind and for constant parameters in the EKF formulation. Therefore, process noise properties are considered constants with a zero mean white noise additive term.

$$\begin{bmatrix} MPN(j+1) \\ \operatorname{diag}(CPN(j+1)) \end{bmatrix} = \kappa * \begin{bmatrix} MPN(j) \\ \operatorname{diag}(CPN(j)) \end{bmatrix} + \begin{bmatrix} w21(j) \\ w22(j) \end{bmatrix} \tag{16}$$

where, $\kappa = \begin{bmatrix} I & 0 \\ 0 & I \end{bmatrix}$ (17) and $w2x(j)$ is a Gaussian white noise vector with properties $w2x(j) \sim N(0, Q2x)$. Compiling (13) and (15), in a matrix to form the measurement model

$$\begin{bmatrix} MPN_m(j) \\ \text{diag}\left(CPN_m(j)\right) \end{bmatrix} = \begin{bmatrix} \hat{x}(j)-\tilde{x}(j) \\ \text{diag}\left(C_{MPN}(j)-\left(\Phi(j)\hat{P}(j-1)\Phi^T(j)-\hat{P}(j)\right)\right) \end{bmatrix} \text{ and consequently} \tag{18}$$

$$\begin{bmatrix} MPN_m(j) \\ \text{diag}\left(CPN_m(j)\right) \end{bmatrix} = \kappa * \begin{bmatrix} MPN(j) \\ \text{diag}\left(CPN(j)\right) \end{bmatrix} + \begin{bmatrix} v21(j) \\ v22(j) \end{bmatrix} \tag{19}$$

where $v2x(j)$ is a Gaussian white noise vector with properties $v2x(j) \sim N(0, R2x)$. Concluding, (16)(19) compose the first parallel filter dynamic and measurement models respectively.

3.2 Estimation of measurement noise covariance

The second parallel filter is used to estimate the main filter measurement noise statistics on the same basis used in the last section, but now residuals used are innovation. This approach is derived from the covariance matching technique. The innovations should have white noise, zero mean characteristics, and be stationary and time uncorrelated. The main filter measurement noise means MMN and covariance CMN are estimated by considering them as states of second parallel filter.

The measurement of the main filter measurement noise means can be obtained from the innovations as

$$MMN_m(i) = z(i) - h(\tilde{x}(i)) \tag{20}$$

The innovation covariance matrix can be calculated as

$C_{MMN}(i) = \left(MMN_m(i) - M\overline{MN}_m(i)\right)\left(MMN_m(i) - \left(M\overline{MN}_m(i)\right)^T\right)$ The theoretical innovation variance

(21) $H(i)\tilde{P}(i)H(i)^T + R(i)$ can be approximated directly by its estimated values as

$$E\ [z(i)-h(\hat{x}(i))][z(i)-h(\hat{x}(i))]^T = H(i)\tilde{P}(i)H(i)^T + R(i) \tag{22}$$

Thereby providing a measurement model for the main filter measurement noise covariance matrix

$$CMN_m(i) = C_{MMN}(i) - H(i)\tilde{P}(i)H(i)^T \tag{23}$$

The same procedure described in the last section was adopted here to consider only diagonal elements and to guarantee the positive semi definite property. Negative values are interpreted as low covariance levels and therefore zeroed. The second parallel filter has taken the same consideration regarding the stochastic nature of the problem. The dynamic model is considered as a random walk, that is,

$$\begin{bmatrix} MMN(j+1) \\ \text{diag}\left(CMN(j+1)\right) \end{bmatrix} = \kappa * \begin{bmatrix} MMN(j) \\ \text{diag}\left(\text{CMN}(j)\right) \end{bmatrix} + \begin{bmatrix} w31(j) \\ w32(j) \end{bmatrix} \tag{24}$$

where $\kappa = \begin{bmatrix} I & 0 \\ 0 & I \end{bmatrix}$ and $w3x(j)$ is a Gaussian white noise vector with properties $w3x \sim N(0, Q3x)$. grouping equations (20) and (23) in matrix form, we obtain

$$\begin{bmatrix} MPN_m(i) \\ \text{diag}\left(CPN(i)\right) \end{bmatrix} = \begin{bmatrix} z(i)-h(\tilde{x}(i)) \\ \text{diag}\left(C_{MMN}(i)-H(i)\tilde{P}(i)H(i)^T\right) \end{bmatrix} \text{(25)and} \begin{bmatrix} MMN_m(i) \\ \text{diag}\left(CMN(i)\right) \end{bmatrix} = \kappa * \begin{bmatrix} MMN(j) \\ \text{diag}\left(CMN(i+1)\right) \end{bmatrix} + \begin{bmatrix} v31 \\ v32 \end{bmatrix}$$

(26)

4 Results and discussions

The proposed adaptive filtering algorithm was applied to the flight simulated data containing turbulence effect. Flight simulation was conducted at the flight conditions of Mach 0.3 and altitude of 1800m for a given rudder pedal input. The effect of turbulence was also incorporated in the simulation. The simulated data were sampled at an interval of 0.025 seconds.

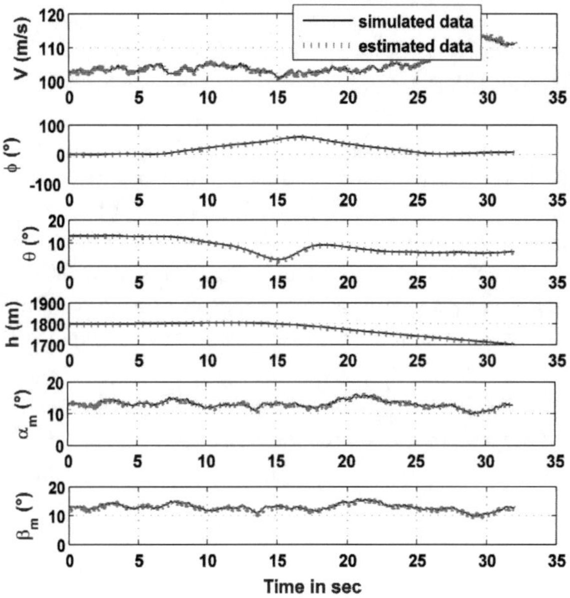

Figure 2. Measured and estimated values for rudder pedal input

In the estimation model, the dynamical representation of atmospheric turbulence is considered by including Dryden model [17] in addition to the aircraft system equations (1). The features that distinguish one turbulence structure from the other are the turbulence intensity σ and integral scale of turbulence L.

$$\dot{y}_u = \frac{-y_u + x_u k_u \sqrt{\frac{\pi}{\Delta t}}}{t_u}$$

$$\dot{y}_{v_2} = y_{v_1}$$

$$\dot{y}_{v_1} = -\frac{y_{v_2}}{t_v^2} - \frac{2y_{v_1}}{t_v} + x_v \sqrt{\frac{\pi}{\Delta t}}$$

$$\dot{y}_{w_2} = y_{w_1}$$

$$\dot{y}_{w_1} = -\frac{y_{w_2}}{t_w^2} - \frac{2y_{w_1}}{t_w} + x_w \sqrt{\frac{\pi}{\Delta}} \tag{27}$$

where x_u, x_v, x_w are the random numbers are generated to simulate the random nature of turbulence; $t_u, t_v, t_w, k_u, k_v, k_w$ are the time constants and are defined as :

$$t_u = L_u / V_t;\quad t_v = L_v / V_t;\ t_w = L_w / V_t \text{ where } V_t = \sqrt{u^2 + v^2 + w^2}\ ,\ \sigma_u = \sigma_v = \sigma_w = \sigma$$

$$\text{and } L_u = L_v = L_w = L\ k_u = \sqrt{\frac{2\sigma_u^2 t_u}{\pi}},\ k_v = \sqrt{\frac{2\sigma_v^2 t_v}{\pi}},\ k_w = \sqrt{\frac{2\sigma_{vv}^2 t_w}{\pi}} \tag{28}$$

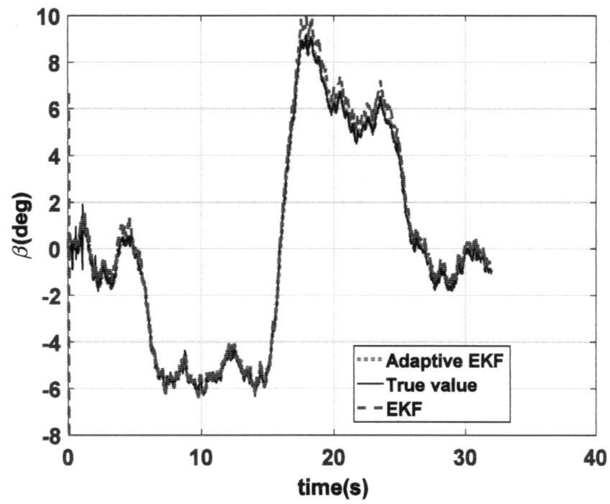

Figure 3. Comparison plot of true values with estimates

The turbulence in velocity components in flight path axes can now be obtained using the relations [17]

$$u_{fT} = y_u$$

$$u_{fT} = \frac{k_v}{t_v}\left[\frac{y_{v_2}}{t_v} + \sqrt{3}\cdot y_{v_1}\right] \tag{29}$$

$$w_{fT} = \frac{k_w}{t_w}\left[\frac{y_{w_2}}{t_w} + \sqrt{3}\cdot y_{w_1}\right]$$

Turbulence generated in flight path axes transformed to body axes are given by:

$$\begin{bmatrix} u_{bT} \\ v_{bT} \\ w_{bT} \end{bmatrix} = A_{\alpha\beta}\cdot\begin{bmatrix} u_{bT} \\ v_{bT} \\ w_{bT} \end{bmatrix} \tag{30}$$

$$A_{\alpha\beta} = \begin{bmatrix} \cos(\alpha)\cos(\beta) & -\cos(\alpha)\sin(\beta) & -\sin(\alpha) \\ \sin(\beta) & \cos(\alpha) & 0 \\ \sin(\alpha)\cos(\beta) & -\sin(\alpha)\sin(\beta) & \cos(\alpha) \end{bmatrix} \tag{31}$$

where, $\beta = \sin^{-1}\left(\frac{v}{V_t}\right)$ and $\alpha = \tan^{-1}\left(\frac{w}{u}\right)$

Figure 3 shows that estimated AOSS by adaptive EKF makes more agreement with its true values compared to the estimate obtained from EKF. This illustrates the superiority of proposed Adaptive EKF to calibrate the aircraft airdata system over EKF in the presence of turbulence.

5 Conclusions

An adaptive version of EKF was applied to dynamic maneuvers to calibrate the AOA and AOSS in real time from the simulated data. Simulations in software were carried out with moderate turbulence effects at 1800 meters and 0.3 Mach to obtain sufficient variations in AOA and AOSS of the aircraft. The investigation was made primarily from simulated data with turbulence effects using simulation software, and it was shown that both AOA and AOSS estimates were accurate. It is observed that adaptive EKF is superior to EKF in terms of accuracy of estimation as it uses instantly estimated noise covariance matrices instead of the tuned value of covariance matrices in EKF. Thus, the consistently good results obtained using adaptive filtering algorithm establish that this approach is great value for online implementation.

6 References

[1] G. Chowdhary, R. V. Jategaonkar, Aerospace S.&T. **14**, pp 106-117, (2010)
[2] C. B. Mendona, E. M. Hemerly, and L. C. Sandoval, J. of aircraft, **44**, pp 1546-1558, (2007)
[3] M. Majeed, I. N. Kar, The Aero. Journal, **115**, pp. 113-122, (2011)
[4] M. Lando, M. Battipede, P. A. Gili, The J. of Aircraft, **44**, pp. 945-953, (2007)
[5] J. Leitne, A. Calise, V. R. Prasad, J. of G.C.D, **20**, pp. 972-979, (1997)
[6] U. J. Pesonen, J. E. Steck, K. Rokhsaz, H. S. Bruner, N. Duerksen, J. of G.C.D, **27**, pp. 434-443, (2004)
[7] D. Lerro, Y. Shalom, IEEE Trans. On Aero. Electro. Sys, **29**, pp 1015-1022, (1993)
[8] J. H. Ahrens and H. Khalil, IEEE Trans. On Auto. Cont., 52, pp. 536-540, (2007)
[9] L. Chanying, L. Guo, Automatica, **46**, pp. 999-1007, (2010)
[10] T. Lombaerts, E. V. Oort, Q. P. Chu, J. A. Mulder, D. Joosten, J. of G.C.D, **33**, (2009)
[11] Q. Song, Z. Jiang, J. Han, IEEE Inter. Conf. on Robo. & Auto., pp. 4164-4169, (2007)
[12] M. Majeed, I. N. Kar, Inter. J. of Aerospace and Innov., **33**, pp. 2, (2010)
[13] R. Jategaonkar, AIAA Progress in AAS, **216**, (2006)
[14] M. Majeed, I. N. Kar, Aircraft Engg. & Aerospace Tech., **85**, pp. 267-279, (2013)
[15] P. S. Maybeck and D. E. Mercier, IEEE Conf. on Decision and Cont. incl. the Sympos. On Adpt. Processes, **2**, pp. 285-289, (1979)
[16] K. K. C. Yu, N. R. Watson, J. Arrillaga, IEEE Trans. On Power Delivery, **20**, pp. 1577-1584, (2005)
[17] H. Dryden, *A review of the statistical theory of turbulence*, InterSci. Pub., (1961)
[18] T. Sderstrm, M. Mossberg, Automatica, **47**, pp. 272-282, (2011)

IOP Conf. Series: Journal of Physics: Conf. Series **1215** (2019) 012011 doi:10.1088/1742-6596/1215/1/012011

Design and Analysis of Rotor/Wing Flap for Canard Rotor/Wing Aircraft

Yangping Deng[1] **and Baigang Mi**[2]

[1]School of Aeronautics, Northwestern Polytechnical University, Xi'an 710072, China
[2]School of Astronautics and Aeronautics, Tsinghua University, Beijing 100084, China

Abstract. In the present study, a kind of blunt trailing edge airfoil with its flap is designed for the purpose of reducing drags of blunt leading and trailing edge airfoils which are used for rotor/wing of CRW. The aerodynamic characteristics of designed flaps and elliptic airfoil which has the same thickness are numerically investigated based on the Reynolds-averaged Navier–Stokes equations and the SST k-ω two-equation turbulence model. The comparisons of aerodynamic force and moment coefficients show that, the lift characteristic is improved for designed flaps, but the drag will not be reduced if the flap has great bending trailing edge, and small bending trailing edge design is suitable for this kind of flap, and it will reduce the drag and improve the longitudinal stability characteristics.

1. Introduction

The Canard Rotor Wing (CRW) aircraft is a high-speed vertical takeoff and landing concept. It combines the low disk loading hover efficiency and low-speed flight characteristics of a helicopter with the high-subsonic cruise speed of a jet powered fixed-wing aircraft. For takeoff and landing, the wing rotates as a reaction-driven rotor. At conversion speed, the canard and lifting horizontal tail reduce the rotor thrust to zero, allowing the rotor to be stopped in a benign condition. After the rotor/wing is stopped, the vehicle continues flying as a three surface aircraft and the jet is used for the required propulsive force [1].

This concept was demonstrated under the X-50A "Dragonfly"program conducted by Boeing Company and Defense Advanced Research Projects Agency (DARPA) in 2003-2005. Although the X-50A program was cancelled in 2006 after the crashes of two prototypes, this concept is still a potential high-speed rotorcraft, and more investigations have been made by China and Korea recently [2-5].

Blunt leading and trailing edge airfoils make up the sections of the rotor/wing of CRW, because the trailing edge of the retreating blade in rotary-wing flight becomes the leading edge in fixed-wing flight. For X-50A prototype, the rotor/wing had an elliptical airfoil section with maximum thickness-to-chord (t/c) ratio varying from 24% at the root to 16% at the tip [6]. This incurs a profile drag penalty in both rotary and fixed-wing modes. The use of drag reduction techniques continues to be investigated, which range from circulation control to reversible airfoils, and the use of deployable flaps [7-9].

In the present study, a kind of blunt trailing edge airfoil with its flap is designed, and a Computational Fluid Dynamics (CFD) based study is also undertaken to evaluate the aerodynamic performance of the designed airfoil with flap in high speed condition, relative to elliptic airfoils which has the same thickness.

2. Design of rotor/wing airfoil and flap

For the purpose of containing flap, and maintaining blunt shape as leading edge, this kind of rotor/wing airfoil is constructed by arcs and spline curves as show in figure 1, and the profile of this airfoil is symmetrical relative to both X and Y axis.

The process of the designing of this airfoil is summarized below:

Firstly, the dimension and position of the arcs at the leading and trailing edge are determined, according to the setting radius of curvature;

Secondly, the reference point P1 of spline curve is determined, according to the setting rotary angle of flap from fold to deploy. This point is on the trailing edge arc, and we can draw a straight line from the center of this arc to point P1. The angle between this line and X axis is θ, and the setting rotary angle of flap from fold to deploy is 2θ.

Thirdly, the reference point P3 of spline curve is determined, according to the setting thickness of this airfoil.

Fourthly, the spline curve is determined according to the position of reference point P2. The tangent of this curve at point P3 is horizontal, and the tangent of this curve at point P1 is vertical to the straight line which has been described in the secondly step.

Finally, the spline curve is mirrored relative to both X and Y axis, and this kind of rotor/wing airfoil including four spline curves and two arcs is constructed after that.

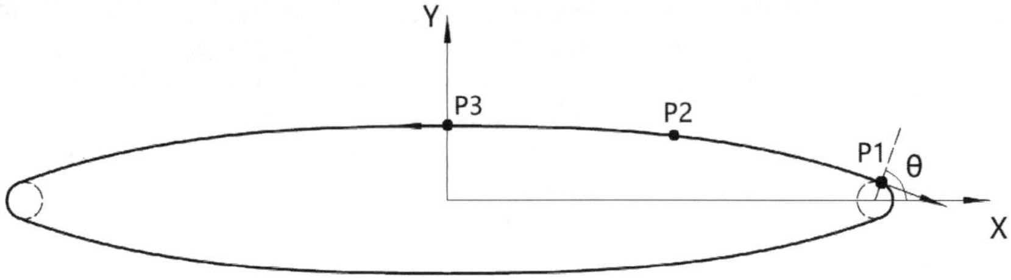

Figure 1. Design of the Rotor/Wing Airfoil

Following the design of the blunt trailing edge airfoil which has been described above, the flap can be designed according to the process below and the key construction points are shown in figure 2.

Firstly, the flap cracking point P4 is determined according to the setting length of the flap.

Secondly, the airfoil curve between P1′and P4 is rotated around the center of the trailing edge arc anticlockwise, the rotary angle is 2θ and this curve is upper surface of the flap. P1′is the symmetrical point of point P1 and it will coincide with P1 after rotation. The point P4 becomes P4′after rotation.

Thirdly, the spline curve which is the lower surface of flap is determined by point P4′and positions of reference point P5 and P6. Point P6 is on the lower curve of this airfoil ahead of point P1′, and the tangents of these two curves at point P6 are coincident.

Finally, the curve between point P4 and P6 is setting as the cover of the groove appeared after the flap is deployed, and this cover can void the vortex flow in the groove, which is helpful for reducing the aerodynamic drag.

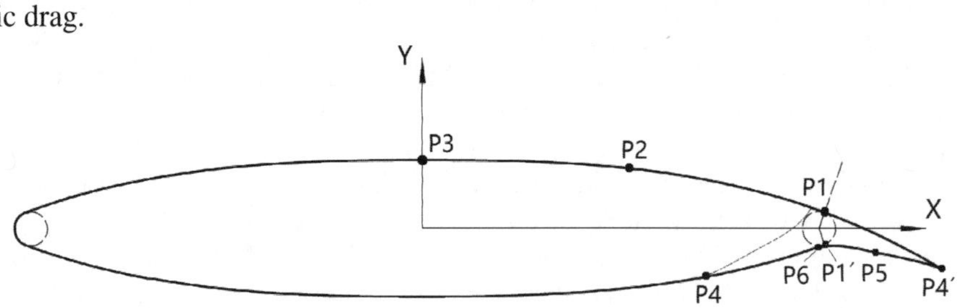

Figure 2. Design of the Rotor/Wing Flap

According to the designs of this kind of rotor/wing airfoil and flap, series of airfoils and flaps can be constructed by setting different radius of arcs, flap rotary angles, positions of point P4 and P5, and so on. The influences of such design parameters on the aerodynamic performance of the rotor/wing airfoil with flap will be analyzed in following section.[10]

3. Analysis of rotor/wing flap based on CFD

3.1. Numerical Method

3.1.1. Governing Equations. The two-dimensional, time-dependent, Reynolds-averaged Navier-Stokes equations for an ideal gas may be written in an integral form for an arbitrary control volume Ω with the boundary $\partial\Omega$ [10]:

$$\frac{\partial}{\partial t}\int_{\Omega} Q dV + \int_{\partial\Omega} F(Q)\cdot ndS = \int_{\partial\Omega} G(Q)\cdot ndS \tag{1}$$

Where Q is the vector of the conservative flow variables; $F(Q)$ and $G(Q)$ are the inviscid and viscous flux vectors in the standard conservation form, respectively; and n is the unit normal vector out from the control volume boundary. The flow variables were nondimensionalized by the free stream quantities, and the length was normalized by the airfoil chord length.

3.1.2. Turbulence model. For turbulence closure, the SST k-ω two-equation turbulence model was employed. The transport equations for the SST k-ω model are [11]:

$$\frac{\partial}{\partial t}(\rho k)+\frac{\partial}{\partial x_i}(\rho k u_i)=\frac{\partial}{\partial x_j}\left[\left(\mu+\frac{\mu_t}{\sigma_k}\right)\frac{\partial k}{\partial x_j}\right]+G_k-Y_k+S_k \tag{2}$$

$$\frac{\partial}{\partial t}(\rho \omega)+\frac{\partial}{\partial x_i}(\rho \omega u_i)=\frac{\partial}{\partial x_j}\left[\left(\mu+\frac{\mu_t}{\sigma_\omega}\right)\frac{\partial \omega}{\partial x_j}\right]+G_\omega-Y_\omega+D_\omega+S_\omega \tag{3}$$

In these equations, G_k represents the generation of turbulence kinetic energy due to mean velocity gradients. G_ω represents the generation of ω. Y_k and Y_ω represent the dissipation of k and ω due to turbulence D_ω represents the cross-diffusion term. S_k and S_ω are source terms.
In the present study, all calculations is performed by assuming that the flow is fully turbulent.

3.1.3. Other Computational Details. A characteristic boundary condition was imposed at the far-field boundary, and the no-slip isothermal boundary condition was applied at the solid surface. In the present study, all calculations were performed by assuming that the flow was fully turbulent.
To simulate the viscous shear layer accurately and to avoid the numerical difficulties involved in using highly stretched triangular cells, hybrid unstructured meshes were adopted in the present study. The hybrid meshes consisted of quadrilateral cells inside the boundary layer and isotropic triangular cells for the rest of the computational domain, as shown in figure 3. More than 161 points are set around the airfoil, and the non-dimensional minimum spacing normal to the solid wall is 10e-6c to capture the vicious effect near the wall, where c is the reference chord length of the air foil. The mesh contains about 57 thousands points to simulate the aerodynamics of the airfoil.

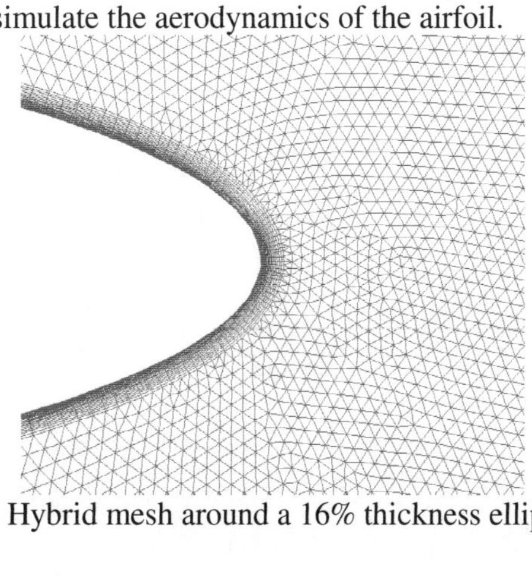

Figure3. Hybrid mesh around a 16% thickness elliptic airfoil

IOP Conf. Series: Journal of Physics: Conf. Series **1215** (2019) 012011 doi:10.1088/1742-6596/1215/1/012011

Because the flow field around an elliptic airfoil is inherently unsteady due to the presence of a pair of vortices at the blunt trailing edge, calculation is performed in a time-accurate manner and the steady-state result is obtained by averaging the periodic time-varying solution.

3.2. Results and Discussion

Five rotor/wing airfoil with flaps are designed for analysis, and they are labelled from E01 to E05. Shapes of these airfoil with flaps are presented in figure 4, and table 1 shows parameters of them.

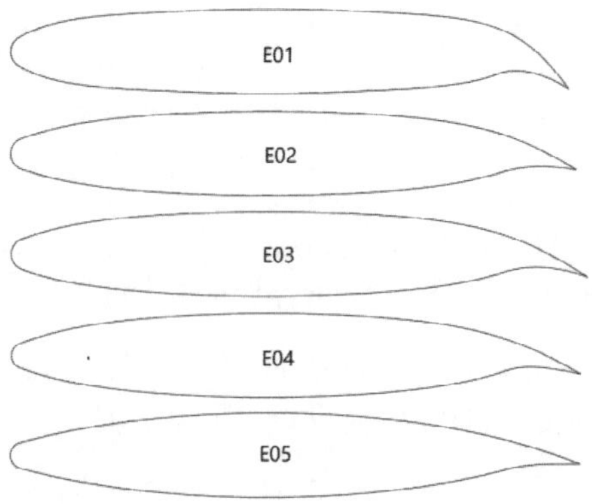

Figure4. Shapes of rotor/wing airfoil with flap for analysis

Calculations are firstly described for several meshes to investigate the mesh dependency of E01. The basic mesh is the same as elliptic airfoil, while a coarse mesh contains about 20 thousands points and two fine meshes with 80 and 100 thousands points are respectively generated to calculate the foil's aerodynamics. Figure 5 and figure 6 show the lift and drag calculated by the meshes with 57, 80 and 100 thousands points are nearly the same, while the results of the coarse mesh present low accuracy. It indicates that the mesh with 57 thousands points is reasonable for the simulations in this study.

Table 1. Parameters of rotor/wing airfoil with flap for analysis

Parameters	E01	E02	E03	E04	E05
Thickness of airfoil(% of airfoil chord length)	16%	16%	16%	16%	16%
Radius of arc(% of airfoil chord length)	3%	3%	3%	2.5%	2.5%
Flap rotary angle(degree)	120	140	140	140	150
Extending length of flap(% of airfoil chord length)	8.6%	10.1%	12.2%	10.8%	11.0%

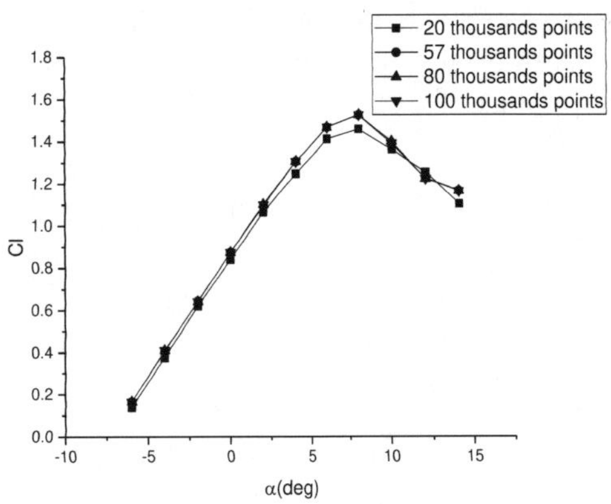

Figure5. Lift coefficients of E01 with different meshes

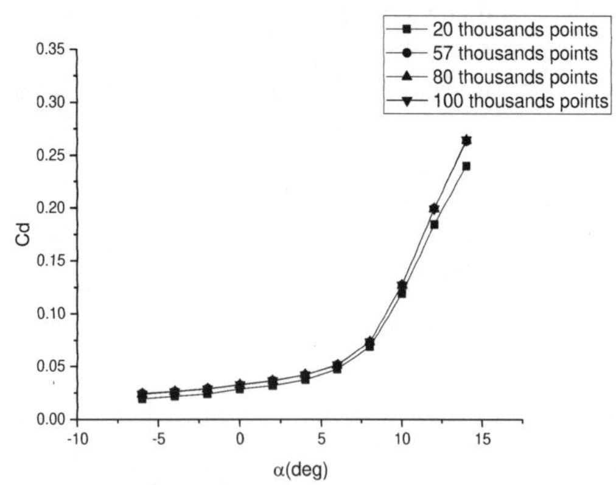

Figure6. Drag coefficients of E01 with different meshes

The aerodynamic forces and moment of a 16%-thick elliptic airfoil are compared with those of rotor/wing airfoil with flaps from E01 to E05 at the same flow condition. Calculations were made for high speed condition at a free stream Mach number of 0.4 and a Reynolds number of 1.9×106. For the purpose of comparing the effects of deployed flaps, the reference length for calculating dimensionless coefficient is the chord length of airfoils with flaps folded, equal to the chord length of elliptic airfoil.

In figure 7, the lift coefficients are compared between the elliptic airfoil and designed rotor/wing airfoil with flaps. It shows that the lift coefficients of all designed flaps are greater than elliptic airfoil in the same angle of attack before stall, and the maximum lift coefficients are also greater than elliptic airfoil except E05. The lift-curve slops of all designed flaps are greater than elliptic airfoil, but the stall angles are all about 8 degree, which are smaller than elliptic airfoil. It can be seen that, the E01 has the least rotary angle and largest bending trailing edge, and the lift coefficient of it is greatest among these designed flaps. The analysis of figure 7 and the differences between designed flaps shows that bending trailing edge, long extending length of flap and small radius of leading edge are beneficial for the improvement of lift performance.

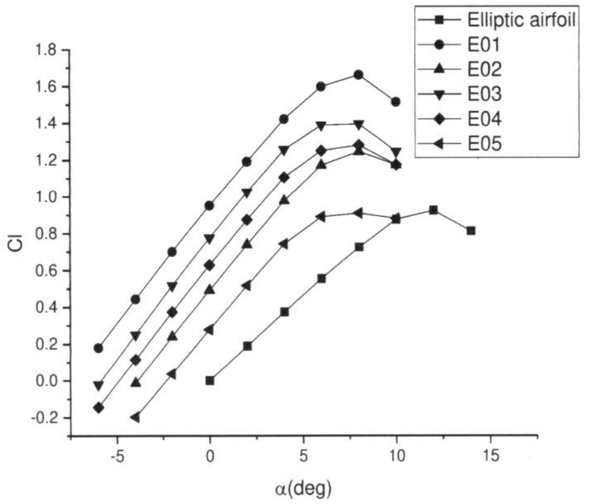

Figure 7. Comparison of lift coefficients

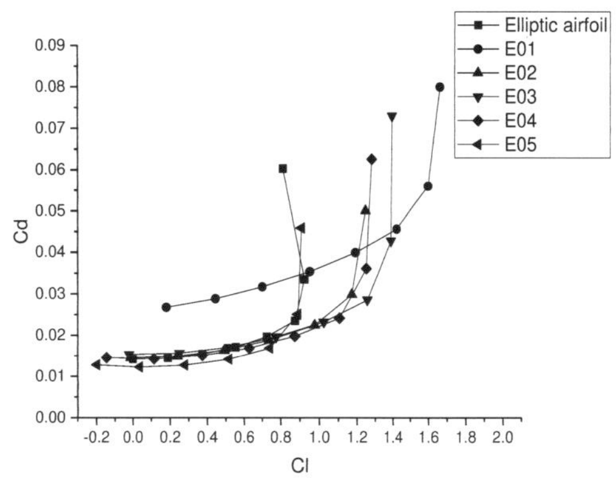

Figure 8. Comparison of drag coefficients

In figure 8, the drag coefficients are compared between the elliptic airfoil and designed rotor/wing airfoil with flaps. It shows that the drag coefficient of E01 is enormously greater than other flaps and elliptic airfoil, and the drag performances of these flaps except E05 are not evidently better than elliptic airfoil, although those flaps have sharp trailing edges.

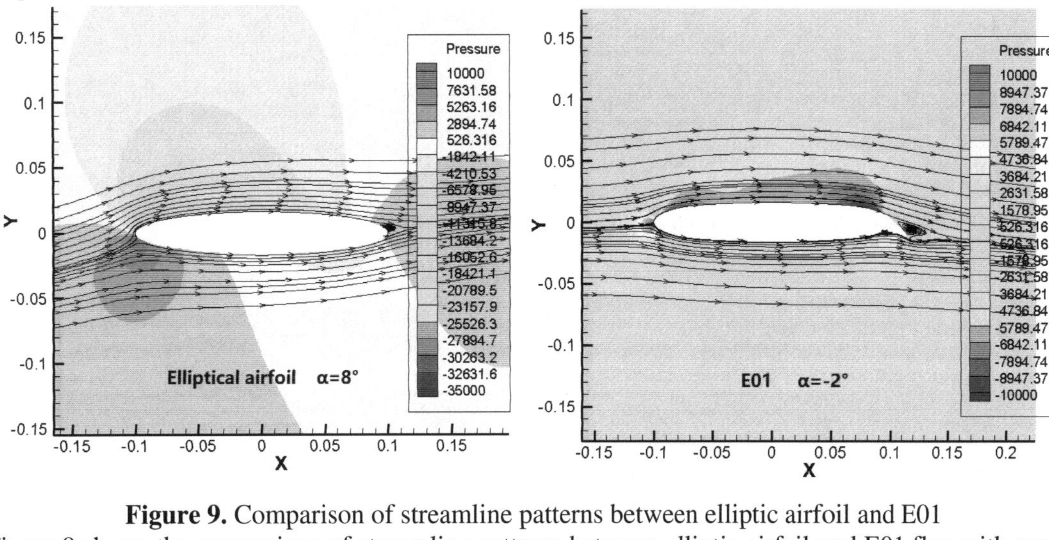

Figure 9. Comparison of streamline patterns between elliptic airfoil and E01

Figure 9 shows the comparison of streamline patterns between elliptic airfoil and E01 flap with nearly same lift coefficient. It is shown that, the flow separating area at the trailing edge of E01 is bigger than elliptic airfoil, and this will produce more pressure drag. Concluding from these comparisons, larger bending trailing edge is beneficial for the improvement of lift performance, but is adverse for the reducing of drag.

In figure 10, the lift-to-drag ratios are compared between the elliptic airfoil and designed rotor/wing airfoil with flaps. It shows that the lift-to-drag ratio of E05 flap is greater than elliptic airfoil at the same lift coefficient before stall, and the lift-to-drag ratios of other flaps except E01 are close to the elliptic airfoil.

In figure 11, the moment coefficients are compared between the elliptic airfoil and designed rotor/wing airfoil with flaps. It shows that, the moment coefficient of the elliptic airfoil continuously increased up to stall as the lift coefficient increased, whereas those of designed flaps decreased down until approaching stall. It is shown that, more lift is generated at the rear section of designed airfoil with flaps than elliptic airfoil. In addition, the moment coefficient of E05 shows a more gradual changes than that of elliptic airfoil, this is beneficial for longitudinal stability characteristics.

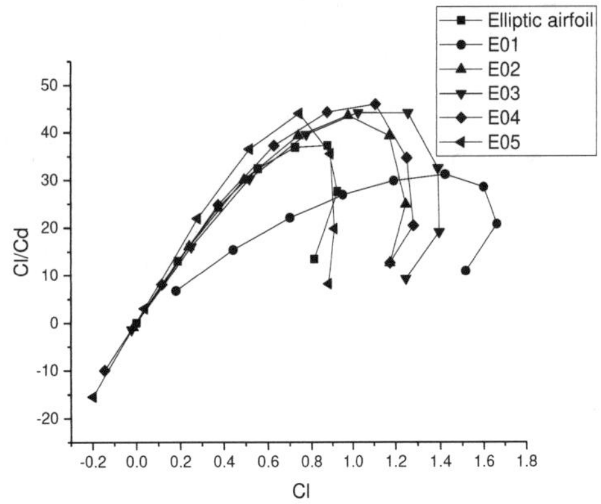

Figure 10. Comparison of lift-to-drag ratio

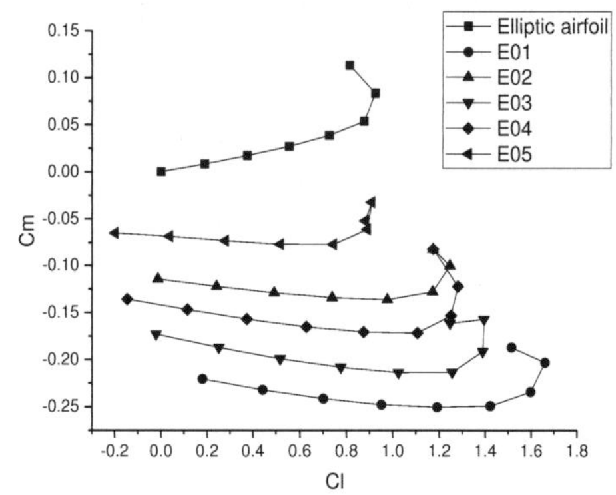

Figure 11. Comparison of moment coefficients about the quarter-chord point

4. Conclusions

For the purpose of reducing drags of blunt leading and trailing edge airfoils which are used for rotor/wing of CRW, a kind of blunt trailing edge airfoil with its flap is designed. For evaluating the effect of drag reduction and the influences of design parameters, five different rotor/wing airfoil with flaps are constructed, and the aerodynamic performances of them are compared to elliptic airfoil which has the same thickness based on CFD.

The results show that, the lift characteristic is improved for designed flaps, but the drag will not be reduced if the flap has great bending trailing edge. It has been proved from the comparison that, small bending trailing edge design is suitable for this kind of flap, and it will improve the drag and longitudinal stability characteristics.

The results also show that, the lift characteristic of flap with small bending trailing edge (for example E05) is not improved compared to elliptic airfoil, and this is a problem need to be studied.

References

[1] Clark A M and Barvara J V, The canard rotor wing (CRW) aircraft – a new way to fly, AIAA 2003-2517

[2] Kim C and Lee J, Hub drag-reduction strategies of canard rotor/wing unmanned aerial vehicles, *Journal of Aircraft*, **44**(4) :1388-91

[3] Lee J W , Jeon K S , et al, Rotor design for the performance optimization of canard rotor/wing aercraft, *Proc. of the 2006 international conference on computational and its applications (Glasgow UK)*, May08-11

[4] Yibo Li and Dongli Ma, Numerical simulation of rotor-aerodynamic surface interaction in hover using moving chimera grid, *Chinese Journal of Aeronautics*, **25**(3) :342-348

[5] Wendong Gai and Honglun Wang, Closed-loop dynamic control allocation for aircraft with multiple actuators, *Chinese Journal of Aeronautics*, **26**(3) :676-686

[6] Thomas L. Thompson, Roger L. Smith, et al, Wind tunnel test results for a canard rotor/wing aircraft configuration, *Proc. of the American Helicopter Society 57th Annual Forum (Washington DC, US)* May 9-11,2001

[7] Gerald Angle II, Jonathan Kweder, et al, Experimental investigation into a 10 :1 elliptical airfoil with circulation control augmentation for rotorcraft use, AIAA 2008-3873

[8] Niemied R, Jacobellis G and Gandhi F, Reversible airfoil for stopped rotors in high speed flight, *Smart Materials and Structures*, **23**(2014)

[9] Barocela E, Ballwin, et al, Low-drag rotor/wing flap, US Patent 7014142B2

[10] Sungyoon C and Joon K Oh, Aerodynamic characteristics of elliptic airfoils at high reynolds numbers, *Journal of Aircraft*, **45**(2) :641-650

[11] Menter F R, Zonal two equation k-ω turbulence models for aerodynamic flows, AIAA 1993-2906

IOP Conf. Series: Journal of Physics: Conf. Series **1215** (2019) 012012 doi:10.1088/1742-6596/1215/1/012012

Observer-Based Backstepping Fault-tolerant Control for Spacecraft with Reaction-wheel Failures

Z Z Zhang, D Ye and Z W Sun

Research Center of Satellite Technology, Harbin Institute of Technology, Harbin 150001, China

Abstract. In this paper, nonlinear satellite system with actuator faults is considered. To solve the problem of attitude stabilization of satellite under external disturbances and actuator faults, an observer is designed firstly. And the output of the observer is proved to be uniformly ultimately bounded. By using the estimated information of the observer, a backstepping fault-tolerant controller is derived. The proposed controller could achieve attitude stabilization in the presence of actuator faults. Finally, a simulation of attitude stabilization is carried out to illustrate effective of the fault-tolerant controller

1 Introduction

With the deepening of deep space exploration missions, makes it put forward higher requirements for spacecraft guidance, navigation and control technology. To this end, the concept of fault-tolerant control system design was proposed[[1]-[3]].As a popular technology, adaptive control has been used in spacecraft control systems widely [[4]-[6]]. One of the advantage of adaptive control is that it could identify uncertain parameters in real-time. In [[7]], for rendezvous and proximity problem of spacecrafts, the author designed an adaptive controller under input saturation. In [[8]], under the gain actuator fault, the author proposed an adaptive controller to deal with it. To solve the problem of actuator transient response, paper [[9]] proposed a control scheme based on technology of variable structure control. But the control schemes mentioned above were not proved to achieve attitude stability in a finite time. To solve this problem[[10]], the author combined adaptive control technique and adding a power integrator to solve the problem of attitude tracking control. In [[11]], an SMC-based control approach is derived under actuator outage fault. The proposed controller could stabilized the attitude quickly. In [[12]], an SMC scheme is proposed. Under the roll or yaw faults, attitude rapid maneuvering was achieved.

Although the adaptive control technology is robust to uncertainties, but it is under the assumption that the uncertainty is bounded. However it will increase the conservation of controller. Under this situation, controller based on fault diagnosis scheme was investigated. In [[13]], a nonlinear unknown input observer was proposed to observe the actuator faults under external disturbances and gyro drift. It could achieve the isolation of actuator faults. Similarly, paper [[13]] designed an observer to detect the time-vary thruster faults. The observer could observe the fault torque quickly. However both of them did not derive controllers for spacecraft. In [[15]], the author proposed a fault-tolerant control scheme based on an observer. The controller could achieve attitude stabilization under gain fault of actuators. In [[16]], an sliding-mode observe is designed to identify the uncertain parameters.

In the following of this paper, attitude kinematics and dynamics of the spacecraft is derived. And problem formulation is given. In section 3, an observer is designed to observer the fault torque of actuators and external disturbances. The boundness of the observer is proved by Lyapunov stability theory. In section 4, a backstepping controller is proposed based on the observer. And the system state is uniform ultimately bounded. Simulation is carried out in section 5.

2 Kinematics and dynamics model

2.1 Mathematical model of rigid spacecraft

MRPs is used to describe the spacecraft attitude in this paper. $\boldsymbol{n} \in \Re^3$ is defined as the Euler principal axis. $\varphi \in \Re$ is the rotation angle around $\boldsymbol{n}$. Define $\boldsymbol{\omega}$ as the angular velocity of satellite. The mathematical model of satellite attitude kinematics are given by

$$\sigma_i = \frac{q_i}{1+q_0}, \quad i = 1,2,3, \tag{1}$$

$$\dot{\boldsymbol{\sigma}} = F(\boldsymbol{\sigma})\boldsymbol{\omega} \tag{2}$$

$$\boldsymbol{F}(\boldsymbol{\sigma}) = \frac{1}{4}\left\{\left(1-\boldsymbol{\sigma}^{\mathrm{T}}\boldsymbol{\sigma}\right)\boldsymbol{I}_{3\times 3} + 2\left[\boldsymbol{\sigma}^{\times}\right] + 2\boldsymbol{\sigma}\boldsymbol{\sigma}^{\mathrm{T}}\right\} \tag{3}$$

where q_0 and q_i, $i = 1,2,3$ denote the unit quaternions.

Assumed that the satellite equipped N (N>3) reaction-wheels. The satellite's dynamics equation is given as follows

$$\boldsymbol{J}\dot{\boldsymbol{\omega}} = -\boldsymbol{\omega}^{\times}\boldsymbol{J}\boldsymbol{\omega} + \boldsymbol{\tau} + \boldsymbol{d} \tag{4}$$

where $\boldsymbol{J}$ represents the satellite inertial matrix, $\boldsymbol{\tau}$ represents the control torque, $\boldsymbol{d}$ represents the disturbance torque, $\boldsymbol{\omega}^{\times}$ represents the skew symmetric matrix of $\boldsymbol{\omega}$.

2.2 Model of reaction wheel faults

When faults occur, the torque generated by reaction-wheels could be modeled as follows

$$\tau_i = e_i(t)\tau_{ci} + \overline{\tau}_{ci} \tag{5}$$

Further, Eq.(5) could be modified as

$$\tau_i = \tau_{ci} + \tau_{fi} \tag{6}$$

where $\tau_{fi} = \left[e_i(t) - 1\right]\tau_{ci} + \overline{\tau}_{ci}$ denotes the fault torque.

Assumption 1 There exists a constant $\tau_{\max}$ that makes following inequality holds $\left|\tau_{fi}\right| \le \tau_{\max}$.

3 Design of observer

In this paper, fault torque is not estimated for one single reaction-wheel. Design the input of observer as $\boldsymbol{v}(t)$. If the observer is stable, the following equation holds

$$\boldsymbol{\tau}(t) = \boldsymbol{\tau}_c(t) + \boldsymbol{v}(t) \tag{7}$$

if $\boldsymbol{T}_c(t) = diag\left(\left[\tau_{c1} \quad \tau_{c2} \quad \tau_{c3}\right]^T\right)$, then $\boldsymbol{\tau}(t) = \boldsymbol{T}_c(t)\boldsymbol{\rho}(t)$, where $\boldsymbol{\rho}(t) = \left[\rho_1(t) \quad \rho_2(t) \quad \rho_3(t)\right]$, the following equation holds

$$\boldsymbol{T}_c(t)\boldsymbol{\rho}(t) = \boldsymbol{\tau}_c(t) + \boldsymbol{v}(t) \tag{8}$$

According to the dynamics model(4), the observer designed as follows

$$\boldsymbol{J}\dot{\hat{\boldsymbol{\omega}}} = -\hat{\boldsymbol{\omega}}^{\times}\boldsymbol{J}\hat{\boldsymbol{\omega}} + \boldsymbol{\Lambda}(\boldsymbol{\omega} - \hat{\boldsymbol{\omega}}) + \boldsymbol{\tau} + \boldsymbol{v}(t) \tag{9}$$

$$\boldsymbol{v}(t) = \boldsymbol{K}_1\boldsymbol{v}(t-T) + \boldsymbol{K}_2\left[\boldsymbol{\omega}(t) - \hat{\boldsymbol{\omega}}(t)\right] + \boldsymbol{K}_3\left[\boldsymbol{\omega}(t-T) - \hat{\boldsymbol{\omega}}(t-T)\right] \tag{10}$$

where $\boldsymbol{K}_1$, $\boldsymbol{K}_2$ and $\boldsymbol{K}_3$ are gain matrices of the observer, T denotes the sampling-time. $\boldsymbol{\Lambda}$ is a positive

defined matrix to be designed.
From Eq.(4) Eq.(6) and Eq.(9), equation of estimation error could be derived

$$\boldsymbol{J}\dot{\tilde{\boldsymbol{\omega}}} = \left(-\boldsymbol{\omega}^{\times}\boldsymbol{J}\boldsymbol{\omega}+\hat{\boldsymbol{\omega}}^{\times}\boldsymbol{J}\hat{\boldsymbol{\omega}}\right) - \boldsymbol{\Lambda}\tilde{\boldsymbol{\omega}} + \boldsymbol{\tau}_c + \boldsymbol{u}_f(t) - \boldsymbol{v}(t) \tag{11}$$

where $\tilde{\omega} = \omega - \hat{\omega}$ and $\boldsymbol{u}_f(t) = \boldsymbol{\tau}_f + \boldsymbol{d}$.
The following assumptions are made to prove the stability of the observer:
Assumption 2. $\boldsymbol{u}_f(t)$ is bounded and the following inequality holds $\boldsymbol{u}_f(t) - \boldsymbol{K}_I\boldsymbol{u}_f(t-T) \le \boldsymbol{k}_d$.

Assumption 3. The nonlinear term $\boldsymbol{\omega}^{\times}\boldsymbol{J}\boldsymbol{\omega}$ satisfies $\left\|-\boldsymbol{\omega}^{\times}\boldsymbol{J}\boldsymbol{\omega}+\hat{\boldsymbol{\omega}}^{\times}\boldsymbol{J}\hat{\boldsymbol{\omega}}\right\| \le \eta\|\tilde{\boldsymbol{\omega}}\|$, $\eta > 0$ is an unknown constant[[18]].
Assumption 4. The angular velocity information is available, and it is continuously differentiable, so $\|\tilde{\omega}(t-T)\| \le \xi\|\tilde{\omega}(t)\|$ for an unknown $\xi > 0$.

4 Design of backstepping fault-tolerant controller

Through the observer designed before, then an adaptive backstepping controller will be designed for achieve the attitude stabilization under actuator failures and disturbances. First, variables are defined as

$$\boldsymbol{x}_1 = \int \boldsymbol{\sigma} dt,\ \boldsymbol{x}_2 = \boldsymbol{\sigma},\ \boldsymbol{x}_3 = \boldsymbol{\omega} \tag{12}$$

Then the dynamics model could be rewritten as

$$\dot{\boldsymbol{x}}_1 = \boldsymbol{x}_2 \tag{13}$$

$$\dot{\boldsymbol{x}}_2 = F(\boldsymbol{x}_2)\boldsymbol{x}_3 \tag{14}$$

$$\boldsymbol{J}\dot{\boldsymbol{x}}_3 = -\boldsymbol{x}_3^{\times}\boldsymbol{J}\boldsymbol{x}_3 + \boldsymbol{\Theta}(t)\boldsymbol{\tau}_c - \boldsymbol{\Delta}(t)\boldsymbol{\tau}_c + \boldsymbol{d} \tag{15}$$

where $\boldsymbol{\Theta}(t) = diag([\rho_1 \ \ \rho_2 \ \ \rho_3])$, $\boldsymbol{\Delta}(t) = diag([\delta_1 \ \ \delta_2 \ \ \delta_3])$, $\delta_i = \hat{\rho}_i - \rho_i$ .To derive the backstepping controller, making the following transformation

$$z_1 = \boldsymbol{x}_1, z_2 = \boldsymbol{x}_2 - \boldsymbol{\alpha}_1, z_3 = \boldsymbol{x}_3 - \boldsymbol{\alpha}_2 \tag{16}$$

Where $\boldsymbol{\alpha}_1$ and $\boldsymbol{\alpha}_2$ are virtual control variables.
Step 1 According to Eq.(16)

$$\dot{z}_1 = \dot{\boldsymbol{x}}_1 = \boldsymbol{x}_2 = z_2 + \boldsymbol{\alpha}_1 \tag{17}$$

Consider a Lyapunov function as $V_1 = 1/2 \cdot z_1^T z_1$, And design $\boldsymbol{\alpha}_1 = -c_1\boldsymbol{x}_1$, c_1 is a constant to be designed, then the time derivative of V_1 could be obtain $\dot{V}_1 = z_1^T \dot{z}_1 = z_1^T(z_2 - c_1 z_1) = -c_1\|z_1\|^2 + z_1^T z_2$, When $z_2 = 0$, $\dot{V}_1 = -c_1\|z_1\|^2$, z_1 is asymptotically convergence, that is $\lim_{t\to\infty} z_1(t) = \boldsymbol{0}$.
Step 2 The time derivative of z_2 is

$$\dot{z}_2 = \dot{\boldsymbol{x}}_2 - \dot{\boldsymbol{\alpha}}_1 = F(\boldsymbol{x}_2)\boldsymbol{x}_3 + c_1\boldsymbol{x}_1 \tag{18}$$

Consider Lyapunov candidate as $V_2 = V_1 + 1/2 \cdot z_2^T z_2$ .Design $\boldsymbol{\alpha}_2$ as $\boldsymbol{\alpha}_2 = \boldsymbol{F}^{-1}(\boldsymbol{x}_2)(-z_1 - c_1\boldsymbol{x}_2 - c_2 z_2)$, Then $\dot{V}_2 = -c_1\|z_1\|^2 - c_2\|z_2\|^2 + z_2^T\boldsymbol{F}(\boldsymbol{x}_2)z_3$, if $z_3 = 0$, $\dot{V}_2 = -c_1\|z_1\|^2 - c_2\|z_2\|^2$. So z_1 and z_2 are asymptotically convergence, that is $\lim_{t\to\infty} z_1(t) = \lim_{t\to\infty} z_2(t) = \boldsymbol{0}$
If the controller could make $\boldsymbol{\tau}_c$ satisfy $\lim_{t\to\infty} z_3(t) = \boldsymbol{0}$, then $\lim_{t\to\infty} z_1(t) = \lim_{t\to\infty} z_2(t) = \boldsymbol{0}$ also satisfied.
Theorem 2 For satellite dynamics system determined by Eq.(2) and Eq.(4) with actuator failures. Under observer determined by Eq.(9) and Eq.(10), design the controller as follows

$$\boldsymbol{\tau}_c = \boldsymbol{Sat}\left(\boldsymbol{\mu}, \tau_{\max}\right) \tag{19}$$

$$\begin{aligned}\boldsymbol{\mu}(t) = \boldsymbol{\Theta}(t)^{-1}\Bigg\{ &\boldsymbol{J}\left[\frac{d\boldsymbol{F}^{-1}(\boldsymbol{x}_2)}{dt}(z_1 + c_1\boldsymbol{x}_2 + c_2 z_2) + \boldsymbol{F}^{-1}(\boldsymbol{x}_2)(\dot{z}_1 + c_1\dot{\boldsymbol{x}}_2 + c_2\dot{z}_2)\right] + \\ &\boldsymbol{x}_3^{\times}\boldsymbol{J}\boldsymbol{x}_3 - c_3 z_3 - G_1\boldsymbol{x}_a - \boldsymbol{F}^T(\boldsymbol{x}_2)z_2 - \frac{\chi_{T_d}^2 z_3}{\chi_{T_d}\|z_3\| + \varepsilon\exp(-\beta t)}\Bigg\}\end{aligned} \tag{20}$$

$$\dot{\boldsymbol{x}}_a = -G_2\boldsymbol{x}_a - \frac{\|\boldsymbol{\Theta}(t)\|^2\|\boldsymbol{\tau}_c - \boldsymbol{\mu}\|^2}{\|\boldsymbol{x}_a\|^2}\boldsymbol{x}_a - \boldsymbol{\Theta}(t)(\boldsymbol{\tau}_c - \boldsymbol{\mu}) \tag{21}$$

where $\chi_{T_d} = \tau_{\max} + d_{\max}$, ε is a small constant, K_1, K_2, β and c_3 are gains of the controller. If the following inequalities satisfied $c_3 - 1 > 0, G_2 - G_1^2/2 - 1/2 > 0$. Then, the close loop of attitude control system is stable. Furthermore, there exists $\varepsilon_1^* > 0$, $\varepsilon_2^* > 0$ and $T^* > 0$, for an arbitrary $t > T^*$, $\|\sigma(t)\| \le \varepsilon_1^*$ and $\|\omega(t)\| \le \varepsilon_2^*$ holds.

Proof. From Eq.(15)

$$\dot{z}_3 = \boldsymbol{J}^{-1}\left[-\boldsymbol{x}_3^{\times}\boldsymbol{J}\boldsymbol{x}_3 + \boldsymbol{\Theta}(t)\boldsymbol{\tau}_c + \boldsymbol{d}\right] + \frac{d\boldsymbol{F}^{-1}(\boldsymbol{x}_2)}{dt}(z_1 + c_1\boldsymbol{x}_2 + c_2 z_2) + \boldsymbol{F}^{-1}(\boldsymbol{x}_2)(\dot{z}_1 + c_1\dot{\boldsymbol{x}}_2 + c_2\dot{z}_2) \tag{22}$$

Where

$$\frac{d\boldsymbol{F}^{-1}(\boldsymbol{x}_2)}{dt} = \frac{16\left(1 + \boldsymbol{x}_2^T\boldsymbol{x}_2\right)^2\left[\frac{d\boldsymbol{F}(\boldsymbol{x}_2)}{dt}\right]^T - 64\left(1 + \boldsymbol{x}_2^T\boldsymbol{x}_2\right)\boldsymbol{x}_2^T\boldsymbol{F}(\boldsymbol{x}_2)\boldsymbol{x}_3\boldsymbol{F}^T(\boldsymbol{x}_2)}{\left(1 + \boldsymbol{x}_2^T\boldsymbol{x}_2\right)^4} \tag{23}$$

$$\frac{d\boldsymbol{F}(\boldsymbol{x}_2)}{dt} = \frac{1}{2}\left\{-\boldsymbol{x}_2^T\left[\boldsymbol{F}(\boldsymbol{x}_2)\boldsymbol{x}_3\right]I_3 + \left[\boldsymbol{F}(\boldsymbol{x}_2)\boldsymbol{x}_3\right]^{\times} + \boldsymbol{F}(\boldsymbol{x}_2)\boldsymbol{x}_3\boldsymbol{x}_2^T + \boldsymbol{x}_2\boldsymbol{x}_3^T\boldsymbol{F}^T(\boldsymbol{x}_2)\right\} \tag{24}$$

Consider a Lyapunov candidate as

$$V_3 = V_2 + \frac{1}{2}z_3^T\boldsymbol{J}z_3 + \frac{1}{2}\boldsymbol{x}_a^T\boldsymbol{x}_a \tag{25}$$

Time derivative of V_3 could be derived

$$\begin{aligned}\dot{V}_3 = &-c_1\|z_1\|^2 - c_2\|z_2\|^2 + z_2^T F(\boldsymbol{x}_2)z_3 - G_2\|\boldsymbol{x}_a\|^2 - \|\boldsymbol{\Theta}(t)\|^2\|(\boldsymbol{\tau}_c - \boldsymbol{\mu})\|^2 - \\ &\boldsymbol{x}_a^T\Theta(t)(\boldsymbol{\tau}_c - \boldsymbol{\mu}) + z_3^T\left[-\boldsymbol{x}_3^{\times}\boldsymbol{J}\boldsymbol{x}_3 + \boldsymbol{\Theta}(t)\boldsymbol{\tau}_c - \boldsymbol{\Delta}(t)\boldsymbol{\tau}_c + \boldsymbol{d}\right] + \\ &\boldsymbol{J}\left[\frac{d\boldsymbol{F}^{-1}(\boldsymbol{x}_2)}{dt}(z_1 + c_1\boldsymbol{x}_2 + c_2 z_2) + \boldsymbol{F}^{-1}(\boldsymbol{x}_2)(\dot{z}_1 + c_1\dot{\boldsymbol{x}}_2 + c_2\dot{z}_2)\right]\end{aligned} \tag{26}$$

According to the following inequalities

$$\begin{aligned}&z_3^T\boldsymbol{\Theta}(t)(\boldsymbol{\tau}_c - \boldsymbol{\mu}) \le \frac{1}{2}\|\boldsymbol{\Theta}(t)\|^2 + \frac{1}{2}\|z_3\|^2, \quad -G_1 z_3^T\boldsymbol{x}_a \le \frac{1}{2}\|z_2\|^2 + \frac{G_1^2}{2}\|\boldsymbol{x}_a\|^2 \\ &-\boldsymbol{x}_a^T\boldsymbol{\Theta}(t)(\boldsymbol{\tau}_c - \boldsymbol{\mu}) \le \frac{1}{2}\|\boldsymbol{\Theta}(t)\|^2\|(\boldsymbol{\tau}_c - \boldsymbol{\mu})\|^2 + \frac{1}{2}\|\boldsymbol{x}_a\|^2\end{aligned} \tag{27}$$

If gains of controller selected as **Theorem 2**

$$\begin{aligned}\dot{V}_3 &\le -c_1\|z_1\|^2 - c_2\|z_2\|^2 - (c_3-1)\|z_3\|^2 - \left(G_2 - \frac{G_1^2}{2} - \frac{1}{2}\right)\|\boldsymbol{x}_a\|^2 + \varepsilon\exp(-\beta t) \\ &\le -m\left(\|z_1\|^2 + \|z_2\|^2 + \boldsymbol{J}_{\min}\|z_3\|^2 + \|\boldsymbol{x}_a\|^2\right) + \varepsilon\end{aligned} \tag{28}$$

Where $m = \min\left(c_1, c_2, (c_3-1)/J_{\max}, G_2 - G_1^2/2 - 1/2\right)$. According to the definition of V_3

$$2V_3 \ge \|z_1\|^2 + \|z_2\|^2 + \boldsymbol{J}_{\min}\|z_3\|^2 + \|\boldsymbol{x}_a\|^2 \tag{29}$$

Furthermore

$$\dot{V}_3 \le -2mV_3 + \varepsilon \tag{30}$$

So V_3 is uniform ultimately bounded. And for $\varepsilon^* > \sqrt{\varepsilon / m}$, there exist a finite time $T_f > 0$, makes $\|z_i\| \le \varepsilon^*$ and $\|\boldsymbol{x}_a\| \le \varepsilon^*$ holds. The proof is completed.

5 Simulation Results

In order to verify the effectiveness of the proposed controller, the controller proposed in [[17]] is compared in this paper.

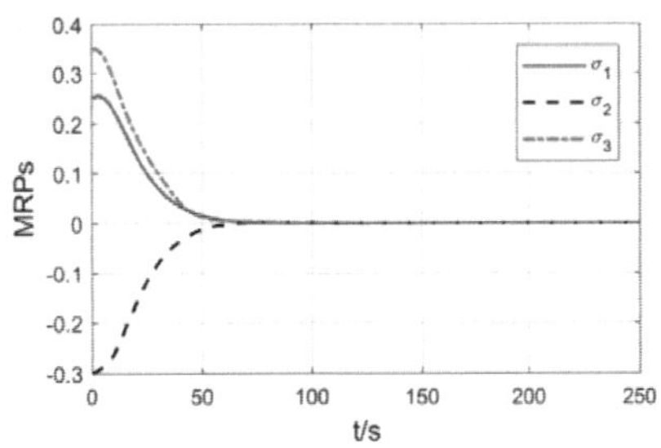

Figure 1 Attitude response of satellite

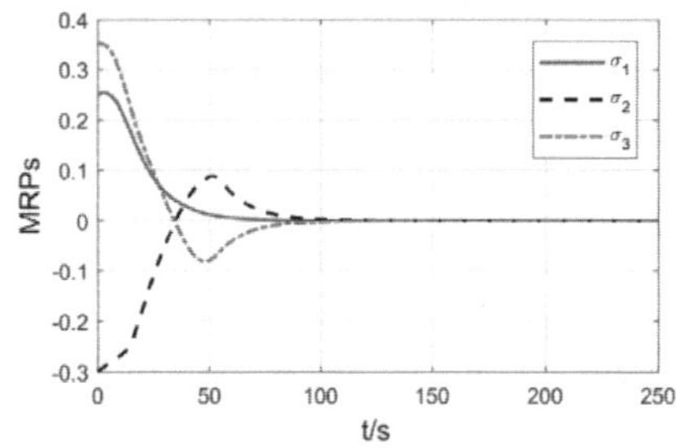

Figure 2 Attitude response under compared controller

The Attitude response of the backstepping fault-tolerant control method is shown in Figure 1. And Attitude response of controller proposed in [[17]] is shown in Figure 2.The Angular velocity response of the backstepping fault-tolerant control method is shown in Figure 3. And Angular velocity response of controller proposed in [[17]] is shown in Figure 4. Figure 5 shows the output signals of the observer. It could be seen in the figure, when faults happened the output signals will show the effectiveness on the system.

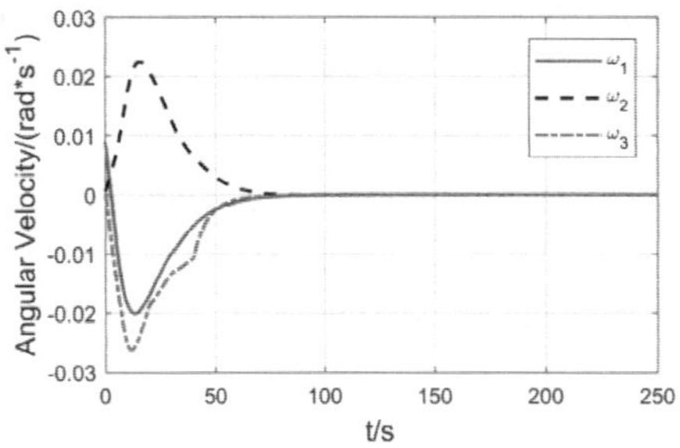

Figure 3 Angular velocity response of satellite

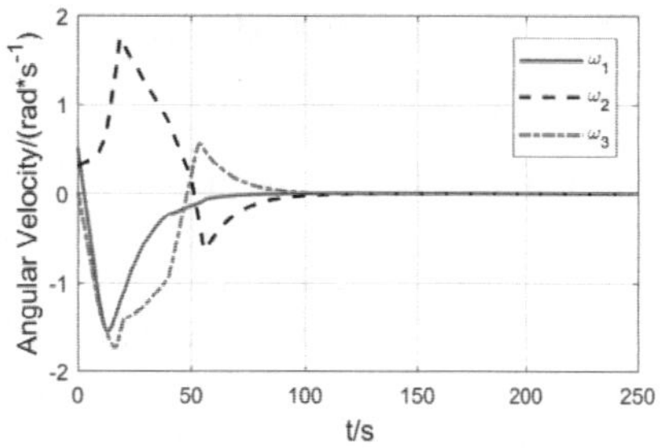

Figure 4 Angular velocity response under compared controller

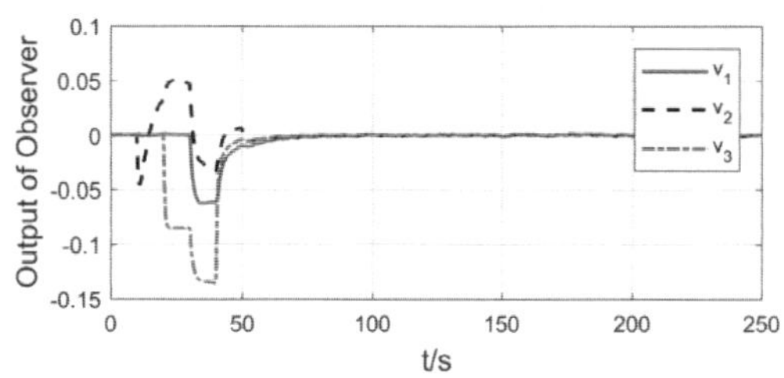

Figure 5 output signals of the observer

6 Conclusion

In this paper, to solve the problem of attitude stabilization of satellite under external disturbances and actuator fault, nonlinear satellite system with actuator fault is and external disturbance is considered. An observer is developed to estimate the actuator fault torque. And the output of the observer is proved to be uniform ultimately bounded. Furthermore, by using the estimated information of the observer, a backstepping fault-tolerant controller is derived. The proposed controller could achieve attitude stabilization in the presence of actuator fault. The simulation results show the robustness to actuator faults and disturbances. And compared to a similar controller. The results show that the proposed controller could achieve attitude stabilization more effectively.

7 References

[1] Blanke M, Izadi-Zamanabadi R and Bøgh S A 1997. Fault-tolerant control systems—a holistic view. *Control Engineering Practice*, **5** 693-702.

[2] Patton R J. Fault-tolerant control: the 1997 situation 1997. *IFAC Proceedings Volumes*, **30** 1029-1051.

[3] Zhang Y, Jiang J 2008. Bibliographical review on reconfigurable fault-tolerant control systems. *Annual reviews in control*, **32** 229-252.

[4] Bustan D, Sani S K H, Pariz N(2014. Adaptive fault-tolerant spacecraft attitude control design with transient response control. *IEEE/ASME Transactions on Mechatronics*, **19** 1404-1411

[5] Yang G H, Ye D(2008. Adaptive fault-tolerant H∞ control against sensor failures. *IET Control Theory & Applications*, **2** 95-107

[6] Qinglei H U, Zhang Y, Xing H U O(2011, et al. Adaptive integral-type sliding mode control for spacecraft attitude maneuvering under actuator stuck failures. *Chinese Journal of Aeronautics*, **24** 32-45

[7] Sun L, Huo W, Jiao Z(2017. Adaptive backstepping control of spacecraft rendezvous and proximity operations with input saturation and full-state constraint. *IEEE Transactions on Industrial Electronics*, **64** 480-492

[8] Han Y, Biggs J D, Cui N(2015. Adaptive fault-tolerant control of spacecraft attitude dynamics with actuator failures. *Journal of Guidance, Control*, and Dynamics, **38** 2033-2042

[9] Bustan D, Sani S K H, Pariz N(2014. Adaptive fault-tolerant spacecraft attitude control design

IOP Conf. Series: Journal of Physics: Conf. Series **1215** (2019) 012012 doi:10.1088/1742-6596/1215/1/012012

with transient response control. *IEEE/ASME Transactions on Mechatronics*, **19** 1404-1411
[10] Zhao L, Yu J, Yu H(2018. Adaptive finite-time attitude tracking control for spacecraft with disturbances. *IEEE Transactions on Aerospace and Electronic Systems*, **54** 1297-1305
[11] Liang Y W, Xu S D, Tsai C L(2007. Study of VSC reliable designs with application to spacecraft attitude stabilization. *IEEE Transactions on Control Systems Technology*, **15** 332-338
[12] Kumar K D(2011. Robust attitude stabilization of spacecraft subject to actuator failures. *Acta Astronautica*, **68** 1242-1259
[13] Liu C, Vukovich G, Sun Z and Shi K(2018, Observer-Based Fault-Tolerant Attitude Control for Spacecraft with Input Delay. *Journal of Guidance, Control, and Dynamics*, **41** 2041-2053
[14] Chen W, Saif M(2007. Observer-based fault diagnosis of satellite systems subject to time-varying thruster faults. *Journal of dynamic systems, measurement, and control*, **129** 352-356
[15] Zhang A, Hu Q, Zhang Y(2014. Observer-based attitude control for satellite under actuator fault. *Journal of Guidance, Control, and Dynamics*, **38** 806-811
[16] Yan X G, Edwards C(2008. Adaptive sliding-mode-observer-based fault reconstruction for nonlinear systems with parametric uncertainties. *IEEE Transactions on Industrial Electronics*, **55** 4029-4036
[17] Cai W, Liao X, Song D Y(2008. Indirect robust adaptive fault-tolerant control for attitude tracking of spacecraft. Journal of Guidance, *Control, and Dynamics*, **31** 1456-1463
[18] Liu C, Sun Z, Shi K and Wang F(2017, Robust dynamic output feedback control for attitude stabilization of spacecraft with nonlinear perturbations. *Aerospace Science and Technology*, **64** 102-121

Multi-missile Coordination High Precision Guidance and Control Method for Beam-riding Guidance

Qi Guodong[1], Yu Jianqiao[1*], Ai Xiaolin[1] and Jiang Jun[1]

[1] School of Aerospace Engineering, Beijing Institute of Technology, Beijing 100081, China

jianqiao@bit.edu.cn

Abstract. Due to the limitation of target indication error and steady-state tracking error, it is difficult for the existing beam-riding guidance weapon system to attack long-range and high-maneuvering targets accurately. In order to solve this problem, this paper proposes a high precision guidance and control method for beam-riding guidance multi missile coordination: through sharing error information between multiple missiles in the same guiding beam, correcting the center line of guided beam and eliminating the steady-state error in the process of missile tracking target, thus the strike capability of the guided missile for long distance and high maneuvering target is enhanced. The six degree of freedom controlled trajectory simulation is used to verify this method. The simulation results show that the guidance and control method is correct and effective. Compared with the existing beam-riding guidance system, it has a higher hit precision and better application prospect.

1. Introduction

In complex battlefield environment, it is always an important direction of ammunition development to be able to hit targets accurately at a long distance[1]. Among many types of guided munition, the guidance technology mainly includes autonomous guidance technology, remote control guidance technology, homing guidance technology and composite guidance technology. The remote control guidance is a guidance technology which sends guidance information to the ammunition by the guidance station and directs the ammunition to the target or the predetermined area[2]. Remote control guidance can be divided into remote command guidance and beam-riding guidance, beam-riding guidance can be divided into radio (radar) beam-riding guidance and optical (laser) beam-riding guidance. Promoted by the development of high-energy continuous-wave solid-state laser technology, ammunition using laser beam-riding guidance mechanism has been equipped with a large number of national troops.

The beam-riding guidance weapon system is widely used because of its strong anti-interference capability, simple structure and low cost[3]. The basic principle of the guided weapon is that the guidance Station emits the guiding beam and directs the center line of the beam to the target. When the missile flies in the guiding beam, it relies on the missile borne equipment to perceive its position in the beam, and then produces the control instruction guided missile to fly along the beam center line, eventually leads the missile to the target[4]. However, it is difficult for the existing beam-riding guidance system to be used to attack long-range and high-maneuvering targets accurately. There are two main factors affecting the system: on the one hand, the guidance station has the target indication error in the process of controlling the guidance beam to track the target; on the other hand, the missile has the steady-state tracking error when tracking the guidance beam center line. In order to solve the

above problems, this paper proposes a multi missile cooperative guidance control method under the beam-riding guidance system. By sharing the error information between the pre launched missile and the following missile, this method can greatly improve the hit precision of the beam-riding guidance system to the long distance and high maneuvering target.
The weapon system currently retrieved to fly multiple missiles in the same guiding beam is the British 'Starstreak' air defense missile, with 3 submunitions in the front end of the main projectile. When the two level engine of the projectile is burned out, the 3 submunitions are separated and fly by laser beam-riding guidance to the target[5]. There is no literature on multi-missile coordination in the same guided beam in China, and the 'Starstreak' air defense missile in Britain is the most mature product at present.

2. Multi missile coordination high precision guidance and control method

As shown in Figure 1, the guidance accuracy of three missiles flying in the same guidance beam is mainly affected by two aspects:
(1) The guidance station has the target indication error in the process of controlling the guidance beam tracking target. The error includes the angle tracking error in the course of tracking the target in the center line of the target angle tracking device of the guidance station, and the angular error between the center line of the guiding beam central line and the target angle tracking device. The principle of target indication error determines that the linear deviation of missile tracking target will be scaled up with the increase of target distance. The principle of target indication error determines that the linear deviation of missile tracking target will be scaled up with the increase of target distance. For example, when tracking a target with a distance of 10 km, the target indication error of 0.5mil will make the tracking line of the missile deviate to 5 meters, for a target with a distance of 3 km, the tracking error of the missile is only 1.5 meters[6].
(2) The beam-riding guidance system is essentially three point guidance. Its guiding principle determines the steady tracking error in the center line of the missile tracking guided beam when the target is high maneuvering. The size of the error depends on the design of the guidance control system and the motion characteristics of the target. This error will not only increase the miss distance of the missile, but also even cause the missile to fly out of the guidance beam.
Therefore, this paper proposes the multi-missile cooperative high-precision guidance and control method to improve the hit accuracy of the missile under the Beam-riding guidance system includes the following steps:
Step1: the guidance station acquires the target and then launches the guided beam to track the target.
Step 2: according to a certain time interval, the guided missile will launch 3 missiles in turn.
Step 3: the follow-up missiles send the tracking error formed in the course of tracking the center line of the guided beam to all the missiles launched in advance.
Step 4: after the missile receives the tracking error of all the subsequent missiles, the tracking error is used as the open loop error compensation instruction, and the tracking error is compensated in its guidance and control system.
Step 5: when the first missile that completes the tracking error compensation is close to the target, the position information of the target in its body coordinate system is obtained by the strapdown detector, then the position of the target in the guiding beam is calculated based on the position information. According to the position of the target in the guiding beam, the target indication error of the guide beam center line is obtained, and the target indication error is sent to all subsequent missiles.
Step 6: the subsequent missile corrects the guide beam center line according to the target indication error and tracks the corrected pilot beam center line, until the next missile approaches the target, repeating step 5 to get the new target indication error. According to the new target indication error, the subsequent missile corrects the guide beam center line again, and so on, until all the missiles hit the target.

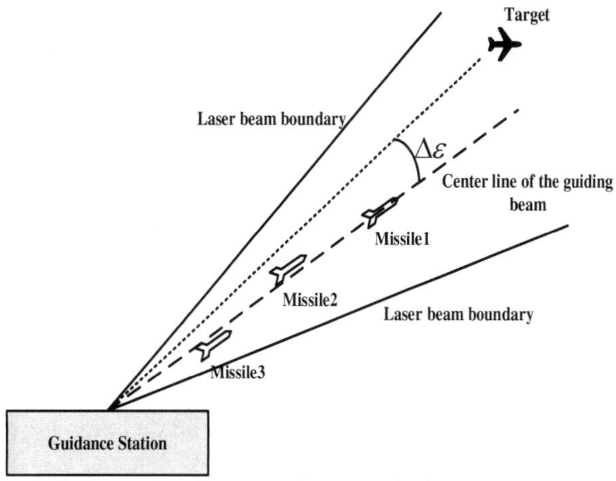

Figure 1. Schematic diagram of three missiles flying in the beam.

2.1. Correction method of steady state tracking error

When the target is maneuvering in high dynamic state, the steady-state tracking error exists when the single missile tracks the center line of the guidance beam, which will cause the missile to miss the target or even fly out of the guidance beam and out of control. However, by means of multi-missile cooperative guidance, the tracking error formed in the process of tracking the beam center line of the guided missile launched subsequently is transmitted to the pre launched missile. By introducing the received tracking error into the pre launched missile's control system as an open-loop error compensation instruction, the type of the guidance control system can be greatly improved, and the problem of missile miss target and even fly out of the guidance beam can be effectively solved when the missile tracks the center line of the guidance beam.

2.1.1. Mathematical model. Taking the 3 missile cooperative operation in the longitudinal plane as an example, establishing the mathematical model of steady state tracking error correction method. As shown in Figure 2, a typical longitudinal guidance control loop of a beam-riding guided missile is taken as an example, the system consists of a lead correction network and an autopilot[7]. It is assumed that the desired position of the missile determined by the moving state of the target is represented by $R(s)=\frac{A}{s^n}$ (n is signal order, A is input signal amplitude, s is complex variable). From the open loop transfer function of the system, it is known that the system is a "II" type system, and under the action of $R(s)$, the steady state tracking error of the guidance and control system is $e_{ss}=\lim_{s\to 0} s\left(R(s)-Y(s)\right)=0$ only when the motion law of the desired position of the missile satisfies $n\leq 2$ [8]. Therefore, when the target is high dynamic maneuver ($n>2$), there is a steady state tracking error when the single missile tracks the centerline of the beam, which can cause the missile to miss the target or even fly out of the guide beam and lose control. By using the multi missile cooperative guidance method, the steady tracking error of the missile is corrected by the steady tracking error of the following missile. It can improve the type of the control system and effectively solve the problem that the missile is out of the target to fly out of the guide beam when the missile tracking guide beam center line.

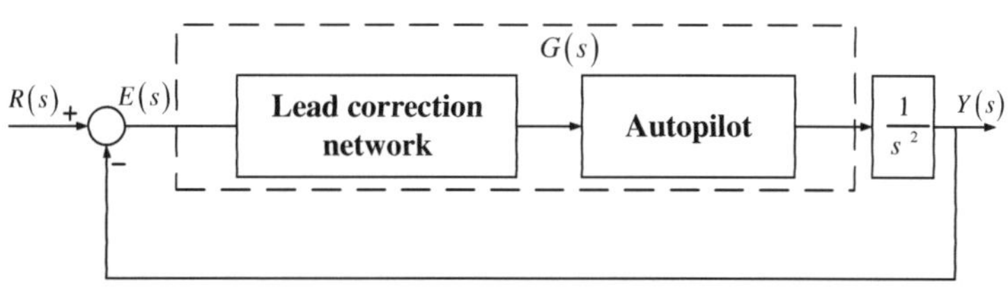

Figure 2. Structure diagram of typical longitudinal guidance and control system for Missile.

During the actual operation, the third missile measure the dynamic tracking error when tracking the center line of the guide beam, and then send them to first, second missile by the data chain on the missile. The second missile receives the dynamic tracking error of the third missile and forms an open loop error compensation command to compensate the guidance and control system error. Then the second missile sends the dynamic tracking error of the tracking guide beam center line to first missile through the data chain on the missile, and the first missile uses the dynamic tracking error of the received second, third missile to form the open loop error compensation instruction to compensate its guidance and control system error, in order to improve the precision of the hit. From Figure 3, we can see that the theoretical tracking error of the third missile is:

$$E_3(s)=\frac{1}{1+G(s)\frac{1}{s^2}}R(s) \tag{1}$$

After the compensation signal is introduced into the guidance and control system of the second missile, the theoretical tracking error of the second missile is:

$$E_2(s)=\frac{1}{\left(1+G(s)\frac{1}{s^2}\right)^2}R(s) \tag{2}$$

After the compensation signal $E_3(s)$ and $E_2(s)$ are introduced into the guidance and control system of the first missile, the theoretical tracking error of the first missile is:

$$E_1(s)=\frac{1}{\left(1+G(s)\frac{1}{s^2}\right)^3}R(s) \tag{3}$$

Therefore, when the target motion law satisfies $n<5$, the steady-state error of the central line of the first, second missile tracking guidance beam is:

$$\begin{aligned} e_{ss,1}&=\lim_{s\to 0} sE_1(s)=\lim_{s\to 0}\frac{A}{\left(s^{(n-1)/3}+G(s)\frac{1}{s^{2-(n-1)/3}}\right)^3}=0 \\ e_{ss,2}&=\lim_{s\to 0} sE_2(s)=\lim_{s\to 0}\frac{A}{\left(s^{(n-1)/2}+G(s)\frac{1}{s^{2-(n-1)/2}}\right)^2}=0 \end{aligned} \tag{4}$$

Thus, under the guidance of multi-missile cooperative guidance strategy, the first and second missile can track the center line of the guidance beam with high precision.

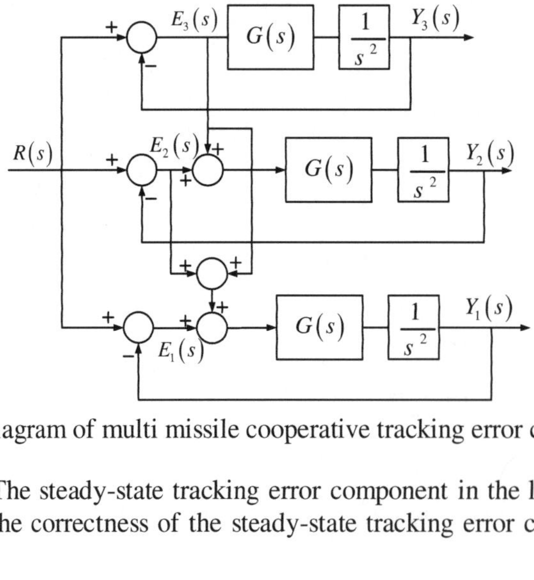

Figure 3. Schematic diagram of multi missile cooperative tracking error compensation structure.

2.1.2. Simulation result. The steady-state tracking error component in the longitudinal plane is taken as an example to verify the correctness of the steady-state tracking error correction method. Steady-

IOP Conf. Series: Journal of Physics: Conf. Series **1215** (2019) 012013 doi:10.1088/1742-6596/1215/1/012013

state tracking error correction method is used in the simulation program, and longitudinal deviation comparison charts of three missile during flight are obtained as shown in Figure 4. From this figure, you can see that the error of the missile 3 is the largest because it did not correct the steady-state error. The missile 2 was corrected once, and the error was reduced a lot. The missile 1 was corrected two times, with the smallest error and the highest hit accuracy of the missile, which accords with the theoretical expectation. It is proved that the error correction method of steady state tracking error is correct and effective.

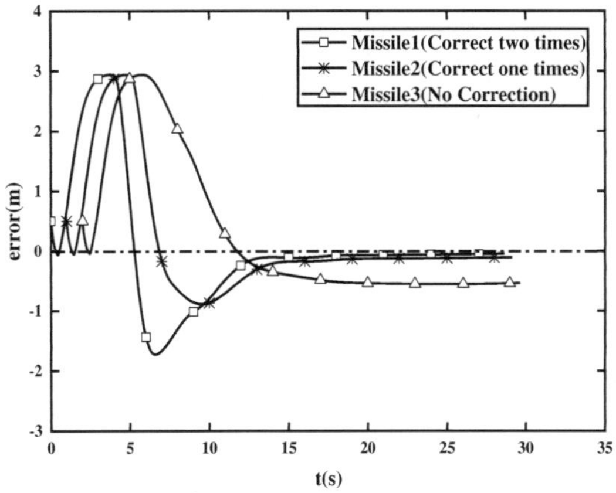

Figure 4. Contrast diagram of three missile longitudinal deviations.

2.2. Estimation method of indicator error

When the missile can accurately track the center line of the guidance beam, the hit accuracy of the missile mainly depends on the target indication error in the process of tracking the target by the guidance station. In this paper, a cooperative operation method is proposed to eliminate the indication error: by installing strapdown detectors on the head of the pre launched missile, the target indication error is measured by the pre launched missile, and then the error information is sent to the following missile to correct the guidance beam of the subsequent missile launched. So as to improve the hit accuracy of following missiles.

2.2.1. Mathematical model. Taking the 3 missile cooperative operation in the longitudinal plane as an example, establishing the mathematical model of target indication error estimation method. For the convenience of modeling, the $\Delta\varepsilon$ is first projected into the longitudinal and lateral planes to obtain two target indication error components $\Delta\varepsilon_y$ and $\Delta\varepsilon_z$, and then the mathematical models of component $\Delta\varepsilon_y$ and $\Delta\varepsilon_z$ are derived respectively.

Taking the component $\Delta\varepsilon_y$ in the longitudinal plane of the target indication error as an example, deducing the mathematical model. The process of deriving the mathematical model of $\Delta\varepsilon_z$ is similar to $\Delta\varepsilon_y$. As shown in Figure 5, it is assumed that the strapdown detector is installed on the missile head, and the detector can measure the distance d and azimuth λ of the target relative to the original point O_1 of the coordinate system in the body coordinate system $O_1x_1y_1$. The missile body coordinate system origin O_1 coincides with the detector center, the O_1x_1 axis coincides with the missile axis and points to the missile in front of the movement is positive, the O_1y_1 axis is perpendicular to the O_1x_1 axis, and is positive to the right[9]. Suppose that in a short time interval $\Delta t = t_2 - t_1$, the relative motion of the target and the missile makes the target moving from the T_1 point to the T_2 point relative to the missile position. As shown in Figure 6, the distance and azimuth angle of the t_1 time target in the body coordinate system $O_1x_1y_1$ are respectively d_1 and λ_1, and the distance and azimuth of the t_2 time target in the body coordinate system $O_1x_1y_1$ are d_2 and λ_2 respectively. When Δt is sufficiently small, it can be considered that the attitude of the missile remains unchanged relative to the guiding beam. At this time, the connection between the T_1 point and the T_2 point can be used as an estimate of the velocity direction of the missile relative to the center line of the guided beam. When the missile can track the beam center line stably, the velocity direction of the missile is basically the same as that of the beam center line. Thus, the relative attitude angle estimation formula of the missile relative to the guidance centerline can be established:

$$\theta = \arccos\left(\frac{d_1^2 + \overline{T_1T_2}^2 - d_2^2}{2d_1\overline{T_1T_2}}\right) - \lambda_1 \tag{5}$$

$$\overline{T_1T_2} = \sqrt{d_1^2 + d_2^2 - 2d_1d_2\cos(\lambda_2 - \lambda_1)} \tag{6}$$

According to the target measurement value of the first missile, the following formula can be used to calculate the position $x_{2,T}$ and $y_{2,T}$ of the target in the guidance beam information field coordinate system $O_2x_2y_2$. As shown in Figure 5, the origin of the $O_2x_2y_2$ coordinate system is located in the guidance station, the O_2x_2 axis is coincided with the guiding beam center line, and the O_2y_2 axis is perpendicular to the O_2x_2 axis. Then:

$$\begin{aligned} x_{2,T} &= R_{M1} + (d\cos\lambda + L)\cos\theta - d\sin\lambda\sin\theta \\ y_{2,T} &= (d\cos\lambda + L)\sin\theta + d\sin\lambda\cos\theta \end{aligned} \tag{7}$$

Where, L is the distance between known strapdown detectors and missile riding guidance information receivers. R_{M1} is the distance between the first missile and the guidance station, and can be estimated by the missile speed plan. Based on the above estimation formula, the formula of target indication error component in longitudinal plane can be further obtained:

$$\Delta\varepsilon_y = \arctan\frac{y_{2,T}}{x_{2,T}} \tag{8}$$

In the same way, the mathematical model of a can be deduced as follows:

$$\begin{aligned} x_{2,T} &= R_{M1} + (d\cos\lambda + L)\cos\theta - d\sin\lambda\sin\theta \\ z_{2,T} &= (d\cos\lambda + L)\sin\theta + d\sin\lambda\cos\theta \end{aligned} \tag{9}$$

$$\Delta\varepsilon_z = \arctan\frac{z_{2,T}}{x_{2,T}} \tag{10}$$

During the actual operation, the strapdown detector begins to work when the first missile approaches the target, and uses the measured data to estimate the target indication error and then uses the data chain on the missile to transmit the indication error to the second missile and the third missile. The central line of the first modified guide beam can be obtained by the original guiding beam center line and the indication error. The second missile and the third missile tracking modified guided beam center line can greatly eliminate the effect of the target indication error on the missile hit precision. When the first missile exploded and the second missile approaches the target, the target indication error is recalculated according to the same method. The target indication error calculated by the second missile is passed to the third missile, and the guidance process of the third missile is further corrected.

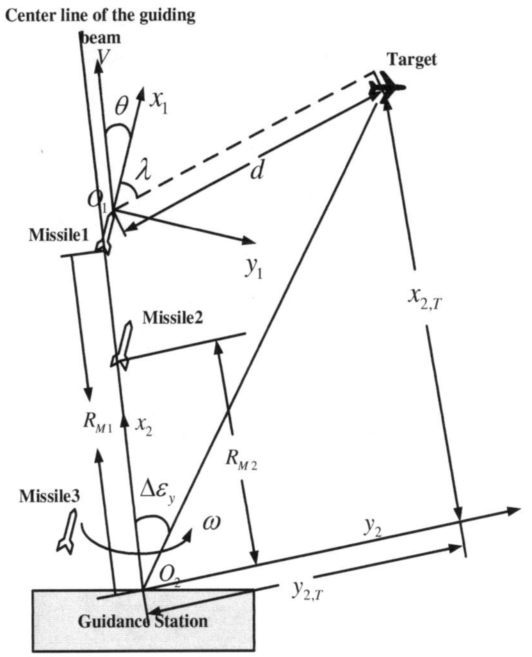

Figure 5. Schematic diagram of multi missile synergistic guidance in a longitudinal plane.

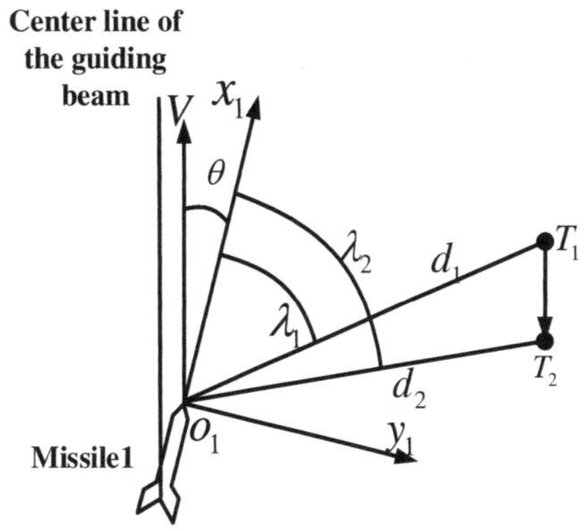

Figure 6. Schematic diagram for estimating the relative attitude angle of a missile.

2.2.2. Simulation result. T A low cost strapdown detector installed on a missile is used to measure the parameters needed to estimate the mathematical model of the target indication error. When the first missile is 300 meters away from the target, the strapdown detector begins to work and calculates the target indication error, which is passed to the second, third missile for the first time correction of the guiding beam center line. After the first missile is detonated, when the second missile is 300 meters away from the target, the target indication error is calculated and transferred to the third missile to correct the guide beam center line again.

Taking the indication error of 0.5mil in the longitudinal plane as an example, the relationship between the estimated value of the indicator error and the distance between the missile and the target is shown as shown in Figure 7. It can be seen from the graph that the smaller the distance between the missile and the target, the more accurate the target indication error estimate is. This deviation is due to the assumption that when Δt is sufficiently small, the relative guided beam of the missile remains unchanged when the mathematical model is derived from the derivation of the indicator error.

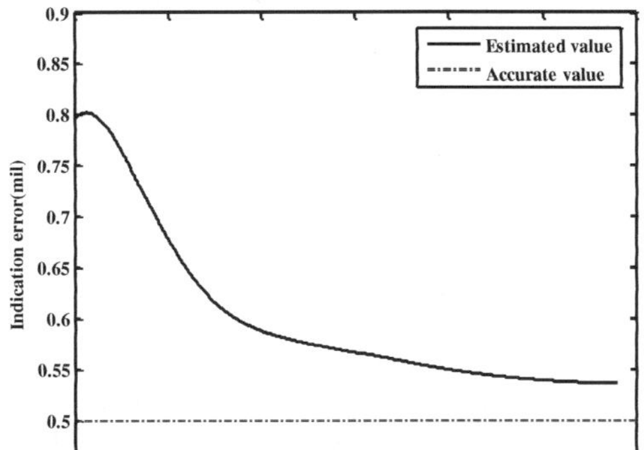

Figure 7. Relationship between the estimated value of the indicator error and the distance between the missile and the target.

The correctness and validity of the mathematical model of the estimation method of target indication error is verified by Monte Carlo simulation with the example of correcting the error component $\Delta\varepsilon_y$ in the longitudinal plane. Multiple mil, such as (-0.5,-0.4, $\cdots$,0, $\cdots$,0.4,0.5), are selected as target indication errors, and the error of the 11 sets of indication error estimates is obtained by simulation as shown in Figure 8. It is divided into 11 groups according to the indication error, each group carries out 200 Monte Carlo simulation, analyzes the data obtained by each group of simulation, and gets the CEP of the falling points of three missiles under the different target indication error as shown in Figure 9.

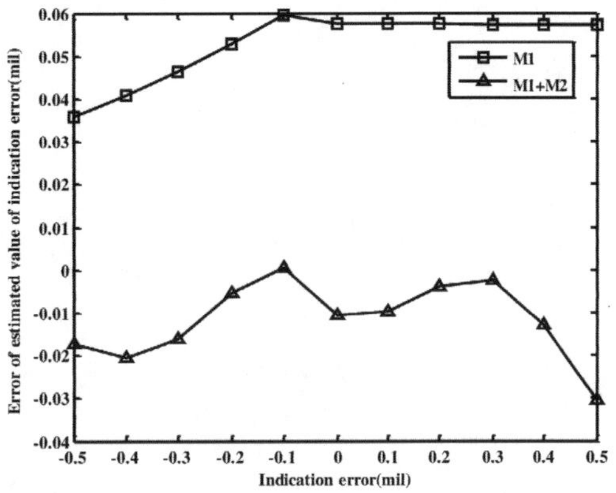

Figure 8. Comparison diagram of the estimated value of the indication error with the accuracy.

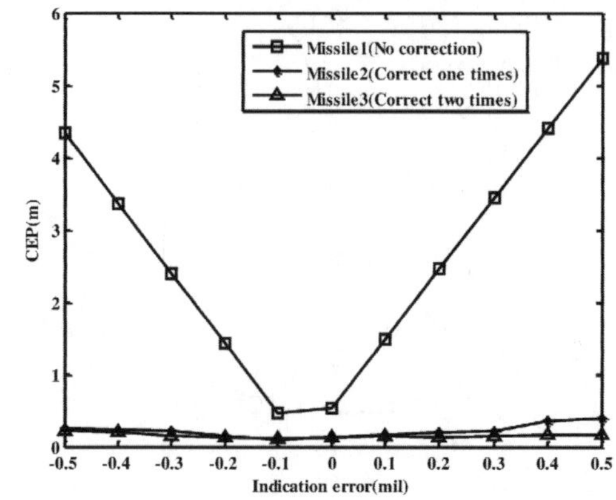

Figure 9. CEP contrast diagram of three missiles with different indication errors.

In Figure 8, M1 represents the indicator error estimate calculated by first missile, which is used for the first correction of the guiding beam center line of the second and third missile. M1+M2 represents the sum of the indicator error estimates calculated by the first and second missile, which is used to revise the guiding beam center line of the third missile. It can be seen from the diagram that the error of the estimated value is very small under different indication errors, and after two calculations of the first and second missile, the error of the estimated value is smaller and the corrected guide beam center line is more accurate, indicating that the mathematical model of indication error estimation method is correct.

In Figure 9, the first missile only calculates the indicator error without correcting the guiding beam center line, the second missile receives the first missile calculation indication error correction 1 times, the third missile receives the first and second missile calculation indication error correction 2 times. As can be seen from the diagram, the first missile follows the guided beam center line with error, and the CEP of the falling points becomes larger as the absolute value of the target indication error increases, and basically satisfies the conclusion that the indication error of the 0.5mil has a 5m deviation at 10 km. The second missile follows the guided beam center line which has been modified one time, and the CEP of the falling points is almost unchanged with the variation of the indication error, and the CEP is very small. The third missile follows the center line of the guided beam after two times correction, and the CEP of the falling points is the smallest. It is consistent with the conclusion obtained in Figure 8, which shows that it is correct and effective to use the indication error calculated by the mathematical model of the indicator error estimation method to correct the guiding beam center line.

3. Controlled ballistic simulation of six degree of freedom

3.1. Setting of simulation conditions

In the six degree of freedom simulation program, the steady state tracking error correction model and the target indication error correction model are used simultaneously to verify the correctness and effectiveness of the multi missile coordination high precision guidance and control method. The simulation conditions are set as follows:

The position coordinates of the guidance Station in the inertial coordinate system are (0, 2, 0) m. The muzzle of the missile launcher is 0.5m below the guidance station and 2m on the right. The target flight height is 50m, and the cruising speed is 200m/s. When the distance between the missile and the target is 15000 meters, three missiles are launched in turn, and the initial velocity of the missile is 700m/s. The simulation should ensure that the target attacked by three missiles is in the same motion state. And the standard deviation of the indication error pulling deviation is 0.3mil.

3.2. Simulation result

The Monte Carlo method is used to carry out 200 simulation tests on the working conditions of the multi missile high precision guidance and control method before and after the correction, and the CEP data of the three missiles are shown in Table 1. The falling points of the first, second, third missile are shown in Figures 10, 11 and 12 respectively, where (a) is the scatter plot before correction, and (b) is the corrected scatter plot.

Table 1. CEP contrast table for three missiles.

	Pre modified CEP(m)	Modified CEP(m)
Missile 1	3.670	3.260
Missile 2	3.570	0.662
Missile 3	3.100	0.972

As shown in Figures 10, 11 and 12 (a), the falling points of 3 missiles are basically consistent before correction, which is consistent with the situation that error of 0.3mil has a 3m deviation at 10 km. From Table 1, it can be seen that after the correction of the steady-state tracking error of the first missile, its CEP is lower than that before the correction, and from Figure 10 it can be seen that the corrected missile drop distribution is closer to the target and the fall points' density increases, but the CEP is still up to 3 meters because the indication error is not corrected. As shown in Figure 11(b), the second missile corrected the indication error and steady state tracking error. Compared with the pre correction, the precision of the missile was greatly improved, and the accuracy of the second missile in the 3 revised missiles was also the highest, consistent with the theoretical expectation. The third missile only corrected the indication error, so the CEP was much lower than that before the correction. However, because the steady state tracking error was not corrected, the falling points was concentrated on the side of the target, as shown in Figure 12 (b).

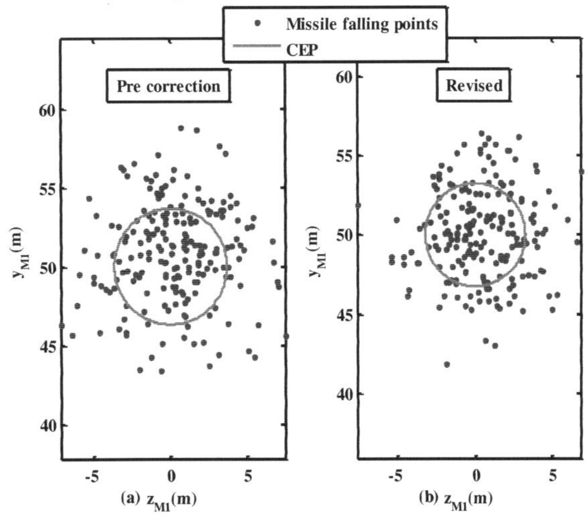

Figure 10. Monte Carlo Simulation Falling Points Chart of the First Missile before and after Correction.

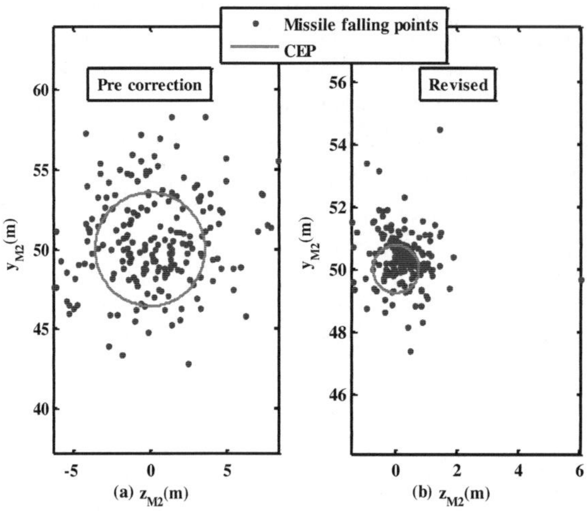

Figure 11. Monte Carlo Simulation Falling Points Chart of the Second Missile before and after Correction.

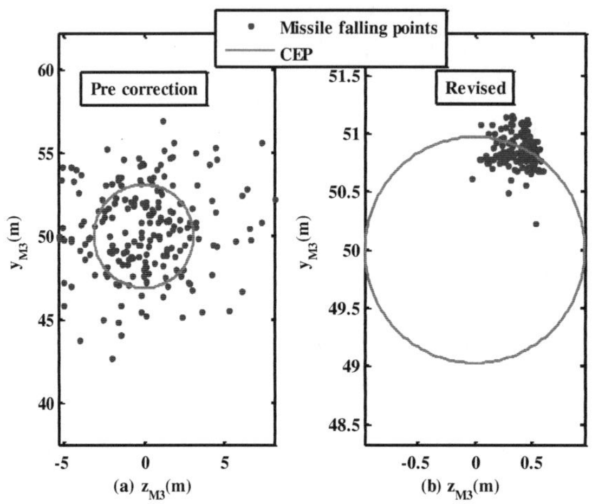

Figure. 12. Monte Carlo Simulation Falling Points Chart of the Third Missile before and after Correction.

4. Conclusions

(1) When the steady-state tracking error of the missile is corrected, the deviation between the missile and the target decreases, the falling point is closer to the target and the density is higher.
(2) After the target indication error of the missile is corrected, the CEP of the falling point of the missile is obviously reduced, and the hit accuracy of the missile is greatly improved.
(3) Through the simulation verification, the multi missile coordination high precision guidance and control method proposed in this paper is correct and effective. Compared with the current beam-riding guidance system, the multi missile cooperative combat system has higher hit precision and better application prospect.

5. References

[1] Jianwei Zhen, Lili Wu and Jinming Li 2017 *Aerodynamic Missile Journal.* **04**, 60-64
[2] Zaikang Qi and Qunli Xia 2005 *Guided Weapon Control Systems*(BEIJING: BEIJING INSTITUTE OF TECHNOLOGY PRESS)
[3] Tengfei Zhang, Hexin Zhang and Junjun Hu 2015 *Electronics Optics & Control.* **10**, 62-67+94
[4] Baohu Fan, Changming Zhao and Guoqiang Ma 2006 *Winged Missiles Journal.* **05**, 47-50

[5] Shukui Wang 1996 *Modern Weaponiy.* **06**, 20-22+49
[6] Xiangren Yu 1977 *Meteorological Monthly.* **09**, 16
[7] Jianqiao Yu, Zhonghui Wen, Yuesong Mei and Teng Long 2010 *ZHANSHU DAODAN ZONGTI SHEJI*(BEIJING: BEIHANG UNIVERSITY PRESS)
[8] Jianmei Song and Zhengjie Wang 2012 *ZIDONG KONGZHI YUANLI*(BEIJING: Publishing House of Electronics Industry)
[9] Xingfang Qian, Ruixiong Lin and Yanan Zhao 2011 *DAODAN FEIXING LIXUE*(BEIJING: BEIJING INSTITUTE OF TECHNOLOGY PRESS)

Modeling and Simulation of Satellite Attitude Dynamics and Control System using Modelica

Y H Liu[1], D S Shangguan[1], J Yao[2], L P Chen[1] and J W Ding[1]

[1]School of Mechanical Science and Engineering, Huazhong University of Science and Technology, 430074 Wuhan, China
[2]Shanghai Institute of Satellite Engineering, 200000 Shanghai, China

Abstract. The control of attitude of a satellite is difficult to verify since several subsystems of different disciplines, e.g., flexible multibody dynamics, machines, electronics and control, are involved and strongly coupled. A traditional method is to model each single discipline subsystem with a domain specific software, and combine them through interfaces. However, the problems are that lots of interfaces between software have to be developed, and manual decoupling of systems should be considered during modelling process. In this Paper, models of the dynamics and the control system are established in a unified modelling software, MWorks, which is an Integrated Development Environment of Modelica. The model can be used to verify the satellite control strategy, and also provide an effective means for satellite attitude control system design. Moreover, on this basis, the whole satellite was modelled to access state parameters changes under the coupling of the dynamic subsystem, the attitude control subsystem, the power supply subsystem, the propulsion subsystem and the integrated electronic subsystem can be synthetically considered.

1. Introduction

With the rapid development of modern technology and increasing competition in satellite market, simulation technology as a means to improve the efficiency in development of satellite product, its importance has become increasingly prominent [1-2]. Satellite simulation involves the coupling of different fields such as mechanics, electricity and control. The current multi-domain simulation [3-4] mainly uses single domain software to establish a single discipline model and through the interface between software or the use of HLA to solve single-disciplinary simulation and co-simulation of some disciplines [5], but cannot support completely multi-system modelling and simulation [6], which has poor reusability and scalability about the models. At the same time, the traditional method is only used for specific satellite simulation, but cannot be directly used for other satellites.

Modelica has been widely used in aerospace because of its support for multi-domain unified modelling, object-oriented physical modelling, non-causal modelling, and continuous discrete mixed modelling [7]. Abroad in the development of satellite systems design and test process has been widely used digital prototyping technology to improve development efficiency and product quality, and achieved remarkable results. In 2009, Zhang H J [8] established the Modelica model of the satellite flywheel, considering the flywheel control, electrical appliances, mechanical properties, validated the Modelica language in satellite multi-domain modelling of the feasibility and effectiveness. In 2011, Xu W F [9] developed a multi-domain unified modelling and simulation platform for spacecraft, and carried out a number of areas of simulation analysis. However, it has not been reported on the simulation of satellite attitude dynamics and control. Therefore, this paper first builds dynamics and control model library of the satellite, and validates the satellite attitude control strategy. On the basis of the above, the satellite state change under the coupling of the dynamic subsystem, the attitude control subsystem, the power

IOP Conf. Series: Journal of Physics: Conf. Series **1215** (2019) 012014 doi:10.1088/1742-6596/1215/1/012014

subsystem, the propulsion subsystem and the integrated electronic sub-system is considered synthetically, which is effective for the overall system-level multi-disciplinary collaborative simulation and verification.

2. Models of Dynamics and Control System

2.1. System decomposition
In the case of a satellite being developed, the attitudes and the control implementation of satellite are verified to meet the design requirements during the unfoldment of the satellite moving parts. According to the principle of structural hierarchy, reusability and extensibility, the satellite dynamics and control System is decomposed of satellite cabin, solar arrays(structure parts), antenna, thruster, earth sensor, sun sensor and star sensor, attitude control and management unit, environment, and so on.

2.2. Main devices
2.2.1. Satellite cabin. Figure 1 shows the satellite cabin model. Based on the Modelica standard library of basic components, in particular multi-body dynamics library whose physical theory have been described very well ,it is helpful to save the modeling engineer the time spent in formula derivation.. According to the actual physical topology of the satellite, the model of the satellite cabin is built by building blocks, and the information of the position of the solar arrays and the antennas is also defined. This model mainly considers the mass and inertia of the cabin.

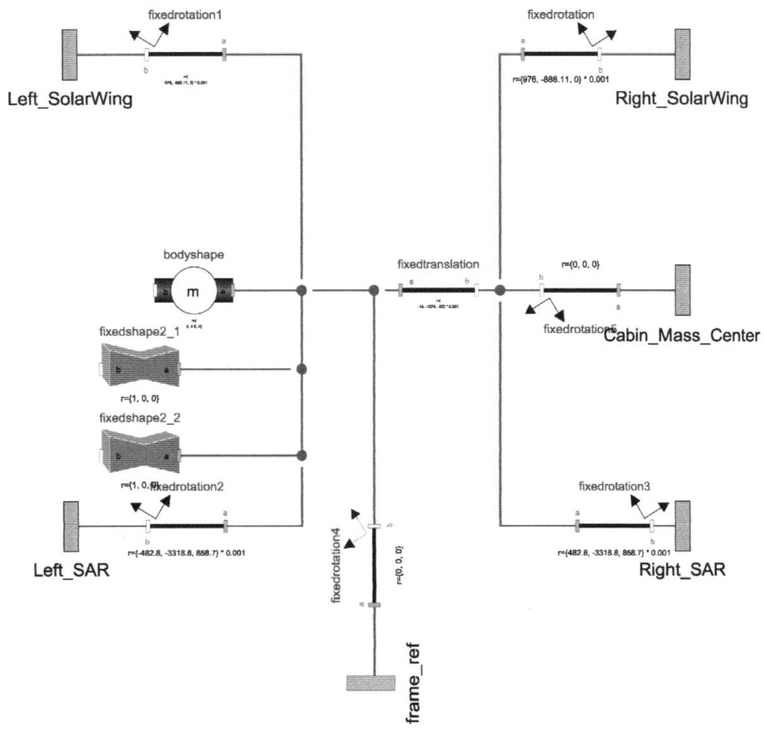

Figure 1. The satellite cabin model.

2.2.2. Solar array wings. Since the solar array in the unfoldment process of satellite attitude a great impact, it is necessary to consider the flexibility of the wings when modeling. The main process is as follows[10]: Firstly, the three-dimensional geometric model is processed into the modal neutral file using the finite element tool, then it is imported into the flexible body library based on MWorks. Finally, the related parameters are set. The main equations of the flexible body are:

$$M\ddot{s} + C\dot{s} + Ks = Fu \quad (1)$$

In this equation, M is the mass matrix, C is the damping matrix, and K is the stiffness matrix. The nu-component vector u represents the nontrivial force inputs to the system, while the ns×nu matrix F maps these nontrivial inputs to the corresponding degrees of freedom.

2.2.3. Antenna. Figure 2 shows the antenna model. The model mainly considers the rigid dynamic characteristics of the antenna during its unfoldment. The drive is driven by a DC motor. Since the antenna unfoldment process takes a long time, it does not need to consider the effect of its flexibility.

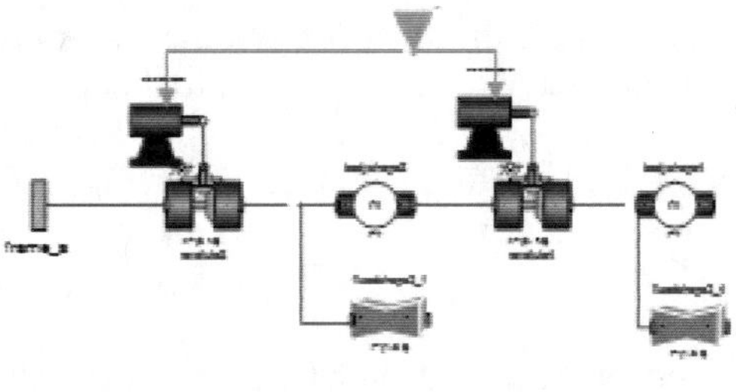

Figure 2. The antenna model.

2.2.4. Thruster model. Thruster model uses a simplified model, without considering the propellant combustion process, the thrust value as constant. The model mainly considers the influence of the number of thruster, installation positions and installation angles on satellite attitudes. The switch of the thruster is determined by the attitude control and management unit.

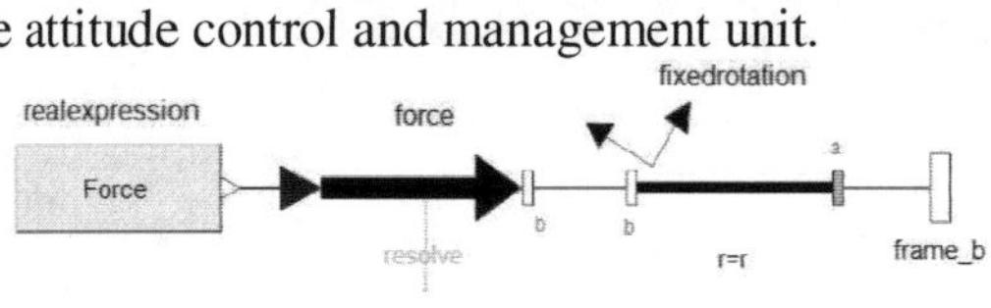

Figure 3. Thruster model.

2.2.5. Earth sensor. Earth sensor model uses an infrared earth sensor. The principle of the model is to obtain the angle between the geocentric vector and the scanning spin axis and the angle between the geocentric vector and the projected X-axis on the plane of the coordinate system by solving the attitude angle information of the input satellite. The angle between the geocentric vector and the scanning spin axis.

$$\eta = \arccos\left(z_{se} \left(x_{se}^2 + y_{se}^2 + z_{se}^2 \right)^{-1/2} \right) \tag{2}$$

The angle between the geocentric vector and the projected X-axis on the plane of the coordinate system:

$$\lambda = \arccos\left(z_{se} \left(x_{se}^2 + y_{se}^2 \right)^{-1/2} \right) \tag{3}$$

Where: x_{se}, y_{se}, z_{se}, respectively, is the earth vector in the measurement of coordinate system representation.

2.2.6. Sun sensor. The principle of the sun sensor model is to compute the angle between the projection of the solar vector in the XOZ plane and the Y-axis, and the angle between the projection of the sun vector in the measurement coordinate system YOZ plane and the Y axis by inputting the right ascension and declination and the attitude angle of the satellite relative orbital system. The angle between the projection of the sun vector in the measurement coordinate system XOZ plane and the Y axis:

$$\xi = \arccos\left(z_s \left(x_s^2 + z_s^2 \right)^{-1/2} \right) \tag{4}$$

The angle between the projection of the sun vector in the measurement coordinate system YOZ plane and the Y axis:

$$\eta = \arccos\left(z_s \left(y_s^2 + z_s^2 \right)^{-1/2} \right) \tag{5}$$

Where: xs, ys, zs, respectively, is the sun vector in the measurement coordinate system representation.

2.2.7. Gyroscope. The model principle of the gyroscope is to input the angular velocity of the satellite on the rotation axis of the gyro, and output the satellite attitude angular velocities considering the constant drift and zero mean white noise.

2.2.8. Attitude and orbit control computer (AOCC). AOCC is the core component of satellite attitude control. It mainly consists of PID controller, control signal conversion, jet time control and control strategy. AOCC calculates the control voltage signal by inputting the attitude angles and attitude angular velocities of the satellite and the control modes.

2.3. System model construction

Figure 4 shows the dynamics system model .Including satellite cabin, solar array wings, antennas, earth sensor, gyroscope and integrated drive controller and other stand-alone components. The main function of the integrated drive controller is to transmit commands according to the flight program, including solar arrays unfolding and driving and unfolding unwinding commands. The earth sensor is used to measure the attitude angles of the satellite. The gyroscope is used to measure the attitude angular velocities of the satellite. The system model is mainly used to simulate the change of the attitudes during the development of moving parts.

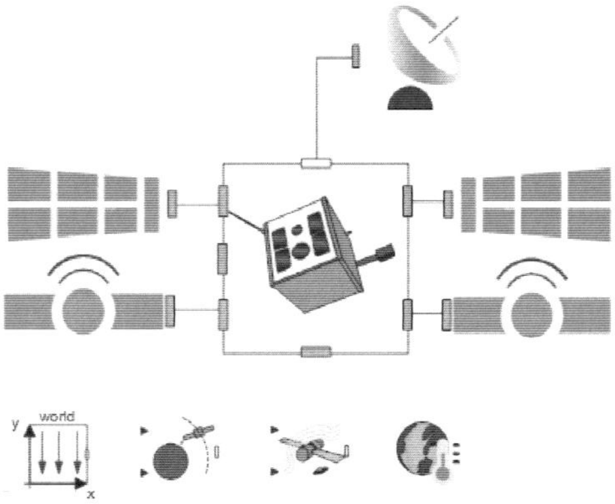

Figure 4. Dynamical system model.

The satellite attitude dynamics and control integrated model is shown as Figure 5.Including antenna, integrated drive controller, attitude control and management unit, earth sensor, solar sensor, star sensor, gyroscope and environment model. The model is mainly used to simulate control strategy validation.

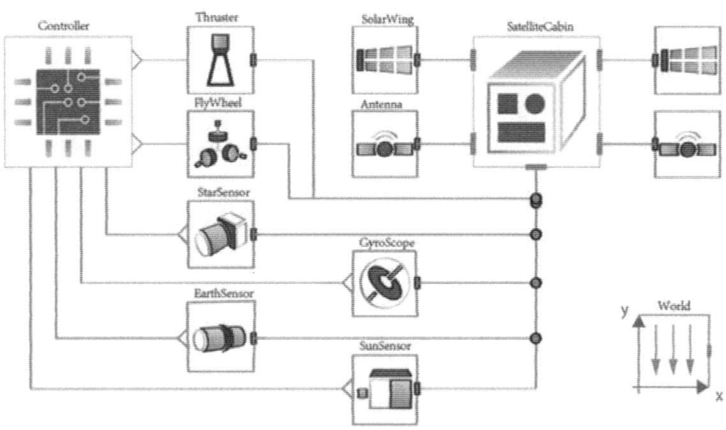

Figure 5. Attitude dynamics and control model.

3. Simulations and Analysis

3.1. Dynamic model validation

In order to verify the correctness of the model with Modelica/MWorks software, the satellite dynamics model based on ADAMS software was compared, which has been a lot of test validation in satellite's designs. Based on the agreement between the model parameters and given the same flight control command, both the results were compared.

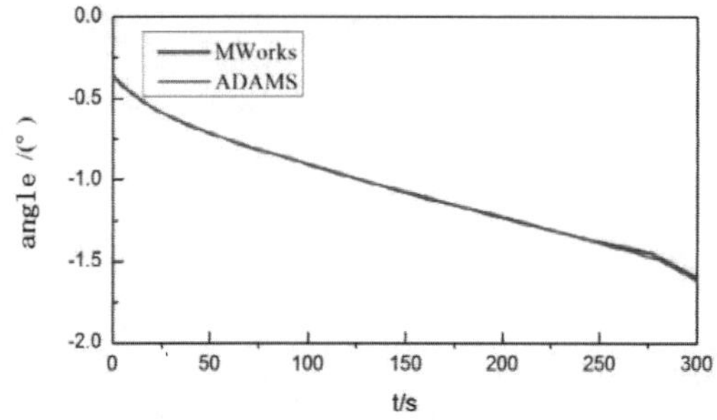

Figure 6. Comparison of yaw angles.

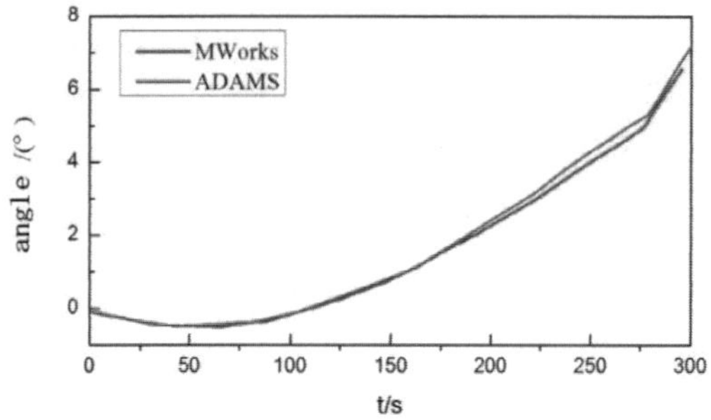

Figure 7. Comparison of roll angles.

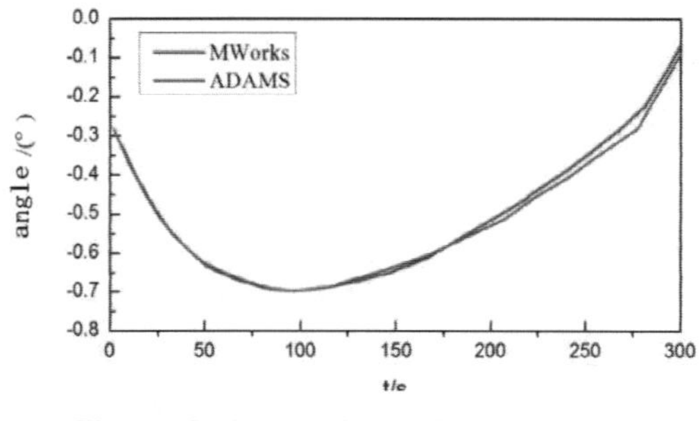

Figure 8. Comparison of pitch angles.

The attitude angles obtained by MWorks/Modelica agrees well with that of ADAMS shown in Figure 6, Figure 7 and Figure 8, the maximum relative error is 0.1%, and the average relative error is 0.05%, thus confirming that it is feasible for Modelica multi-domain unified modelling technology to simulate process in the satellite system.

3.2. Effect of flexible solar wings

Since the separation of the satellites and the rocket into the states of the solar arrays completely unfolds will cause the inertia to vary widely, it is very difficult to plan the satellite attitude control algorithm and the flight program. Therefore .In this section, the effects of the presence of rigid solar wings and flexible solar wings on simulation results were compared. The results were shown in Figure 9. It could be seen that replacing rigid solar wings with flexible wings has little effect on the overall attitude variation of the satellite, but it could be seen from the simulation result curves and the deformation cloud that there is obvious vibration when the moving parts start to expand and stop.

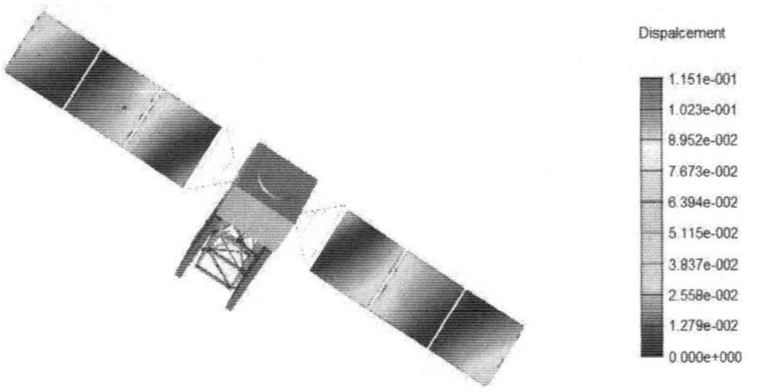

Figure 9. Deformation cloud.

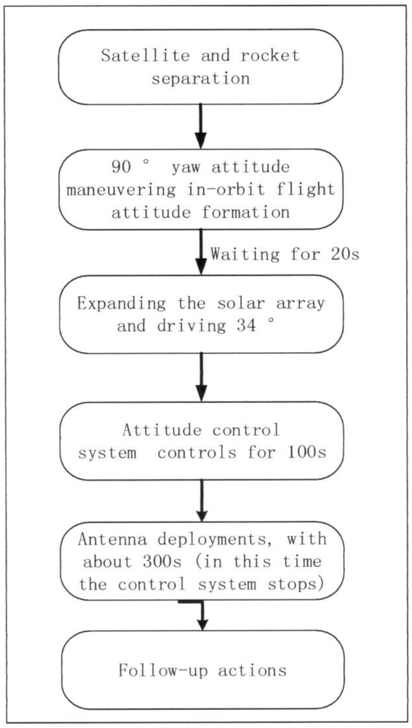

Figure 10. Flight procedure.

After separated from the rocket, the satellite was to complete the following actions, as shown in Figure 10. The model of the satellite attitude dynamics and control system constructed in section 2.3 was set up parameters according to the flight program, and the whole flight process simulation was carried out .The simulation process was mainly concerned with the effect of the control and whether the design requirements were met about the satellite attitude changes to verify the correctness of the control strategy.

The satellite attitude and attitude angular velocity curves are shown in Figure 11 and Figure 12, respectively. As can be seen from the figures, in 0~100 s the satellite worked in eliminating the initial deviation mode, and at 150 s in the yaw-capture mode, when attitude angle and attitude angular velocity have a great change, the attitude angle and attitude angular velocity satisfy the design requirements at the end of the yaw acquisition,. at 700 s the antenna begins to expand when control program is in the standby mode, the maximum attitude angle deviation caused by 0.01 rad within the design specifications. Figure 14 shows the thrust switch control signal, it can be seen from the working state of the thruster during the control implementation. Through the above analysis, it is convenient to provide reference and basis for the development of attitude control strategy.

Figure 14 shows a three-dimensional animation display of the satellite. It is very intuitive to reflect the satellite attitude changes.

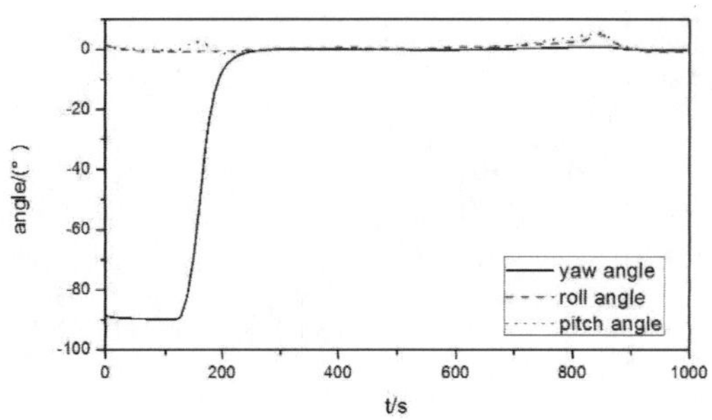

Figure 11. Attitude angle curves of satellite.

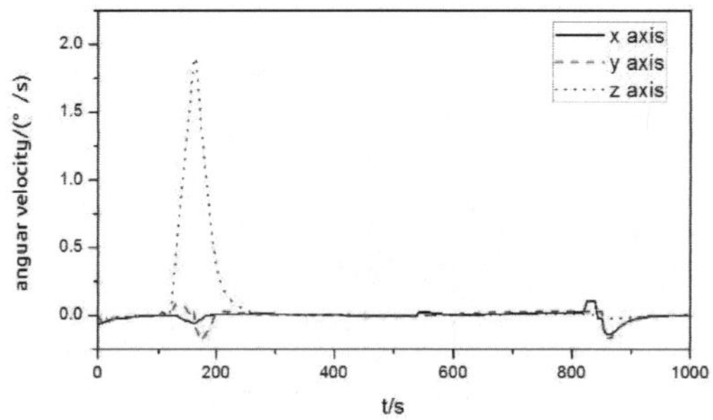

Figure 12 Attitude angle velocity curves of satellite.

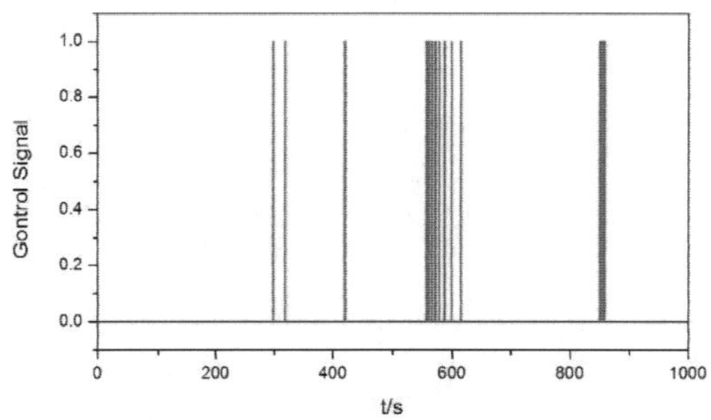

Figure 13. The thrust switch control signal.

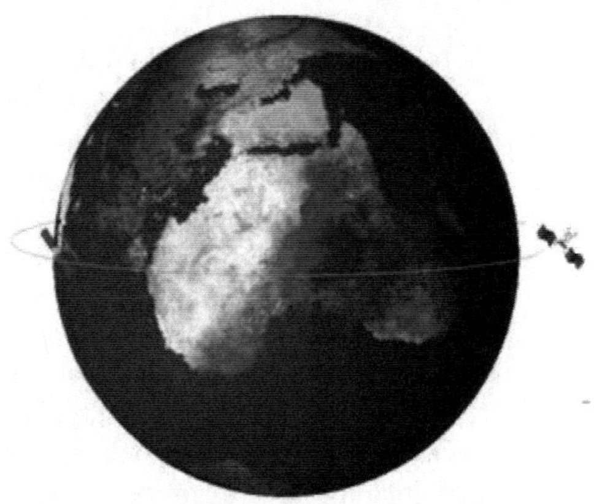

Figure 14. A three-dimensional animation display of the satellite.

4. Satellite multi-system modelling and coupling analysis

Based on the model of satellite dynamics and control model and the multi-professional features of the single device taken into accout, six model libraries including power supply, propulsion, measurement and control, communication, data management and transmission, and integrated electronics were extended and constructed.

Based on the construction of the six professional model library, the integrated satellite System Model is built. The coupling of dynamics and control subsystem, power subsystem, propulsion subsystem, measurement and control subsystem, data management and transmission sub-system and integrated electronic subsystem is considered to obtain satellite attitude changes with different control modes, the dynamic balances among the solar array, the battery and the device loads, the real-time output characteristics of the thruster under different control modes and the propellant remaining amount, satellite uplink and downlink calculation, information storage Balance characteristics as well as sending and receiving characteristics of flight procedures in the program instructions and remote control. Through the coupling analysis of satellite multi-system, it can provide reference for the designer to design the satellite system and verify the index in the stage of the plan.

5. Satellite multi-system modelling and coupling analysis

Through analysis of satellite dynamics and control system, the following conclusions are drawn:

- It is helpful to verify the correctness of the control strategy by comparing the satellite attitude dynamics and control simulation results with the design indexes;
- After replacing the rigid solar wings with the flexible wings, although the influence on the attitude of the satellite is not significant, both the simulation results and the Deformation cloud show that the satellite has a significant vibration in the process of solar wings and antennas unfolding.
- Modelica technology solves the coupling problem of satellite attitude dynamics and control system. The system of satellite dynamics and control is modelled on the same platform based on the unified physical modelling standard, and the natural coupling integration helps to validate the control strategy.
- On this basis, the satellite state changes under the coupling of the dynamic subsystem, the attitude control subsystem, the power subsystem, the propulsion subsystem and the integrated electronic subsystem are obtained synthetically, which provides an effective way of system-level multi-disciplinary collaborative simulation and verification for the whole department. It also lays the foundation for the satellite system-wide digitalization and multidisciplinary optimization.

6. References

[1] Pulecchi T, Casella F, Lovera M 2006 A Modelica Library for Space Flight Dynamics.Proceedings of the 5th International Modelica Conference. Vienna, Austria.4-5:107-116.

[2] Casella F, Lovera M 2008 High-Accuracy Orbital Dynamics Simulation through Keplerian and Equinoctial Parameters.Proceedings of the 6th International Modelica Conference.University of Applied Sciences Bielefeld, Bielefeld, Germany, (2):505-514.

[3] Ye X,Pan Q,Dong Z H,et al 2014 A Survey of Multi - Domain Modeling and Simulation.SOFTWARE,35（3）：233~236.

[4] Averill M 2009 Simulation Modeling and Analysis. Fourth Edition.Beijing: Tsinghua University Press.

[5] Sinha R, Liang V.C 2001 Paredis C.J.J. et al. Modelling and Simulation Methods for Design of Engineering Systems. Journal of Computing and Information Science in Engineering, 1: 84~91.

[6] Otter M,Elmqvist H,Mattsson S E 2013 The new modelica multiubody library// Proceedings of the 3rd International Modelica Conference.Link0oing,The Modelica Association and Linkoping University.

[7] Fritzson P 2003 Principles of object-oriented modeling and simulation with Modelica 2.1 New York. IEEE Press.

[8] Zhang H J 2009 Modelica-based satellite multi-domain modeling and simulation methods.Harbin University of Science and Technology:30-50.
[9] Xu W F, Liang B, Li C 2011 Multi-domain unified modeling and simulation platform for three-axis stabilized spacecraft. Journal of Harbin Institute of Technology,09: 74~80.
[10] Xie G,Zhao Y,Zhou F L,et al 2012 Modelling and Simulation of the Coupled Rigid-flexible Multibody Systems in MWorks. Proceedings of the 9th International Modelica Conference. Munich Germany. September 3-5:405~416.

IOP Conf. Series: Journal of Physics: Conf. Series **1215** (2019) 012015 doi:10.1088/1742-6596/1215/1/012015

An Alternative Tuning Scheme for Simple Adaptive Flight Control System

C Orlando*, A Esposito and A Alaimo

Kore University of Enna, Faculty of Engineering and Architecture, 94100 Enna, Italy

calogero.orlando@unikore.it

Abstract. The occurrence of system failure or of external disturbances are problem that must be considered in the design of safe flight control systems. Adaptive control schemes show the ability to handle such issues without any a priori knowledge about the fault or disturbance. The main condition for obtaining a stable adaptive controller is the passivity of the plant but most real system do not satisfy this condition. The use of properly defined Parallel Feedforward Compensator (PFC) can allow the augmented system to meet the passivity requirements. An alternative design approach for tuning the PFC and the invariant gains of the Simple Adaptive Controller (SAC) is presented in this work. The tuning procedures is a modification of the PSO and is called Population Decline Swarm Optimization (PDSO) since it takes into account a decline demographic model to speed up the tuning steps. The method is applied to adaptively control the pitch response of an aeroplane during level flight. Tuning results are presented along with flight mechanics simulation taking into account different command laws.

1. Introduction

Modern control systems, as the fly by wire, have become a standard for most aircrafts leading to an increase in term of safety and performance. With the increase of flight control systems complexity the requirement for safety and reliability has become more and more stringent [1, 2]. It has stemmed the need to design control systems capable of tolerating potential faults and disturbances as well. For such reason, robust, fault-tolerant or adaptive flight control systems are usually investigated [3, 4]. Different Adaptive Control techniques have been developed and applied to manage the changes that the aircraft may undergo during its flight. Among these, the following may be cited: Viswanathan and Lakshmi [5] that redesign an autopilot system using Sliding Mode Control; Lee et al. [6] that study an L1 adaptive control capable of recovering nominal performance in the presence of failures and disturbances; Nivison and Khargonekar [7] that present a Model Reference Adaptive Control system based on Neural network to take into account uncertainties; and Nishiyama et al. [8] that propose the use of Simple Adaptive Control to realize fault tolerant flight systems.

The use of the adaptive control requests for the passivity of the system which, for the SAC problem, can be identified by the Almost Strictly Positive Realness condition for the transfer functions that model the plant [9]. However, real systems generally do not satisfy such condition. For this reason Barkana [9] has proposed the use of a Parallel Feedforward Compensator to render the augmented plant ASPR.

An alternative design procedure for synthesizing the PFC and tuning the SAC invariant gains is studied in this work. The procedure, called PDSO - Population Decline Swarm Optimizer, is a modification of the particle swarm optimization (PSO) [10] that takes into account a population decline model [11] allowing the use of a large swarm for the exploration steps and then reducing it during the exploitation. The tuned adaptive scheme is then applied to control aircraft pitch motion, whose model is taken from literature [12]. Outline on the simple adaptive flight control system is given in section 2; the population decline swarm optimizer is briefly described in section 3 while tuning and simulation results are commented in section 4.

2. Simple Adaptive Flight Control System

Let us assume that the aircraft longitudinal behaviour can be represented by a simplified controllable and observable SISO linear model through the following transfer function [12] that links the elevator deflection δ to the aircraft pitch angle θ as

$$G_0(s) = \frac{4.22s^2 + 4.31s + 0.212}{s^4 + 3.01s^3 + 6.96s^2 + 0.232s + 0.224} \tag{1}$$

Such model is obtained by linearizing the aircraft longitudinal dynamic about the steady level flight trim conditions given by true airspeed 66.5 *m/s* and angle of attack of 4.98 *deg* at an altitude of 1524 *m* with constant thrust. Let now assume that the plant has to follow a stable reference model that can be written as a first order transfer function whose output y_m gives the state x_m while the bounded input scalar signal is u_m.

The SAC scheme is similar to the MRAC (Model Reference Adaptive Control) but it requests for an output feedback term that allows to build the input control signal as a linear combination of the model reference input, of the model reference state vector and of the output tracking error $e(t) = y(t) - y_m(t)$ and writes as

$$u(t) = K_e(t)e(t) + K_{um}(t)u_m(t) + K_{xm}(t)x_m(t) \tag{2}$$

The terms $K_i(t)$, with $i=\{e, um, xm\}$ are the adaptive gains of the SAC, obtained by summing a proportional and an integrative terms, and each is properly defined as

$$\begin{aligned} K_i &= K_{i,Prop} + K_{i,Int} \\ \dot{K}_{i,Int} &= e\Gamma_i i \\ K_{i,\mathrm{Prop}} &= e\Gamma_i i \end{aligned} \tag{3}$$

where Γ_i are the invariant gains of the SAC algorithm. The condition requested for stability of the adaptive controller is the passivity of the plant. This means that the plant transfer function must be at least Almost Strictly Positive Real (ASPR) which, for a SISO transfer function, implies that all zeroes are negative, the relative degree is 1 and that the leading coefficient is positive. It is to be said, however, that almost all real systems do not satisfy the ASPR. Such problem is tackled by adding to the plant a PFC such that the augmented system meets the passivity requirements. Moreover the PFC transfer function must be such that the output of the augmented plant $G_a=G_0+G_{PFC}$ is almost equal to the original plant output, namely $y_a=y+y_{PFC} \approx y$. For more detail about the SAC algorithm, the ASPR and PFC and for the proof of stability, the interested reader is referred to the literature [9]. The block diagram representation of the plant augmented by the parallel feedforward compensator and controlled by SAC to follow the reference model is shown in Figure 1.

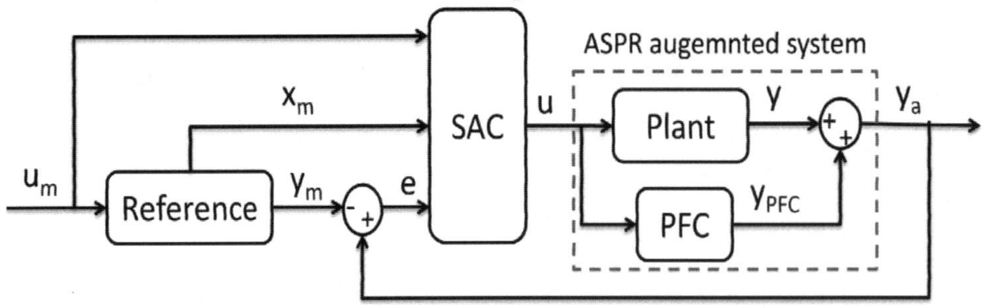

Figure 1. Block diagram representation of the controlled plant.

3. Population Decline Swarm Optimizer

In order to synthesize the parallel feedforward compensator and to properly select the invariant parameters of the SAC scheme the population decline swarm optimizer is used. The P_DSO is a modification of the particle swarm optimization and it relies upon the simplified social model that describes a swarm of birds (the particles) looking for food (minimum/maximum value of the objective function) while flying on field (the search space). More in detail, each particle **p** represents a position vector in the n-dimensional search space and, after a random initialization, it moves according to the following law

$$\mathbf{p}^i_{\lambda+1} = \mathbf{p}^i_\lambda + \mathbf{v}^i_{\lambda+1} \tag{4}$$

where $\lambda=\{1,2,\ldots,\Lambda\}$ is the iteration step being Λ the max number of iteration while $i=\{1,2,\ldots,P_\lambda\}$ labels the i-th particle of the swarm and P_λ is the number of particle present in the swarm at the iteration λ. The updated particle velocity in equation (4) writes as

$$\mathbf{v}^i_{\lambda+1} = \chi\left[\mu_\lambda \mathbf{v}^i_\lambda + c_c r_1\left(\mathbf{p}^i_b - \mathbf{p}^i_\lambda\right) + c_s r_2\left(\mathbf{p}^g_b - \mathbf{p}^i_\lambda\right)\right] \tag{5}$$

being c_c the cognitive acceleration constant, c_s the social constant, $\mathbf{p}^i_b$ is the current best position attained by any of the particle of the swarm while $\mathbf{p}^g_b$ is the best position of all the swarm; r_1 and r_2 are random numbers in the interval [0; 1]; μ_λ and χ are the linear decreasing inertia weight and the constriction factor [11], respectively. Last, the decline population model introduced to modify the standard PSO read as

$$P_\lambda = \begin{cases} P_{\max} & \lambda = 1 \\ \lceil P_{\lambda-1}\xi \rceil & if \quad \lambda/\Delta = \lfloor \lambda/\Delta \rfloor \\ P_{\lambda-1} & otherwise \end{cases} \tag{6}$$

being Δ the number of iterations during which the swarm size is constant, ξ is the percentage reduction of the population and $\lceil x \rceil$ is the integer part of the variable x and $\lfloor x \rfloor = -\lceil -x \rceil$. It is worth noting that the use of the decline population model allows to use a large size swarm during the initialization to increase the exploration capability of the swarm and allows to continuously reduce the number of particles in the swarm during the iterations to speed up the procedure with the objective of minimizing an *a priori* chosen fitness function by reducing the overall computational cost.

More in detail, with reference to the problem described in the previous Section 2, the P_DSO is applied twice to design both the parallel feedforward compensator transfer function G_{PFC} and to select the SAC invariant parameters Γ_i. The fitness function to be minimized to synthesize the PFC transfer function is a modified norm of the discrepancy between the augmented y_a and the actual y systems outputs undergoing a unitary step input excitation, u. The PFC is written as a first order transfer function having DC gain κ_{PFC} and time constant τ_{PFC}, that are the element of the particle vector **p** while the fitness function is

$$\Im_{PFC}(\mathbf{p}) = \phi(\mathbf{p})\left\{\int_{T_W}\left[y_a(\mathbf{p},t) - y(t)\right]^2 dt\right\}^{\frac{1}{2}} \tag{7}$$

being $\phi(\mathbf{p})$ a penalty function that equals 1 if G_a meets the ASPR condition and becomes numerically infinite otherwise. This approach allows to constraint the optimization problem, that specifies as in equation. (8), in such a way to look for ASPR augmented plant only.

$$\begin{gathered} \min \Im_{PFC}(\mathbf{p}) \\ s.t.\, \mathbf{p} \in \left[\mathbf{p}_{PFC,\min}, \mathbf{p}_{PFC,\max}\right] \end{gathered} \tag{8}$$

On the other hand, in order to define the SAC invariant parameters, the particle position vector is defined as p=[Γ_e, Γ_{um}, Γ_{xm}] and the minimization problem write as

$$\min \Im_{SAC}(\mathbf{p})$$
$$s.t.\mathbf{p} \in \left[\mathbf{p}_{SAC,\min}, \mathbf{p}_{SAC,\max}\right] \tag{9}$$

where the objective function specifies as

$$\Im_{SAC}(\mathbf{p}) = \int_{T_W} \left[\alpha e^2(\mathbf{p},t) + \beta \delta^2(\mathbf{p},t)\right] dt \tag{10}$$

In equation 10 α and β are weight parameters while T_W is the integration time window.

4. Results

Simulation results are summarized in this section. At first PFC and SAC tuning results by mean of the Population Decline Swarm Optimizer are presented and commented, successively flight mechanics simulation results are given.

4.1. PFC tuning results

In order to synthesize the PFC by using the P_DSO, the particle is defined as **p**=[κ_{PFC}, τ_{PFC}], the minimum and maximum value of the search space are set to 0.01 and 1 for both particle elements (in accord with [13]) while the time window extent to compute the index to be minimized, namely equation 7, is set to 600 s. The parameters of the optimization algorithm are: number of swarm = 1; cognitive constant c_c = 2.05; social constant c_s = 2.05; minimum and maximum of inertia weight are 0.4 and 0.9, respectively; maximum number of iterations Λ = 20; time for population decline Δ = 3; percentage reduction of the population ξ=0.5; initial population size P_{max} = 10. To account for the stochastic nature of the P_DSO, 50 optimizations have been run. Figure 2 shows convergence history of the fitness function equation 7. It can be appreciated the random initialization and that after the fifth iteration the convergence is met.

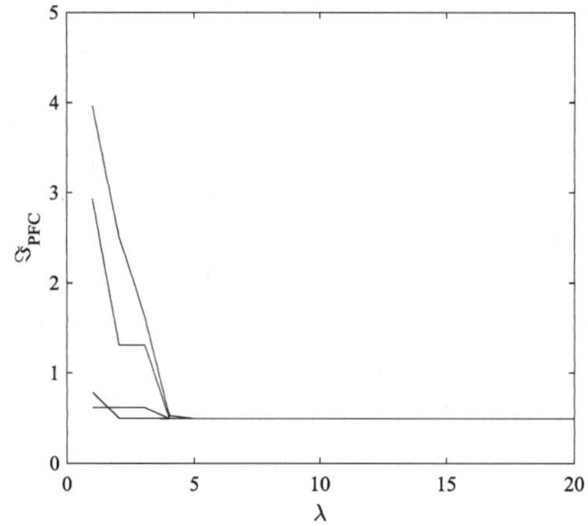

Figure 2. PFC tuning convergence (five representative curves).

Table 1 collects the results of the minimization procedure carried out using the proposed population decline swarm optimizer and the standard PSO using the same optimizer parameters previously given. It can be noted that the proposed optimizer, that takes into account a decline model for the population of the swarm, manages to find the same results of the standard (constant population) particle swarm optimizer in terms of minimum value of the objective function equation 7 and values of the best position particle.

Table 1. PFC tuning results.

	P_DSO	PSO
min $\Im_{PFC}$	0.495	0.495
$\mathbf{p}_{PFC}$	[0.01,1]	[0.01,1]
t / t_{PSO}	0.27	1

On the contrary it is worth noting that, in the present case, the P_DSO requests for a computational time corresponding to the 27% of the one needed by the standard PSO and resulting in an increased computational efficiency. The comparison of step responses of original G_0 and PFC augmented $G_a=G_0+G_{PFC}$ plants is shown in Figure 3 evidencing that the parallel feedforward compensator does not modify the plant response. This is true at low frequencies, below 100 rad/s where phugoid and short period dynamics lie. However the PFC makes the plant ASPR allowing for the application of the simple adaptive control technique.

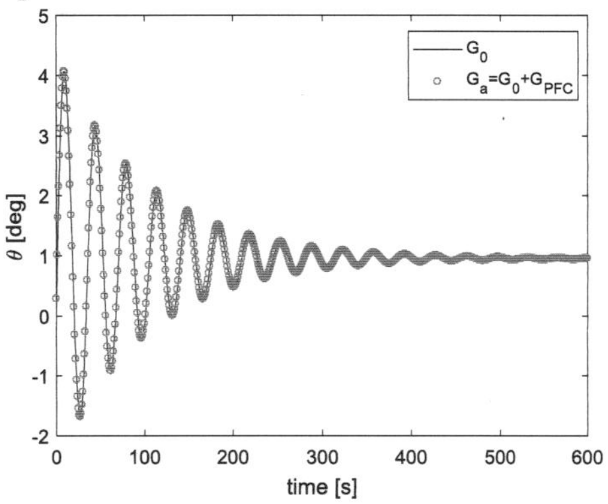

Figure 3. Unitary step response: comparison of original and PFC augmented plant.

4.2. SAC tuning results

To synthesize the SAC controller, a smoothed square wave defined in equation 11 is passed as command u_m to the reference model. The amplitude of the square wave is A_{ssw}=2 deg, the circular frequency is ω_{ssw}=0.1 rad/s while smoothing parameter ρ_{ssw} is set to 50.

$$\delta_c = \frac{2A_{ssw}}{\pi}\tan^{-1}\left[\rho_{ssw}\sin\left(\omega_{ssw}t\right)\right] \tag{11}$$

The reference model transfer function is defined as $G_{ref}=(0.05s+1)^{-1}$ while the computation time window is T_W=10 s. The P_DSO parameters are set as: number of swarm = 1; cognitive constant c_c=2.05; social constant c_s=2.05; minimum and maximum of inertia weight are 0.4 and 0.9, respectively; maximum number of iterations $\Lambda = 20$; time for population decline $\Delta = 5$; percentage reduction of the population ξ=0.8; initial population size $P_{max} = 20$. To account for the stochastic nature of the P_DSO, 50 optimizations have been run. The minimum and maximum values of the search domain are set to 0.001 and 100, respectively, for all SAC parameters while the weights of the objective function equation 10 are set as α=1 and β=0.04.

The convergence history of the objective function equation 10 is given in figure 4 showing the random initialization and the convergence after about five iteration. The optimization results are collected in Table 2 and, even in this case, the P_DSO results match well with the standard PSO solution, showing a speed up ratio t/t_{PSO}=0.71 that confirms a reduced computational effort of the proposed decline population scheme with respect to the standard procedure.

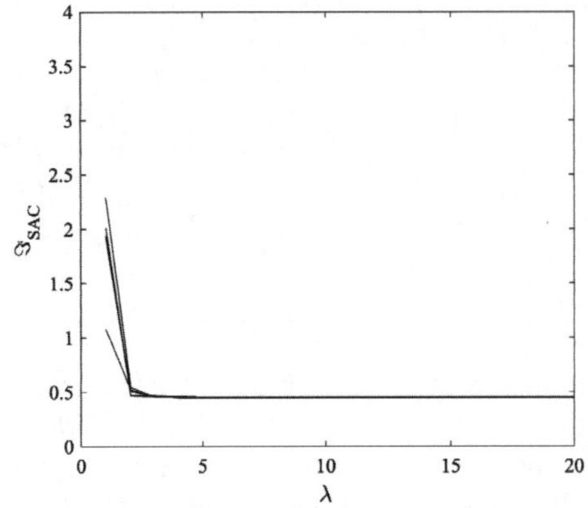

Figure 4. SAC tuning convergence (five representative curves).

Table 2. SAC tuning results

	P_DSO	PSO
min $\Im_{PFC}$	0.447	0.447
$\mathbf{p}_{SAC}$	[24.7,0.001,0.277]	[25.0,0.001,0.272]
t / t_{PSO}	0.71	1

Time history results for the ten second analyzed are given in figure 5 in terms of the aircraft pitch angle variation θ commanded by the elevator deflection δ. It can be seen that the pitch angle never exceed the value of 2 deg and, after a transient response, reaches the reference model at about t=4 sec.

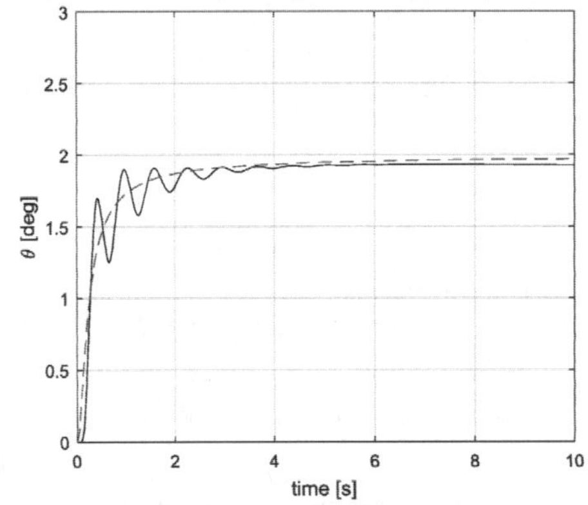

Figure 5. Time history of pitch angle (solid: plant response; dashed: reference).

4.3. Flight simulation results

Once the controller parameters have been set by using the proposed alternative optimization procedure that has allowed to minimize the chosen objective functions, the pitch dynamic behaviour of the controlled aircraft is investigated. Again, the reference model transfer function that the plant is request to follow is $G_{ref}=(0.05s+1)^{-1}$, the smoothed square wave equation 11 is used as input while the computation time window is set to T_W=300 s.

The effects of the input command frequency and of the smoothing parameters on the aeroplane pitch and elevator deflection angle are investigated. In such study it is taken into account that plant transfer function models the behaviour of the experimental aircraft called MuPAL-α [12]. The phugoid and short period frequencies are 0.0287 Hz and 0.343 Hz, respectively, while the elevator actuator (modelled as a first order transfer function) has a cut-off circular frequency of 1.59 Hz [14], maximum deflection angle of 25 deg and minimum deflection angle of -30 deg [15]. The system transient response in terms of pitch angle θ and elevator deflection δ is plotted in figure 6 for the case ω_{ssw}=0.1 rad/s and ρ_{ssw}=75.

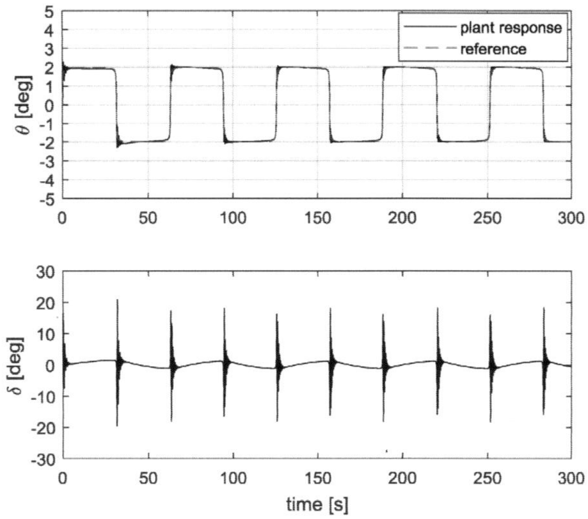

Figure 6. Time history of pitch angle and elevator deflection.

Looking at the figure 6 it can be concluded that the simple adaptive controller tuned by means of the P_DSO is capable of leading the plant in such a way to closely follow the reference. The maximum pitch value in absolute value is 2.28 deg at about 31 seconds while the frequency of oscillation of pitch variable is about 1.51 Hz which is below the plant cut-off frequency of 1.59 Hz; on the other hand the maximum deflection reached by the elevator is about 20 deg and the transient behaviour is characterized by a frequency of 1.58 Hz that is close but below the cut-off one. A closeup representation of the transient behaviour of the system is given in Figure 7.

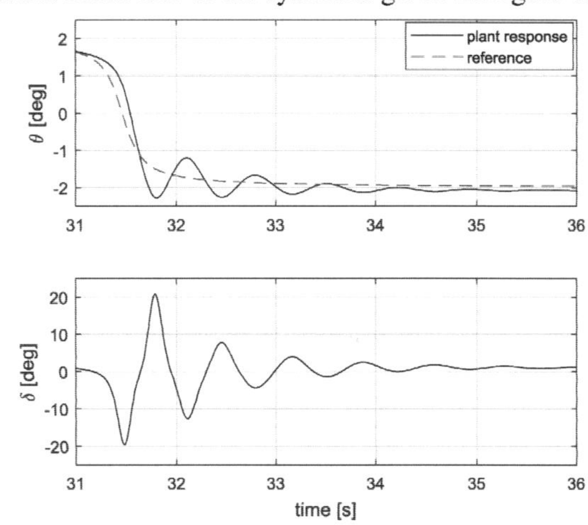

Figure 7. Particular of transient response.

Last, the elevator command parameters ω_{ssw} and ρ_{ssw} are let vary to study the influence of the command signal on the plant response. Results collected in Table 3 and 4 highlights that as the smoothing parameter ρ_{ssw} increases, which means that the square wave becomes sharper, maximum value of the pitch angle increases over 2 deg. The maximum elevator deflection increases as well with the smoothing parameter and the same stands for the transient frequency of both the input δ and output θ variables of the plant. The effect of the actuation frequency ω_{ssw} is similar since all the investigated variables increase with it. In addition, it can be noted that the transient frequency of both δ and θ

move towards the cut-off frequency in both cases. Moreover, looking at the maximum elevator deflection results it can be noted that when ω_{ssw}=0.1 rad/s and ρ_{ssw}=100 the elevator deflection exceeds the limit value of 25 deg suggesting the use of a lower smoothing parameter when an actuation frequency higher or equal to 0.1 rad/s is requested.

Table 3. Influence of command parameters ρ_{ssw} for ω_{ssw}=0.05 rad/s

ρ_{ssw}	$\|\theta\|_{max}$ [deg]	f_θ [Hz]	$\|\delta\|_{max}$ [deg]	f_δ [Hz]
25	1.99	0.04	2.6	1.1
50	2.08	0.08	5.7	0.94
75	2.12	1.17	8.7	1.15

Table 4. Influence of command parameters ρ_{ssw} for ω_{ssw}=0.1 rad/s

ρ_{ssw}	$\|\theta\|_{max}$ [deg]	f_θ [Hz]	$\|\delta\|_{max}$ [deg]	f_δ [Hz]
25	2.05	1.07	5.7	1.07
50	2.11	1.25	11.5	1.28
75	2.28	1.51	20.8	1.58

5. Conclusions

An alternative method based on a modification of the particle swarm optimization has been presented in this work with the aim of tuning the PFC and the invariant gains of the SAC scheme. The simple adaptive controller has been employed to control the pitch dynamic of an aircraft during horizontal flight. Tuning results have shown that the search capability of the Population Decline Swarm Optimizar – PDSO is comparable with the standard PSO but have also evidenced a reduced computational effort of the proposed method. Numerical simulation have also been performed to investigate the influence of the input to the reference model on the aircraft response taking into account physics limitation of the system. Next steps of present work will deal with the analysis of the fault tolerant and disturbance rejection capabilities of the proposed controller taking into account the deflection limits of the elevator surface.

6. References

[1] Favre C 1994 *Int. J. Contr.*, **59**(1), 139-157
[2] Liu Y, Ren Z and Cooper JE 2018, *J. Guid., Contr. Dyn.*, **41**(6), 1423-34
[3] Eterno JS, Weiss JL, Looze DP and Willsky A 1985 *24th IEEE Conf. on Decision and Control*, 24, 900-905
[4] Steinberg M 2005 *Proc. Inst. Mech. Eng., Part G: J. Aer. Eng.*, **219**(4), 263-275
[5] Viswanathan R and Lakshmi PS 2018 *J. Inst. Eng. India Ser. C* **99**, 369
[6] Lee H, Snyder S and Hovakimyan N 2017 *J. Guid. Contr. Dyn.*, **40**(4), 1013-26
[7] Nivison SA and Khargonekar P, 2018 *AIAA Guid., Nav. and Contr. Conf.*, 0842
[8] Nishiyama T, Suzuki S, Sato M and Masui K 2016 *AIAA Infotech@ Aerospace Conf.*, 0132
[9] Barkana I 2014 *Int. J. Adap. Contr. Sig. Proc.*, **28**(7-8), 567-603
[10] Shi Y and Eberhart R 1998 *Evolutionary Computation Proc. IEEE World Congress on Computational Intelligence*, 69-73
[11] Orlando C and Alaimo A 2017 *Mech. Syst. Sig. Proc.* **84**, 551-569
[12] Tokunaga D, Masui K and Suzuki S 2015 *Proc. Eng.*, **99**, 1035-43
[13] Omori Y, Suzuki S and Masui K 2013 *AIAA Infotech@ Aerospace Conf.*, 5210
[14] Sato M 2006 *J. Gui. Contr. Dyn.*, **29**(6), 1476-82
[15] RUAG Aerospace Services GmbH, Dornier 228 Advanced Commuter (AC) Facts & Figures, accessed 12.09.2018 via dornier228.ruag.com

A Piece-Wise Linearization PID Control of Aero-engine Software Defined Control System

Xiaoxiang Ji[1] *, Jianghong Li[1] , Jiao Ren[1] and Yafeng Wu[1]

[1] School of Power and Energy, Northwestern Polytechnical University, 710129 Xi'an, China

jixiaoxiang@mail.nwpu.edu.cn

Abstract. A scheme of piece-wise linearization PID control of aero-engine software defined control system based on wireless network is proposed in this paper. The characteristic of the system is that the PID controller is separated and distributed among the nodes of the wireless network ,each node has the function of computing, memory and wireless communication, no core control unit in the system and the whole wireless network acts as a distributed PID controller. The linearization model and corresponding tuning parameters at different working points in flight envelope of the aero-engine are established offline using the memory resource of each node to make the node task parameters switch when the aero-engine working at different points. Finally, MATLAB/Simulink software tool is used to conduct the digital simulation analysis, the result shows that the proposed scheme achieves good dynamic and steady-state performance, and achieves good control effect.

1. Introduction

The aero-engine electronic control system undergoes analog electronic control, digital electronic control and distributed control[1][2]. In the above electronic control system, the components of the control system are connected by cables. During the flight, cables and connecters are prone to failure due to vibration and other factors, the thrust-weight ratio is reduced due to numerous cables and connecters in the system. In addition, the damage of the center control unit which covers the whole or main functions of the system has a great impact on the control system.

In the field of industrial control, with the development of wireless communication technology, Wireless Sensor Networks (WSN) and Wireless Networked Control System (WNCS) have appeared. WSN is a multi-sensor system, which is characterized by intelligent sensing nodes and shared communication links[3][4]. In WNCS, wireless network acts as a communication medium to undertake data communication tasks between nodes and has the characteristics of low cost, self-organization and easy maintenance[5][6]. [7] proposes the concept of EVM for industrial process control, the characteristic of which is to make use of the assignment and scheduling of the control tasks in nodes that the computing, communication, memory and energy are limited to complete complex control task.

Wireless sensor research for the harsh environment of aero-engine has achieved great results. For example, NASA Johnson Space Center[8], the University of Maine has conducted extensive research on passive wireless sensors in harsh environment. Environetix Technologies Corporation has developed a series of wireless sensors suitable for the harsh environment of aero-engine, of which the wireless temperature sensor EVHT-100 has been commercialized[9][10].

Linearization control of aero-engine is still a common control method, but the aero-engine is strongly nonlinear, it is difficult to guarantee the real-time requirement for online calculation of real-time linear model and tuning of controller parameters, for the offline calculation of linear model and tuning of controller parameters, with the improvement of the model accuracy of aero-engine, the number of aero-engine models and the corresponding controller parameters will be very large, this poses a high requirement for local memory of control system, and is a great challenge to the existing aero-engine control system.
The above research indicates that the aero-engine control based on wireless network is an important development direction of aero-engine control research in the future. In this paper, we proposed a aero-engine Software Defined Control System (SDCS) based on PID controller which is the simplest and most commonly used[11]. SDCS realizes a fully distributed controller within wireless network using linearization model of aero-engine at different working points in the flight envelope[12] to make the corresponding controller parameters switched. The control scheme can overcome the impact of the cables, connecters and core control unit in traditional electronic control system on aero-engine control system, and the powerful memory capacity of SDCS is suitable for high-precision aero-engine piecewise linearization control. The effectiveness of the scheme is analyzed by MATLAB/Simulink environment.

2. Architecture of Aero-engine SDCS

Figure 1 shows the overview of an aero-engine SDCS, which includes two parts: the Control Platform and the Support Platform. For the Control Platform, each node has the function of computing, memory and wireless communication, the PID controller is virtualized and dispersed in the wireless nodes in Figure 1, and the control system is a combination of wireless node tasks. The Support Platform is the manager of the control system (i.e. the whole wireless network), which mainly includes node resource monitoring, node scheduling and the management of node task mapping.
The wireless network based on the IEEE 802.11 protocol standard is used to build a Ad-Hoc network that is peer-to-peer and un-center. In order to simplify the research, this paper limits the scope of study to a small jet engine, and all nodes in the network can jump to each other.

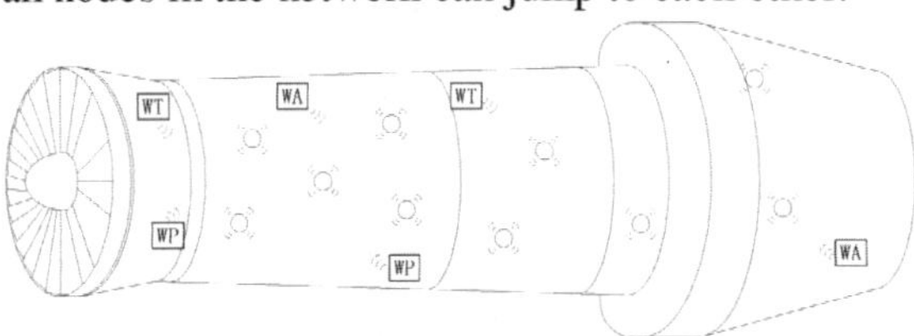

Figure 1. Overview of a aero-engine SDCS.

2.1. Control task virtualization based on MATLAB/Simulink

SDCS requires that control tasks be divided(i.e. virtualized) and then distributed among the nodes in wireless network. The process of task segmentation is called control task virtualization, and each segmented task is called Virtual Control Task (VCT).

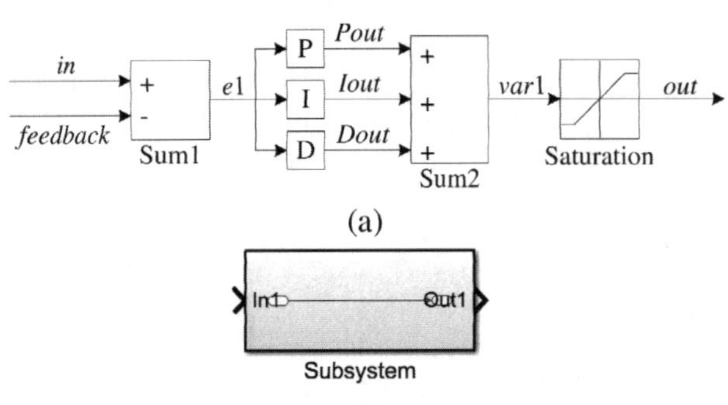

Figure 2. (a)Model of a PID controller; (b)Subsystem Block of Simulink software.

The scheme of virtualization of the control task designed in this paper is based on MATLAB/Simulink, a PID controller Simulink model is shown in Figure 2(a). The Subsystem Block (as shown in Figure 2(b)) in Simulink is used to package the blocks that make up the controller, and each packaged subsystem is a VCT. For example, the Figure 3 is a virtualization scheme of the PID controller shown in Figure 2.

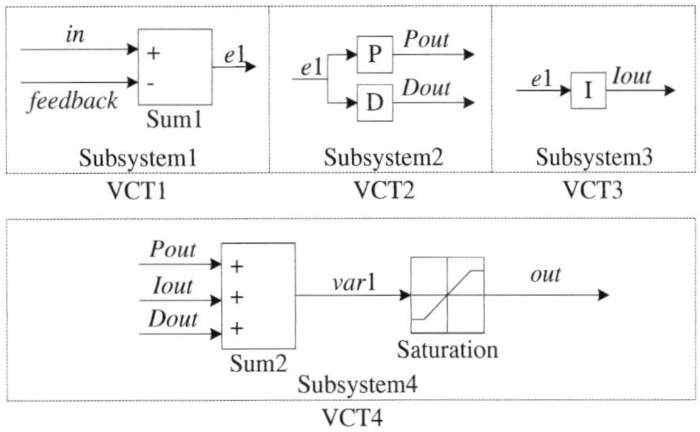

Figure 3. VCTs from Figure 2(a).

2.2. Introduction of Support Platform Architecture

The Support Platform architecture is shown in Figure 4.The Support Platform Manager (SPM) realizes all the functions of the Support Platform. Each node runs a Local Operating System (LOS) which executes Local Tasks (LT) of the node and VCT mapped to the node.

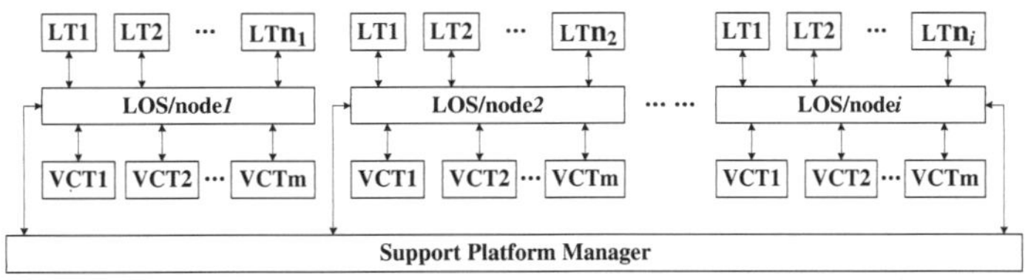

Figure 4. Architecture of the Support Platform.

3. Piece-wise Linearization operation of aero-engine SDCS

Linear control method is a mature and common method in aero-engine control, but the operation process of aero-engine is a complicated aerothermodynamics process. Within the full flight envelope, the aerothermodynamic process of the aero-engine varies with environment condition and working condition. The piece-wise linear model is required in the design of control system by using linear control law because aero-engine has strong nonlinearities, multimodality and high-order characteristics. For the SDCS scheme proposed in this paper, due to each node in the wireless network have capacity of memory, the whole control system can achieve powerful memory capacity to meet the requirement of aero-engine piecewise linear model accuracy improvement on the memory capacity of the control system. The implementation of piecewise linearization control scheme of aero-engine based on software defined control system is discussed in this section.

3.1. The operation analysis of Support Platform

In order to describe and schedule the whole wireless network and the control tasks, two data structures named SPM Description Table (SPMDT) and Nodes Control Table (NCT) are introduced. SPM gives the description of wireless network, VCT status and operating points of the aero-engine in flight envelope, including CPU utilization and memory usage of each node and communication status between nodes. NCT includes mapping relationship between VCT and network nodes, data transmission relationship between wireless network nodes, and the switching of VCT parameters of aero-engine at different working points.

The scope of SPM spans the entire wireless network and supports the following online operations: (1) Each node in the wireless network has a local backup of SPMDT and NCT, and keeps the SPMDT and NCT of the nodes in the wireless network consistent though periodic broadcast; (2) Each VCT is

IOP Conf. Series: Journal of Physics: Conf. Series **1215** (2019) 012016 doi:10.1088/1742-6596/1215/1/012016

mapped to 1 primary node and B backup nodes, the primary node and backup nodes can receive and process data synchronously, but the backup nodes do not send the process calculating data. (3) Switch all the parameters of VCT when working points in the aero-engine flight envelope changed.
During the operation of system, the following operating constraints should be satisfied:
(1) Each VCT is mapped to 1 primary node and B backup nodes

$$\sum_{i=1}^{m} v_{i,j}^{p} = 1, \sum_{i=1}^{m} v_{i,j}^{b} = B, v_{i,j}^{p} + v_{i,j}^{b} \leq 1$$

$$\forall i \in \{1,2,\cdots m\}, j \in \{1,2,\cdots,n\}$$

(2) The number of all primary nodes is equal to the number of all VCTs exactly, the number of all backup nodes is great than or equal to the number of all VCTs

$$\sum_{i=1}^{m} v_{i,j}^{p} = 1, \sum_{i=1}^{m} v_{i,j}^{b} = B, v_{i,j}^{p} + v_{i,j}^{b} \leq 1$$

$$\forall i \in \{1,2,\cdots m\}, j \in \{1,2,\cdots,n\}$$

(3) The IP address and MAC address in the network have a one-to-one mapping relationship

$$\sum_{j} ip_{i}^{j} = 1, \forall i \in \{1,2,\cdots m\}$$

Where, the decision variables in the above constraints are:
(1) $v_{i,j}^{p}$, the assignment variable of the primary node of VCT_j, where $i \in \{1,2,\cdots,m\}, j \in \{1,2,\cdots,n\}$

$$v_{i,j}^{p} = \begin{cases} 1, node_i \text{ is the primary node of } VCT_j \\ 0, else \end{cases}$$

(2) $v_{i,j}^{b}$, the assignment variable of backup node of VCT_j, where $i \in \{1,2,\cdots,m\}, j \in \{1,2,\cdots,n\}$

$$v_{i,j}^{b} = \begin{cases} 1, node_i \text{ is the backup node of } VCT_j \\ 0, else \end{cases}$$

(3) ip_i^j, the mapping variable of IP address, where $i \in \{1,2,\cdots n\}$, $j \in \{1,2,\cdots,a\}$, $IPADD = \{ipadd_1, ipadd_2, \cdots, ipadd_a\}$ is the set of all IP addresses

$$ip_i^j = \begin{cases} 1, ipadd_j \text{ is mapped to } node_i \\ 0, else \end{cases}$$

3.2. VCT parameters switching
Computer memory technology has developed rapidly in these years, more than 100GB high-speed FLASH chips can be carried in a few square centimeters. There are many wireless nodes in the aero-engine SDCS, and the memory capacity of the whole control system is equal to the sum of all the nodes' memory capacity in wireless network.

A set of nodes in the system can be used to store the parameters of controller at different working points in aero-engine flight envelope, which is called Parameter Nodes. Each node executing VCTs contains the working points of aero-engine at the present and local backup of controller parameters at nearby working points.

The working points of aero-engine flight envelope will be numbered. Assume that the aero-engine in flight envelope is divided into v working points, $\mathrm{WP} = \{\mathrm{wp}_1, \mathrm{wp}_2, \cdots, \mathrm{wp}_\mathrm{v}\}$ is a set of all working points in flight envelope, $\mathrm{PAR} = \{\mathrm{par}_1, \mathrm{par}_2, \cdots, \mathrm{par}_\mathrm{v}\}$ is a set of controller parameters at v working points after the aero-engine is piece-wise linearized, where par_i is a set of controller parameters at working point wp_i of aero-engine. In addition, assume that wp_j is the current working point of aero-engine, $\mathrm{LPAV} = \mathrm{U}(\mathrm{par}_\mathrm{j}, \delta)$ is a set of local backup of controller parameters of all VCT nodes, where $\mathrm{U}(\mathrm{par}_\mathrm{j}, \delta)$ is the area with radius δ of the element par_j in the set PAR. With the change of the working point of aero-engine, VCT nodes apply for the controller parameters of corresponding working point from the parameter node, and discard the controller parameters that do not belong to $\mathrm{U}(\mathrm{wp}_\mathrm{j}, \delta)$ working point.

During the switching of VCT parameters above, the parameter set LPAV of local controller for the primary and backup nodes of all VCTs is dynamically adjusted as the aero-engine operating at different working points in flight envelope. When the working state of aero-engine is switching between the adjacent working point, the VCT nodes obtain the corresponding controller parameters directly from the local and update the parameter set of local controller by using the idle time of the nodes to reduce the time consumption of the switching of the controller parameters and ensure that the local storage resource of VCT nodes is sufficient.

In addition to the constraints of the section 3.1, the system also needs to meet the constraint below during the operation

(4) the range of the local controller parameters is not exceeding the neighborhood with center of the current working point and radius of δ

$$\sum_j lmp_i^j \le 2\delta + 1, j - i \le \delta, i, j \in \{1,2,\cdots v\}$$

where the decision variable $\mathrm{lmp}_\mathrm{i}^\mathrm{j}$ is

$$lmp_i^j = \begin{cases} 1, \text{parameters of } wp_j \text{ belongs to } U(par_i, \delta) \\ 0, else \end{cases}$$

4. Simulation analysis

In order to verify the effectiveness of the scheme, a analysis of digital simulation is made by MATLAB/Simulink software tool in this section. The piecewise linearization of a twin-shaft turbofan aero-engine at zero height and zero Mach number conditions from [13] is taken as the research object, and 5 working points running on the flight envelope of aero-engine are adopted for the purpose of simplifying the study. The PID controller and its virtualization plan are shown in Figure (2) and Figure (3).

The transfer function model of aero-engine is:

$$G_{AE} = \frac{K(T_z s + 1)}{(T_{p1} s + 1)(T_{p2} s + 1)}$$

Table 1 shows the model parameters at 5 working points in the flight envelope, where WP denotes the working point.

IOP Conf. Series: Journal of Physics: Conf. Series **1215** (2019) 012016 doi:10.1088/1742-6596/1215/1/012016

Table 1. Mode parameters of 5 working points in flight envelope.

Working Points	1	2	3	4	5
K	0.0454	0.0472	0.0475	0.0476	0.0476
T_{p1}	0.2156	0.2205	0.2190	0.2082	0.1717
T_{p2}	0.1947	0.1860	0.1625	0.2072	0.1021
T_z	0.2684	0.2305	0.2448	0.2726	0.1410
T_z	0.2684	0.2305	0.2448	0.2726	0.1410

Given the system input as a Step input, the aero-engine is sequentially switched from the working point 1 to the working point 5, the linear model of aero-engine at 5 working points is obtained from the parameters in Table 1. Assume that the switching time of aero-engine at 5 working points is 5s, 10s, 15s and 20s respectively, since the system has to judge the current working point, the switching of VCT parameters lags behind that of working points of aero-engine. In this simulation, the lag time is set as 2 sampling periods, and the system response is shown in Figure 5. It can be seen from the figure that the switching of VCT parameters caused by the change with working point of aero-engine has little impact on the control system, the system maintains good dynamic tracking performance and steady state performance.

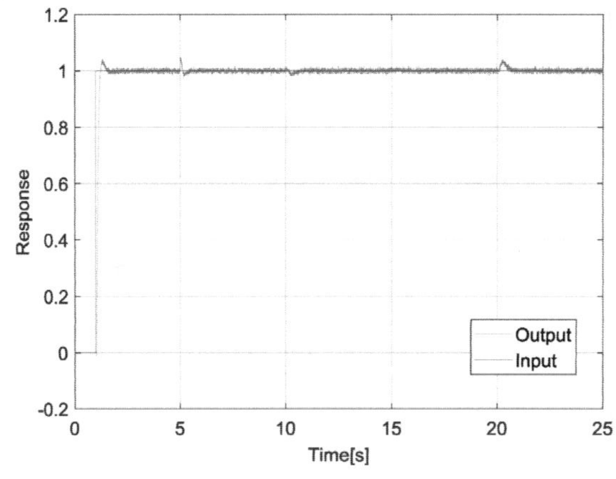

Figure 5. System response when parameters of VCT switched in Step input

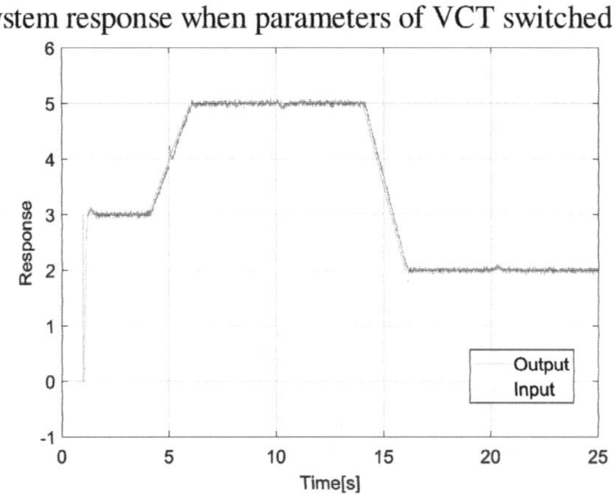

Figure 6. System response when parameters of VCT switched in a series of different Step and Ramp inputs

IOP Conf. Series: Journal of Physics: Conf. Series **1215** (2019) 012016 doi:10.1088/1742-6596/1215/1/012016

Given the system input as a series of different Step and Ramp input, other conditions are consistent with that in Figure 5, the system response diagram is shown in Figure 6. It can be seen from the figure that the switching of VCT parameters caused by the change with working point of aero-engine has little impact on the control system, the system maintains good dynamic tracking performance and steady state performance.

It can be concluded from the above two simulation results that the proposed scheme achieves good dynamic and steady state performance in the piecewise linearization of aero-engine control, which shows the effectiveness of the proposed control scheme.

5. Conclusion

Footnotes should be avoided whenever possible. If required they should be used only for brief notes that do not fit conveniently into the text.

In this paper, a concept of aero-engine SDCS based on wireless network is proposed. The virtualization scheme of PID controller which is the commonly used is introduced, the virtualized PID controller is distributed in wireless network and a fully distributed controller based on wireless network is realized. Aero-engine has strong nonlinear characteristics, in this paper, a controller parameters switching scheme is design by using the linearized model of aero-engine at different working points in flight envelope. This control scheme makes full use of the powerful memory function of SDCS.

According to the piece-wise linearization control scheme of aero-engine proposed in this paper, the digital simulation analysis is carried out. The simulation results show that the control scheme proposed in this paper achieves good control effect when the aero-engine works at different working points in the flight envelope, and the VCT parameter switching has little effect on the output response of the control system. Therefore, the scheme proposed in this paper is suitable for the subsection linearization control of aero-engine flight envelope, and reduces the impact of cables, connectors and core control unit on aero-engine control system at the same time.

6. References

[1] PL Shaffer. J ENG GAS TURB POWER,121 (1999)
[2] J Zang. Modern aircraft engine control technology (Aviation Industry Press, Beijing China, 2016)
[3] S Bera, S Misra, SK Roy, MS Obaidat. IEEE SYST J,99 (2016)
[4] M Yigit, VC Gungor, E Fadel, et al. J NETW COMPUT APPL,71 (2016)
[5] B Wu, MD Lemmon, H Lin. IEEE T CONTR SYST T, 26,(2018)
[6] H Yang, M Shi, Y Xia, et al. IEEE T CYBERNETICS (published online, 2018)
[7] R Mangharam, M Pajic. IEEE ICDCS (Montreal, QC, Canada, 2009)
[8] G Studor. PWST 2012 Workshop Plan, NASA/JSC (2012)
[9] SC Moulzolf, R Behanan, T Pollard, et al. IEEE IUS (Prague, Czech Republic, 2013)
[10] MPD Cunha, A Maskay, RJ Lad, et al. IEEE IUS (Tours, France, 2016)
[11] MX Di. Research on robust PID controller for aero-engine (Nanjing University of Aeronautics and Astronautics, Nanjing China, 2012)
[12] B Wang, X Wang, YL Shi, et al. Journal of Aerospace Power, 29 (2014)
[13] YT Tang. Research on frequency domain design method of PID control for aeroengines (Nanjing University of Aeronautics and Astronautics, Nanjing China, 2012)

IOP Conf. Series: Journal of Physics: Conf. Series **1215** (2019) 012017 doi:10.1088/1742-6596/1215/1/012017

Stability Analysis of PID Control of Aero-engine Software Defined Control System under Nodes Faults

Jiao Ren[1*], Jianghong Li[1] , Xiaoxiang Ji[1] and Yafeng Wu[1]

[1] School of Power and Energy, Northwestern Polytechnical University, 710129 Xi'an, China

renjiao@mail.nwpu.edu.cn

Abstract. A distributed aero-engine PID controller implementation scheme based on nocenter and peer to peer wireless network is proposed in this paper. The basic method is to make the PID controller virtualized and then distribute them in the nodes of the wireless network, the control task of the system is the synthesis of node tasks of multiple wireless network, and there is no special controller node in the system. In order to deal with the potential faults of nodes in the wireless network, tasks have the ability to migrate among nodes. Based on the analysis of complex frequency domain theory, task migration has less impact on the system response, Finally, the influence of task migration on the system is simulated and analyzed in the MATLAB/Simulink environment, simulation results are consistent with the theoretical analysis, the availability of the designed control system is proved.

1. Introduction

At present, the aero-engine control system is mainly based on the electronic control connected by cables[1][2]. There is a central control unit in the control system, all or main functions of the system are concentrated in the core unit. The damage of the core unit has a great influence on the system. In addition, the cables and connectors used in large quantities increase the failure probability of the control system, and reduce the thrust-to-weight ratio of the aero-engine at the same time.

With the development of low-cost sensing, computing and wireless communication technologies, Wireless Sensor Networks (WSN) have emerged[3][4], it is composed of several sensors, which take intelligent sensor nodes and shared communication links as the main topological features. Inspired by the WSN, Wireless Networked Control System (WNCS) appears in the control field[5]. WNCS has the characteristics of low cost, self-organization and easy maintain. In WNCS, wireless network is mainly used as communication medium and only takes on the task of data communication among the nodes in the wireless network[6]. In [7], a preliminary concept of EVM is proposed for industrial process control, the characteristics of EVM is to use the assignment and scheduling of control tasks in nodes with limited resources such as computing, communication and memory to complete relatively complex control task.

At present, research of aero-engine wireless sensor has achieved certain results, for example, NASA Johnson Space Center introduced the research status of passive wireless sensors in aero-engines and rocket-engines at the Report of Passive Wireless Sensor Technology 2012 Workshop Plan[8]. Environetix has developed a full suite of wireless sensors including high-temperature sensors, pressure sensors, vibration sensors and strain sensors for taking measurements in harsh and extreme environments with the supporting of the US Air Force and other organizations, and the wireless high-temperature sensor EVHT-100 has been commercialized and was available for customized installation[9]. The above research shows that the control scheme based on wireless network is the development direction of future aero-engine control system. The aero-engine Software Defined Control System SDCS control system was proposed based on the development of the above mentioned related supporting technologies.

Aiming at the challenge of modern aero-engine control system, based on the development of wireless communication technology, computer technology, memory technology, and other supporting technologies, this paper proposes the concept of SDCS, which is a fully distributed control system based on IEEE802.11 protocol standard wireless network.

2. Architecture of Aero-engine SDCS

Figure 1 is overview of a aero-engine SDCS on a turbine engine, where WT denotes the wireless temperature sensor, WP denotes the wireless press sensor, WA denotes the wireless actuator, denotes nodes in wireless network, denotes the node have the abilities of wireless communication. The SDCS consists of the Support Platform and the Control Platform, every node in the Control Platform has the abilities of computation, memory and wireless communication, virtualized control tasks are distributed in the wireless nodes, the whole wireless network forms complete controller to complete the control function of the system; the Support Platform manages the control system, mainly including the monitoring and analysis of nodes and control tasks, and task migration among the wireless nodes.

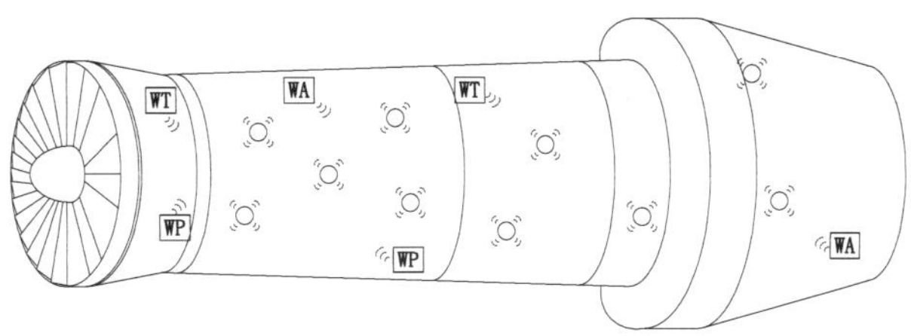

Figure 1. Overview of a SDCS on a turbine engine.

2.1. Virtualization of PID controller for the Control Platform based on MATLAB/Simulink

The Control Platform requires that control tasks be distributed in nodes of the wireless networks, and the overall control tasks of the system need to be divided into multiple sub-tasks. The division of the control task is named control task virtualization, and each sub-task is named Virtual Control Task (VCT).

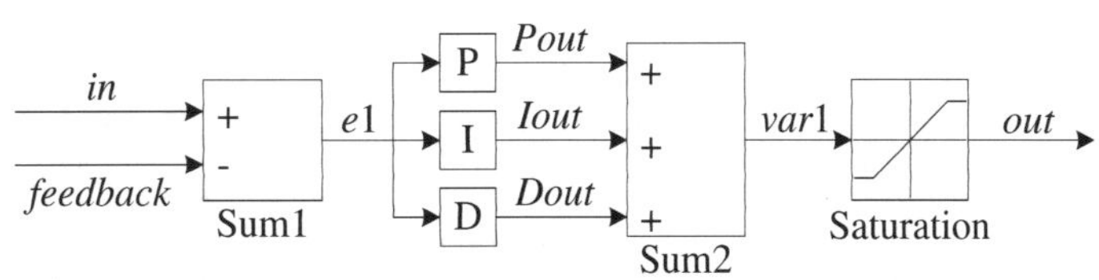

Figure 2. Simulink model of a closed-loop PID controller.

As shown in Figure 2, it is the Simulink Model of the PID controller, the closed-loop PID controller consists of several Simulink Blocks, each VCT is a Simulink Block or a combination of Simulink Blocks, each Block or combination of Blocks can be encapsulated as a subsystem, each subsystem is a VCT. An example of implementation of virtualization of PID controller is shown in Figure 3. The function model of each VCT can be generated by using the Real-Time Workshop, it is compiled under node platform to generate executable code of node platform.

IOP Conf. Series: Journal of Physics: Conf. Series **1215** (2019) 012017 doi:10.1088/1742-6596/1215/1/012017

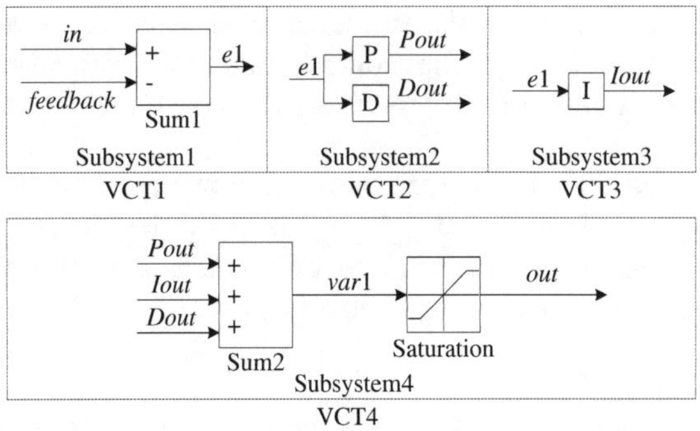

Figure 3. Virtualization of controller in Figure 2.

2.2. Architecture of the Support Platform

Figure 4 is the architecture of the Support Platform, the function of the Support Platform is realized by the Support Platform Manager (SPM), the scope of the SPM span the entire wireless network, every node of the wireless network runs a Local Operating System (LOS), LOS executes Local Task (LT) and VCT assigned to the node, SPM realized the mapping relationship between VCT and nodes in the wireless network, and monitors the local resources (CPU, memory, etc.). In addition, SPM also complete the assignment and migration of VCT, and manages the entire network.

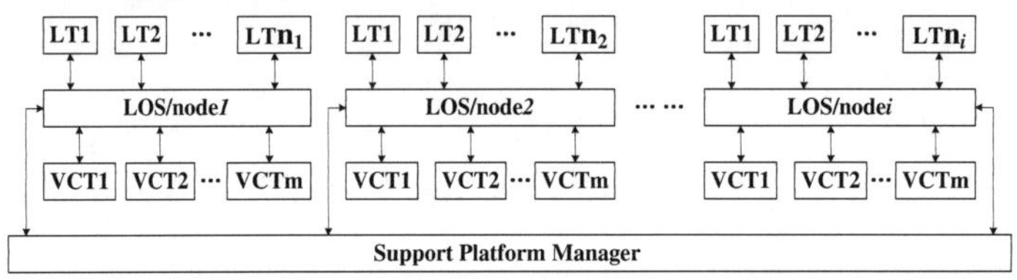

Figure 4. Architecture of the Support Platform.

3. Support Platform handle & node fault analysis

3.1. Support Platform handle

The operation of the Support Platform is described in this section. The set $\mathrm{SVCT} = \{\mathrm{VCT}_1, \mathrm{VCT}_2, \cdots, \mathrm{VCT}_n\}$ denotes set of all VCTs, the set $\mathrm{NODE} = \{\mathrm{node}_1, \mathrm{node}_2, \cdots, \mathrm{node}_m\}$ denotes set of all physical nodes in the wireless network, each VCT is assigned to $\mathrm{B}+1$ nodes, one of which is the primary node and the other B node(s) is(are) the backup node(s). When the primary node of $\mathrm{VCT_j}$ is at fault, the corresponding backup node is selected as the new primary node according to the ascending order rule of the IP address. When a backup node of $\mathrm{VCT_j}$ is at fault, the corresponding idle node (the node does not execute any VCT) is selected as the new backup node according to the ascending order rule of the IP address. The state data is periodically exchanged among the primary node and the backup node and broadcast to the whole wireless network for node status monitoring.

The following decision variables is introduced to describe the operating process:

(1) Primary node assignment variables $\mathrm{v}_{\mathrm{i,j}}^{\mathrm{p}}$, where $\mathrm{i} \in \{1,2,\cdots,\mathrm{m}\}$, $\mathrm{j} \in \{1,2,\cdots,\mathrm{n}\}$

$$\mathrm{v}_{\mathrm{i,j}}^{\mathrm{p}} = \begin{cases} 1, \mathrm{node_i} \text{ is the parmary node of } \mathrm{VCT_j} \\ 0, \text{else} \end{cases}$$

(2) Backup node assignment variables $\mathrm{v}_{\mathrm{i,j}}^{\mathrm{b}}$, where $\mathrm{i} \in \{1,2,\cdots,\mathrm{m}\}$, $\mathrm{j} \in \{1,2,\cdots,\mathrm{n}\}$

$$\mathrm{v}_{\mathrm{i,j}}^{\mathrm{b}} = \begin{cases} 1, \mathrm{node_i} \text{ is the backup node of } \mathrm{VCT_j} \\ 0, \text{else} \end{cases}$$

(3) IP address mapping variables ip_i^j, where $i \in \{1,2,\cdots,m\}$, $j \in \{1,2,\cdots,a\}$, the set $\mathrm{IPADD} = \{\mathrm{ipadd}_1, \mathrm{ipadd}_2, \cdots, \mathrm{ipadd}_a\}$ denotes all IP address

$$\mathrm{ip}_i^j = \begin{cases} 1, \mathrm{ipadd}_j \text{ have been mappped to } \mathrm{node}_i \\ 0, \text{else} \end{cases}$$

During the running of the system, the following operating constraints must be satisfied:
(1) Every VCT have 1 primary node and B backup node(s), $\forall i \in \{1,2,\cdots m\}, j \in \{1,2,\cdots,n\}$

$$\begin{cases} \sum_{i=1}^{m} v_{i,j}^p = 1 \\ \sum_{i=1}^{m} v_{i,j}^b = B \\ v_{i,j}^p + v_{i,j}^b \le 1 \end{cases} \tag{1}$$

(2) The number of all primary nodes is exactly equal to the number of all VCTs, and the number of all backup nodes is greater than or equal to the number of all VCTs, $\forall i \in \{1,2,\cdots m\}, j \in \{1,2,\cdots,n\}$

$$\begin{cases} \sum_{i=1}^{m}\sum_{j=1}^{n} v_{i,j}^p = n \\ \sum_{i=1}^{m}\sum_{j=1}^{n} v_{i,j}^b \ge n \end{cases} \tag{2}$$

(3) The IP address and MAC address have one-to-one mapping relationship, $\forall i \in \{1,2,\cdots m\}$

$$\sum_j \mathrm{ip}_i^j = 1 \tag{3}$$

3.2. Node fault analysis

When a node fault causes a VCT migration, the output of the VCT is 0 during the migration, which can be regarded as adding a square wave disturbance to the VCT's output when the migration action occurs, the width of the square wave is the duration of the migration action, the amplitude of the square wave is the same as the amplitude of the output of the VCT when the migration occurs, and the direction is opposite, i.e., disturbance $d(t) = -R \cdot \varepsilon(t - t_1) + R \cdot \varepsilon(t - t_2)$, where $\varepsilon(t)$ is the Step Function, R is the amplitude, $t_2 - t_1$ is the duration of the migration action, and $t_2 > t_1$.

In this section, we mainly discuss the migration of VCT_2 and VCT_3 in Figure 3. The disturbance transfer function of the system is:

$$\frac{Y(s)}{D(s)} = \frac{G_{AE}(s)}{1 + G_{PD}(s)G_{AE}(s) + G_I(s)G_{AE}(s)} \tag{4}$$

where $G_{AE}(s)$ is the aero-engine model, $G_{PD}(s)$ is the model of VCT_2 and $G_I(s)$ is the model of VCT_3. From the disturbance transfer function, the impact of migration on system response mainly depends on the amplitude R, the greater the amplitude R, the greater the fluctuation of system response during task migration.

The steady-state error of the system under the disturbance is:

$$e_{ss} = \lim_{s \to 0} \frac{sD(s)}{1 + G_{PD}(s)G_{AE}(s) + G_I(s)G_{AE}(s)} \tag{5}$$

when the disturbance is Step $\mathrm{r(t) = R \cdot \varepsilon(t - t_i)}$, where R is the amplitude, $\mathrm{t_i}$ is the Step time, and the stead-state error of the system under this disturbance is:

$$e_{ss} = \frac{R}{1 + \lim_{s \to 0}[G_{PD}(s)G_{AE}(s) + G_I(s)G_{AE}(s)]} \tag{6}$$

Considering the disturbance $\mathrm{d(t)}$ as a superposition of two Step signals with opposite amplitudes. For the transfer function model of PID controller, the order of the numerator is two and the order of the denominator is one, the order of the numerator is greater than the denominator, it is easy to conclude that the stead-state error of the system under disturbance input $\mathrm{d(t)}$ is 0.

4. Simulation analysis

In order to verify the effectiveness of this scheme, this section is based on MATLAB/Simulink software tool for digital simulation analysis. In this simulation, the fuel-high-pressure rotor speed linear model of a twin-shaft turbofan aero-engine at zero height and zero Mach condition[10] is used as the plant model. The controller and virtualization scheme are shown in Figure 2 and Figure 3.

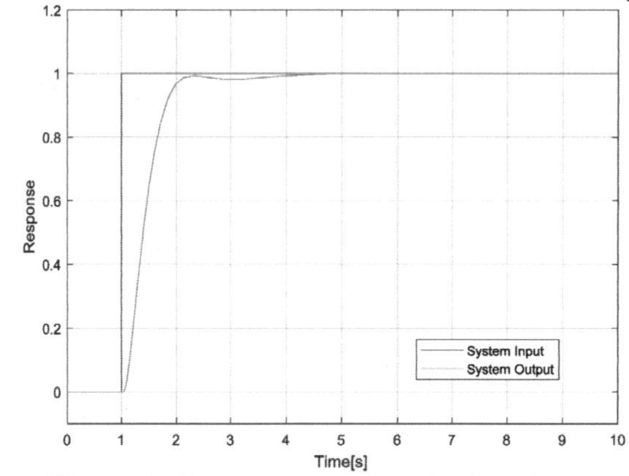

Figure 5. System response under Step input

The system response under Step input is shown in Figure 5, it shows that the system response has good stead-state characteristics and dynamic tracking performance. The scheme designed in this paper meets the requirements of the control system.

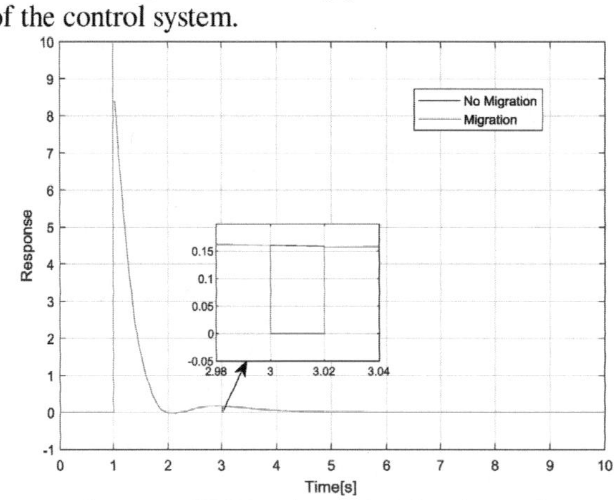

Figure 6. Output of VCT_2 when VCT_2 is migrated at $t = 3s$.

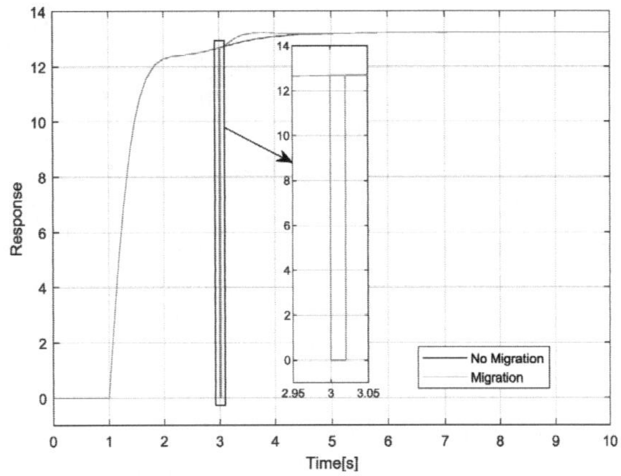

Figure 7. Output of VCT_3 when VCT_3 is migrated at $t = 3s$.

Figure 6 is the output of VCT_2 when VCT_2 is migrated at $t = 3s$ and there is no migration of VCT_2. Figure 7 is the output of VCT_3 when VCT_3 is migrated at $t = 3s$ and there is no migration of VCT_3. It can be concluded that when migration happened, the output of VCT_2 is smaller than the output of VCT_3. From the perspective of disturbance, it can be seen that the amplitude of square wave disturbance when migration of VCT_2 occurs is smaller than that of VCT_3.

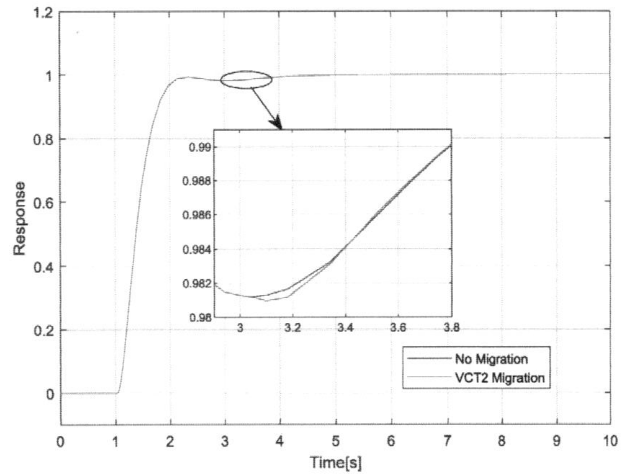

Figure 8. System response when VCT_2 is migrated at $t = 3s$.

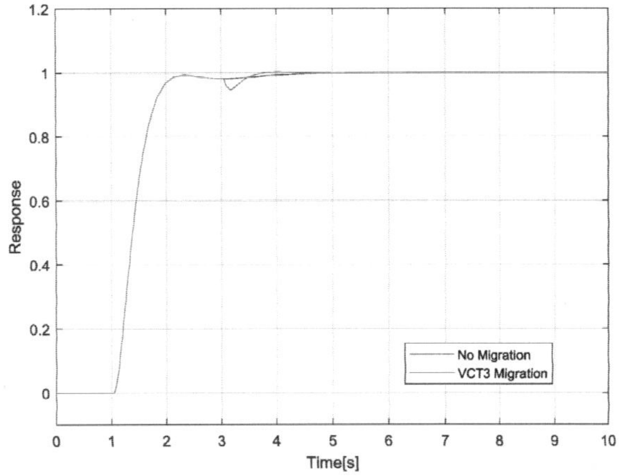

Figure 9. System response when VCT_3 is migrated at $t = 3s$.

Figure 8 is the system response when VCT_2 is migrated at $t = 3s$ and there is no migration of VCT_2, and Figure 9 is the system response when VCT_3 is migrated at $t = 3s$ and there is no migration of

IOP Conf. Series: Journal of Physics: Conf. Series **1215** (2019) 012017 doi:10.1088/1742-6596/1215/1/012017

VCT_3. From Figure 8 and Figure 9, we can conclude that when VCT_2 and VCT_3 is migrated at same time separately, the impact of the migration on system response is different. The migration of VCT_3 has greater impact on system response than that of VCT_2. From the simulation results in Figure 6 and Figure 7, the difference of the system response is mainly due to the difference of the amplitude of the equivalent perturbation of the migration of VCT_2 and VCT_3.

From Figure 6 to Figure 9, it can be concluded that when VCT_j is migrated, the greater the output of VCT_j at the time of migration (it can be seen as the larger the amplitude of the equivalent disturbance is), the greater the impact of migration on system response, this is consistent with the theoretical analysis above, which shows that the analysis of the impact of VCT migration on system response is correct.

Different from the existence of the core control unit hot backup scheme, the controller of the control scheme proposed in this paper is distributed. Compared with the scheme with the core control unit, when the controller node fails and performs backup switching or task migration, the proposed scheme has less impact on system control performance, and risk caused by node fault is reduced due to the distributed structure of the controller.

5. Conclusion

This paper introduces a distributed aero-engine control system based on wireless network, and presents a virtualization scheme of PID controller based on MATLAB/Simulink software, the controller in the scheme is distributed. VCT(s) can be migrated between nodes to cope with the potential faults of nodes, and the impact of migration of VCT(s) on system response is analyzed.

Finally, the digital simulation analysis is carried out. The simulation results show that the control system designed in this paper has good control performance, the impact of migration of different VCT on the system are simulated and compared, and the simulation results are consist with the analysis in this paper. It can be obtained that in the distributed control scheme proposed in this paper, while the control task are distributed in the wireless network, the risks brought by the node fault(s) to the control system is also distributed in the wireless network.

6. References

[1] WR Schley. J ENG GAS TURB POWER,**121** (1998)
[2] J Zang. *Modern aircraft engine control technology* (Aviation Industry Press, Beijing China, 2016)
[3] S Bera, S Misra, SK Roy, MS Obaidat. IEEE SYST J,99 (2016)
[4] H Kaschel, J Ortega. *IEEE ICA-ACCA2016*(Curico, Chile, 2016)
[5] P Park, SC Ergen, C Fischione, C Lu, KH Johansson. IEEE COMMUN SURV TUT, **20** (2018)
[6] B Wu, MD Lemmon, H Lin. IEEE T CONTR SYST T, **26**,(2018)
[7] M Pajic, R Mangharam. *IEEE RTAS* (Stockholm, Sweden, 2010)
[8] G Studor. PWST 2012 Workshop Plan, NASA/JSC (2012)
[9] MPD Cunha, A Maskay, RJ Lad, et al. IEEE IUS (Tours, France, 2016)
[10] YT Tang. *Research on frequency domain design method of PID control for aeroengines* (Nanjing University of Aeronautics and Astronautics, Nanjing China, 2012)

Optimal Formation Control for Quadrotors with Collision Avoidance Based on Dynamic Constraints

P F Jin[1], J Q Yu[1], Z Y Jia[1] and D Yang[1]

[1]School of Aerospace Engineering, Beijing Institute of Technology, Beijing 100081, China

Abstract. This paper studied the optimal formation control of quadrotor UAV based on the dynamic model, and the collision avoidance between quadrotors is considered. By constructing the problem into a standard convex quadratic programming problem, we hope to improve the solving efficiency of the formation control problem. Firstly, the nonlinear dynamic model of quadrotor is linearized and the prediction model is established. Then, the safety zone constraints are transformed from a circular zone to a half-plane zone, making the optimization problem be a standard convex quadratic programming problem. Finally, the quadratic programming problem is solved using distributed receding horizon optimization. Numerical simulations in three-dimensional space show that this method can obtain the optimal formation trajectory with collision avoidance, and can improve the solving efficiency.

1 Introduction

In the field of optimal formation control for multiple unmanned aerial vehicles (UAVs), model predictive control (MPC) is often used to solve optimization problems, usually based on kinematic models. Researchers studied the formation trajectory optimization of multiple UAVs based on the particle kinematics model [[1]]. The dynamic characteristic of quadrotor UAV is quite different from fixed-wing aircraft. It is necessary to study the formation control of quadrotors according to its 6-DOF dynamic model. Quadrotor's dynamics has strong nonlinearity, so the direct applying of nonlinear dynamic model will increase the difficulty of solving optimization. Linearizing the dynamic model can help the optimization problem be solved quickly under the linear MPC framework.

Collision avoidance should be considered during formation control. Researchers adopt some intelligent optimization algorithms, such as A* algorithm[[2]], Ant Colony Algorithm (ACA) [[3]], Particle Swarm Optimization (PSO) [[4]], the Artificial Potential Field (APF) [[5]], searching for cooperative formation control of multiple UAVs. However, the trajectories solved by these intelligent optimization algorithms are generally not optimal, and they usually consume large computing resources. Another feasible solution for the formation control problem is the Mixed Integer Programming (MILP) [[6]], which transforms the problem into a mixed integer linear programming model by introducing auxiliary decision variables, and then uses CPLEX, a relatively mature optimization tool, to solve the problem centrally. In [[7]], the collision avoidance constraints are linearized into the range of abscissa and ordinate coordinates, and the receding horizon control is used to reduce the calculation time, which makes the real-time trajectory planning with collision avoidance be realized.

When the performance index function is a quadratic function with linear constraints, the quadratic programming algorithm can be used to solve the optimization, which is more efficient than intelligent algorithm. However, the circular safety zone constraints of quadrotors are nonlinear constraints and cannot be accepted by quadratic programming algorithm. In this paper, the safety zone constraints are transformed from a circular zone into a half-plane zone, which can be expressed by a linear inequality, and the optimal variables belong to a nonempty closed convex set. Then the optimal problem is transformed into a standard convex quadratic programming problem, which can be solved by mature quadratic programming algorithms.

2 Modelling

2.1 Linear dynamic model of a quadrotor UAV

For a single quadrotor UAV, the following continuous time particle kinematics model is used to describe it [[8]]:

$$\begin{cases} \ddot{x} = (\sin\phi\sin\psi + \cos\phi\sin\theta\cos\psi)u_z \\ \ddot{y} = (\cos\phi\sin\theta\sin\psi - \sin\phi\cos\psi)u_z \\ \ddot{z} = (\cos\phi\cos\theta)u_z - g \\ \ddot{\phi} = u_\phi \\ \ddot{\theta} = u_\theta \\ \ddot{\psi} = u_\psi \end{cases} \tag{1}$$

Based on the following operations: 1) small angle assumption; 2) omitting high-order minor terms; 3) applying $u_z = g$ to horizontal motion, we can obtain

$$\dot{s}_i = \bar{A}s_i + \bar{B}u_i + \bar{C} \tag{2}$$

In the formula (2),

$$\bar{A} = \begin{bmatrix} 0 & 1 & 0 & 0 & 0 & 0 & 0 & 0 & 0 & 0 & 0 & 0 \\ 0 & 0 & 0 & 0 & 0 & 0 & 0 & 0 & 9.81 & 0 & 0 & 0 \\ 0 & 0 & 0 & 1 & 0 & 0 & 0 & 0 & 0 & 0 & 0 & 0 \\ 0 & 0 & 0 & 0 & 0 & 0 & -9.81 & 0 & 0 & 0 & 0 & 0 \\ 0 & 0 & 0 & 0 & 0 & 1 & 0 & 0 & 0 & 0 & 0 & 0 \\ 0 & 0 & 0 & 0 & 0 & 0 & 0 & 0 & 0 & 0 & 0 & 0 \\ 0 & 0 & 0 & 0 & 0 & 0 & 0 & 1 & 0 & 0 & 0 & 0 \\ 0 & 0 & 0 & 0 & 0 & 0 & 0 & 0 & 0 & 0 & 0 & 0 \\ 0 & 0 & 0 & 0 & 0 & 0 & 0 & 0 & 0 & 1 & 0 & 0 \\ 0 & 0 & 0 & 0 & 0 & 0 & 0 & 0 & 0 & 0 & 0 & 0 \\ 0 & 0 & 0 & 0 & 0 & 0 & 0 & 0 & 0 & 0 & 0 & 1 \\ 0 & 0 & 0 & 0 & 0 & 0 & 0 & 0 & 0 & 0 & 0 & 0 \end{bmatrix}, \bar{B} = \begin{bmatrix} 0 & 0 & 0 & 0 \\ 0 & 0 & 0 & 0 \\ 0 & 0 & 0 & 0 \\ 0 & 0 & 0 & 0 \\ 0 & 0 & 0 & 0 \\ 1 & 0 & 0 & 0 \\ 0 & 0 & 0 & 0 \\ 0 & 1 & 0 & 0 \\ 0 & 0 & 0 & 0 \\ 0 & 0 & 1 & 0 \\ 0 & 0 & 0 & 0 \\ 0 & 0 & 0 & 1 \end{bmatrix}, \bar{C} = \begin{bmatrix} 0 \\ 0 \\ 0 \\ 0 \\ 0 \\ -9.81 \\ 0 \\ 0 \\ 0 \\ 0 \\ 0 \\ 0 \end{bmatrix} \tag{3}$$

s_i is the state vector of the quadrotor i, and $s_i = [x_i; v_{xi}; y_i; v_{yi}; z_i; v_{zi}; \phi_i; p_i; \theta_i; q_i; \psi_i; r_i]$, including the current position, speed, attitude angle and angular velocity of the quadrotor. u_i is the control input, and $u_i = [u_{zi}; u_{\phi i}; u_{\theta i}; u_{\psi i}]$. Discretize (2) and the following discrete model is obtained:

$$s_i(k+1) = As_i(k) + Bu_i(k) + C \tag{4}$$

In (4), there is

$$A = I + \bar{A}\Delta t, \quad B = \bar{B}\Delta t, \quad C = \bar{C}\Delta t \tag{5}$$

2.2 Prediction model and performance index function

The core idea of MPC is to use a known model, the current state of the system and future control sequence to predict the future output of the system. The output length is an integral multiple of the control period [[9]]. Since the future control is unknown, it is necessary to optimize the problem according to certain optimization conditions to obtain the future control sequence.

In the prediction time domain, at time k, the state sequence and control sequence of the quadrotor i domain are defined as

$$\tilde{s}_i(k) = [s_i(k+1 \mid k); s_i(k+2 \mid k); \cdots ; s_i(k+N \mid k)] \tag{6}$$

$$\tilde{u}_i(k) = [u_i(k \mid k); u_i(k+1 \mid k); \cdots ; u_i(k+N-1 \mid k)] \tag{7}$$

N is the prediction time domain length. The prediction model during flight can be expressed as

$$\tilde{s}_i(k) = Es_i(k) + F\tilde{u}_i(k) + G \tag{8}$$

In (8), there is

$$E = \begin{bmatrix} A \\ A^2 \\ \vdots \\ A^N \end{bmatrix}, F = \begin{bmatrix} B & 0 & \cdots & 0 \\ AB & B & \cdots & 0 \\ \vdots & \vdots & & 0 \\ A^{N-1}B & A^{N-2}B & \cdots & B \end{bmatrix}, G = \begin{bmatrix} I \\ A+I \\ \vdots \\ A^{N-1} + A^{N-2} + \cdots + 1 \end{bmatrix} C \tag{9}$$

The control objective of the problem is to minimize the control energy under various constraints, so in the prediction time domain, the performance index function of the quadrotors system with the control $\tilde{u}_i(k)$ can be

$$J(\tilde{u}_i(k)) = \sum_{i=1}^{N_V} \tilde{u}_i'(k) \tilde{Q} \tilde{u}_i(k) \tag{10}$$

In (10), $\tilde{Q} = I_N \otimes Q, Q = (diag(1,0,0,0))$. N_V is the number of quadrotors. Equation (10) is a standard quadratic function, and constructing the standard quadratic programming problem will facilitate the quick solving of the optimization.

2.3 Constraints

There are many constraints that must be satisfied in the formation configuration of quadrotors. This paper takes three types of constraints into consideration, including terminal state constraints, platform performance constraints, and inter-aircraft collision avoidance constraints.

(1) Terminal state constraints require that each quadrotor must reach the reference target point at the terminal time to form and maintain a specific formation, which is expressed as

$$s_i(k+N \mid k) = \theta + \Delta\theta_i \tag{11}$$

In (11), θ is reference state vector for quadrotors formation in target point, and $\Delta\theta_i$ represents the formation structure, including the relative position and velocity vector.

(2) The platform performance constraints require that the values of velocity, attitude angle and control input during flight do not exceed their limits. In order to simplify the processing, only the components in the horizontal are considered. The velocity vector is projected to different directions to obtain $\forall m \in [1, \cdots, M]$, $\forall i \in [1, \cdots, N_V]$

$$v_{xi} \cos(\frac{2m\pi}{M}) + v_{yi} \sin(\frac{2m\pi}{M}) \le v_{\max} \tag{12}$$

M is a positive integer. As for the attitude angle limits and control input limits, there are

$$-\phi_{\max} \le \phi_i \le \phi_{\max}, -\theta_{\max} \le \theta_i \le \theta_{\max}, -\psi_{\max} \le \psi_i \le \psi_{\max} \tag{13}$$

$$-u_{\phi\max} \le u_{\phi i} \le u_{\phi\max}, -u_{\theta\max} \le u_{\theta i} \le u_{\theta\max}, -u_{\psi\max} \le u_{\psi i} \le u_{\psi\max}, -u_{z\max} \le u_{zi} \le u_{z\max} \tag{14}$$

(3) Collision avoidance constraints: In the 2D case, define $pos_i(k)$ and $pos_j(k)$ as the position vector of the quadrotor i and quadrotor j at time k, and $pos_i(k) = [x_i(k); y_i(k)]$. $pos_i(k+1)$ represents the position vector of the quadrotor i at the next time. The safety zone is generally a circle whose center is on the quadrotor, with a safety distance R. The constraint of a circle is a quadratic inequality, and the quadratic programming problem with quadratic inequality constraints and non-convex optimization variables is not easy to find the optimal solution. Consider changing the safety zone from a circle constraint to a half-plane constraint, as shown in figure 1.

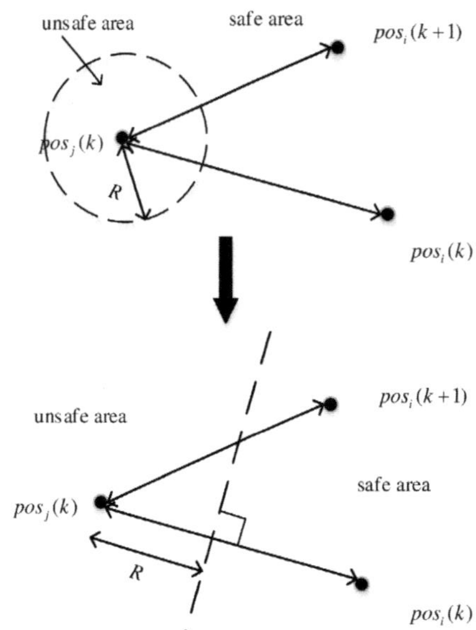

Figure 1. The collision avoidance constraints in 2D case

For the half-plane constraints in figure 1, the position of the quadrotor must satisfy

$$(pos_i'(k+1) - pos_j'(k))(pos_i(k) - pos_j(k)) \ge \| pos_i(k) - pos_j(k) \|_2 R, \ \ i, j = 1, \cdots N_V, j \ne i \tag{15}$$

Based on the above analysis, the performance index function of trajectory optimization for the quadrotors formation with collision avoidance is expressed as

$$J(\tilde{u}_i(k)) = \sum_{i=1}^{N_V} \sum_{n=0}^{N-1} u_i'(k+n \mid k) Q u_i(k+n \mid k) \tag{16}$$

Therefore, the optimization model is

$$\begin{aligned} & \min_{u_1, u_2, \cdots, u_{N_V}} \ J(\tilde{u}_i(k)) \\ & \text{s.t.} \\ & \begin{cases} h(\tilde{s}_i(k), \tilde{u}_i(k)) = 0, & \forall i = 1, 2, \cdots, N_V \\ g(\tilde{s}_i(k), \tilde{u}_i(k)) \le 0, & \forall i = 1, 2, \cdots, N_V \\ f(\tilde{s}_i(k), \tilde{s}_j(k), \tilde{u}_i(k)) \le 0, \forall i, j = 1, 2, \cdots, N_V, i \ne j \end{cases} \end{aligned} \tag{17}$$

In (17), $h(\tilde{s}_i(k), \tilde{u}_i(k))$ is the terminal state constraints, $g(\tilde{s}_i(k), \tilde{u}_i(k))$ is the platform performance constraints, and $f(\tilde{s}_i(k), \tilde{s}_j(k), \tilde{u}_i(k))$ is the collision avoidance constraints between quadrotors. The above performance index function is a quadratic function, constrained by some linear equalities and linear inequalities, and can be optimized by mature quadratic programming algorithm to obtain the optimal solution.

3 Optimal formation control algorithm

The optimization model (17) is a centralized model. As the number of quadrotors increases, the number of optimization decision variables and constraints increases, resulting in a sharp increase in computational complexity and unsatisfactory on-line computational efficiency. Moreover, for quadrotors system, there is no centralized optimization node in many cases. In order to avoid the system failure caused by centralized node's problem and reduce the scale of the optimization problem, a distributed algorithm is proposed for quadrotors.

Due to the possible model mismatch and disturbance, the actual state of the quadrotor in flight will deviate from the predicted state. In order to further improve the computational efficiency, the distributed trajectory optimization and receding horizon control are combined to carry out distributed receding horizon optimization, that is: at any time, based on the current state, plan the future trajectory over a period of time, solve the optimal control sequence. Only the first item of the control sequence is applied to the platform, and this solution process will be repeated the next time.

3.1 Model of the distributed receding horizon optimization

A communication-based distributed receding horizon optimization model is used to decompose the centralized optimization problem into a local trajectory optimization problem for each quadrotor. Specifically, for the quadrotor i, the performance index function of local optimization is

$$J_i(\tilde{u}_i(k)) = \sum_{n=0}^{N-1} u_i'(k+n \mid k) Q u_i(k+n \mid k) \tag{18}$$

The distributed optimization model is

$$\begin{aligned} &\min_{u_i} J_i(\tilde{u}_i(k)) \\ &\text{s.t.} \\ &\begin{cases} h(\tilde{s}_i(k), \tilde{u}_i(k)) = 0 \\ g(\tilde{s}_i(k), \tilde{u}_i(k)) \le 0 \\ f(\tilde{s}_i(k), \tilde{s}_j(k), \tilde{u}_i(k)) \le 0, \forall j \in N_i \end{cases} \end{aligned} \tag{19}$$

In (19), N_i is a set of the quadrotor nodes within the communication distance of the quadrotor i. From the above model, it can be seen that the optimization problem of the quadrotor i is only related to the local decision variable $\tilde{u}_i(k)$, which means the scale of the optimization problem is greatly reduced. In the process of the distributed solution, each quadrotor need to communicate with other quadrotors and exchange their solution information. The information transmitted by each quadrotor includes the state sequence and control sequence in prediction time domain.

3.2 Optimal formation control based on MPC

The cooperative collision avoidance of a quadrotors system is essentially a cooperative game problem of multi-objective and multi-player [[10]]. The Nash optimal trajectories for a quadrotors system with collision avoidance is defined as: for an arbitrary quadrotor i in the system, there is a control sequence $\tilde{u}_i^* = [u_i^*(k \mid k); \cdots; u_i^*(k+N-1 \mid k)]$, can satisfy the following inequality for any control sequence

$$J_i(\tilde{u}_i^*, \tilde{u}_{j \ne i}^*) \le J_i(\tilde{u}_i, \tilde{u}_{j \ne i}^*) \tag{20}$$

Then $\tilde{u}_i^*$ is the Nash optimal control sequence of quadrotor i for the quadrotors system.

The Nash optimal control sequence can be obtained by the following iterative process: First, at time k, each quadrotor estimates the initial control sequence and state sequence and sends them to other quadrotors through communication network. On the basis of obtaining the control sequence and state sequence of other quadrotors, each quadrotor can obtain the needed control sequence at present iteration by solving the standard quadratic programming problem. If the control sequence does not satisfy (20), it indicates that the whole system does not reach Nash equilibrium. Further iteration can make the local and global performance better until the system reaches Nash equilibrium. After that, the control sequence of each quadrotor is the Nash optimal control sequence.

4 Simulations

In order to verify the rationality of this algorithm, three unmanned aerial vehicles are taken as an example to conduct in-plane formation simulation experiments. Solve the optimal trajectories in the process of formation configuration without/with collision avoidance respectively. The simulation hardware performance is: Intel Core i5-2400 CPU, 3.10GHz, 4GB RAM. The simulation software environment is MATLAB.

The initial state vectors of three UAVs are

$$s_1(0) = [50\text{m}, 20\text{m/s}, 50\text{m}, 0\text{m/s}, 0\text{m}, 0\text{m/s}, 0\text{rad}, 0\text{rad/s}, 0\text{rad}, 0\text{rad/s}, 0\text{rad}, 0\text{rad/s}]^T$$
$$s_2(0) = [450\text{m}, -20\text{m/s}, 450\text{m}, 0\text{m/s}, 0\text{m}, 0\text{m/s}, 0\text{rad}, 0\text{rad/s}, 0\text{rad}, 0\text{rad/s}, 0\text{rad}, 0\text{rad/s}]^T$$
$$s_3(0) = [50\text{m}, 0\text{m/s}, 450\text{m}, -20\text{m/s}, 0\text{m}, 0\text{m/s}, 0\text{rad}, 0\text{rad/s}, 0\text{rad}, 0\text{rad/s}, 0\text{rad}, 0\text{rad/s}]^T$$

The formation reference target point is (322 m, 250 m, 50 m), and the positions of three quadrotors relative to the reference point are (-20 m, 20 m, 0 m), (-20 m, -20 m, 0 m), (0 m, 0 m, 0 m) limited by the formation. All three quadrotors are required to have terminal speed vector (20 m/s,0 m/s, 0 m/s), with maximum speed not exceeding 25 m/s and maximum accelerations not exceeding 3.5 m/(s^2) in flight. The safe distance between aircraft to avoid collision is set to 25 m. Sampling interval is 0.5 sec. The decision-making interval for trajectory optimization is 1 sec. The prediction time domain length is 20 sec, that is, every predictive trajectory consist of 40 sampling points. Figure 2 show the optimal formation trajectories with considering collision avoidance.

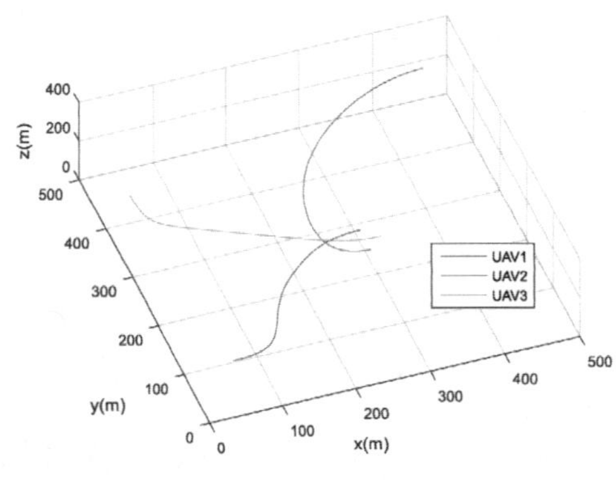

Figure 2. The optimal formation trajectories with considering collision avoidance

It can be seen that the optimal formation control can be obtained by solving the quadratic programming problem after changing the safety zone from a circle to a half-plane. Three quadrotors reach the reference target point at the designated time and constitute the required formation successfully.

Figure 3 and figure 4 show the distance between each two quadrotors in the two simulation cases. The minimum distances in flight are 11.57 m, 15.53 m and 17.05 m without considering collision avoidance. After adding the half-plane collision avoidance constraint, the distance between any two quadrotors is not less than 25 m, which means the collision avoidance is realized successfully.

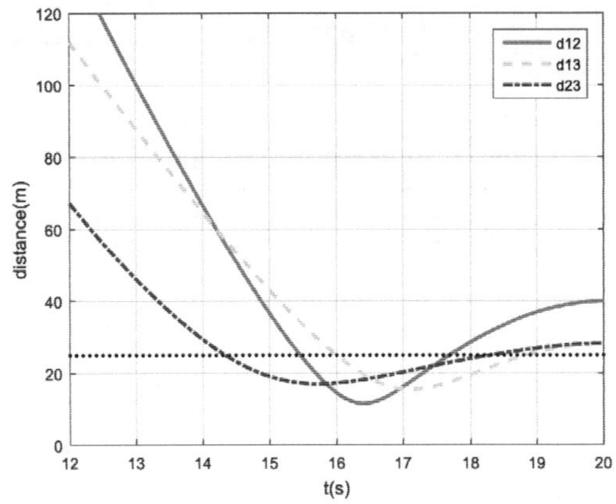

Figure 3. The distances between the quadrotors during flight without collision avoidance

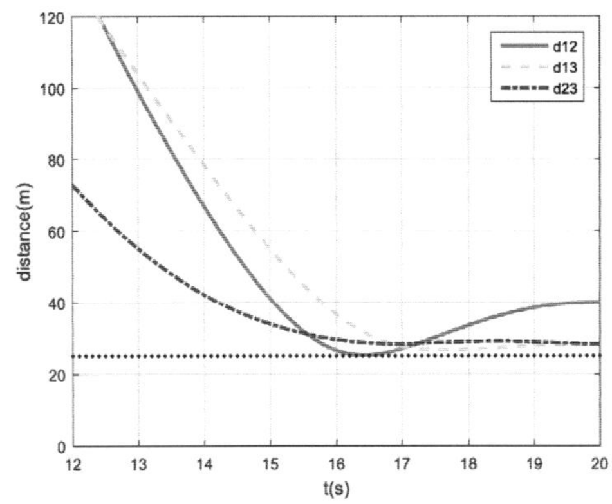

Figure 4. The distances between the quadrotors during flight with collision avoidance

In the iteration process, the performance index function value using distributed algorithm changes as shown in figure 5. The Nash optimal state is soon reached, with the performance index value being 11773.5. The first performance index value is lower because the iteration process takes the optimal control sequence of not considering collision avoidance constraints as the initial control sequence, which saves control energy needed for collision avoidance.

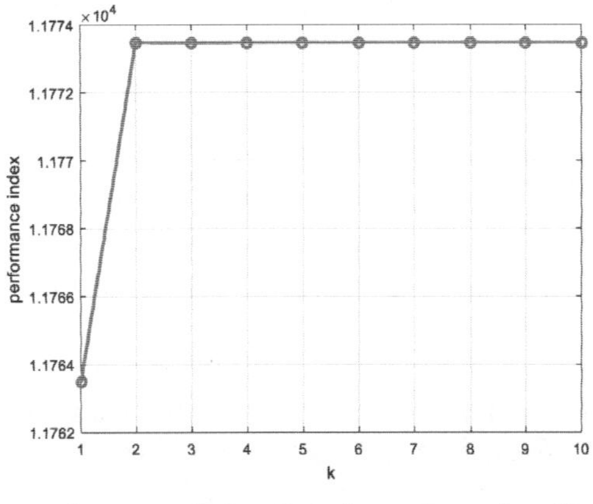

Figure 5. The performance index function value using distributed algorithm

Table 1 is the running time of centralized and distributed algorithm for three quadrotors, and table 2 is the running time of convex quadratic programming and Nonlinear programming respectively used in the centralized algorithm.

Table 1. Running time comparison

Solving Algorithm		Running Time (sec)
Centralized		6.532
Distributed	UAV1	2.676
	UAV2	2.357
	UAV3	2.793

Table 2. Running time comparison

Distributed Algorithm	Average Running Time (sec)
Convex quadratic programming	2.608
Nonlinear programming	4.174

It's obvious that when the formation scale is three, the distributed solution time is less than half of the centralized solution time. As the formation scale increases with the number of quadrotors, it is foreseeable that the computational burden of distributed algorithm will be greatly reduced compared with centralized algorithm. And from table 2, if the circular safety zone is adopted, the average solving time of nonlinear programming is 4.174s, which is much longer than that of convex quadratic programming. It shows us that using the half-plane safety zone to solve the collision avoidance is more efficient

5 Conclusions

This paper studied the optimal formation control with collision avoidance between quadrotors. Firstly, the nonlinear dynamic model of quadrotor is linearized. MPC is used to establish the prediction model of the optimization problem based on the linear dynamic model. Then, three kinds of constraints are analyzed, and among them, the safety zone constraints are transformed from the quadratic inequality of the circular zone into the linear inequality of the half-plane zone. A distributed optimization model of quadrotors formation control is presented. The model is decomposed into several local optimization models of single quadrotor, and we use the distributed receding horizon optimization to solve them. Finally, a formation configuration flight is simulated to validate that the algorithm can obtain Nash optimal formation trajectories under the condition of cooperative collision avoidance.

6. References

[1] Shen L, Niu Y and Zhu H 2013 Theories and methods of autonomous cooperative control for multiple UAVs. (Beijing: National Defense Industry Press)

[2] Liang X, Meng G and Xu Y 2018 A geometrical path planning method for unmanned aerial vehicle in 2D/3D complex environment. *Intelligent Service Robotics*, 11.3, pp 301-312

[3] Rathbun D, Kragelund S, Pongunwattana A and Capozzi B 2002 An evolution based path planning algorithm for autonomous motion of a UAV through uncertain environments. *AIAA 21st Digital Avionics System conference*, pp 551-607

[4] Pongpunwattana A 2004 Real-time planning for teams of autonomous vehicles in dynamic uncertain environments. (Washington: University of Washington)

[5] Zhu X, Yan M, Zhang C, Lin H and Qu Y 2017 UAV formation collision avoidance control method based on improved artificial potential field. *Journal of Harbin Engineering University*, 38.6, pp 961-968

[6] Richards A and How J P 2002 Aircraft trajectory planning with collision avoidance using mixed integer linear programming. *American Control Conference*, 3.3, pp 1936-1941

[7] Kamal W A, Gu D W and Postlethwaite I 2005 Real time trajectory planning for UAVs using MILP. *IEEE Conference on Decision and Control, and the European Control Conference*, pp 3381-3386

[8] Cowling I 2008 Towards autonomy of a quadrotor UAV. (Cranfield: Cranfield University)

[9] Gong J, Jiang Y and Xu W 2014 Model predictive control for self-driving vehicles. (Beijing: Beijing Institude of Technology Press)

[10] Nash 1950 The bargaining problem. Econometrica, 18.2, pp 155-162

7. Acknowledgment

This work was supported by the National Natural Science Foundation of China under Grant No.11532002.

Antlion Optimization for Bending of Arrow Honeycomb

JIANG Hengkun[1], FANG Houfei[2], HOU Yangqing[2], WU Ke[1]

[1] China Academy of Space Technology (Xi'an), Xi'an 710000, China;
[2] Shanghai YS Information Technology Co., Ltd, Shanghai 200240, China

Abstract. In view of the single curved deformation characteristics of zero Poisson's ratio (ZPR) honeycomb, Arrow honeycomb with ZPR is applied to the reflecting surface of the spatial parabolic cylinder antenna in this paper. First of all, the loading method for cylinder formation of honeycomb with ZPR is analysed. Then, Abaqus analysis, surface fitting and precision calculation are integrated into the objective function. Antlion Optimization (ALO) algorithm is used to optimize the RMS error of the deformed surface. The analysis results show that it is feasible to apply the honeycomb with ZPR to the reflector of cylindrical antenna, and the accuracy of deformed profile is improved by 45.5% through the ALO algorithm, which proves the effectiveness and efficiency of the ALO algorithm to solve the problem with unknown initial value. The research results are certainly significant for the optimization design of honeycomb with ZPR and the application of ALO algorithm.

1. Introduction

The structure of parabolic cylinder antenna has been preliminarily studied in the late 20th century and further developed in the early 21st century. In 2003, John k. Lin et al. [1] demonstrated the feasibility of technology which meet the precision requirements through the development of the membrane structure and the precision test of the molded surface. In 2004, British O.Soykasap et al. [2] conducted the research of the bidirectional deployable elastic shell parabolic cylinder antenna. In 2004, Christopher g. Meyer et al. [3] discovered the influence of boundary support conditions and gravity acceleration on the reflector of a thin film parabolic cylinder in orbit. In 2007, Eastwood Im et al. [4] designed a half-size prototype PR-2 parabolic cylinder antenna. In 2010, Jack Leifer et al. [5] studied the shape variation of parabolic films under different gravity conditions.

At the early stage of the research, antenna reflector is made of metal materials which are heavy and stiff, and the membrane is easy to fold to influence the precision. All of these limit the development of diameter and the scope of applications [6,7].Compared with other traditional materials, honeycomb is characterized by light weight, high strength, stiffness and shear modulus, shock resistance, heat insulation, designability and stress concentration reduction. The deformation of the structure with ZPR will not produce coupling in the orthogonal direction. Similarly, when the structure with ZPR is bending, it usually appears as cylinder [8].Honeycombs such as Silicomb [9, 10], Chevron[11] and Accordion[12] which usually have a Poisson's ratio of zero. ZPR structures under external bending deform like a single curve, as shown in Figure 1, which makes the structure more suitable for the single curvature bending deformation [8, 13].

In this paper, Arrow honeycomb will be applied to fit parabolic antenna reflector to demonstrate the feasibility of the ZPR honeycombs used for parabolic antenna and the effectiveness of ALO algorithm, and accuracy will be optimized through ALO algorithm.

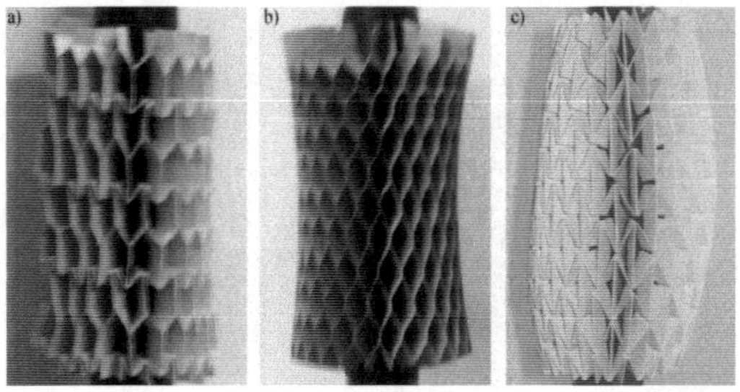

Figure 1. Images of ''tubes'' formed from honeycombs with
(a) ZPR-single curve (b) PPR-concave (c) NPR-convex

2. Cylinder formation

In this paper, the ZPR honeycomb Arrow is selected as the research object and the ideal surface is parabolic cylinder. The model (122*100*2.5(mm)) of honeycomb panel is established by UG-NX, as shown in Figure 2. ABAQUS is adopted in the analysis and calculation, and the element is S4R.

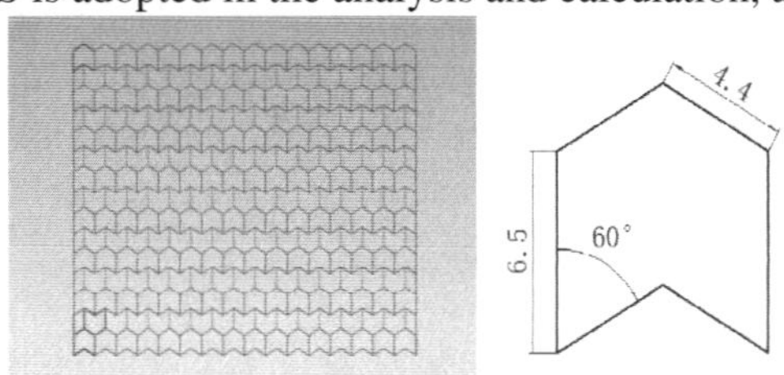

Figure 2. Geometry of Arrow honeycomb

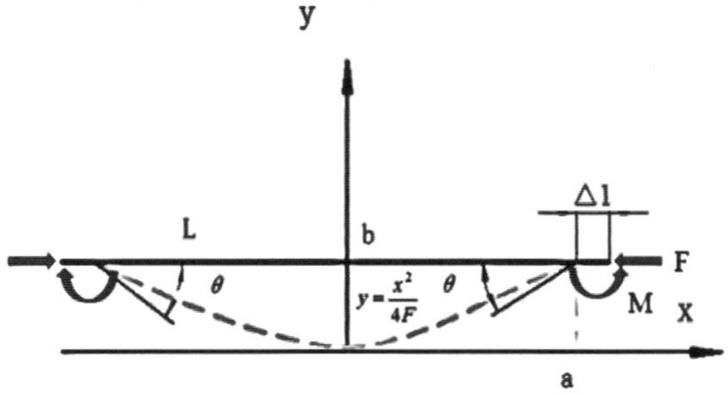

Figure 3. Illustration of loading method

Assuming that the length of the honeycomb panel will not change before and after deformation, symmetric rotation and displacement loads are applied to both sides of the panel, as shown in Figure 3. The parameters in the figure are expressed as follows:

$$L = \int_{-a}^{a} \sqrt{1 + y'^2}\, d \tag{1}$$

$$\Delta l = \frac{L}{2} - a \tag{2}$$

$$tan\theta = y'(a) \tag{3}$$

where L is the original length of honeycomb panel, y is the target curve.

In fact, the influence of thickness cannot be ignored due to the dimension limitation of the panel. So the Eqs.(2) should be amended as follows:

$$\Delta l = \frac{L}{2} - a + \frac{h}{2} \times cos\theta \tag{4}$$

where h is the thickness of panel.

FE analysis is carried out according to the calculated loads. The plot contour shows the consistency of the deformation along width direction, as shown in Figure 4. The bending deflection proves the effectiveness of the parabolic cylinder and the feasibility of ignoring the influence of width. The value of required loads is obtained through geometric calculations above. Although the influence of thickness of the panel has been considered, the precision of the cylinder isn't the highest because of the nonlinear deformation during bending process. So the value of loads need to be further optimized.

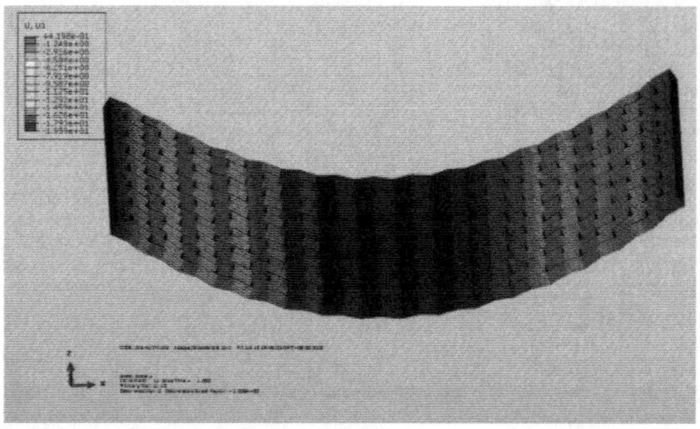

Figure 4. FE simulation of honeycomb deformation

3. Overview of Antlion Optimization algorithm

The Antlion Optimization algorithm is a novel optimization algorithm which mimics the hunting process of antlions in nature. It was first proposed by Australian scholar Mirjalili [14] in 2015. An antlion larvae digs a cone-shaped pit by rotating in the sand as it drills down [15-17]. It hides under the bottom of the funnel and waits for insects to be trapped [18]. When ants or other insects climb into traps, they slide down because of loose sands, antlion will cast sands out continuously. Then antlions with large jaws will clamp their prey and eat it under the sand. Then the corpse will be thrown out of the trap for the next hunting [19].

3.1 Operators of the ALO algorithm

The ALO algorithm mimics the interaction between antlion and ant in the process of predation. Ants are required to move in the search space, while antlions capture them and become more adaptable by building traps for prey. To model the interaction process, a random walk of ants is simulated as follows:

$$X(t) = [0, cumsum(2r(t_1) - 1), cumsum(2r(t_{12}) - 1), \cdots, cumsum(2r(t_n) - 1)] \quad (5)$$

where n is the maximum number of iterations, t is the step of random walk, *cumsum* calculates the cumulative sum, and $r(t)$ is expressed as follows:

$$r(t) = \begin{cases} 1 \; if \; rand > 0.5 \\ 0 \; if \; rand \leq 0.5 \end{cases} \quad (6)$$

In the optimization process, the location of ants are stored in following matrix:

$$M_{Ant} = \begin{matrix} Ant_{1,1} & Ant_{1,2} & \cdots & Ant_{1,d} \\ Ant_{2,1} & Ant_{2,2} & \cdots & Ant_{2,d} \\ \vdots & \vdots & \ddots & \vdots \\ Ant_{n,1} & Ant_{n,2} & \cdots & Ant_{n,d} \end{matrix} \quad (7)$$

where $Ant_{i,j}$ is the *j-th* value of the *i-th* ant, n is the number of ants, and d is the number of variables. Each ant's location represents an optimal parameter value. The fitness matrix of ants is calculated by using the objective function:

$$M_{OA} = \begin{bmatrix} F([Ant_{1,1}, Ant_{1,2}, \cdots, Ant_{1,d}]) \\ F([Ant_{2,1}, Ant_{2,2}, \cdots, Ant_{2,d}]) \\ \vdots \\ F([Ant_{n,1}, Ant_{n,2}, \cdots, Ant_{n,d}]) \end{bmatrix} \quad (8)$$

where F is the objective function. Similarly, the antlions are hiding in the search space, the location matrix and fitness matrix can also be expressed as follows:

$$M_{AntL} = \begin{matrix} AntL_{1,1} & AntL_{1,2} & \cdots & AntL_{1,d} \\ AntL_{2,1} & AntL_{2,2} & \cdots & AntL_{2,d} \\ \vdots & \vdots & \ddots & \vdots \\ AntL_{n,1} & AntL_{n,2} & \cdots & AntL_{n,d} \end{matrix}$$

$$M_{OAL} = \begin{bmatrix} F([AntL_{1,1}, AntL_{1,2}, \cdots, AntL_{1,d}]) \\ F([AntL_{2,1}, AntL_{2,2}, \cdots, AntL_{2,d}]) \\ \vdots \\ F([AntL_{n,1}, AntL_{n,2}, \cdots, AntL_{n,d}]) \end{bmatrix} \quad (9)$$

where $AntL_{i,j}$ is the *j-th* value of the *i-th* antlion, n is the number of antlions, d is the number of variables, and F is the objective function.

3.1.1 Random walks of ants. Ants change their location randomly through Eqs. (5). In order to ensure the ants walk inside the search space, they need to be normalized using following equation:

$$X_i^t = \frac{(X_i^t - a_i) \times (d_i^t - c_i^t)}{(b_i - a_i)} + c_i^t \quad (10)$$

where a_i is the minimum of random walk of *i-th* variable, b_i is the maximum of random walk of *i-th* variable, c_i^t is the minimum of random walk of *i-th* variable at *t-th* iteration, d_i^t is the maximum of random walk of *i-th* variable at *t-th* iteration.

3.1.2 Trapping in pits. Random walks of ants are affected by traps of antlions. This behavior can be described as follows:

$$c_i^t = Antlion_j^t + c^t \quad (11)$$

$$d_i^t = Antlion_j^t + d^t \quad (12)$$

where c^t is the minimum of all variables at *t-th* iteration; d^t is the maximum of all variables at *t-th* iteration; c_i^t is the minimum of all variables for *i-th* ant; d_i^t is the maximum of all variables for *i-th* ant; $Antlion_j^t$ is the position of *j-th* antlion at *t-th* iteration.

3.1.3 Building trap. Antlion optimization algorithm mimics the behavior of the antlion predation in nature by a roulette wheel. Antlion is selected based on their fitness to optimization progress. The mechanism gives more chances to the fitter antlions for catching ants.

3.1.4 Sliding ants towards antlions. Once the antlion realizes that an ant is in the trap, it will throw sands out of the center of the trap to make the ant trying to escape slide down slowly. The following equations are presented in this regard:

$$c^t = \frac{c^t}{I} \quad (13)$$

$$d^t = \frac{d^t}{I} \quad (14)$$

where I is the ratio. In Eqs. (13) and (14), $I = 10^W \times \frac{t}{T}$. Where t is the current iteration, T is the maximum number of iteration, and W is a constant based on the current iteration as follows:

$$W = \begin{cases} 2 & t > 0.1T \\ 3 & t > 0.5T \\ 4 & t > 0.75T \\ 5 & t > 0.9T \\ 6 & t > 0.95T \end{cases}$$

Basically, the constant W adjust the accuracy level of optimization algorithm.

3.1.5 Catching prey and re-building the trap. If the ant has a better fitness than the selected antlion then it changes its position to the largest position of the hunted ant to improve its chance of catching new one. The following equation is illustrated in this regard:

$$Antlion_j^t = Ant_i^t \ if \ F(Ant_i^t) > F(Antlion_j^t) \quad (15)$$

where t shows the current iteration, $Antlion_j^t$ shows the position of *j-th* antlion at *t-th* iteration, Ant_i^t indicates the position of *i-th* ant at *t-th* iteration.

3.1.6 Elitism. In this algorithm, the best antlion obtained so far in each iteration is saved as an elite. Since the elite is the fittest antlion, it can affect all the ants during the iterations. Thus, it is assumed that every ant randomly walks around a chosen antlion by the roulette wheel and the elite simultaneously as follows:

$$Ant_i^t = \frac{R_A^t + R_E^t}{2} \tag{16}$$

where R_A^t is the random walk around the antlion selected by roulette wheel at *t-th* iteration, R_E^t is the random walk around the elite at *t-th* iteration.

4 Results

As mentioned in section 1, the loads are obtained by geometric calculation. When the thickness of the panel is not taken into account (Eqs. (2) and (3)), the results are:

θ= 0.616rad, Δl = 4.4mm.

Whereas, if the thickness of the honeycomb panel is taken into account, the results derived from Eqs. (4) are: θ=0.616 rad, Δl =5.12 mm.

The loads are respectively applied to the honeycomb panel in ABAQUS, and then the RMS error of the deformed surface is calculated. The results are shown in Figure 5.

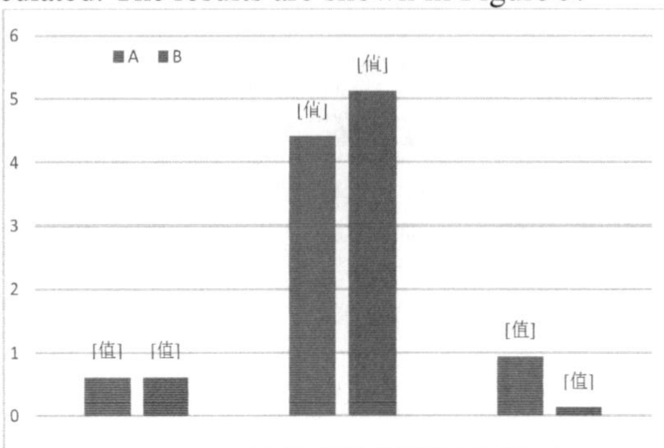

Figure 5. Result of geometric loads

In the figure, series B represents that the thickness of the panel is considered, while series A represents the opposite situation. It can be seen that in the geometric calculation, the thickness only influences the displacement. When the loads derived from Eqs. (4) are applied, the precision of the surface is greatly improved, and the RMS error is reduced by 86.0% in comparison. That means when the thickness of the panel is not negligible compared with its dimension, the thickness must be considered in the loads calculation.

In this paper, the Antlion Optimization algorithm (ALO) is applied to optimize the precision of the deformed surface, then to validate the effectiveness and excellent performance of ALO. The algorithm iteration process is shown in Figure 6.

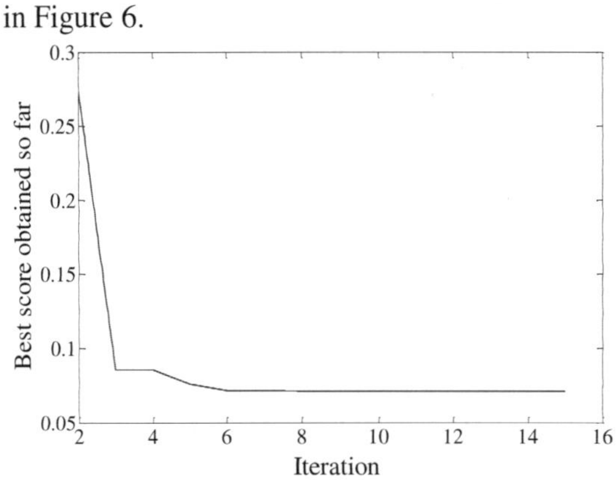

Figure 6. Convergence curve of ALO algorithm

The results show that the Arrow honeycomb with ZPR can form a parabolic cylinder with a certain precision. Meanwhile, when ALO algorithm is used to solve the optimization problem, the iteration tends to be converged quickly. The results are shown as follows:

θ=0.5883 rad; Δl=5.0530 mm; RMS=0.0713mm.

The results show that ALO algorithm is effective. Compared with geometric deformation results, RMS error decreases by about 45.5%.The ALO algorithm can solve the optimization problem efficiently and quickly. The variation trend of its target value will not fluctuate greatly during the iteration, which also proves the high computational efficiency of ALO.
Figure 7.shows the fitness of each ant during the iteration. It can be seen that as the iteration proceeding, all ants are gradually approaching, which indicates that the random walk of all ants is decreasing. This figure proves the convergence of ALO algorithm. The position of the ant with the highest fitness in each iteration will accordingly update the position of the elite.

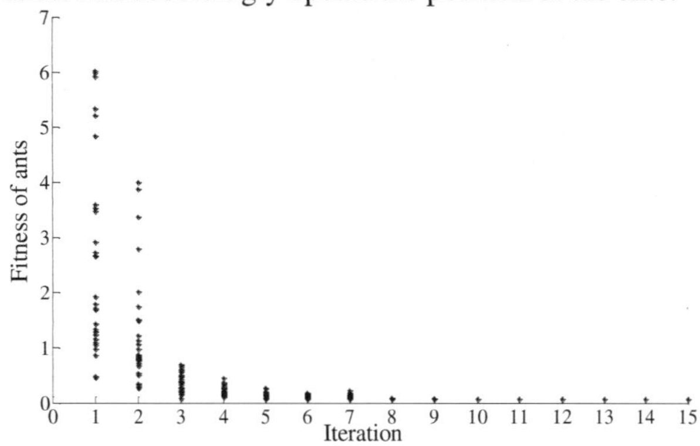

Figure 7. Search history of ALO algorithm

5. Conclusion

In this paper, the Arrow honeycomb with ZPR is applied to form a parabolic cylinder by means of the combination of rotation and displacement. The RMS error is further reduced by using the ALO algorithm, which further improves the precision of the shaped surface of the parabolic cylinder. The optimization results fully prove the feasibility of applying the Arrow honeycomb to the parabolic cylinder antenna to make a high-precision reflector. Meanwhile, the ALO algorithm adopted in the optimization process also shows excellent performance, which can avoid the local optimal result very well. The algorithm itself and the target problem are independent from each other, and are not related to the initial value, which proves the feasibility of ALO algorithm for solving various optimization problems in reality.

References

[1] K.L. John, H.S. George.Ⅲ, E.S. Stephen, et al. Advanced precipitation radar antenna singly curved parabolic antenna reflector development [C]//44th AIAA/ ASME/ASCE/AHS Structures, Structural Dynamics, and Materials Conference, Norfolk:7-10 (2003).
[2] O. Sovkasap, A.M. Watt, S. Pellegrino. New deployable reflector concept[C]// 45th AIAA/ASME/ ASCE/AHS/ASC structures, Structural Dynamics & Materials Conference, Palm Springs: 19-22 (2004).
[3] G. Greschik, M.M. Mikulas, A. Palisoc. Torus-less inflated membrane reflector with an exact parabolic center [J]. *AIAA Journal*, 42.12: 2579-2584 (2004).
[4] E. Im, S.L. Durden. Next-generation spaceborne precipitation radar instrument concepts and technologies[C] // 45th AIAA Aerospace Sciences Meeting and Exhibit, Reno, Reno: 8-11 (2007).
[5] J. Leifer, D.C. Jones, A.M. Cook. Gravity-induced wrinkling in subscale, singly curved parabolic gossamer membrane [J]. *Journal of Spacecraft & Rockets*, 47.1:214-219 (2012).
[6] YUAN Sichen, YANG Bingen, FANG Houfei. Form-finding of large deployable mesh reflectors with elastic deformations of supporting structures[C]// AIAA Spacecraft Structures Conference, AIAA SciTech Forum: 1198 (2018).
[7] SHI Hang, YUAN Sichen, and YANG Bingen. New methodology of surface mesh geometry design for deployable mesh reflectors [J], *Journal of Spacecraft and Rockets*, 55.2: 266-281 (2018).

[8] J.N. Grima, L. Oliveri, D. Attard, et al. Hexagonal honeycombs with zero Poisson's ratios and enhanced stiffness [J]. *Advanced Engineering Materials*, 12.9:855–862 (2010).
[9] R.M. Neville, A. Monti, K. Hazra, et al. Transverse stiffness and strength of Kirigami zero-ν PEEK honeycombs [J]. *Composite Structures*, 114.1:30-40(2014).
[10] F. Scarpa, Y. Chen, C. Remillat, et al. Curved Kirigami SILICOMB cellular structures with zero Poisson's ratio for large deformations and morphing [J]. *Journal of Intelligent Material Systems & Structures*, 25.6:731-743(2014).
[11] J.N. Grima, L. Oliveri, D. Attard, et al. Hexagonal honeycombs with zero Poisson's ratios and enhanced stiffness [J]. *Advanced Engineering Materials*, 12.9:855-862(2010).
[12] R. PAINTER, C.D.L. Remillat, F. Scarpa, et al. A morphing mechanical and electromagnetic antenna reflector concept based on curved corrugated structures [C]// ICAST 25th International Conference on Adaptive Structures and Technologies, Hague (2014).
[13] K.R. Olympio, F. Gandhi. Zero Poisson's ratio cellular honeycombs for flex skins undergoing one-dimensional morphing [J]. *Journal of Intelligent Materials Systems & Structure*, 21.21:1737-1753 (2010).
[14] S. Mirjalili, The ant lion optimizer [J].*Advanced in Engineering Software*, 83.C:80-98(2015).
[15] M.M. Nischal, S. Mehta. Optimal load dispatch using ant lion optimization [J]. *Journal of Engineering Research and Applications*, 5.8:10-19 (2015).
[16] I. Scharf, A. Subach, O. Ovadia. Foraging behaviour and habitat selection in pit building antlion larvae in constant light or dark conditions[J].*Animal Behaviour*,76.6:2049-2057 (2008).
[17] D. Griffiths, Pit construction by ant-lion larvae: A Cost-Benefit Analysis [J]. Journal of Animal Ecology, 55.1:39-57(1986).
[18] J. Goodenough, B. Mcguire, E. Jakob, Perspectives on animal behavior[J], *Wiley & Sons*,45.95(1993).
[19] I. Scharf, O. Ovadia. Factors influencing site abandonment and site selection in a sit-and-wait predator: a review of pit-Building antlion larvae [J]. *Journal of Insect Behavior*, 19.2:197-218(2006).

Research and Implementation of Key Techniques for Indoor Movement Object Trajectory Prediction

Ping Zhang[1], Qianqian Jiang[2], Ruilin Zhang[2] and Bo Li[2*]

[1]The Second Research Institute of CAAC, South No.17 Second Ring Road, 610041 Chengdu, Sichuan, P.R. China
[2]School of Aeronautics and Astronautics, University of Electronic Science and Technology of China, No.2006 Xiyuan Ave, West Hi-Tech Zone, 611731 Chengdu, Sichuan, P.R. China

libo@uestc.edu.cn

Abstract. With the popularity of smart mobile terminals and Wi-Fi signals, people's demand for indoor location services has also increased. However, in the indoor space, GPS positioning is inaccurate, and the Wi-Fi signal may also have signal instability even no signal in some areas. This paper proposes a prediction method based on improved HMM model combined with historical trajectory clustering. The experimental results on UJIIndoorLoc data set show that the predicting trajectories can greatly improve the real-time performance of location services and the proposed method has great improvement in accuracy and scalability comparing with another model.

1. Introduction

With the rapid development of wireless interconnect technology (Wi-Fi), indoor positioning technology and smart mobile terminals, Wi-Fi based indoor location services have been widely used. Although Wi-Fi technology can provide location information for mobile devices, if there are large indoor environments with a high population density, it is difficult to accurately obtain the position of the moving object, so it is necessary to predict the position to which the moving object may move in the future. In order to satisfy the user experience, it is necessary to improve the real-time positioning of indoor positioning online. At present, some achievements have been made for the research of indoor mobile object trajectory prediction. The main schemes are as follows:

(1) establishing an efficient index for trajectory data of indoor moving objects in order to query quickly [1];

(2) combining the features of indoor space model, the trajectory data is projected and converted, and then compressed. The aim is to save storage space [2].

Some scholars use spatial density clustering such as DBSCAN algorithm [3] and the STING algorithm based on grid analysis [4] to excavate the activity characteristics of moving objects according to the historical trajectory of mobile objects; Song et al. proposed a state-based moving object model [5], using Markov model performs state changes. Shaojie Qiao segmented the trajectory by using segmented segment information as the hidden state of the Markov model [6], and then performed position prediction. At the same time, he proposed improved parameters on the basis of the original Markov model to adapt to trajectory prediction methods. However, these methods are based on the classical HMM model, the classical HMM model method for the location prediction of indoor moving objects when there is a state of stay may lead to prediction failure.

IOP Conf. Series: Journal of Physics: Conf. Series **1215** (2019) 012020 doi:10.1088/1742-6596/1215/1/012020

Meantime, the indoor space contains a large number of spatial entity elements such as walkway elements, corridors, and elevators [7]. Therefore, it is necessary to construct an efficient model to express the interior space, which is the basis for improving all related technologies of indoor positioning. At present, methods for modelling indoor space mainly include object-based feature model [8], geometric model method [9] and symbol model method [10]. This paper mainly uses the symbol model to represent the indoor space model. Therefore, the indoor space model needs to be gridded.

In view of the shortcomings of the above methods, this paper proposes a prediction method for indoor moving trajectory data based on HMM model to achieve trajectory prediction of indoor moving objects. The grid track time series is formed and based on the matching moving trajectory data of Wi-Fi signal. The historical track data is mined and analyzed, and the trajectory location point is predicted with HMM model, and then the trajectory path sequence is compressed. This prediction can solve the problem of prediction failure.

The main work of this paper is as follows:

Firstly, the space model is divided into meshes, and then the trajectory data is projected on the model to generate the grid trajectory sequence according to the spatial model. The generated grid locus sequence is pre-processed and the historical grid locus eigenvalues are extracted and clustered to form an important location cluster database.

Secondly, the HMM model is used to predict the grid trajectory sequence according to the clustering database using the clustered cluster object as the hidden state of HMM. At the same time, the HMM model is used to generate the most probable hidden state sequence corresponding to the grid time series, in other words, the grid trajectory sequence represented by the cluster sequence, the grid trajectory sequence is compressed, and the compressed code table is constructed. Optimize the trajectory database of moving objects.

Finally, a simulation platform based on Canvas and NODE is established to verify the accuracy of trajectory location prediction and compression efficiency, which provides a new way of thinking and method for indoor trajectory prediction.

2. Framework

Prediction model based on HMM is divided into five modules, as shown in Figure1.

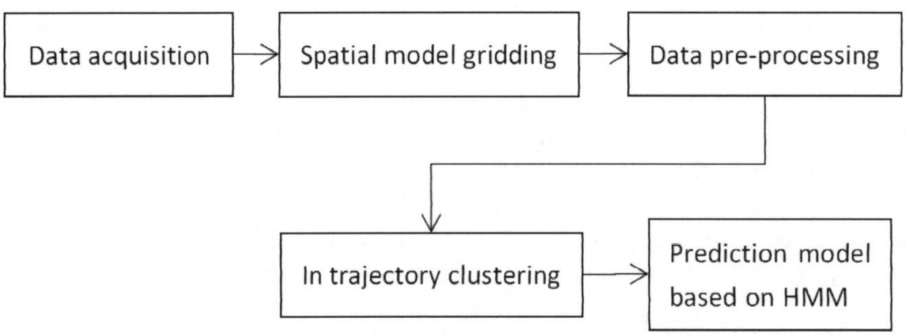

Figure 1. Trajectory prediction model

Data acquisition collects trajectory data of users by client devices. The spatial model is based on the historical trajectory data to calculate the unit edge length and the grid division. Data pre-processing module filter and formats data which projected by the grid space model and provides reasonable data for trajectory clustering. In trajectory clustering module, we identify trajectory clusters based on the historical data of users to prepare for trajectory prediction model based on HMM. Trajectory prediction module predicts the next trajectory according to the user's current location. In this paper, the classical HMM model is improved by combining the state residence time, and the problem of predicting failure in the classical HMM model is solved. This paper mainly focuses on trajectory data in an indoor environment where the signal is unstable pre-processing, trajectory clustering, and trajectory prediction algorithms.

3. Pre-processing and clustering

3.1. Definition

In order to make a better description, the related definitions are given as follows before data processing.

Definition 1 Moving locus. Describes a single-track point of an indoor moving object whose data format can be represented by an object (1).

$$p = \{ p \mid p = (x, y, time, angle, level) \} \tag{1}$$

In this formula,(x, y) is latitude and longitude of the trajectory point; angle is deflection Angle; Level is barometer value; time is the time stamp of the collection time is the Unix timestamp.

Definition 2 Trajectory sequence. A sequence composed of a single-track point in chronological order.

$$originalTrail = \{ < p_1, ..., p_n > \mid \forall p_i, p_j (\exists i < j, p_i.time < p_j.time) \} \tag{2}$$

In this formula p_i is a single-track point, N is the number of trace points contained in a trajectory.

Definition 3 Unit length. Based on the historical trajectory data, the shortest distance between two adjacent non-repeat locus points in the interior space.

Definition 4 grid sequence. According to the length of the unit, the interior space model is meshed to generate the grid interior space model, and the points in a grid can be represented by its centre. For the sake of simplicity, take a square to represent a grid.

According to the grid model, the trajectory data is projected and the grid sequence (3) is generated.

$$gridSequence = \{ g_1, g_2, ..., g_m \}, g_i = \{ id, stoptime, angle, level \} \tag{3}$$

In this formula, id is grid number. Stop time is the duration of stay; angle is interval deviation angle; level is floor;

Definition 4 stop time. If a grid sequence is given such as $< ..g_j, g_{i1}, g_{i2}, ..., g_{ik}, ... >$, g_{i1} and g_{ik} of them have the same grid number, different time of grid track points, g_j, g_{i1} is a grid of different grid number track points, time Settings, such as type (4).

$$stoptime = \begin{cases} 0, g_j[id] \neq g_{i1}[id] \,||\, g_{ik}[time] - g_{i1}[time] < \delta_t \\ g_{ik}[time] - g_{i1}[time], g_{ik}[id] = g_{i1}[id] \end{cases} \tag{4}$$

In this formula, δ_t is the time threshold. When the time interval between the continuous mesh point is greater than that, it is considered that the grid point has a stop state.

3.2. Data pre-processing

Users' location information can be depicted with grid sequence. In this paper, trajectory data is pre-treated according to the above definition.

Step1 according to the measurement value of the internal barometer of the MEMS inertial sensor and air pressure formula, the floor of the track point is calculated. The angle of deflection is obtained through the angle between two trajectory points. Traverse the generated grid sequence, calculate the dwell time of grid point according to definition 4.

Step2 the generated grid trajectory sequences are complemented by redundancy and grid trajectory sequences.

Step3 the angle in the grid trajectory sequence formats. The Angle interval is divided into eight sub-intervals that do not intersect. The statistics are mapped to the number of angles of each sub-interval and the angle information object is generated.

3.3. Trajectory clustering

Before the trajectory of clustering, this paper first presents the following definition

Definition 5 corner. An angle that belongs to the range of $[\frac{\pi}{4},\frac{7\pi}{4}]$

Definition 6. important location. The grid point meets the following condition:

(1) Stop time is greater than time threshold.

(2) The number of corners is greater than threshold.

(3) The number of track points through the grid point is greater than threshold.

In this paper, based on the definition of important points, the historical trajectory data is clustered by using DBSCAN algorithm, and finally different clusters are formed.

4. Trajectory prediction

In this paper, the trajectories were predicted by combining trajectory clustering and HMM model. Before trajectory prediction, this paper first presents the following definition.

Definition 7 Hidden states. In this paper, the trace cluster is used as a hidden state.

Definition 8 Observed sequence. The Observed sequence is grid Sequence (Def 4)

Definition 9 Initial state probability matrix.

$$\lambda = [0,..0,\frac{1}{k},0,\frac{1}{k},...,0]^T, \tag{5}$$

Definition 10 Initial state transfer probability matrix A.

$$A=\begin{bmatrix} a_{11} & a_{12} & \cdots & a_{1m} \\ a_{12} & \cdots & \cdots & a_{12} \\ \cdots & a_{ij} & a_{ii} & \cdots \\ a_{m1} & a_{m2} & \cdots & a_{mm} \end{bmatrix}, \sum_{j=1}^{m} a_{ij} = 1 \tag{6}$$

$$a_{ij} = \left[\beta_a^{(ij)}, \beta_{st}^{(ij)}, \beta_{dis}^{(ij)}\right] \begin{bmatrix} \omega_a \\ \omega_{st} \\ \omega_{dis} \end{bmatrix} \tag{7}$$

In this formula, a_{ij} is the state transfer probability of hidden state C_i and C_j; $\beta_a^{(ij)}$ is the characteristic value of hidden state C_i and C_j; $\beta_{st}^{(ij)}$ is the state transfer probability of hidden state C_i and C_j; $\beta_{dis}^{(ij)}$ is the distance of C_i and C_j; $\begin{bmatrix} \omega_a \\ \omega_{st} \\ \omega_{dis} \end{bmatrix}$ is the is the weight of the three eigenvalues.

Definition 11 confusion matrix B. The confusion matrix B is like equation (8), which is the transfer probability matrix of hidden state and grid sequence, m is the number of hidden states, n is the state observation length, b_{ij} Is the state transfer probability of the observed sequence and the hidden state, b_{ij} is derived from b'_{ij}.

$$B=\begin{bmatrix} b_{11} & b_{12} & \dots & b_{1n} \\ b_{21} & \dots & \dots & b_{2n} \\ \dots & b_{ij} & b_{ii} & \dots \\ b_{m1} & b_{m2} & \dots & b_{mn} \end{bmatrix}, \sum_{j=1}^{n} b_{ij} = 1 \tag{8}$$

$$b_{ij}^{'} = \begin{cases} 1, g_j \in C_i \\ \left[\gamma_a^{(ij)}, \gamma_{dis}^{(ij)}\right]\begin{bmatrix} w_a \\ w_{dis} \end{bmatrix}, g_j \notin C_i \end{cases} \tag{9}$$

In this formula, b_{ij} is the state transfer probability of hidden state C_i and observed sequence g_j; $\gamma_a^{(ij)}$ is the normalized angular eigenvalues; $\gamma_{dis}^{(ij)}$ is the distance of C_i and g_j; $\begin{bmatrix} w_a \\ w_{dis} \end{bmatrix}$ is the is the weight of the two eigenvalues.

In this paper, the moving object mesh sequence is segmented by sliding window. Because of the moving objects movement patterns of randomness, can be thought of in the mobile mode, every point and turning point is the beginning of a new model, the input to the HMM model, which accords with the HMM model of short-term memory, and can avoid the excessive fitting model training. Based on the segmented processing of the mobile object grid sequence, the specific trajectory prediction steps are as follows.

Step1 Segmenting the grid track points based on important points. Step2 The HMM initial model is calculated according to definitions 7 to 11. Step3 use the forward-backward algorithm and the Baum-Welch algorithm training model. Because there may be multiple consecutive grid sequence trajectories projected into the same cluster, generating successive cluster sequences. If the classical HMM definition of the state transfer matrix is adopted, the probability of the hidden state rotation is zero, which leads to the failure of prediction.

In the training model phase, when the rotation probability of matrix A is 0, use a linear smoothing method.as equation (10), (11).

$$\varepsilon_a = \text{Logisitic}(stoptime(C_t)), 0 < \varepsilon_a \le \max(a_{tz}) \tag{10}$$

$$stoptime(C_t) = \max_{1 \le v \le k}(stoptime(g_{tv})) \tag{11}$$

In those formulas, ε_a is the value of the logarithmic trajectory of the maximum of stop time in the cluster. a_{tz} is the value of row t, column z in the state transition matrix A.

Step 4 Based on the HMM model trained in Step3, the Viterbi algorithm is used for prediction. Step 5 Based on the implicit state sequence group returned in Step4, the current shift Angle is calculated based on the mobile terminal MEMS inertial navigation unit, and the Angle of the deflection Angle is standardized. Gets the most likely implicit state in this Angle interval, returning the nearest grid point in the cluster that is the most likely implicit state representation to the current grid location.

5. Experimental results and analysis

The experiment adopts the data-set that public data provided by UJIIndoorLoc. According to the data-set, the grid side length is 1m. The error threshold accuracy calculation method defined in this section is calculated as follows.

Definition 12 The acceptance value of the prediction position $H(i)$. When the prediction position is less than the actual position, the European distance (dis(i)) is less than the error threshold and the value is 1, otherwise it is 0.

$$H(i)=\begin{cases}1, dis(i)\le \delta_{mis}\\ 0, dis(i)>\delta_{mis}\end{cases} \tag{12}$$

Definition 13 Error threshold accuracy (AC). Denotes the accuracy of the prediction position within the acceptance range of the pre-set error threshold.

$$AC=\frac{\sum_{i=1}^{N}H(i)}{N} \tag{13}$$

In this formula, N is the predicted points.

Based on the offline and Wi-Fi signal environment, we compared the improved HMM model, the classical HMM model, the probability method and the HMM (HMM-KF) prediction algorithm based on Kalman filter. Based on the method of calculating prediction accuracy of 13, we compare and test three algorithms with different prediction error thresholds. Figure2 shows the test results. When the improved HMM algorithm is below 1.5m in error value, it is 20% higher than the HMM-KF, and is more than 60% higher than the probability method. In addition, the improved HMM algorithm is more accurate than the other two algorithms.

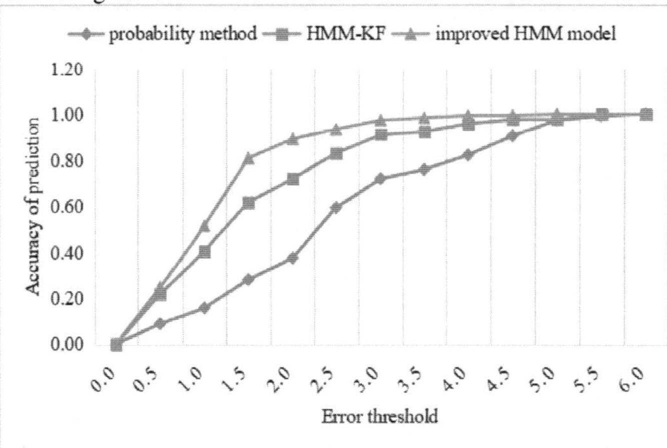

Figure 2. The comparison of prediction accuracy with different prediction error thresholds.

6. Conclusion

This paper presents a scheme to analyse the indoor moving trajectory data, build the interior space, and predict trajectories based on trajectory cluster. The prediction model improves the classical HMM model through the time-dependent characteristic value of historical trajectory. Experiment shows that the method proposed has great improvement in accuracy and scalability as compared traditional methods. Besides, the scheme has a high practical value with its flexible application.

7. References

[1] Dequn Z, Chuanjie X and Tao P 2017 Semantic cell oriented indoor moving objects index based on Hbase *Journal of Geo-Information Science*

[2] Hao L, Shukui Z and Penghui S 2018 Trajectory compression algorithm for adaptive parameters *Computer application research 35* 685-688,716

[3] Ester M, Kriegel H P and Xu X 1996 A density-based algorithm for discovering clusters a density-based algorithm for discovering clusters in large spatial databases with noise *International Conference on Knowledge Discovery & Data Mining* **1996** 226-231

[4] Amini A, Wah T Y and Saboohi H 2014 On density-based data streams clustering algorithms: a survey *Journal of Computer Science & Technology* **29** 116-141
[5] Song M B, Ryu J H, Lee S K, et al 2003 Considering mobility patterns in moving objects database *International Conference on Parallel Processing* IEEE
[6] Qiao S J, Li T R, Han N, et al 2015 Self-adaptive trajectory prediction model for moving objects in big data environment A*cm Sigcomm Computer Communication Review* **26** 2869-2883
[7] Namiot D, Sneps-Sneppe M 2013 *Discovery of Convoys in Network Proximity Log* (Berlin: Springer Berlin Heidelberg) pp 139-150
[8] Kolbe T H, Gerhard Gröger, Lutz PlümerCity 2005 GML – Interoperable Access to 3D City Models *Geo-information for Disaster Management* **2005** 21-23
[9] Elfes A 1989 Using occupancy grids for mobile robot perception and navigation *in Computer* **22** 46-57
[10] Lee J 2004 A spatial access-oriented implementation of a 3-D GIS topological data model for urban entities *GeoInformatica* **8** 237-264

Acknowledgments

This work is supported by Sichuan Science & Technology Program under Grant 2017GZ0034 & 2018GZ0077 & 2018GZ0069.

Image Processing Techniques for Shock Wave Detection and Tracking in High Speed Schlieren and Shadowgraph Systems

Guoshuai Li[1*], Muhammed BurakAgir[1], Konstantinos Kontis[1], Takahiro Ukai[2], Sriram Rengarajan[3]

[1] School of Engineering, University of Glasgow, Glasgow G12 8QQ, UK
[2] Osaka Institute of Technology, Osaka, 535-8585, Japan
[3] India Institute of Technology, Madras Chennai, India

g.li.1@research.gla.ac.uk

Abstract. Schlieren and shadowgraph photography has been widely used to offer insight into the flow field in aerospace engineering due to their ease of application. The high-speed schlieren and shadowgraph techniques are typically applied to investigation of unsteady shock wave structure including shock reflection patterns and shock wave/boundary layer interactions, etc. Generally, qualitative analysis is performed based on the schlieren and shadowgraph image sequences. To process and analyze the large data set of the high-speed schlieren and shadowgraph images quantitatively, especially for the shock wave detection and tracking, a software was developed based on MATLAB GUI and its image processing toolbox. In this study, the image processing techniques exploited in the software, such as background subtraction, filter, threshold, edge detection, and shock tracking are presented. A case study on the phenomena of shock wave reflection from a solid surface was conducted. The results show that the proper filter method and the background image subtraction can effectively eliminate the image noises in frequency domain, which makes it easier to analyze the flow structure. Moreover, the instantaneous locations of shock waves are detected accurately, and the shock wave propagation speed calculated using the developed software are consistent with those of previous studies.

1. Introduction

Schlieren and shadowgraph techniques have been used since the early 19th century to visualize the optical density. Most commonly, they are applicable to visualize diverse subjects such as striations in blown glass, inhalation in humans and animals, shock waves from a plane in flight, and heat emanating from a system. In the experimental fluid mechanics field, schlieren and shadowgraph techniques are widely applied to study compressible flow phenomena, such as shock waves, expansion fans, and variable-density wakes and boundary layers in the supersonic and hypersonic wind tunnels [1-12]. However, as a dynamic and straightforward visualization tool, these systems are primarily applied to conduct qualitative visual studies.

In recent years, investigations on extracting and quantifying the information from the schlieren and shadowgraph image sets using digital imaging processing and computer vision techniques have been conducted. Estruch et al. [13] employed a number of image processing algorithms such as background image subtraction and Canny edge detection methods to quantify the shock wave positions and unsteadiness. Cui et al. [14] evaluated different image edge detection algorithms including Roberts, Prewitt, Sobel Canny, and Laplacian of Gaussian for accuracy of the shock wave detection. Dehghan et al. [15] proposed a two-level scheme involving schlieren image processing and classification to study the effects of speed change on the airplane surface.

The aim of this study is to process and quantify the large data set of the high speed schlieren and shadowgraph images more efficiently and accurately, especially for the shock wave detection and tracking. Based on improved new functions and image processing algorithms, a more versatile software system based on MATLAB GUI and its image processing toolbox was developed.

2. The proposed software system

In this section, the main functions and components of the proposed software system are described:

2.1. Basic image processing

Basic image processing functions such as image reading, saving, zooming, rotating, dragging, data cursor, image format conversion and so on are developed in the menu bar and tool bar. These functions are quite crucial and very useful for other image processing modules.

2.2. Image filter

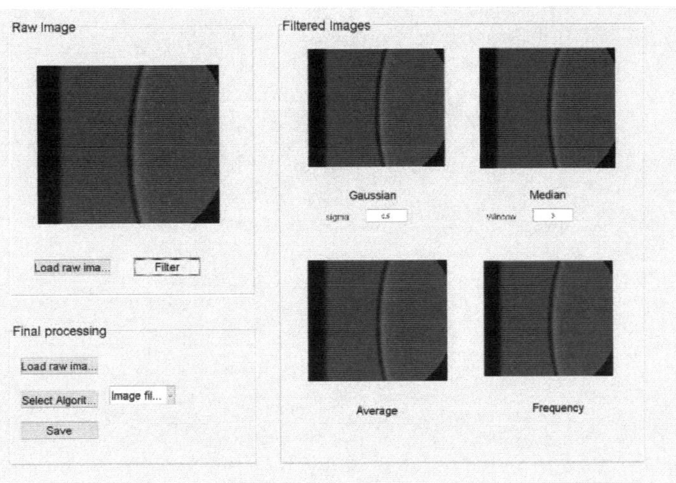

Figure 1. Image filter interface

In high speed schlieren and shadowgraph photography, the images usually contain salt-and-pepper noise and other noises due to high shutter speed and gain (ISO sensitivity). To improve the image quality and the accuracy of image threshold and shock wave detection, several image filter algorithms including Gaussian, median, average, and frequency domain filters are provided. The interface of image filtering is given in Fig. 1. Applying different filter methods and parameters, users can compare the filter effects on an image so that the optimal filer and parameters can be selected. However, it is worth noting that no algorithms can make this judgement perfectly (for all cases) since there is often a trade-off made between noise removal and preservation of fine.

2.3. Background image subtraction

The shock wave and other flow features are usually discernible as distinct dark or light regions in comparison with the background light intensity. To distinguish an image intensity more clearly and eliminate the contaminated area, an image processing routine including background image subtraction and object area restoration is applied. Fig. 2(a) shows the interface for background image subtraction processing. It is clear that the background image subtraction in the space domain and the frequency domain are both listed in the software. However, a lot of cases show that the image conversion into frequency domain using FFT can earn better results.

Unfortunately, the object area which is crucial for identifying the boundary of the flow is also deducted during the process of the background image subtraction. For restoring the object area, a threshold and restoring algorithm is proposed. The interface and the results are shown in Fig. 2 (b).

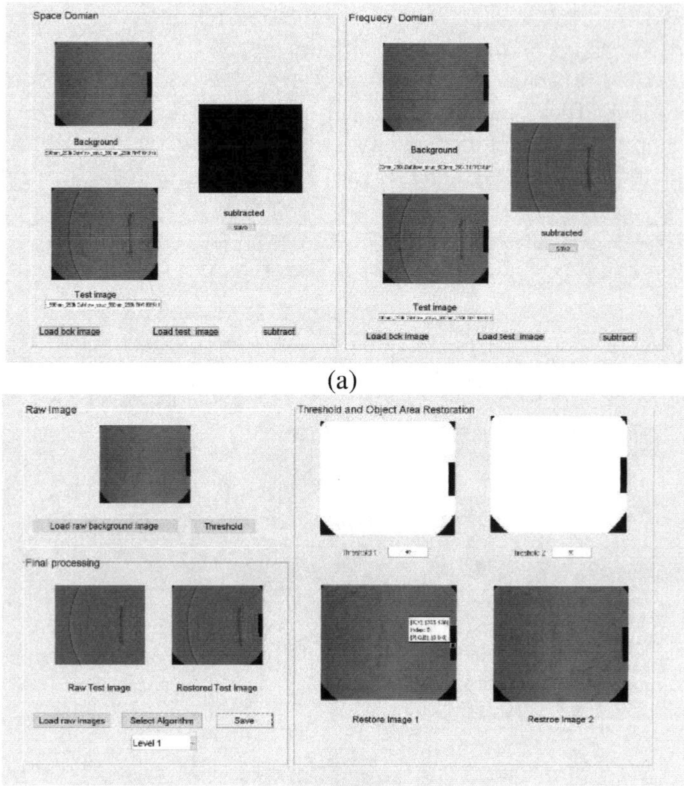

(a)

(b)

Figure 2. Background image subtraction: (a) subtracting the background image; (b) threshold and object area restoration

2.4. Shock wave detection and tracking

In order to detect the shape of the shock waves and other flow features, a number of edge detection algorithms are applied. The interface of edge detection in this software system is shown in Fig. 3(a). The Sobel, Canny, and Roberts edge detection algorithms have been described in the previous studies [13, 14]. To capture the steady shape unsteady change of the shock wave more accurately, a modified Canny algorithm is developed. This modified edge detection algorithm is more robust when a large amount of unsteady shock wave images need to be analyzed.

Through processing a time series of images, shock wave distortion, motion, and unsteadiness can be determined. Fig. 3 (b) shows the interface and some results of shock wave tracking. With these functions, quantitative information the schlieren and shadowgraph images contain can be extracted and quantified effectively and accurately.

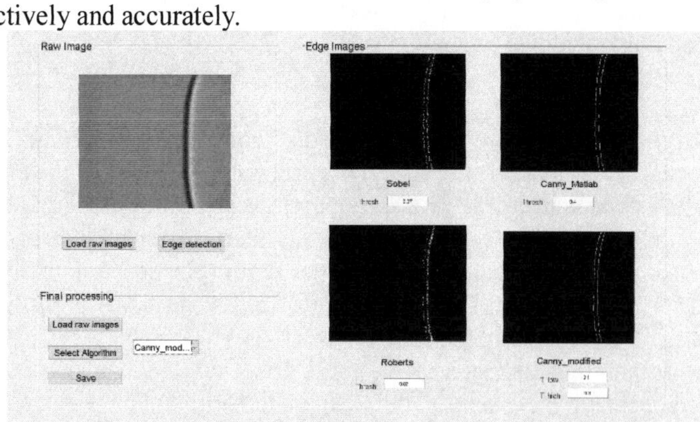

(a)

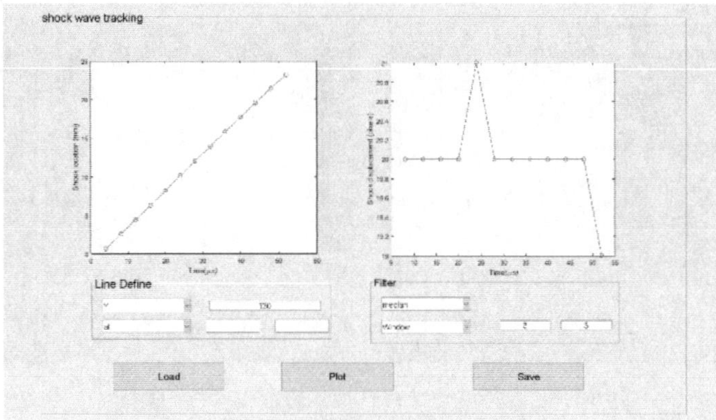

(b)

Figure 3. Shock wave detection and tracking: (a) edge detection; (b) shock wave location tracking

3. Experimental setup

The software and corresponding image processing techniques are applied in a shadowgraph visualization of the blast shock wave impingement on a solid surface. The schematic of the experimental setup is shown in Fig. 4. The open-end shock tube with the diameter D=22mm consists a driven section of 200mm in length and a blasted section. A non-electric tube (Dyno Nobel, model: NONEL DynoLine) [16, 17] was used to induce a blast wave which impinges on the plate model. The detonation was initiated within the NONEL tube by an electric blasting device (Dyno Nobel, model: Dyno Start 2, output voltage: 2500 V). This explosion generates a blast wave from the NONEL tube end, and the blast wave impinged on the plate model.

Three flush-mounted pressure transducers (Kulite Semiconductor Products, Inc., model: XTE-190M, natural frequency: 175 kHz) were flush mounted along the centreline of the square plate model. The shock wave impingement and reflection process were visualised using high-speed shadowgraph photography. The shadowgraph system consists of a 450–1000W continuous light source with an Xe-Hg arc lamp (Newport, model: 66921), a pair of 203.3 mm diameter concave lens with a focal length of 1829 mm, and a high-speed camera (Shimadzu, model: HPV-1). As for the synchronization, the output of the Kulite transducer was utilized to trigger the high-speed camera. The images were acquired at a frame rate of 250 kfps with an exposure time of 1 μs.

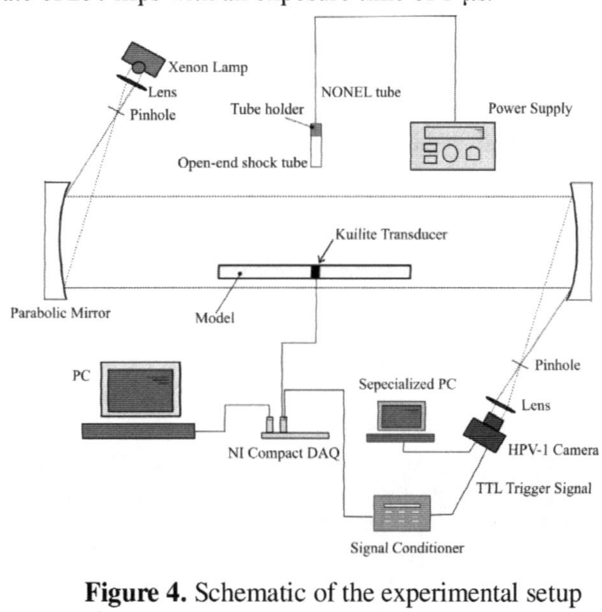

Figure 4. Schematic of the experimental setup

4. Results and analysis

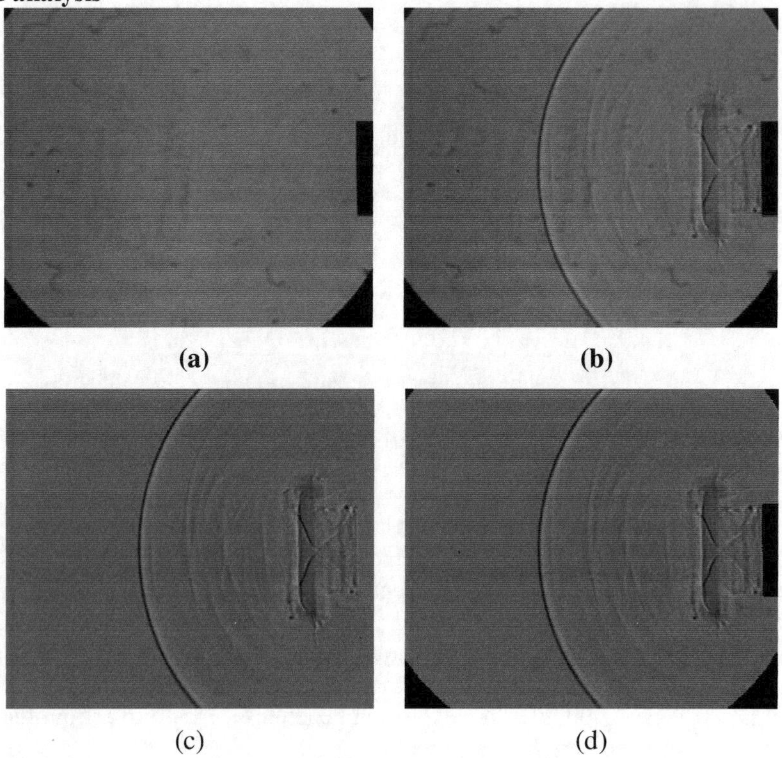

Figure 5. Background image subtraction and object area restoration of a shadowgraph image (Δt = 144 μs): (a) background image; (b) shadowgraph image; (c) background image subtraction; (d) object area restoration

To evaluate the visualization process and shock structure of the blast shock wave exiting from the open-end shock tube, a test case without the plate was conducted in the condition of NONEL tube of 500 mm in length. Fig. 5 shows the background image subtraction and object area restoration results of a shadowgraph image when Δt = 144 μs (the elapse time after the shock wave emission from the shock tube end is defined as t_0). As shown in the Fig. 5, the original shadowgraph images have a number of image noises due to the contamination of the lens/mirror surfaces and some other reasons. This problem can be mitigated by cleaning all the optical components or improving the quality of the system hardware. In the other hand, it can be eliminated in a digital image processing viewpoint. Fig. 5 (c) shows the image processed after the background image subtraction in frequency domain. Obviously, most of the image noises are subtracted from the background image. Unfortunately, the shock tube and the regions out of measurement area in the test image are also classified as noises. The image algorithms based on threshold are developed in the software system to restore the object area. The image processing results in the subtracted shadowgraph image and object area restoration are shown in Fig. 5 (d).

The original and background image subtracted shadowgraph images are given in Fig. 6. It is apparent that subtracted shadowgraph images can distinguish the light gradient and visualize the flow structure better. In Fig. 6 (f-j), several weak shock waves behind the blast wave as well as a contact surface between the emerging and the external fluid are clearly visible. The contact surface moves away and rolls up to from a vortex ring. After the rolling up process, the vortex ring propagates and expands. In the other hand, a rearward facing shock wave with the low pressure upstream and high pressure downstream is generated. This shock wave propagates along with the vortex ring. It is worth noting that sequences of the multiple weak shock waves were observed which was also found in a previous studies [12]. Their origin is not fully understood and needs more investigations.

IOP Conf. Series: Journal of Physics: Conf. Series **1215** (2019) 012021 doi:10.1088/1742-6596/1215/1/012021

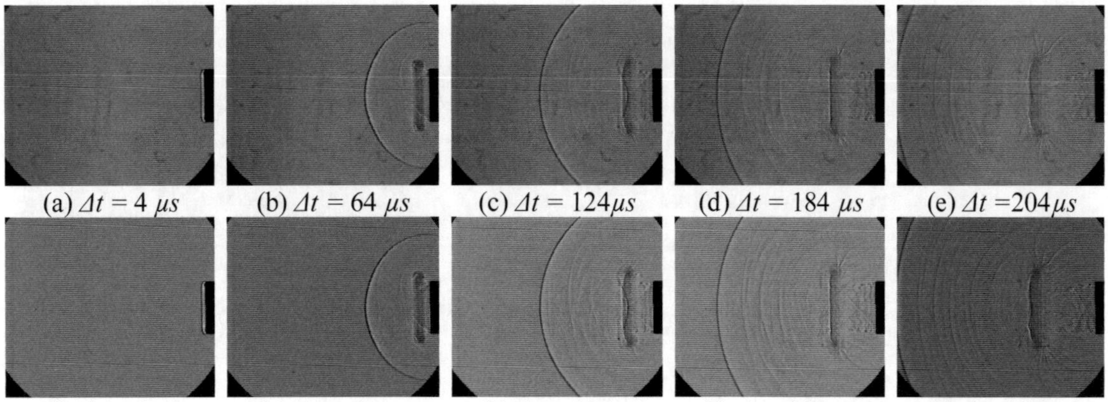

(a) Δt = 4 μs (b) Δt = 64 μs (c) Δt = 124μs (d) Δt = 184 μs (e) Δt =204μs

(f) Δt = 4 μs (g) Δt = 64 μs (h) Δt = 124μs (i) Δt = 184 μs (j) Δt =204μs

Figure 6. Time resolved shadowgraph images: (a-e) original images, (f-j) background image subtracted images

Fig. 7 gives the intensity profiles of the original and processed shadowgraph images (Δt =184μs) in their centrelines. The intensity of the density gradient was normalized by the brightness of the image background. The lateral axis denotes the normalized distance *x/d* from the tube end (*D* =22 mm denotes the diameter of the shock tube). As shown in Fig. 7 (a), the locations of the blast wave, the multiple weak shock waves, the rearward facing shock wave and the vortex ring can be recognized in the intensity profile curve of the original shadowgraph image. However, there are a lot of image noises along the centreline of the shadowgraph image. Moreover, the normalized intensity varies from 0.81 to 1.01 along the background area upstream of the initial shock wave. This means that the parallel light of the shadowgraph system is not ideally uniform.

Fig.7 (b) shows the normalized intensity profile by the background image subtraction process. The level of the noises is decreased significantly, and the locations of the flow features mentioned previously are more salient. In the meantime, the normalized intensity in the background area upstream of the initial shock wave changes from 0.962 to 1.074. The problem of the inhomogeneous incident parallel light is mitigated to some extent by background image subtraction. The normalized intensity profiles by a Gaussian filter with the standard deviation σ =0.5 and 1.1 are shown in Fig. 7 (c) and (d), respectively. With a Gaussian filter, the random noise of the shadowgraph image is eliminated, making it easier to distinguish the main flow features such as the initial shock wave, multiple weak shock waves, rearward facing shock wave, and the vortex ring. A greater standard deviation of the Gaussian filter leads to a lower noise level, however, the boundaries of features are smoother. Namely, there is a trade-off made between noise removal and preservation of fine structure.

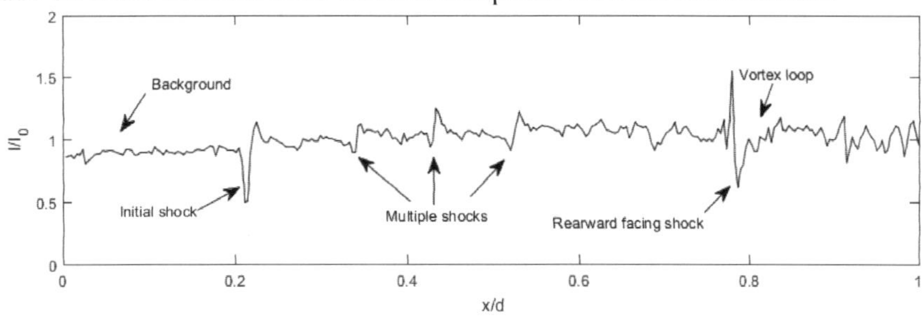

(a) Original shadowgraph image

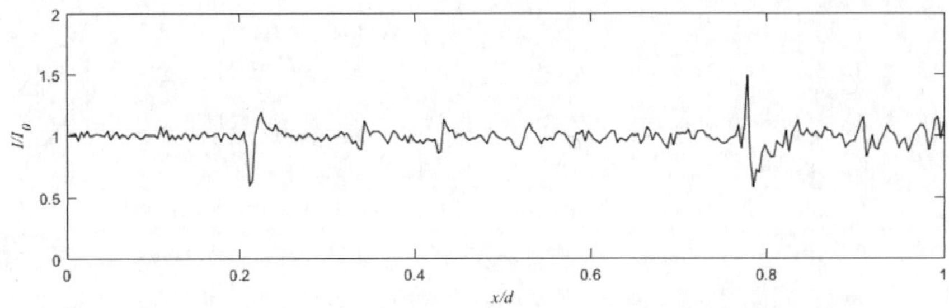

(b) Background image subtracted shadowgraph image

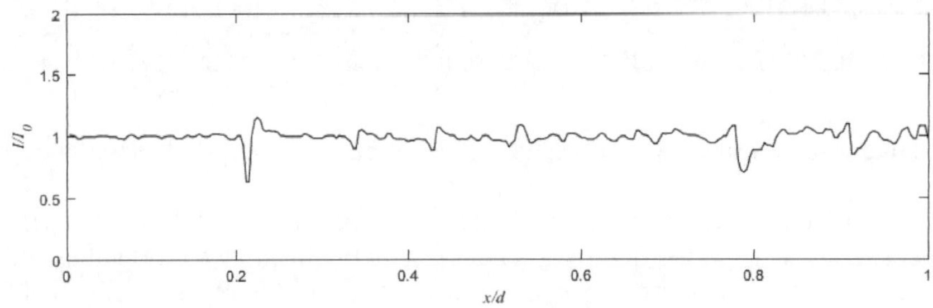

(c) Gaussian filtered shadowgraph image (σ=0.5)

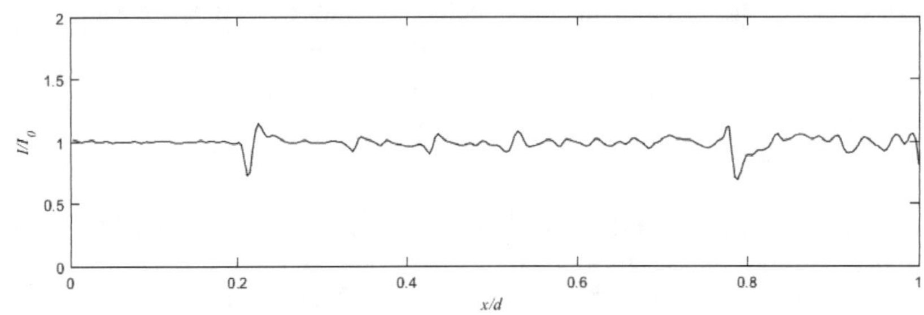

(d) Gaussian filtered shadowgraph image ($\sigma = 1$)

Figure 7. Intensity profiles of the original and processed shadowgraph images (Δt =184 μs) in their centrelines

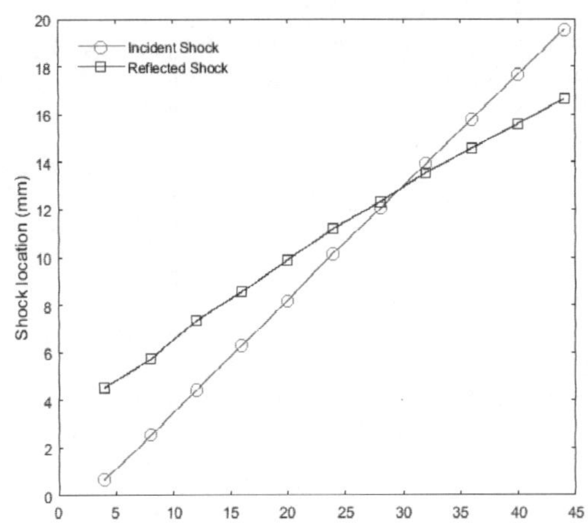

Figure 8. Instantaneous horizontal locations of the initial incident shock wave and the reflected shock wave

Furthermore, experiment regarding shock wave reflection from a solid surface was conducted based on an open-end shock tube. The distance between the tube end and the plate surface is 30 mm and the length of the NONEL tube is 300 mm. Based on a modified Canny edge detection algorithm described in the section 2, the shadowgraph images are processed to detect and track the instantaneous locations of the shock waves. The instantaneous horizontal locations of the initial incident shock and a reflected shock wave (before the reflected shock and vortex ring impingement) along the centrelines of the shadowgraph images are calculated (Fig. 8). Note that elapsed time for the incident shock is defined as t_0 when the shock wave is released from the shock tube, and the start time for the reflected shock was the time when the incident shock reaches the plate surface. Based on the shock wave movement and elapsed time, the velocity of shock wave is calculated. As shown in the Fig. 8, the velocity of the incident shock generally keeps the same and is higher than that of the reflected shock wave. The mean velocity of the incident shock wave with a 300 mm NONEL tube length is 469.8 m/s or Mach number 1.37 and the speed of the reflected shock wave is 333.5 m/s, which is a Mach number of less than 1.0. The result is very close to that of a previous study conducted by Ukai et al. [17]. Based on the shock wave detection and tracking algorithms proposed in this paper, more investigations such as shock wave movement, distortion, and unsteadiness can be performed more effectively and accurately.

5. Conclusions

A software system based on digital image processing techniques for shock wave detection and tracking in the high speed schlieren and shadowgraph techniques was developed. The primary processes incudes background image subtraction, object area restoration, filter, threshold, and edge detection. To validate the image processing algorithms proposed in this paper, experiment regarding shock wave impinging on a solid surface were conducted. The result shows that image noises are eliminated effectively by background image subtraction in frequency domain and proper filter method, which makes it easier to analyse the flow structure. Moreover, the instantaneous locations of shock waves were detected accurately and the velocities of the shock waves calculated using the software were consistent with those of previous studies. Overall, the image processing techniques proposed in this paper make it more effectively to deal with the large data set of the schlieren and shadowgraph images. Investigations such as shock wave movement, distortion and unsteadiness can be conducted using the software more effectively and accurately. Next, more functions and image processing options will be extended in this software to enhance its applicability.

References

[1] Settles, G.S. and M.J. Hargather, A review of recent developments in schlieren and shadowgraph techniques. Measurement Science and Technology, 2017. 28(4).

[2] Nel, L., B. Skews, and K. Naidoo, Schlieren techniques for the visualization of an expansion fan/shock wave interaction. Journal of Visualization, 2014. 18(3): p. 469-479.

[3] Daub, D., S. Willems, and A. Gülhan, Experiments on the Interaction of a Fast-Moving Shock with an Elastic Panel. AIAA Journal, 2016. 54(2): p. 670-678.

[4] Desjouy, C., et al., Irregular reflection of weak acoustic shock pulses on rigid boundaries : Schlieren experiments and direct numerical simulation based on a Navier-Stokes solver. Physics of Fluids, 2016. 28(2).

[5] Wolfram, J. and J. Martinez Schramm. Pattern Recognition in High Speed Schlieren Visualization at the High Enthalpy Shock Tunnel Göttingen (HEG). 2010. Berlin, Heidelberg: Springer Berlin Heidelberg.

[6] Smith, N.T., M.J. Lewis, and R. Chellappa, Detection, Localization, and Tracking of Shock Contour Salient Points in Schlieren Sequences. AIAA Journal, 2014. 52(6): p. 1249-1264.

[7] Saravanan, S., et al., Schlieren Visualization of Shock Wave Phenomena over a Missile-Shaped Body at Hypersonic Mach Numbers. Proceedings of the Institution of Mechanical Engineers, Part G: Journal of Aerospace Engineering, 2011. 225(1): p. 26-34.

[8] Cyril Mauger, L.M., Marc Michard, Alexandre Azouzi, Stéphane Valette, Shadowgraph, Schlieren and interferometry in a 2D cavitating channel flow. 2012.

[9] Zare-Behtash, H., et al., Shock wave-induced vortex loops emanating from nozzles with

IOP Conf. Series: Journal of Physics: Conf. Series **1215** (2019) 012021 doi:10.1088/1742-6596/1215/1/012021

singular corners. Experiments in Fluids, 2010. 49(5): p. 1005-1019.
[10] Mowbray, D.E., The use of schlieren and shadowgraph techniques in the study of flow patterns in density stratified liquids. Journal of Fluid Mechanics, 2006. 27(03).
[11] Clem, M.M., K. Zaman, and A.F. Fagan, Variation of shock-spacing during screech stage-jumps. International Journal of Aeroacoustics, 2016. 15(3): p. 324-335.
[12] Médici, E.F. and G.P. Waite, Experimental laboratory study on the formation of multiple shock waves observed during volcanic eruptions. Geophysical Research Letters, 2016. 43(1): p. 85-92.
[13] Estruch, D., et al., Measurement of shock wave unsteadiness using a high-speed schlieren system and digital image processing. Rev Sci Instrum, 2008. 79(12): p. 126108.
[14] Cui, S., et al., Image Processing Techniques in Shockwave Detection and Modeling. Journal of Signal and Information Processing, 2013. 04(03): p. 109-113.
[15] Dehghan Manshadi, M., et al., Speed detection in wind-Tunnels by processing schlieren images. Vol. 29. 2016. 965-970.
[16] Samuelraj, I.O., G. Jagadeesh, and K. Kontis, Micro-blast waves using detonation transmission tubing. Shock Waves, 2012. 23(4): p. 307-316.
[17] Ukai, T., et al., Three-dimensional shock wave distortion in shock-square vortex loop interaction. Experimental Thermal and Fluid Science, 2016. 79: p. 85-90.

Design for Guidance Method of UAV Autonomous Landing on Mobile Platform Based on Prediction of Intersection Points

Jun Jiang[1] **, Guodong Qi**[1] **and Guoxian Huang**[1]

[1]School of Aerospace Engineering, Beijing Institute of Technology, Beijing, China

Abstract. The development of the technology for UVAs (Unmanned Aerial Vehicle) autonomous landing on mobile platform is introduced in this article. And the existing problems of UAVs autonomous landing on mobile platforms are also analysed. In order to solve these problems, a guidance method of UAV autonomous landing on mobile platform based on prediction of intersection points is proposed in this paper. In this method, the intersection point of the trajectory of UAV and the trajectory of the mobile platform is predicted based on the speed plan for UAV and the motion state of mobile platform. Then, the UAV will be guided to the intersection point by proportional guidance law. And at each time of the landing of the UAV, the location of the intersection point will be corrected in real time. In order to verify this method, a mathematical simulation experiment was designed. The result of this experiment shows that UAVs are able to land more accurately on the mobile platform with less overload using this method.

1. Introduction

These guidelines, written in the style of a submission to *J. Phys.: Conf. Ser.*, show the best layout for your paper using Microsoft Word. If you don't wish to use the Word template provided, please use the following page setup measurements.

In Modern Warfare, the UAV is playing an increasingly important role in both land and naval battlefields. The landing stage of UAV is the last stage of the whole flight process, which determines the success of UAV recovery. At present, the way of UAV recovery includes landing gear recovery, net-catch-landing recovery and parachute recovery[1]. Landing gear recovery is suitable for landing platforms with relatively long runways, such as aircraft carriers or other large surface ships[2]; parachute recovery often makes it difficult to control the falling points of UAVs[3]. The net-catch-landing recovery method is more suitable for mobile platforms which have the need to recover UVAs but have no relatively long landing runway, such as small surface ships and land vehicles[4].

Small surface ships and ground vehicles have relatively great maneuverability, which is difficult for UAVs to land on it automatically. Because, if the platform has too much maneuverability, the UAV which need to land on it has to generate greater normal acceleration to intersect the platform. However, limited by many factors, such as aerodynamic shape, flight speed or structural strength of airframe, the normal acceleration the UAVs generating is limited. If the normal acceleration needed by UAV during landing is less than the maximum normal acceleration it can produce, it may lead to UAV recovery failure[5].

However, If the platform is not mobile, it will increase the possibility of platform being under attack, which reduces the viability of platforms and UAVs on the battlefield. Therefore, it is necessary to design a guidance method which can make the UAV landing successfully even when the platform maneuver amplitude is large[6].

2. The realization principle of guidance method

In order to solve the problem of UAV autonomous landing on mobile platforms with relatively great maneuverability, a guidance method for UAV autonomous landing based on prediction of intersection point is presented in this paper presents. In this method, the time-to-go of UVAs landing is calculated by dichotomy, and the data base of this calculation is the velocity variation law of UAV in landing phase set in advance and the motion state of mobile platform calculated by GPS data. Based on the time-to-go, the location of the intersection point of the trajectory of UAV and the trajectory of the moving platform can be predicted. Then, the UAV will be guided by the proportional guidance method to the intersection point. Because of the influence of wind, air temperature, engine parameters and other factors, it is impossible for UAV to fly completely according to the preset velocity variation law in the landing process[7]. Therefore, in each control cycle of the flight process, the velocity variation law of UAV will be modified according to the UAV's flight speed at this control cycle measured by GPS and the UAV's acceleration at this control cycle measured by INS. Due to the velocity variation law of UAV being modified, the time-to-go of the UVA's landing process and the location of the intersection point are also modified at each control cycle. At a time the location of the intersection point modified, the UAV will be guided to the modified intersection point until it successfully lands on the mobile platform. The implementation of the method is as follows.

2.1. Guidance method for UVAs autonomous landing

The implementation of the method is based on the following three hypotheses:
Firstly, the recovery method of UAV is net-catch-landing recovery, that is, as long as the UAV can intersect the mobile platform, it is considered that the UAV lands on the mobile platform successfully, and the attitude angle of UAV landing is not required;
Secondly, the motion form of mobile platform is uniform acceleration motion.
Last, it is considered that the GPS data and the INS data obtained by UAV are accurate, that is, the GPS positioning and velocity measurement errors and the acceleration measurement errors of inertial devices are neglected.
The relative motion of UAV and platform is discussed only in the vertical plane in this paper. First, spatial coordinate system of UAV and mobile platform has to be established. Take the position of UAV to enter the landing phase as the origin of coordinates. Take the plumb line at the origin of the coordinate system as the Y axis, and the positive direction of it is pointing upward. The X axis and the Y axis are in the same vertical plane, and the X axis is perpendicular to the Y axis. The positive direction of the X axis is pointing to the direction of the mobile platform. The schematic diagram of the coordinate system is shown in Figure 1.

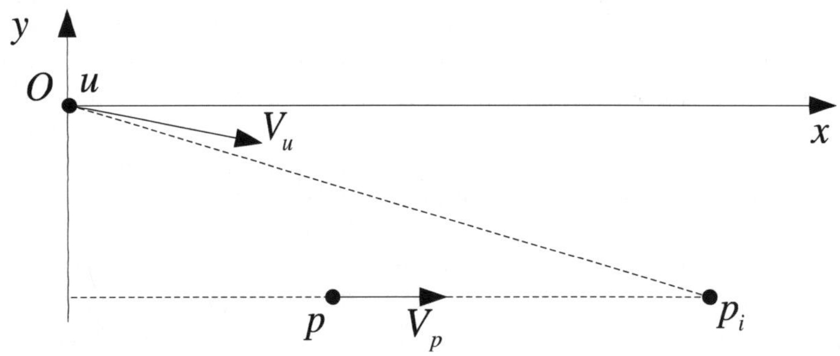

Figure 1. The schematic diagram of the coordinate system.

In Figure 1, point u represents the position of the UAV, point p represents the position of the mobile platform, point p_i represents the position of the predicted intersection point, vector $\boldsymbol{V}_u$ represents the speed of the UAV, vector $\boldsymbol{V}_p$ represents the speed of the mobile platform.
The dynamic model of UAV moving in the vertical plane is shown in Formula 1[8].

$$\begin{cases} \frac{dV_u}{dt} = \frac{F_T \cos\alpha - F_D}{m_u} - g\sin\theta_u \\ \frac{d\theta_u}{dt} = \frac{F_T \sin\alpha + F_L}{m_u V_u} - \frac{g\cos\theta_u}{V_u} \\ \frac{dx_u}{dt} = V_u \cos\theta_u \\ \frac{dy_u}{dt} = V_u \sin\theta_u \end{cases} \quad (1)$$

In Formula 1, g represents the acceleration of gravity, x_u and y_u respectively represents the position of UAV in the coordinate system, mu represents the mass of the UAV, θ_u represents the trajectory inclination angle of the UAV, F_T represents the thrust of UAV's engine, F_D and F_L respectively represents the drag force and the lift force that air acts on the UAV, α represents the angle of attack of the UAV and it is the variable which controls the UAV.
The α can be calculated through the normal acceleration required of during the UAV landing. Set N_r as the normal acceleration required of during the UAV landing, and set N_a as the normal acceleration that the UAV generates during the landing process. The expression of N_a is shown in Formula 2.

$$N_a = \frac{F_T \sin\alpha + F_L}{m_u} - g\cos\theta_u \quad (2)$$

In Formula 2, F_L can be expressed as a function related to α, the expression of F_L is shown in Formula 3[9].

$$F_L = \frac{1}{2}\rho V_u^2 S C_L(\alpha) \quad (3)$$

In Formula 3, ρ represents air density, S represents the reference area of the UVA, $C_L(\alpha)$ represents Lift coefficient related to α. Therefore, N_a can be expressed as a function related to α. The condition for effective and successful control of UAV is that the normal acceleration required of during the UAV landing is equal to the normal acceleration that the UAV actually generates during the landing process. The condition can be expressed as which the Formula 4 shows.

$$N_r = N_a = \frac{F_T \sin\alpha + \frac{1}{2}\rho V_u^2 S C_L(\alpha)}{m_u} - g\cos\theta_u \quad (4)$$

From the above analysis, it can be found that the essence of the UAV landing process control is to design the variation law of normal acceleration required of during the UAV landing, which is the innovation of this article. in this paper, the guidance law of UAV landing phase is proportional navigation method. In the theory of proportional navigation method, N_r can be expressed as:

$$N_r = V_u K\dot{q} \quad (5)$$

In Formula 5, K represents the proportional coefficient of proportional guidance method for UAV landing, $\dot{q}$ represents the rate of change of sight angle between the UAV and the target. Notably, in this article, the target is not the mobile platform, but to the predicted intersection point of the trajectory of UAV and the trajectory of the mobile platform. The predicted intersection point can be calculated by the following procedure.
Set the initial location of the mobile platform is (x_{p0}, y_{p0}), the initial velocity of the mobile platform is (V_{xp0}, V_{yp0}), the acceleration of the mobile platform is (a_x, a_y). The motion state of the mobile platform can be expressed as the Formula 6.

$$\begin{cases} V_{xp}(\mathrm{t})=V_{xp0}+a_x \cdot t \\ V_{yp}(\mathrm{t})=V_{yp0}+a_y \cdot t \\ x_p(t) = x_{p0} + \int_0^t V_{xp}(\mathrm{t})dt \\ y_p(t) = y_{p0} + \int_0^t V_{yp}(\mathrm{t})dt \end{cases} \tag{6}$$

Set the velocity variation law of UAV is $\hat{V}_u(t)$. Therefore, the flight distance of UAV during landing can be expressed as the Formula 7.

$$\hat{R}_u(t) = \int_0^t \hat{V}_u(t)dt \tag{7}$$

In order to calculate the location of the intersection point, the time-to-go of UAV intersecting the mobile platform must be calculated. Set that t_{go} represents the time-to-go of UAV intersecting the mobile platform. At t time, t_{go} can be obtained by solving equations shown in Formula 8.

$$\hat{R}_u(t+t_{go}) = \left(x_p^2(t+t_{go}) + y_p^2(t+t_{go})\right)^{1/2} \tag{8}$$

Therefore, At t time, the location of the intersection point can be calculated by the following formula.

$$\begin{cases} x_{pi} = x_p(t+t_{go}) \\ y_{pi} = y_p(t+t_{go}) \end{cases} \tag{9}$$

Set q as the sight angle between the UAV and the intersection point and it can be expressed as the Formula 10.

$$q = \arctan\frac{y_u - y_{pi}}{x_u - x_{pi}} \tag{10}$$

The rate of change of sight angle between the UAV and the intersection point can be calculated by doing differential calculus for formula 10. Set the rate of change of sight angle in the k control cycle as $\dot{q}_k$, the time span of each control cycle is T, so $\dot{q}_k$ can be calculated as the Formula 11.

$$\dot{q}_k = \frac{q_k - q_{k-1}}{T} \tag{11}$$

Introduce Formula 11 into Formula 5, the expression of the normal acceleration required in the k control cycle can be obtained.

2.2. Correction method of velocity variation law

Mentioned at the beginning of this section, it is impossible for UAV to fly completely according to the preset velocity variation law in the landing process and the velocity variation law has to be modified in each control cycle.

Set the velocity of the UAV measured by GPS in the k control cycle is V^k_{um}, the tangential acceleration of the UAV measured by GPS in the k control cycle is akum. According to the preset velocity variation law, the preset velocity of the UAV in the k control cycle is $\hat{V}_u^k$. The preset tangential acceleration of the UAV is expressed as $\hat{a}_u^k$ and it can be obtained by doing differential calculus for $\hat{V}_u(t)$. In the k control cycle, the deviation between the actual velocity and tangential acceleration of the UAV and their preset values can be expressed as ΔV^k_u and Δa^k_u. Their expression is shown as Formula 12.

$$\begin{cases} \Delta V_u^k = V_{um}^k - \hat{V}_u^k \\ \Delta a_u^k = a_{um}^k - \hat{a}_u^k \end{cases} \tag{12}$$

IOP Conf. Series: Journal of Physics: Conf. Series **1215** (2019) 012022 doi:10.1088/1742-6596/1215/1/012022

The velocity variation law of the UAV can be modified based on ΔVku and Δaku. The modified correction rule is shown as Formula 13.

$$\begin{cases} \hat{V}_u(t) = \hat{V}_u(t) + \Delta V_u^k & (if \;\; \Delta a_u^k = 0) \\ \hat{V}_u(t) = \hat{V}_u(t) + \Delta V_u^k + \Delta a_u^k \cdot t & (if \;\; \Delta a_u^k \neq 0) \end{cases} \tag{13}$$

Introduce the modified velocity variation law into Formula 7, and calculate the time-to-go and the location of the intersection point based on this.

3. Numerical simulation

In order to verify the rationality of the guidance method proposed in this article, numerical simulation experiment is designed. The numerical simulation parameters are shown in Table 1 and Table 2.

Table 1. Parameters of the UAV.

Parameters	Values
Initial velocity	200m/s
Initial trajectory inclination angle	0°
Initial location	(0 m, 0 m)
Mass	18 kg
Thrust	30 N
Reference area	0.0087 m^2
Lift coefficient	$C_L(\alpha)=1.28*\alpha$
Drag coefficient	$C_D(\alpha)=0.186*\alpha$

Table 2. Parameters of the mobile platform.

Parameters	Values
Initial location	(1500 m, -300 m)
Initial velocity	(10 m/s, 0 m/s)
Initial acceleration	(10 m/s^2, 0 m/s^2)

The air density ρ is 1.225 kg/m3. Set the proportional coefficient K is 3. Set the thrust deviation value is $\pm$10% and the drag deviation value is $\pm$20%. The numerical simulation results are shown in the following figures.

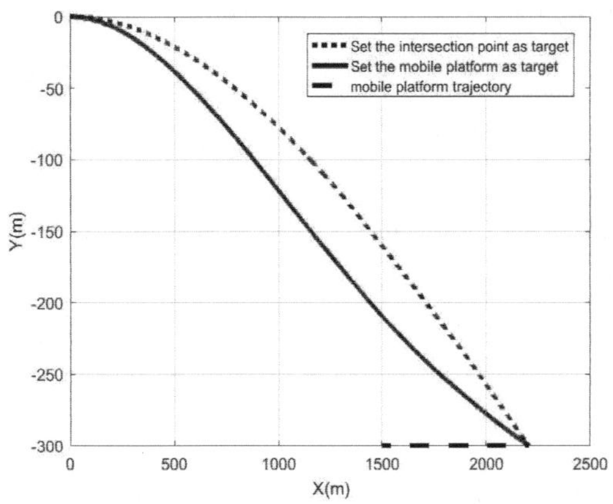

Figure 2. The trajectory of different guidance method.

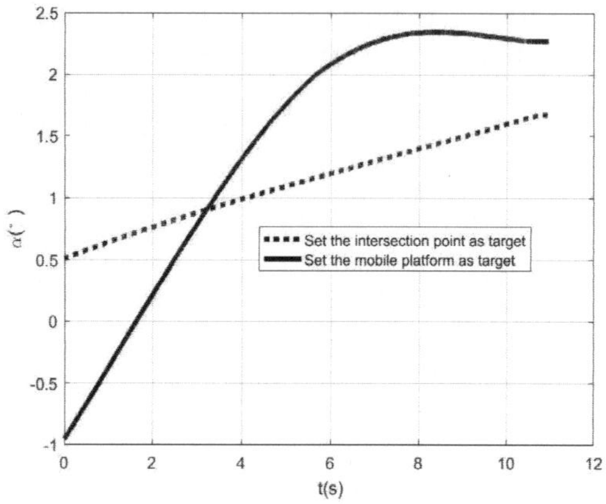

Figure 3. The angle of attack of different guidance method.

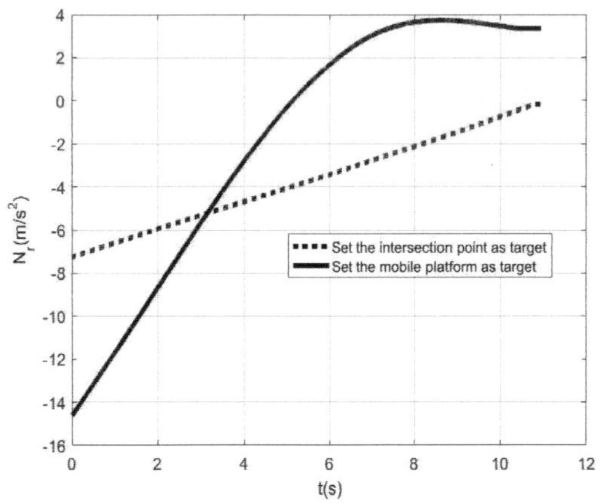

Figure 4. The required normal acceleration of different guidance method.

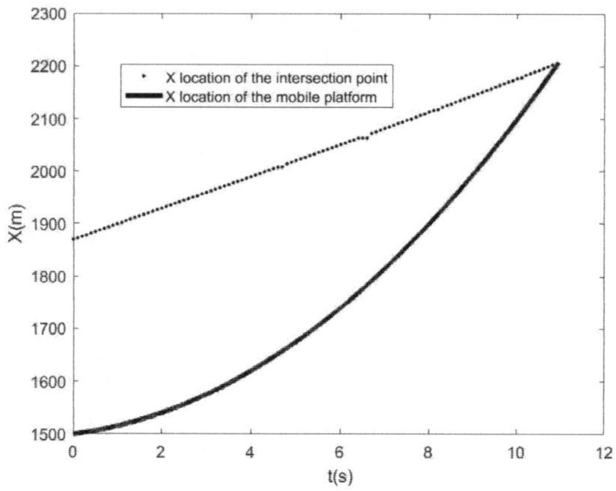

Figure 5. The change of locations of the intersection point and the mobile platform.

As can be seen from Figure 2, the flight trajectory of the UAV is more straight than that of the direct tracking mobile platform when it makes autonomous landing using the guidance method proposed in this article. As shown in Figure 3 and Figure 4, the required angle of attack and the required normal acceleration of the UAV landing with the guidance method presented in this article are both smaller than those of the direct tracking mobile platform. Referring to the results shown in Figure 5, this reason that the required normal acceleration of the UAV landing with the guidance method presented in this article are smaller than those of the direct tracking mobile platform can be explained. Due to the change range of the location of the predicted intersection point is much smaller than that of the mobile platform, the guidance method proposed in this paper is essentially to make UAV track a very small maneuvering target rather than the mobile platform which has greater maneuvering range. Obviously, according to the characteristics of proportional navigation, the normal acceleration required to track a small maneuvering target is less than that required to track a large maneuvering target.

4. Conclusion

Numerical simulation shows that the UAV can land precisely on the mobile platform when it lands with the guidance method based on predicting intersection point proposed in this article. Moreover, the required normal acceleration of this method is less than the traditional guidance method based on mobile platform as tracking target. The autonomous landing guidance method proposed in this article can ensure that the UAV can successfully land with a small normal acceleration while the mobile platform maintains a high maneuverability, which is of great importance in improving the survivability of UAV and mobile platform in the battlefield. The next step is to improve the algorithm of the guidance method so that it can realize the UAV autonomously landing on the platform with more complex maneuvering forms. It can lay a theoretical foundation for the UAV autonomously landing on the mobile platform in air.

5. Reference

[1] Daquan Tang, Bo Bi, Xushang Wang, Fei Li and Ning Shen J 2010 *J China Inertial Tec* **18** 550
[2] Pradeep S A 2002 *Aerospace Sci AIAA* 250
[3] Jing Liu J 2012 *Command Control Sim* **6** 109 Jinhua. Pei
[4] Jinhua Pei J 2009 *J NUAA* **41** 6
[5] D Lee, T Ryan and HJ Kim A 2012 *IEEE Trans Rob Autom* **162**, 971
[6] JL Sanchez Lopez, J Pestana, S Saripalli and P Campoy J 2014 *J Intell Rob* **74** 113
[7] Rui Zhou, Xiaojing Chen, Zhouyi Yu, Pei Chi and Zongji Chen J 2005 *Control Decis* **2** 222
[8] Jianqiao Yu, Zhonghui Wen, Yuesong Mei and Teng Long M 2009 *Overall Design of Tactical Missile* (Beijing)
[9] Xingfang Qian, Ruixiong Lin and Yanan Zhao M 2015 *Missile flight dynamics* (Beijing)

IOP Conf. Series: Journal of Physics: Conf. Series **1215** (2019) 012023 doi:10.1088/1742-6596/1215/1/012023

Vision-Based Hazard Detection with End-to-end Spatiotemporal Networks for Planetary Landing

Cheng Huang[1,2] **and Hongmei Wang**[1,2]

[1]National Key Laboratory of Aerospace Flight Dynamics
[2]School of Astronautics, Northwestern Polytechnical University, Xi'an 710072, Shaanxi, Province, China

Abstract. Hazard detection is the key technique for autonomous landing during a planetary exploration mission. This paper proposed an end-to-end spatiotemporal network to detect all hazards in the image sequences captured by optical camera. The spatial stream processes the colour image sequences and the temporal one learns features from optical flow image sequences. The spatial features and temporal features are fused by the proposed metric fusion method, and then the spatiotemporal features become more discriminative through triplet loss. A hazard map can ultimately be obtained in the testing phase after processing on the full-size image sequences. The evaluation results prove the effectiveness of the proposed network, and the testing results show the feasibility of practical application.

1. Introduction

In recent years, with the development of the planetary and comet exploration technology, autonomous and intelligent probing technique is more and more widely used. And it is truly a great challenge to autonomously land on an unknown celestial body. Therefore, avoiding all hazards and selecting a safety landing area during the decent process is the key to the success of the space exploration mission. Fig.1 shows that there are mainly four hazards that may influence the landing safety including crater, discontinuities, rock and slopes[[1]]. For traditional researches, different methods are adopted for different hazards.

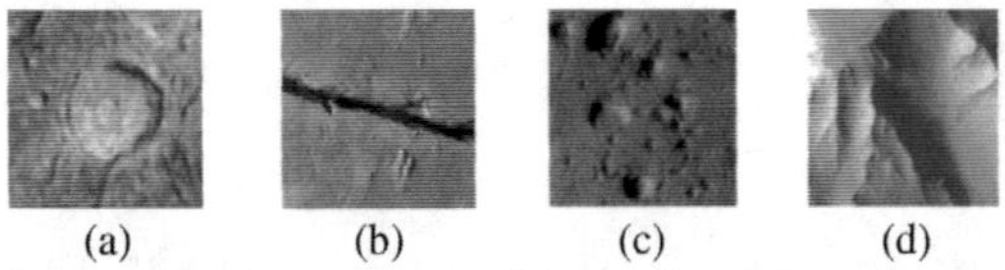

Figure 1. Four types of hazards: (a) crater; (b) discontinuities; (c) rock; (d) slope.

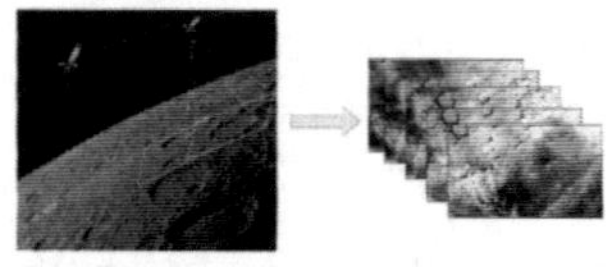

Figure 2. Captured image sequence during landing phase.

For automatic detection of craters, the geometric information is usually utilized. Since the edge of the crater is simply like a circle, circles are used to fit craters edge[[2]][[3]] to detect the existence of craters. And texture features [[4]] are also distinguished information that can be combined with supervised learning classifiers to discriminate craters and non-craters[[5]]. There is one more good method to model the craters according to the photos taken from different angles[[6]]. These methods are simple and easy to be implemented, but the detection accuracies are not enough to meet expectations. Nowadays, with the rapid development of machine learning methods, there are many more intelligent architectures for craters detection, such as simple convolutional neural networks for craters feature learning [[7]-[8]]. Artificial neural networks can also interpret the hazard features and generate a hazard map at different scales[[9]].

There are mainly two methods for the rock detection, including edge contour detection and shadow detection. Edge fragments can be concatenated with image pyramid to detect the rock zones under different scales[[10]]. The shadow detection is mainly based on the shining angle of the sunlight [[11]], and it has the advantages of real-time and high stability.

The studies of slope and discontinuities estimation are relatively few. Early slope estimation techniques for landing of detectors are based on distance information or digital elevation model (DEM) from laser scanning radar[[12]]. But compared with passive vision sensors such as optical cameras, such active vision sensors have the backwards of high cost, heavy weight. So it's meaningful to estimate slope based on optical vision method. The discontinuities detection is relatively simple, there nearly no research to study on it specifically.

In this paper, only optical camera is considered for the hazards detection during planetary landing phase. And the image sequences captured by the camera and their corresponding optical flow [[13]] results are used for the detection algorithms. The Fig.2 shows the way of image sequence captured during landing phase.

A spatiotemporal network is proposed in this paper to uniformly process all types of hazards. And a metric fusion method is also proposed to fuse the spatial features and temporal features, thus form stable spatiotemporal features embedding. In training phase, the features of hazards are learned without pertinently modelling and the classes are correctly distinguished based on triplet loss. In testing phase, full-size images are fed into the network to get the end-to-end hazards distribution results in which the locations and classes are clearly marked. The evaluation results prove the effectiveness of the proposed network.

2. Spatiotemporal networks architecture

The architecture of proposed spatiotemporal networks is shown in Fig. 3. It's a triplet of two-stream architecture.

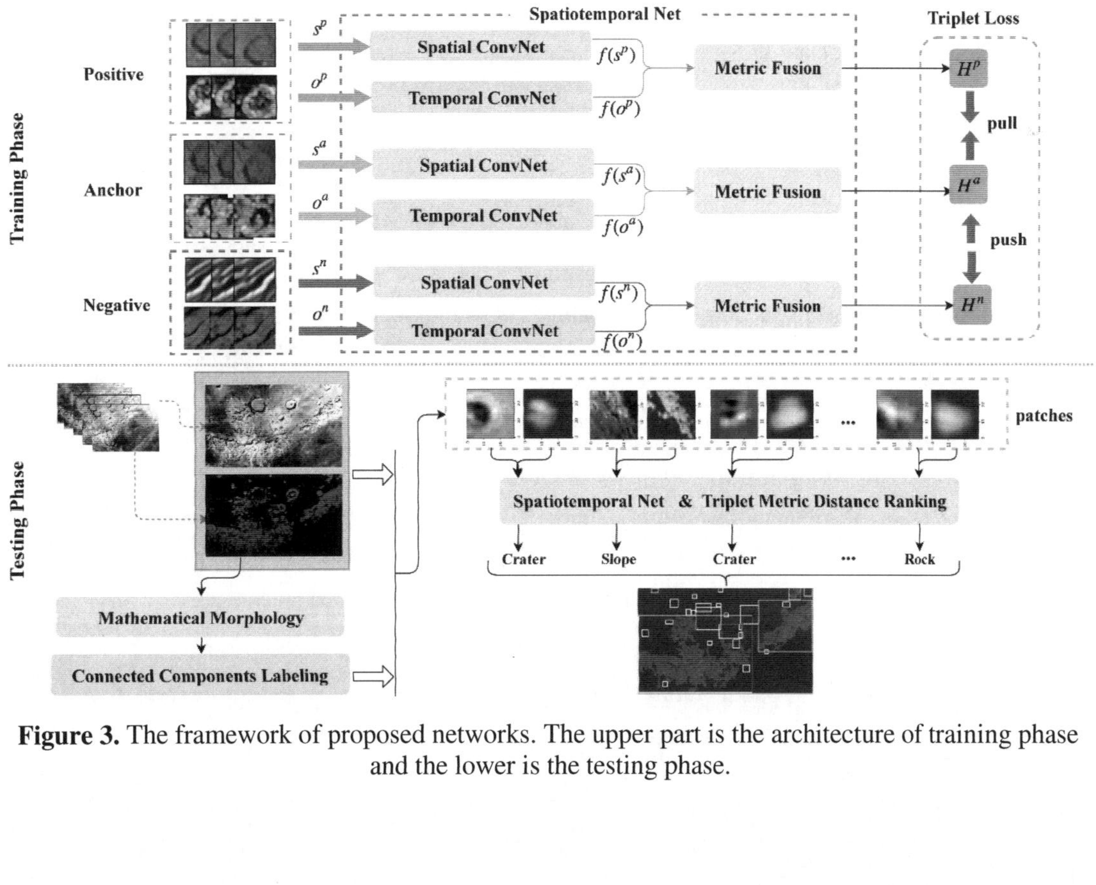

Figure 3. The framework of proposed networks. The upper part is the architecture of training phase and the lower is the testing phase.

Each two-stream branch consists of a spatial convolution network and a temporal convolution network, which takes the colour image sequences and optical flow image sequences as inputs, respectively. The spatial part of the two-stream branch carries information about contents and scenes, and the temporal part conveys the motion and edge information. The features learned by these two parts are fused in the metric space to collaboratively represent the spatial-temporal information. And the triplet loss is utilized to make these features more discriminative.

2.1. Convolutional network

The spatial and temporal convolution networks in this paper are the same layer configurations as in Fig.4. It is basically a simple structure for feature learning, which consists of three convolution layers, two max-pooling layers and ReLU activation function.

Supposing the spatial image sequence is represented as $s^i = \{s_t^i \mid s_t^i \in \mathbb{R}^D\}_{t=1}^{T_1}, i = p,a,n$. T1 is the length of image sequence and D is the dimension of images. p means the positive sample branch, a is the anchor sample branch and n is the negative sample branch. Each spatial stream is a CNN architecture which can be denoted as $f(\bullet)$, and the output can be denoted as $f(s^i)$.

The only difference between these two streams is that the spatial stream operates on the colour image sequence, but the temporal one processes the optical flow image sequence.

The temporal image sequence is expressed by $o^i = \{o_t^i \mid o_t^i \in \mathbb{R}^D\}_{t=1}^{T_2}, i = p,a,n$. T2 is the length of optical flow image sequence. Thus the output of each temporal stream can be denoted as $f(o^i)$.

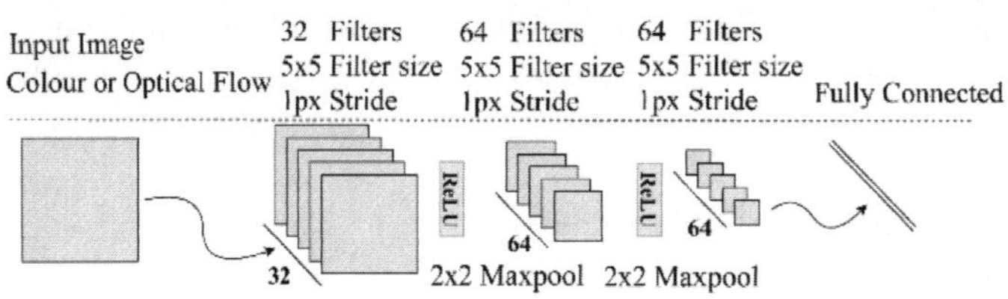

Figure 4. Convolutional network layers.

2.2. Metric fusion

Each branch of the triplet net is a two-stream architecture, in which the features learned by spatial stream and temporal stream are supposed to be fused to represent a spatiotemporal information. Our method is to embed the temporal feature $f(o^i)$ into the metric space of spatial feature. The mapping relations can be expressed by: $H^i(f(s^i), f(o^i)) : (\mathbb{R}^{T_1 \times D}, \mathbb{R}^{T_2 \times D}) \mapsto \mathbb{R}^{T_2 \times D_1}$, $i = p,a,n$,where D1 is the dimension of fused feature.

Specifically, we take use of polar coordinates transformation $R(\bullet)$ to calculate this mapping relationship. Thus, spatial feature $f(s^i)$ and temporal feature $f(o^i)$ are all supposed to be transformed into polar coordinates space, whose radius and polar angles can be expressed by:

$$r(f(s^i)), theta(f(s^i)) = R(f(s^i)) \tag{1}$$

$$r(f(o^i)), theta(f(o^i)) = R(f(o^i)) \tag{2}$$

Where $r(\bullet)$ is the radius and $theta(\bullet)$ is the polar angle.

The mapping result from temporal feature space to spatial feature space can be formulated as following:

$$f'(o^i) = R^{-1}\left(\frac{r(f(s^i))}{r(f(o^i))}, theta(f(s^i))\right) \tag{3}$$

Where $f'(o^i)$ is the metric in spatial feature space and $R^{-1}(\bullet)$ is the polar inverse transformation.

So the final feature fusion result of the two-stream branch is calculated as

$$H^i = \frac{1}{2}\left(R(f(s^i)) + f'(o^i)\right) \tag{4}$$

2.3. Triplet loss

Within the triplet of $\langle H^p, H^a, H^n \rangle$, we usually expect features of the positive sample H^p is more similar to anchor sample H^a than the features of the negative sample H^n. So the triplet loss is adopted in this paper to pull similar pairs and push dissimilar pairs away. And the expression is formulated as following [[14]]:

$$L_{trp} = \frac{1}{N}\sum_{t}^{N}\left[\left\|H_t^a - H_t^p\right\|_2^2 - \left[\left\|H_t^a - H_t^n\right\|_2^2\right] + \alpha\right]_+ \tag{5}$$

Where $[\bullet]_+ = max(\bullet, 0)$, α is a margin that is enforced between positive and negative pairs. $\|\bullet\|$ is the metric distance. But the number of possible triplets grows cubically with the number of examples, it's infeasible to process them all and the training converges slowly. So it's crucial to select triplets to ensure fast convergence. Thus the strategy we adopt is online triplet selection [[15]] method that can randomly select hard negative for each positive pair within a mini-batch.

3. Implementation details

3.1. Network training

The training procedure takes six image sequences as inputs at a time, since the training framework is a triplet of anchor sample, the positive sample and the negative sample. And the meanwhile, each sample branch contains a spatial ConvNet stream and a temporal ConvNet stream. The spatial ConvNet processes the colour image sequence to learn the contextual features while the temporal one takes the optical colour image sequence as input to capture the motion features. Each image sequence just contains one kind of terrain. The training phase aims to make all features more discriminative and classify different target correctly.

3.2. Network testing

The testing phase aims to obtain a hazard map directly through the end-to-end network. There are many image sequences with several scenes taking as test inputs. Each image contains complex terrains, in order to generate a hazard map, we need to extract different terrains to identify the corresponding class and mark them. The mathematical morphology and connected components labelling are utilized to operate on the optical flow image to obtain candidate patches. Then the corresponding patches in colour image and optical flow image are fed into the spatiotemporal net to get the attribute of each patch. All patches can get their labels after ranking with training results. The label of background is set to -1, the labels of terrains are set to 0, 1, 2, 3, respectively. So there is a hazard map obtained, in which the locations and classes of the terrains are marked.

4. Evaluation and discussions

4.1. Dataset

We test the networks on our own dataset which is called Dynamic Planet Terrains (DPT) dataset. The colour image sequence samples are captured in different light intensity, different angles and different height from System Tool Kit (STK). And the corresponding optical flow image sequences are calculated by dense flow method. Parts of samples are listed in Fig.5 below. The number of sequence (NoS) of each type is 40 in train samples and 20 in validation samples. The number of frames (NoF) in each image sequence varies from 15 to 176, in which the image size is 182×184. The detail information of the DPT dataset is shown in table 1.

IOP Conf. Series: Journal of Physics: Conf. Series **1215** (2019) 012023 doi:10.1088/1742-6596/1215/1/012023

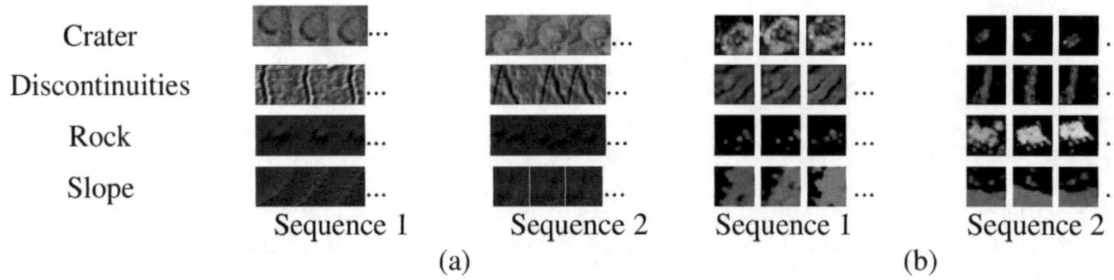

Figure 5. Part of samples in DPT dataset: (a) colour image samples; (b) optical flow image samples.

Table 1. Details of DPT dataset

Phase	Sequence	Types	NoS	NoF
Train	Colour	crater	40	37~176
		discontinuities	40	24~98
		rock	40	23~120
		slope	40	20~77
	Optical Flow	crater	40	33~157
		discontinuities	40	18~86
		rock	40	18~115
		slope	40	15~42
Validation	Colour	crater	20	35~76
		discontinuities	20	65~159
		rock	20	30~94
		slope	20	40~74
	Optical Flow	crater	20	15~91
		discontinuities	20	19~154
		rock	20	26~102
		slope	20	35~69

4.2. Evaluation of the proposed method

The experiments are conducted with PyTorch and OpenCV environment. The spatiotemporal network is trained on the DPT dataset to learn the spatiotemporal features of the six-stream inputs. To evaluate the effectiveness of the proposed network, the classification accuracies of the following three schemes are compared:

I. Single CNN stream
II. Spatiotemporal Net+triplet loss (all possible triplets)
III.Spatiotemporal Net+triplet loss(online triplet selection)

Table 2. The comparison of classification accuracy

Scheme	crater	discontinuities	rock	slope	Average
I	0.62	0.71	0.81	0.96	0.76
II	0.76	0.81	0.89	0.94	0.85
III	0.93	0.94	0.95	0.98	0.95

From the comparison results in the table 2, it can be clearly seen that CNN has not enough ability to divide all targets apart only by the extracted features. The triplet loss can pull similar targets and push dissimilar targets away, so the method II and III have the better classification ability. But the selection method of triplet is the key of improving the effectiveness of algorithm. And it shows that online triplet selection within a mini-batch in method III is more efficient than process them all in method II.

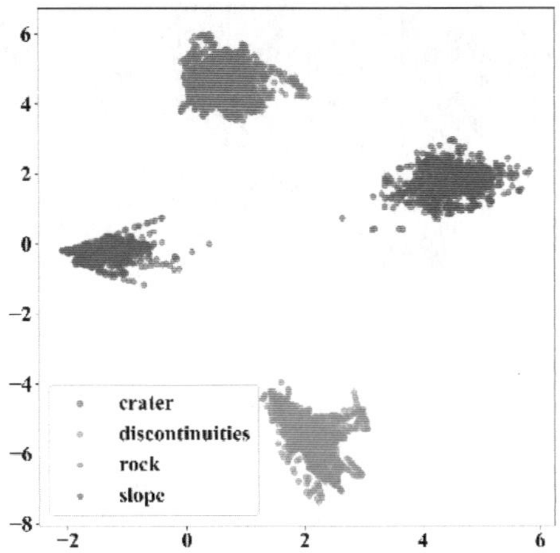

Figure 6. Triplet embedding result

Fig. 6 shows the triplet embedding result in training phase. The learned metrics are close to each other within class and far away from different classes. As is shown in Fig.7, if spatial stream and temporal stream obtain correct metrics, the metric fusion result is undoubtedly correct. But if one of the stream gets wrong results, like the temporal result of slope in Fig. 8, it's embedded into the ranges of rock wrongly. Though the metric result of another spatial stream is also far from the centre of the correct class, our metric fusion method also has the ability to calibrate the fusion result to a correct metric result. It may infer that the spatiotemporal net model has been trained as a good terrain classifier.

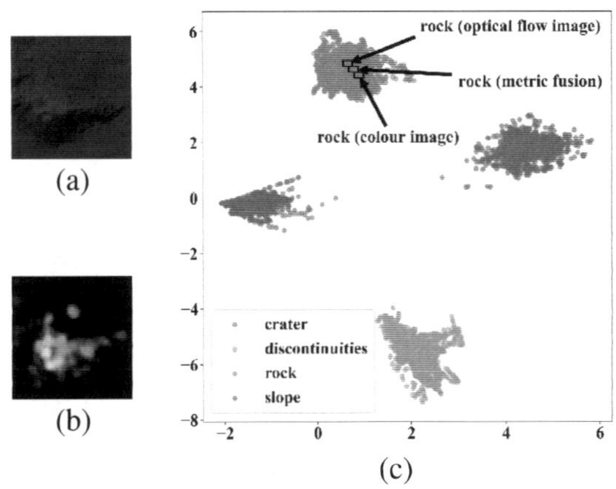

Figure 7. The metric fusion result: (a) The colour image of rock; (b) The optical flow image of rock; (c) The metric embedding result before fusing and after fusing.

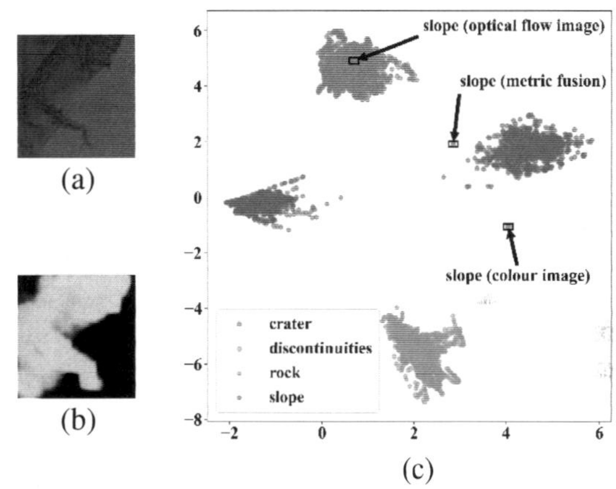

Figure 8. The metric fusion result: (a) The colour image of slope; (b) The optical flow image of slope; (c) The metric embedding result before fusing and after fusing.

In the test phase, the input of the algorithm is the sequence of 1226×734 full-size actual images that contains different terrains which is shown in Fig.9(a). The optical flow images in Fig.9(b) are calculated based on the visual image sequence. And the image patches in Fig.9(c) are cropped from colour and optical flow images and then fed into the spatiotemporal network to get their classes. We mark all hazards in the original image with different colours and draw white rectangular boxes according to the labels result. From the hazard map in Fig.9(d), the spatiotemporal net assigns the correct locations and classes of all terrains in the whole image.

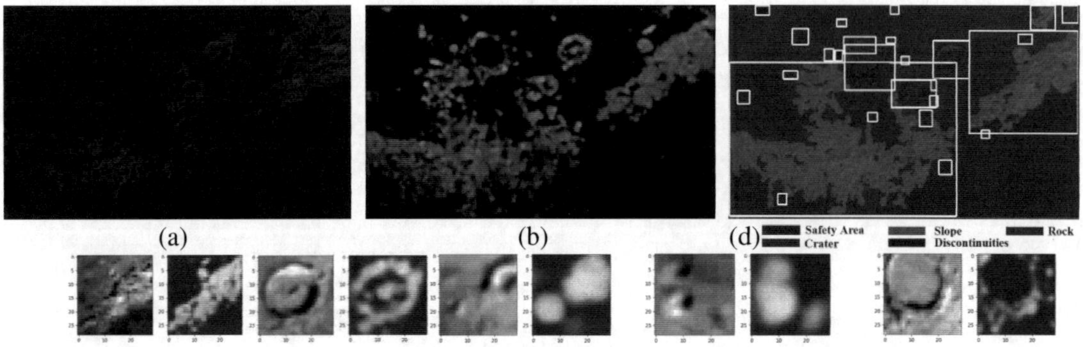

(c)

Figure 9. The testing results of the network: (a) Full-size colour image;(b) Optical flow image calculated by dense flow; (c) Parts of the cropped candidate patches; (d) hazard map generated by the network.

5. Conclusion

The proposed vision-based spatiotemporal network is a triplet of two-stream architecture, which needs 6 inputs at once. The spatial steam operates on the colour image sequence for contextual features and the temporal stream utilize optical flow to obtain motion and edge information. The metric fusion method can make a good balance between these features and form a good spatiotemporal feature. The network is finally trained as a good terrain classifier via triplet loss. And the hazard map can be generated in the test phase through predicting the candidate patches in the full-size images. The experiment results show that the spatiotemporal network with online triplet selection has a better classification performance and the generated hazard map based on full-size image obtains the precise locations and correct classes of terrains. Future work is to improve the computational performance on practical on-board equipment. And the hyperspectral imaging is taken into consideration for the multispectral information fusion.

6. References

[1] Huertas, Cheng Y, and Madison R 2006 Passive imaging based multi-cue hazard detection spacecraft safe landing *IEEE Aerospace Conference*

[2] Chen J, Cui P, Cui H 2011 Automated detection and classification for craters based on geometric matching *International Symposium on Photoelectronic Detection and Imaging*

[3] Sawabe Y Matsunaga T and Rokugawa S 2006 Automated detection and classification of lunar craters using multiple approaches *Advances in Space Research* 21-27

[4] Ding W 2010 Automatic detection of craters in planetary images: An embedded framework using feature selection and boosting *Proceedings of the 19th ACM international conference on Information and knowledge management* 749-758

[5] Di K 2014 A machine learning approach to crater detection from topographic data *Advances in Space Research* 2419-2429

[6] Zhu S Cui P and Hu H 2012 Hazard detection and avoidance for planetary landing based on Lyapunov control method *Intelligent Control and Automation* 2822-2826

[7] Emam E 2015 Automatic crater detection using convex grouping and convolutional neural networks *International Symposium on Visual Computing* 213-224

[8] Campbell T 2017 A deep learning approach for optical autonomous planetary relative terrain navigation *27th AAS/AIAA Space Flight Mechanics Meeting*

[9] Lunghi P Ciarambino M and Lavagna M 2015 Vision-based hazard detection with artificial neural networks for autonomous planetary landing *13th ESA/ESTEC Symposium on Advanced Space Technologies in Robotics and Automation* 1-8

[10] Thompson D R and Castano R 2007 Performance comparison of rock detection algorithms for autonomous planetary geology *Aerospace Conference* 1-9

[11] Huertas A Cheng Y and Matthies L H 2008 Automatic hazard detection for landers

[12] Johnson A E Klumpp A R, Collier J B 2002 Lidar-based hazard avoidance for safe landing on Mars *Journal of guidance control and dynamics* 1091-1099
[13] Hassner T and Liu C 2016 Dense Image Correspondences for Computer Vision *Springer International Publishing* 3-14
[14] Schroff F Kalenichenko D and Philbin J 2015 Facenet: A unified embedding for face recognition and clustering *Proceedings of the IEEE conference on computer vision and pattern recognition* 815-823
[15] Hermans Beyer L and Leibe B 2017 In defense of the triplet loss for person re-identification *arXiv preprint arXiv:1703.07737*

Acknowledgement

This work is jointly supported by the National Natural Science Foundation of China (No.61771400), the Key Research and Development Program of Shaanxi Province (No.2017GY-066) and the Seed Foundation of Innovation and Creation for Graduate Students in Northwestern Polytechnical University (ZZ2018061).

IOP Conf. Series: Journal of Physics: Conf. Series **1215** (2019) 012024 doi:10.1088/1742-6596/1215/1/012024

Path Planning of Unmanned Air Vehicles Relay Communication Based on Lowest Power Loss

Ruiying Xiao[1], Fan Yang[1] and Zhenghong Dong[1]

[1] Space Engineering University, Beijing, 101416, People Republic of China

Abstract. Unmanned aerial vehicle (UAV) relay communication is an important means to realize long-range wireless communication, where the lowest power loss could prolong the air residence time and increase the duration of the relay communication of UAV. In this paper, we present a UAV relay communication path planning based on the minimum encircling circle algorithm for the UAV working at the lowest power loss. We acquired the minimum radius of UAV relay communication based on for moving platform positions, and then determined the 3D position and flight path of UAV. Compared with the centroid enclosing circle algorithm, our simulation can reduce the signal transmitting power, realize the path planning of UAV relay communication under the lowest power loss.

1. Introduction

Aiming at the problem that the communication distance in radio wave transmission is limited and easily affected by environmental factors such as topography and meteorology, many kinds of wireless communication relay methods are put forward in the world at present to expand the communication distance and improve the communication quality. Herein, the UAV platform has been widely used in civil and military fields because of its high flexibility and low cost [1-3]. In recent years, the researches focus on the impact of UAV path planning on communication performance and optimal layout based on the centroid enclosing circle algorithm. Kramer et al. [4] analysed the network capacity of different relay forwarding strategies. Fu et al. [5] shown the optimized search path planning and communication performance of UAV relay communication. Han et al. [6] gave four kinds of network connectivity criteria. In reference [7], UAV optimal path planning based on space-time block coded communication protocol is proposed. However, in practice, the long working time and the large relay range require sufficient power capacity of UAV for the process of mobile and communication. The lowest power loss could lead to prolong the air residence time and increase the duration of the relay communication of UAV. In this paper, we present a UAV relay communication path planning algorithm based on the lowest power loss for the motion law of the ground moving platforms, which ensures the duration of UAV relay operation is extended while the ground moving platforms are within the relay communication range at the same time.

2. Problem Description

The main two objectives of this paper are as follows: Firstly, make the anticipant ground moving platforms included in relay range, keep wireless link unblocked during the course of motion. Secondly, the transmit power of UAV antenna should be reduced as much as possible in order to prolong the single flight time and reduce the power loss of UAV.

The initial ground platforms are assumed randomly distributed in a rectangular range, with the random moving direction. Here, the range of motion is not necessarily limited to the initial rectangle. If the UAV adopts directional antenna, its relay communication range can be simplified to a circle with radius varying with flight altitude. The relay process of UAV to ground moving platforms is divided into two steps: First, search the ground platforms at a certain time to get their position coordinates. Second, on the basis of making sure that all the ground platforms are within the relay communication range, determine the position of the UAV at this time.

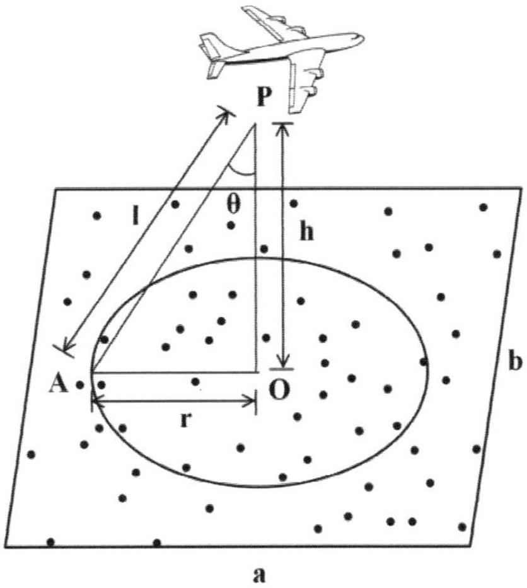

Figure 1. Relay communication scenario diagram.

The relay communication scenario diagram is shown in Figure 1. The black spots represent the ground communication platforms. The *n* platforms are randomly distributed in the bounded rectangular region before the motion starts. The communication platform moves randomly at a fixed speed. The UAV carries a directional antenna, projecting a circular center and radius on the ground at a fixed angle. To ensure that all platforms can communicate with the UAV anywhere, the coverage circle should include all ground communication platforms. In order to prolong the single flight relay operation time, the antenna power loss of UAV should be minimized.

3. Problem Modelling

3.1. Modelling of motion platforms

It is assumed that each ground platform is a vehicle moving at a constant speed and that the direction of movement of each communication vehicle changes once every time (*t* minutes). According to the actual situation, the instantaneous change angle of the driving direction of the vehicle should be $-45° \sim 45°$. Assuming that the speed of the communication vehicle be *v*, and take the random value $-45° < \alpha_1 < 45°$ for each direction change angle, then the mode of the changing direction angle of the vehicle per minute is shown in Figure 2.

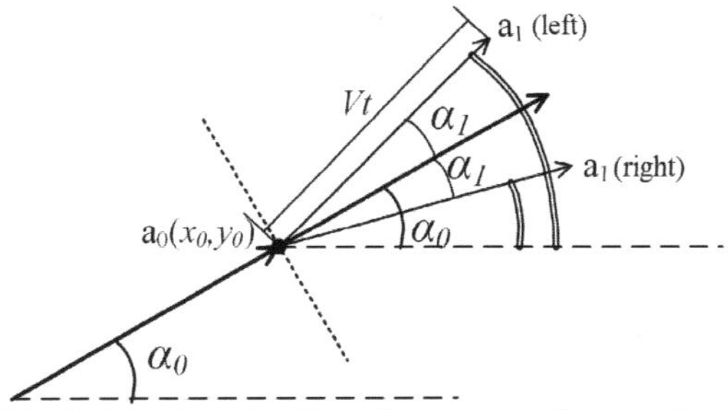

Figure 2. The model of changing the direction angle of ground moving platform.

Compared to the original direction of the vehicle, it turns to the left to be positive, that is $0° < \alpha_1 < 45°$, to the right to be negative, that is $-45° < \alpha_1 < 0°$. The position of the platform is calculated in *t* minutes, and the platform is located at the point $a_0(x_0, y_0)$ before moving, and the angle is α_0 relative to the plane coordinate system. The coordinate of the platform moving after t minutes is

$$a_l(x_0 + vt \times \cos(\alpha_0 + \alpha_1), y_0 + vt \times \sin(\alpha_0 + \alpha_1)) \tag{1}$$

By analogy, the motion paths of each communication vehicle can be obtained in *t* minutes.

3.2. Modelling of optimal flight path for UAV

The optimal flight path of UAV should ensure minimize the transmit power of UAV while the ground platforms located in the relay coverage area. Assuming that the UAV relay communication performance indicators are known, the average received signal power of the ground user is [8]:

$$P_r = P_t G_t G_r (A_f \Delta L)^{-1} \tag{2}$$

$$A_f = (4\pi d / \lambda)^2 \tag{3}$$

Here, the transmit power of UAV is P_t, the antenna gain is G_t, the receiving antenna gain is G_r, the extra loss caused by weather, feeder and other factors is ΔL ,the signal free space propagation path loss is A_f, the propagation distance of unmanned signal is d and the communication wavelength is λ.

According to the model of UAV relay communication coverage area proposed by Zhu et al. [9], the radius of UAV relay communication coverage circle is obtained as follows:

$$r \le (R\left(\frac{P_t G_t G_r \lambda^2}{16\pi^2 \Delta L K R_b T \gamma_0} - H^2\right)(R+H)^{-1})^{1/2} \tag{4}$$

Here, the radius of the earth is R, the altitude of the UAV is *H*, the Polzeman constant is *K* and the noise temperature at the receiver end of the signal is *T* and the minimum threshold of receiver SNR is γ_0.

Equation (4) shows that the transmit power P_t of UAV with the same performance index is positively correlated with its relay coverage radius *r*. In order to reduce the transmit power of UAV, the radius of coverage circle should be minimized. As all ground platforms are located within the relay coverage area of UAV, the circle coverage area should include all the coordinate positions of communication vehicle. Therefore, the lowest transmit power of UAV can be converted into the minimum radius of coverage circle. According to the minimum enclosing circle algorithm [10], the minimum enclosing circle can be obtained by the following four steps according to the coordinates of *n* points in the plane:

The first step is the random selection of three points in the *n*-point set.

The second step is to make a minimum circle containing the three points A, B and C either on the circle (as shown in Figure 3) or two points on the circle at both ends of the diameter of the circle and the other point in the circle.

The third step is to traverse the point D which is the farthest point from the second step in the residual point of the point set. If D is located in or around a circle, the second step is to obtain the circle, which is the minimum encircling circle of the point set, otherwise.

The fourth step is to generate a minimum circle of three points in each of the four points ABCD and select the smallest circle from all the circles. The selected three points are set to the new points A, B, C and return to the second step.

If only two points of ABCD are located on the circumference of the circle generated in the fourth step, then the two points are set to the new point A and point B, and any of the remaining two points is taken as the new C point.

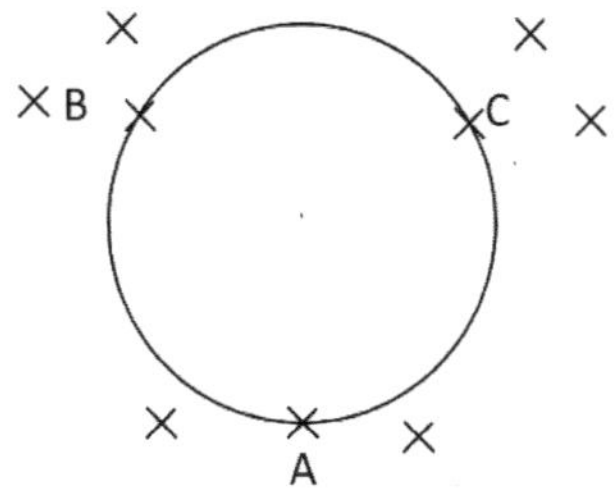

Figure 3. Minimum enclosing circle algorithm.

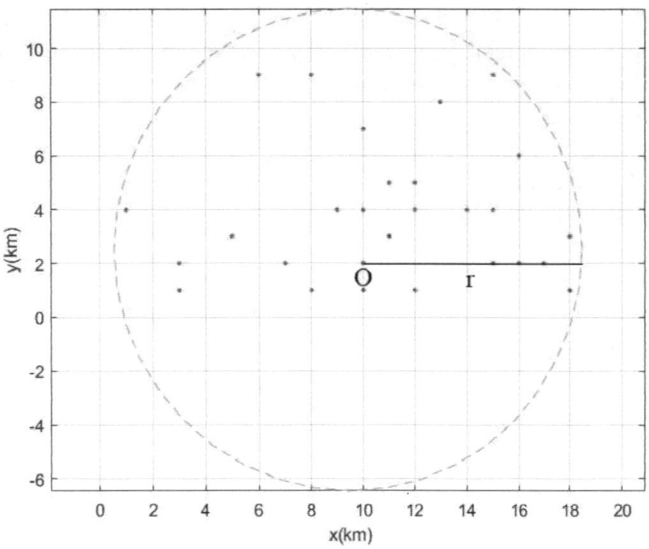

Figure 4. The minimum enclosing circle of the n points set.

For the position coordinates of n communication platforms in the initial rectangle, the minimum radius of relay communication for the initial UAV can be obtained by using the minimum bounding circle algorithm, in which the center position is the plane position of the UAV. As shown in Figure 4. At this point, according to the communication scene Figure 1, the altitude of UAV is

$$h = r\tan\theta \tag{5}$$

The random motion of n-th point set is the motion of ground vehicle. Therefore, the motion model of n-th point set is the same as that of section 3.1, and the direction of motion of each point changes every minute. The coordinates of n points are updated every minute according to the moving platform model, and the center position of the minimum enclosing circle is obtained again. The minimum radius of UAV relaying to the moving platform and its flying height can be obtained for a period of time.

4. Analysis of Simulation Result

In the simulation, the initial vehicle range is set to a rectangle of $20km \times 10km$.

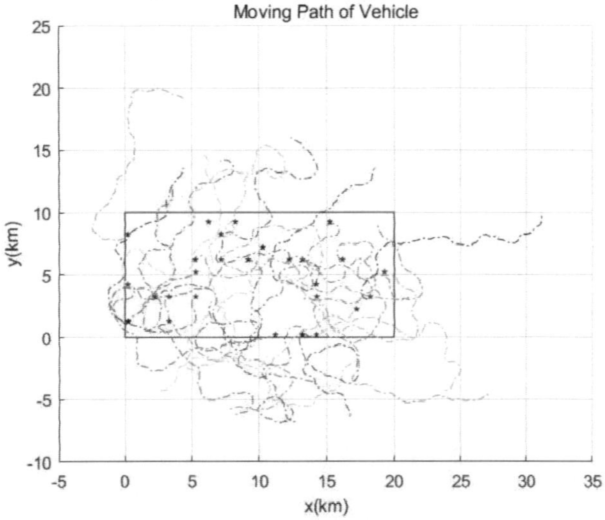

Figure 5. Moving path of 30 vehicles.

Suppose there are 30 ground communication vehicles, the initial position of the vehicle is a random point in the rectangle, and each vehicle moves at the speed of $20km/h$ at a uniform speed. If the direction of movement of the vehicle is randomly changed at $-45° \sim 45°$ per minute, all vehicle tracks are shown in Figure 5 after an hour.

Different color is used to describe every track of 30 vehicles. The rectangular frame is the random range of the initial vehicle, and the asterisks represent the random position of the vehicles in the original rectangle.

Assuming that the antenna direction angle of UAV is $\theta = 60^{\circ}$, the minimum enclosing circle position can be obtained according to the position of ground communication vehicle per minute. The initial position of the vehicle relay circle and the position of the relay circle after one minute of motion of the vehicle are obtained as Figure 6.

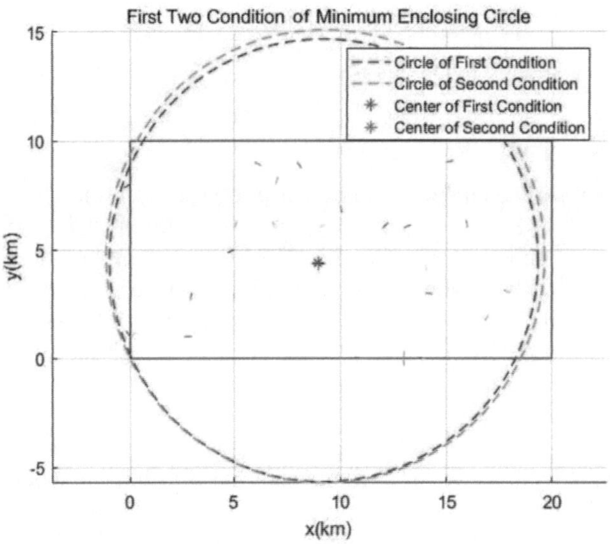

Figure 6. First two-condition of minimum enclosing circle.

The moving path of UAV can be obtained as Figure 7, with the changes of the center of every circle in 60 minutes.

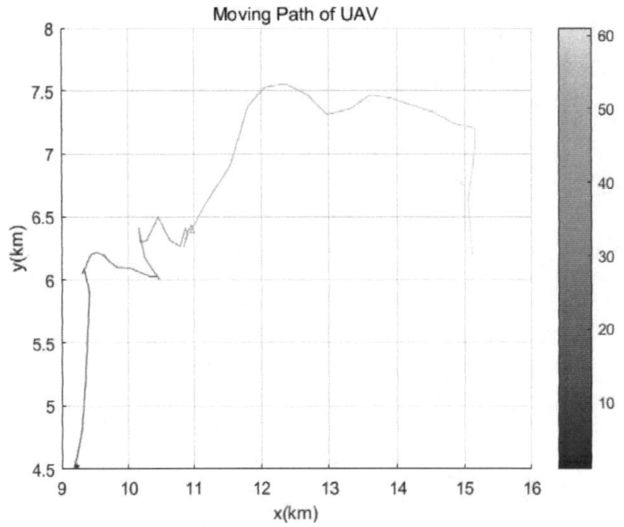

Figure 7. Moving path of UAV.

Different colors are used to represent the changes in the moving path of UAV from 0 to 60 minutes. According to the minimum enclosing circle position in the first 60 minutes, the three-dimensional position diagram of UAV flight can be obtained.

IOP Conf. Series: Journal of Physics: Conf. Series **1215** (2019) 012024 doi:10.1088/1742-6596/1215/1/012024

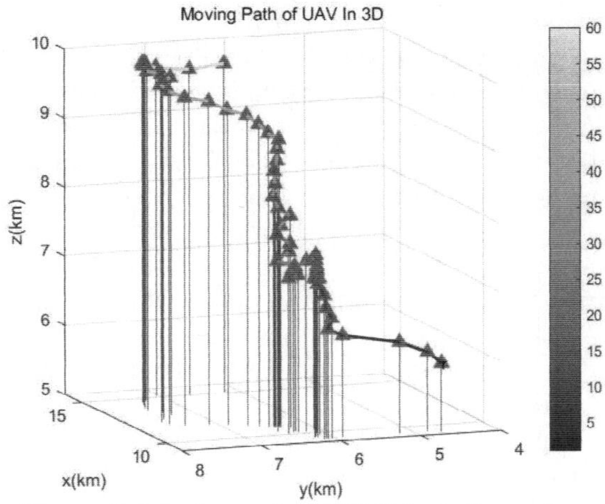

Figure 8. Moving path of UAV in 3 dimensions.

Different colors are used to represent the change in position of the UAV's flight path from 0 to 60 minutes.

Figure 8 shows how the UAV's flight path changes over time due to the random motion of the ground platform. Note that the overall flight height is increasing.

In order to improve the connectivity between UAV and the ground nodes, Xu et al. [11] proposed a weighted centroid UAV flight model. The research object of his model is fixed communication nodes on the ground. After giving different weights to each communication node, the centroid position is calculated to cover the circular center as the relay communication of UAV. In order to ensure that all platforms are located in the UAV relay communication circle, the radius of the circle is the distance between the center and the furthest communication platform.

If n communication platforms are regarded as the same weight nodes, the centroid enclosing circle algorithm can be used to calculate the position and radius of the center of relay communication coverage of UAV. According to formula (5), the altitude of UAV can be further obtained. Using one minute as the simulation step, comparing the circle radius of the UAV relay communication and the flight height calculated by the centroid enclosing circle algorithm and the minimum enclosing circle algorithm within one hour, plotted as Figure 9.

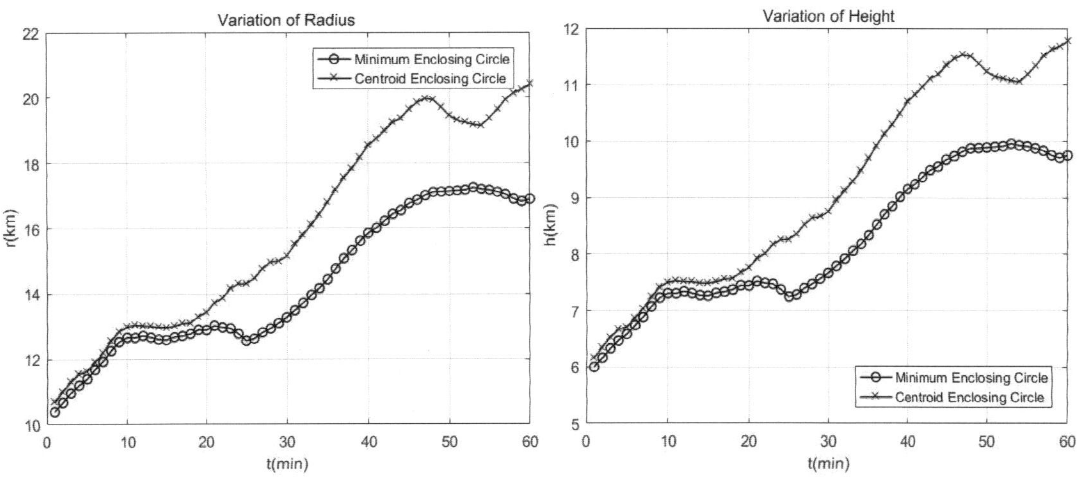

Figure 9. The comparison of circular radius variation and the altitude variation of UAV between the simulation results from the centroid enclosing circle algorithm and the minimum enclosing circle algorithm.

As can be seen from the comparison in the figure, compared with the enclosing circle UAV relay communication path planning method, the UAV relay communication path method based on the lowest power loss is more effective in reducing the radius of the relay circle and reducing the height of the drone. When relaying the mobile communication platform with initial position in the $20km \times 10km$ rectangle, the altitude is between 6km and 10km. Due to the positive correlation between the transmit power and the relay communication radius, the method of this paper is superior to the centroid enclosing circle algorithm in reducing the total power consumption. The simulation results show that the UAV relay communication path method based on the lowest power loss can achieve the goal of extending the air stay time and the relay operation time of a single UAV after take-off.

5. Conclusion

In this paper, the path planning and UAV performance optimization are studied in the course of relay communication between UAV and ground moving platforms. Aiming at the random moving law of the ground motion platform, we present a method of UAV relay communication path planning based on the lowest power loss, and adopts the minimum enclosing circle algorithm. According to the dynamic changes of each vehicle, the UAV flight path and altitude are adjusted in time, so that the UAV can reduce the signal transmission power on the basis of supporting all ground motion platforms to communicate, thus reducing the total power loss of the UAV signal emission. It is shown that the proposed method can achieve the lowest power loss by reduce the radius of the relay communication coverage circle and reduce the UAV flight altitude, then prolong the air residence time and increase the duration of the relay communication of UAV.

References

[1] Gu D L, Ly H, Hong X Y, Gerla M, Pei G Y, Lee Y Z 2000 *IEEE: Wireless Communications and Networking Conference* **2** 879
[2] Xu K X, Hong X Y, Gerla M, Ly H, Gu D L 2001 *IEEE: Military Communications Conference* **1** 230
[3] Perumal S, Baras J S, Graff C J, Yee D G 2008 *IEEE: Military Communications Conference* **1** 7
[4] Kramer G, Gastpar M, Gupta P J 2005 *IEEE Trans. Inf. Theory* **51** 3037
[5] Fu X W, Cheng S M, Gao X G 2014 *J. Syst. Eng. Electron.* **36** 890
[6] Han Z, Swindlehurst A L, Liu K J R 2009 *IEEE Trans. Veh. Technol.* **58** 3533
[7] Liu H T, Zhao W Q, Lee C M, Lee D X 2017 *J. Aeronaut.* **38** 274
[8] Feng Q, McGeehan J, Nix A R 2007 *IEEE: Vehicular Technology Conference* **65** 954
[9] Zhu Q M, Zhou S K, Huo S K, Chen X M, Xu D Z 2014 *Acta Aeronaut. Astronaut. Sin.* **35**, 223
[10] Wang W, Wang W P 2000 *J. Software* **11** 1237
[11] Xu X Z, Yuan J, Wang Y, Zhang Y D, Yi L T, Huo J H, Huo Z M 2011 *J. Tsinghua Univ.: Nat. Sci. Ed* **2** 150

Acknowledgments

We acknowledge the funding from the National Natural Science Foundation of China (No. 61602516).

Neural Partial Differentiation Based Nonlinear Parameter Estimation from Noisy Flight Data

Majeed Mohamed[1], Madhavan G[2], R Manikantan [1] and PS Lal Priya[2]

[1]Senior Scientist, National Aerospace Laboratories, Bangalore, India
[2]Department of Electrical Engineering, College of Engineering, Trivandrum, India

Abstract. This paper focuses on the application of neural partial differentiation (NPD) approach to estimate the longitudinal parameters of an aircraft HFB 320 from noisy flight data. By exciting both short period and phugoid modes of an aircraft with thrust variation, the aircraft system dynamics becomes highly nonlinear and aerodynamic parameters appears nonlinear to the state trajectories of velocity, AOA, pitch rate and pitch angle. This paper highlights the application of NPD for such a class of nonlinear dynamics; previously it was used only for the estimation of parameter appearing linear to the states. The extracted the nonlinear longitudinal parameters of HFB 320 aircraft are compared with the parameters estimated by using adaptive Unscented Kalman Filter (UKF) approach. Finally, the estimation results are validated by comparing with flight data and the responses obtained from the estimates by adaptive UKF.

1 introduction

Aircraft modeling and parameter estimation play a vital role in determining its stability and control characteristics of the complex aircraft system. Presently online aircraft modeling method has attracted more and more research attention for the real-time estimation of aircraft parameters in the design of an adaptive controller to maintain flight stability and high performance in an uncertain environment [1], [2], [3], [4]. However, the recently introduced neural partial differential approach is capable to provide theoretical insight into the statistical information (standard deviation and relative standard deviation) of parameter estimates from noisy data. In the presence of an adverse atmospheric condition, aircraft aerodynamic parameters can be estimated accurately from flight data using adaptive unscented Kalman filter [5][6][7]. But this algorithm requires high computational power and initial values of the parameters.

Online estimation of the aircraft aerodynamic parameters from noisy flight data is challenging since aircraft system dynamics are nonlinear and stochastic in nature. Most of the established algorithms like Filtering Methods (FM) and Output Error Method (OEM) are needed prior knowledge of initial values of the parameters and system dynamic models for estimating aircraft system parameters. The Equation Error Method (EEM) does not require initial values of parameters but it restricted to linear system dynamics. Neural networks are widely used to model the nonlinear aircraft system dynamics [8], [9], [10], [11], [12]. Consequently, Delta and Zero methods are used as estimation approaches from neural model [8], [13], [14]. Even though these approaches provide parameter values through the offline analysis, online estimation of aerodynamic parameters from noisy is not feasible with statistical information of the estimated parameters.

Neural Partial Differentiation (NPD) approach is recently introduced to extract derivatives in real time from noisy flight data with their statistics of Relative Standard Deviation (RSTD) [15], [16]. In the application of NPD to flight data, a neural model of an aircraft is primarily established from flight data and estimated aerodynamic derivatives from the neural model by partial differentiation. When we consider the excitation of both modes of longitudinal motion namely short period mode and phugoid mode, the velocity of an aircraft varies by the employment of thrust variation. In such a case, aircraft system parameters are appearing nonlinear to the state trajectories of the aircraft and estimating those parameters are hard in nature. In this paper, we explain how estimate aircraft aerodynamic derivatives accurately from noisy can flight data of short period and phugoid mode where aerodynamic derivatives considered as parameters appearing nonlinear to the state trajectories. The main contributions of this paper are:

- Longitudinal flight stability and control parameters of HFB-320 aircraft are estimated from flight data. Neural partial differentiation method is used for this purpose.
- For the precise excitation of the system dynamics, both short period and phugoid modes of aircraft are excited during the maneuvering of flight for the estimation of aircraft aerodynamic parameters in longitudinal dynamics.
- In the case of velocity variation, aircraft system dynamics are highly nonlinear and aerodynamic parameters are appearing nonlinear to the state trajectories of velocity, AOA, pitch rate and pitch angle. Neural partial differentiation method is applied to multi- input single out aircraft system (MISO) to extract the nonlinear parameters of HFB 320 aircraft. The results are compared with the estimates obtained from adaptive UKF.
- The estimated Neural Model of aircraft is verified by comparing the time history of a normal acceleration and pitch rate of flight data with their estimated model responses.

The rest of the paper is organized as follows. Section II presents the problem formulation for the aircraft parameter estimation using Neural Partial Differentiation. The estimation results of aerodynamic derivatives and their validation are discussed in section III. Section IV describes concluding remarks of the paper.

2 Aircraft parameter estimation

As the aircraft system dynamics are nonlinear and stochastic in nature, the linearized models without having accurate knowledge of noise characteristics to them may lead to inaccurate results and poor performance. Thus, the feasibility study of adaptive UKF for aircraft parameter estimation is carried out from flight data [5] to addressing the issues related to nonlinearity and unknown noise statistics of covariance. But adaptive UKF algorithm requires initial values of the estimates and shows high computational power. This motivates the use of neural partial differentiation (NPD) method that can provide accurate estimates of aircraft parameters without their initial values. The following section describes how can we apply NPD to estimate nonlinear aircraft parameters from noisy flight data of HFB 320 aircraft (Figure:1). The details of the NPD method are not given in this paper in order to avoid the repetition, and they are referred in [17].

Figure 1. HFB 320 aircraft

2.1 Application of NPD to HFB 320 Aircraft.

The model structure of HFB 320 aircraft for the longitudinal dynamics is given by:

$$\begin{cases} \dot{V} = -\dfrac{\bar{q}S}{m}C_D + g\sin(\alpha-\theta) + \dfrac{F_e}{m}\cos(\alpha+\sigma_T) \\ \dot{\alpha} = -\dfrac{\bar{q}S}{mV}C_L + q + \dfrac{g}{V}\cos(\alpha-\theta) - \dfrac{F_e}{mV}\sin(\alpha+\sigma_T) \\ \dot{\theta} = q \\ \dot{q} = \dfrac{\bar{q}s\bar{c}}{I_y}C_m + \dfrac{F_e}{I_y}(l_{Lx}\sin\sigma_T + l_{Lz}\cos\sigma_T) \end{cases} \tag{1}$$

Where the lift, drag and pitching moment coefficients (C_L, C_D, C_m) are modeled as

$$\begin{cases} C_D = C_{D0} + C_{DV}\dfrac{V}{V_0} + C_{D\alpha}\alpha \\ C_L = C_{L0} + C_{LV}\dfrac{V_0}{V_0} + C_{L\alpha}\alpha \\ C_m = C_{m0} + C_{mV}\dfrac{V}{V_0} + C_{m\alpha}\alpha + C_{mq}\dfrac{q\bar{c}}{2V_0} + C_{m\delta e}\delta_e \end{cases} \tag{2}$$

Where V_0 is initial value of V. The dynamic pressure $\bar{q}(=0.5\rho V^2)$ contains state variable V which introduces nonlinearity in estimates process as it multiplies with all unknown parameters. To obtain aerodynamic parameters $C_{m0}, C_{mV}, C_{m\alpha}$, C_{mq} and C_{me}

Substitute C_m from (2) in $\dot{q}$ in (1).

Thus,

$$\dot{q}I_y - F_e(l_{Lx}\sin\sigma_T + l_{Lz}\cos\sigma_T) = V^2C_{m0} + C_{m\alpha}\alpha V^2 + C_{mq}\frac{q\bar{c}V^2}{2V_0} + C_{m\delta e}\delta_e V^2 + C_{mV}\frac{V^3}{V_0} \tag{3}$$

The left hand side of the above equation can be computed from the measured flight data and it can be used as the output of the neural network by considering model structure of the right hand side of equation (3). With use of the states V, α, q and δ_e the inputs to the same neural network are $V^2, \alpha V^2, \dfrac{q\bar{c}V^2}{2V_0}, \delta_e V^2$ and $\dfrac{V^3}{V_0}$ according to (3). For obtaining the parameters $C_{L0}, C_{LV}, C_{L\alpha}, C_{D0}, C_{DV}$ *and* $C_{D\alpha}$, the equation for acceleration along z-axis is given below:

$$a_{zm} = \frac{\bar{q}S}{m}C_z + \frac{F_e}{m}\sin\sigma_T \tag{4}$$

Where,

$$C_Z = -C_L\cos\alpha - C_D\sin\alpha \tag{5}$$

Therefore, from (2) and (5) in (4):

$$\frac{ma_{zm} - F_e\sin\sigma_T}{0.5\rho S} = -C_{L0}V^2\cos\alpha - C_{LV}\frac{V^3}{V_0}\cos\alpha - C_{L\alpha}V^2\alpha\cos\alpha - -C_{D0}V^2\sin\alpha - C_{DV}\frac{V^3}{V_0}\sin\alpha - C_{D\alpha}V^2\alpha\sin\alpha \tag{6}$$

The left hand side of the above equation can be computed from flight data and used as the output of other neural network with model structure of the expression in right hand side of equation (6). With use of the states are V and α, the inputs to the same neural network are $V^2, \alpha V^2, qV^2 \delta_e V^2$ and V^3 according to (6).

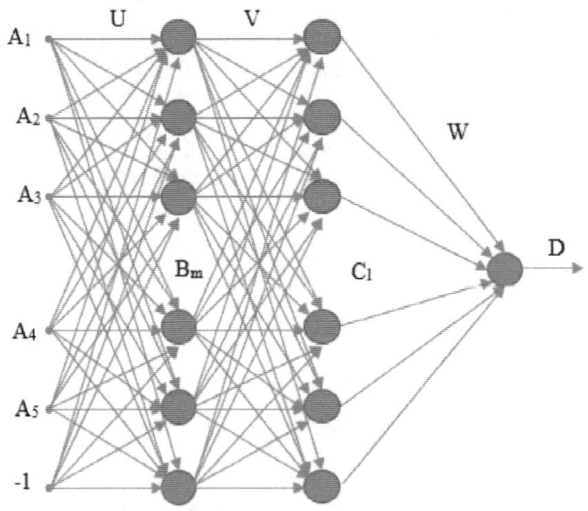

Figure 2. The Schematics of Neural Network

The schematics of such a neural network is given in fig.2 contain two hidden layers with inputs $A_0, A_1 ... A_5$ where A_0 is the bias [17].

$$\begin{cases} A_1 = V^2 \\ A_2 = \alpha V^2 \\ A_3 = \dfrac{q\bar{c}}{2V_0} V^2 \\ A_4 = \delta_e V^2 \\ A_5 = \dfrac{V^3}{V_0} \end{cases} \tag{7}$$

From the Figure. 2, the output D is the sum of products of total inputs from hidden layer 2 and their respective weights of each neuron. Thus,

$$D = W^T C \tag{8}$$

Where,

$$W = \begin{pmatrix} b_{w1} \\ w_1 \\ \vdots \\ w_l \end{pmatrix} \tag{9}$$

Similarly,

$$\begin{cases} C = f\left(V^T B\right) \\ B = g\left(U^T A\right) \end{cases} \tag{10}$$

f and g are activation functions of first and second hidden layers respectively $f(x)$ is expressed as:

$$f(x) = \frac{1 - e^{-\lambda x}}{1 + e^{-\lambda x}} \tag{11}$$

The weight matrices are represented as:

$$V = \begin{pmatrix} b_{v1} & \cdots & b_{vl} \\ v_{11} & \cdots & v_{1l} \\ \vdots & \ddots & \vdots \\ v_{m1} & \cdots & v_{ml} \end{pmatrix} \tag{12}$$

$$U = \begin{pmatrix} b_{u1} & \cdots & b_{um} \\ u_{11} & \cdots & u_{1m} \\ \vdots & \ddots & \vdots \\ u_{n1} & \cdots & u_{nm} \end{pmatrix} \tag{13}$$

The input and output values are scaled to the range of
-0.9 to 0.9. From the above equations, the output can be expressed as:

$$D = \left\{ W^T f \left[V^T g \left(U^T A \right) \right] \right\} \tag{14}$$

Partial differentiation of output of each layers are given by

$$\frac{\partial D}{\partial C} = W^T \tag{15}$$

$$\frac{\partial C}{\partial B} = f'\left(V^T\right) \tag{16}$$

$$\frac{\partial B}{\partial A} = g'\left(U^T\right) \tag{17}$$

Multiplying eq. (15), eq. (16) and eq. (17) gives,

$$\begin{cases} \dfrac{\partial D}{\partial C} \cdot \dfrac{\partial C}{\partial B} \cdot \dfrac{\partial B}{\partial A} = W^T \cdot f'\left(V^T\right) \cdot g'\left(U^T\right) \\ \dfrac{\partial D}{\partial A} = W^T \cdot f'\left(V^T\right) \cdot g'\left(U^T\right) \end{cases} \tag{18}$$

Where, $f' = \text{diag}\left[0\, f_1' \ldots f_l'\right]$ and $g' = \text{diag}\left[0\, g_1' \ldots g' m\right]$
If the input and output of neural network are normalized, then

$$\frac{\partial D}{\partial A} = \frac{\partial D}{\partial D_{\text{norm}}} \times \frac{\partial D_{\text{norm}}}{\partial A_{\text{norm}}} \times \frac{\partial A_{\text{norm}}}{\partial A} \tag{19}$$

The normalized output of neural network can be denormalized by eq. (19). Where,

$$\frac{\partial A}{\partial A_{\text{norm}}} = \begin{bmatrix} 1 & 0 & \cdots & 0 \\ 0 & \dfrac{\partial A_1}{\partial A_{1,\text{norm}}} & \cdots & 0 \\ \vdots & \vdots & \ddots & \vdots \\ 0 & 0 & \cdots & \dfrac{\partial A_n}{\partial A_{n,\text{norm}}} \end{bmatrix} \tag{20}$$

The terms associated of eq. (15) to (20) be intermediate terms of neural networks while getting it trained. Therefore, there is no extra computation required to compute the aerodynamic derivatives, and they are directly given as:

$$\frac{\partial D}{\partial A} = \begin{bmatrix} \dfrac{\partial D}{\partial A_1} & \cdots & \dfrac{\partial D}{\partial A_n} \end{bmatrix} \tag{21}$$

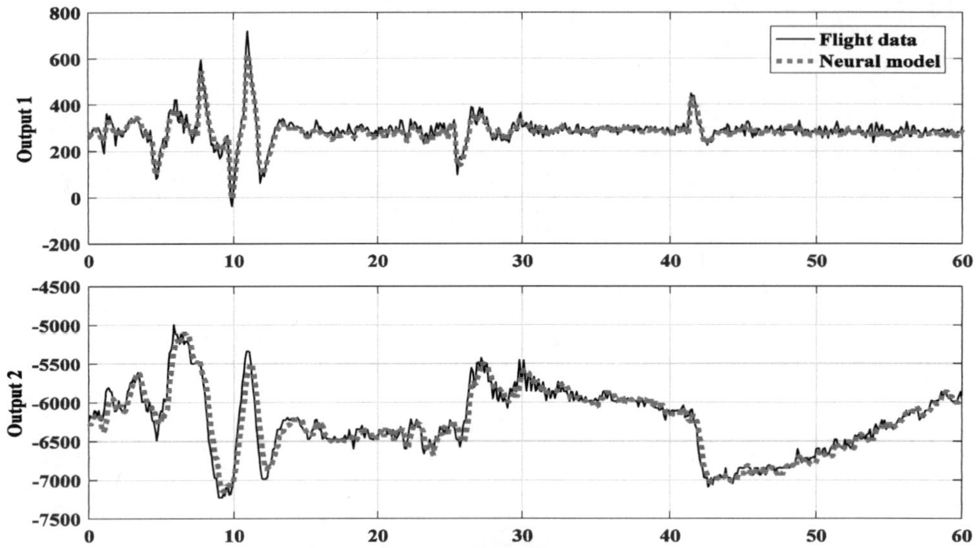

Figure 3. Neural Model of an Aircraft

From, eq. (3) and eq. (7), the output of neural network is:
an advantage over adaptive UKF to estimate aerodynamic parameters from flight data.

$$D = C_{m0}A_1 + C_{m\alpha}A_2 + C_{mq}A_3 + C_{m\delta_e}A_4 + C_{mV}A_5 \tag{22}$$

Therefore,

$$\frac{\partial D}{\partial A} = \begin{bmatrix} C_{m0} & C_{m\alpha} & C_{mq} & C_{m\delta e} & C_{mv} \end{bmatrix} \tag{23}$$

Similarly, the parameters $C_{L0}, C_{LV}, C_{L\alpha}, C_{D0}, C_{DV}$ and $C_{D\alpha}$ can be estimated easily by designing a neural network based on eq. (6). The standard deviation of estimated parameters in eq. (21) is computed by

$$STD = \sqrt{\frac{\sum_{p=1}^{P}\left[\sum_{m=1}^{M}\left(\sum_{l=1}^{L} C'_{lp} v_{lm} w_{kl} D'_{kl}\right) B'_{mp} u_{mi} - AVG\right]^2}{P}} \tag{24}$$

Where,

$$AVG = \frac{\sum_{p=1}^{P}\sum_{m=1}^{M}\left(\sum_{l=1}^{L} C'_{lp} v_{lm} w_{kl} D'_{kp}\right) B'_{mp} u_{mi}}{P} \tag{25}$$

Where, STD and AVG are standard deviation and average of data points, respectively. The relative standard deviation of estimates is given by:

$$RSTD = \frac{STD}{AVG} \times 100\% \tag{26}$$

2.2 Adaptive Unscented Kalman Filter

Parameter estimation of a nonlinear system can be accomplished using Unscented Kalman Filter (UKF) [18]. But the process noise covariance is computed beforehand through Filter Error Method (FEM) [5]. FEM is an offline algorithm which has the potential to estimate noise covariance in addition to system parameters. Whereas, the selection of process noise covariance is carried out in an adaptive UKF algorithm, which consists of two parallel filters referred to as master and slave. The master UKF estimates both states and aerodynamic parameters of aircraft while the slave one estimates the diagonal elements the noise covariance matrix for the master UKF. The adaptive UKF approach has been introduced for the purpose of estimating aerodynamic derivatives from flight data of HFB 320 [5].

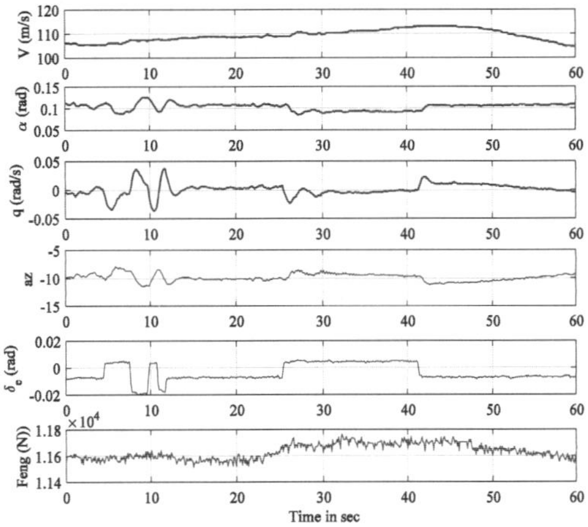

Figure 4. Time history of input signals to the neural network

3 Parameter estimation results and discussion

Aircraft parameter estimation was carried out with open accessible flight test data of the research aircraft HFB- 320 aircraft. The flight tests were conducted to excite the longitudinal motion of research aircraft through a multistep elevator input of 3-2-1-1 resulting in short period motion and a long duration pulse input leading to phugoid motion [5], [19]. The resulting flight data is corrected the flow angle sensor corrections through the flight path reconstruction [20], and ready to use for estimating aircraft longitudinal parameters to quantify the aircraft lift, drag, and pitching moment coefficients.

Table 1. Estimated parameter values

Parameters	NPD	Adaptive UKF
C_{m0}	0.1227 (1.2757)*	0.1115 (4.29)
$C_{m\alpha}$	-1.067 (0.9246)	-0.9703 (1.54)
C_{mq}	-38.899 (1.1909)	-35.3628 (2.82)
$C_{m\delta e}$	-1.6934 (0.930)	-1.5395 (1.65)
C_{mV}	0.0051 (7.3542)	0.0045 (92.19)
C_{l0}	-0.0783 (4.1629)	-0.0871 (22.91)
C_{lV}	0.1321 (2.0504)	0.1468 (11.42)
$C_{l\alpha}$	3.8604 (0.4076)	4.2894 (1.14)
C_{d0}	0.1103 (1.4113)	0.1226 (2.64)
C_{dV}	-0.0578 (3.3394)	-0.0642 (4.27)
$C_{d\alpha}$	0.2877 (0.8204)	0.3197 (2.40)

* The values in parenthesis denote relative standard deviation values in percentage.

A prior information of the postulated model structure of aircraft longitudinal dynamics is required to extract the parameters using neural partial differentiation method from a neural model of an aircraft. For estimating aerodynamic derivatives of C_{m0}, C_{mV}, C_{m}, C_{mq} and $C_{m\delta e}$ we can use eq.(22) as the model structure for estimation algorithm, and neural model of aircraft can be established with inputs of $A_i, i = 1,2,3,4,5$ and output of network computed from the left hand side of the eq.(3). Similarly, we can establish another neural model of the aircraft with output computed from left hand side of eq.(6) and inputs of $-V^2\cos\alpha, -V^3\frac{\cos\alpha}{V_0}, -V^2\alpha\cos\alpha, -V^2\sin\alpha, -V^3\frac{\sin\alpha}{V_0}$ and $-V^2\alpha\sin\alpha$. Subsequently, aerodynamic parameters $C_{L0}, C_L V, C_{L\alpha}, C_{D0}, C_{DV}$ and $C_{D\alpha}$ are estimated by the application of NPD from the established neural model.

Figure 3 shows the established neural models with respective eq. (3) and (6), and the extracted aircraft longitudinal parameters are tabulated in Table 1 with their relative standard deviations. The estimated parameter values were found to attain a stable value as the number of iterations increased as shown in figures 7, 8, where the variation of parameters with number of iterations is given. Aircraft estimated parameter variation with respect to the flight data points are also plotted in figures 5, 6. The estimated results are compared with those obtained from the adaptive UKF algorithm to demonstrate the efficacy of proposed approach. Figure 9 shows the comparative responses of pitch rate and normal acceleration measurements with those reconstructed from the parameters estimated by NPD and adaptive UKF. This estimation results confirms that NPD approach has potential to estimate aerodynamic derivatives from flight data where parameters are appearing nonlinear to the aircraft states in nonlinear dynamics.

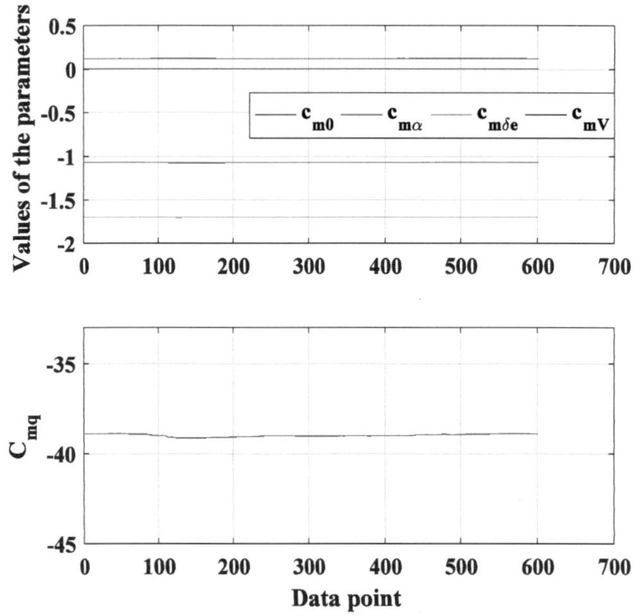

Figure 5. Variation of Parameters w.r.t. data points

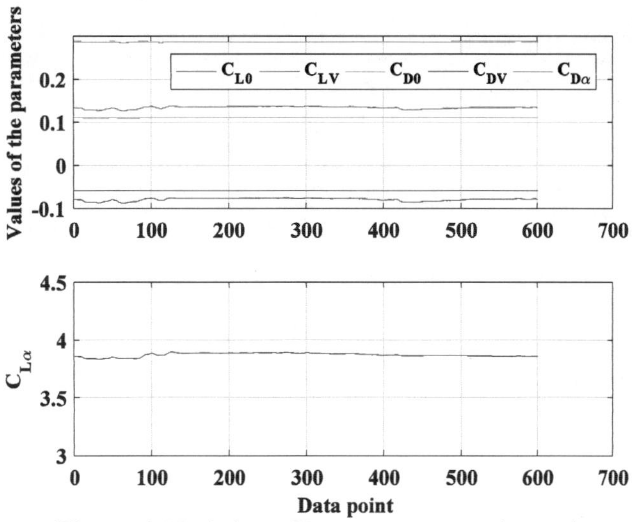

Figure 6. Variation of Parameters w.r.t. data points

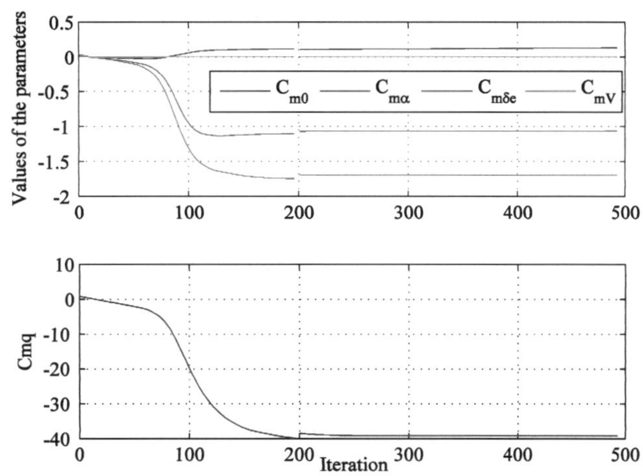

Figure 7. Variation of Parameters w.r.t. number of iterations

4 Conclusions

The NPD method is applied to the flight data of high performance HFB 320 aircraft for the estimation of the longitudinal aerodynamic derivatives. For this, the aircraft was inspired by the multistep input and long duration pulse to the elevator for the excitation of short period and phugoid modes of aircraft. Therefore, the velocity of the aircraft varies with thrust variation to accommodate the phugoid mode of aircraft. This makes the aircraft system dynamics highly nonlinear and aerodynamic parameters appear nonlinear to the state trajectories of velocity, AOA, pitch rate and pitch angle.

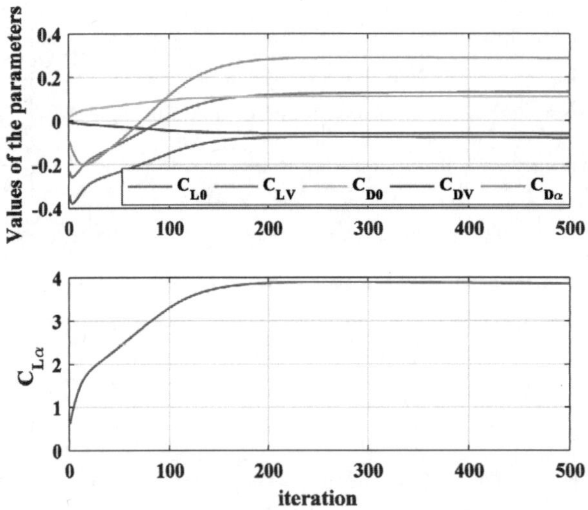

Figure 8. Variation of Parameters w.r.t. number of iterations

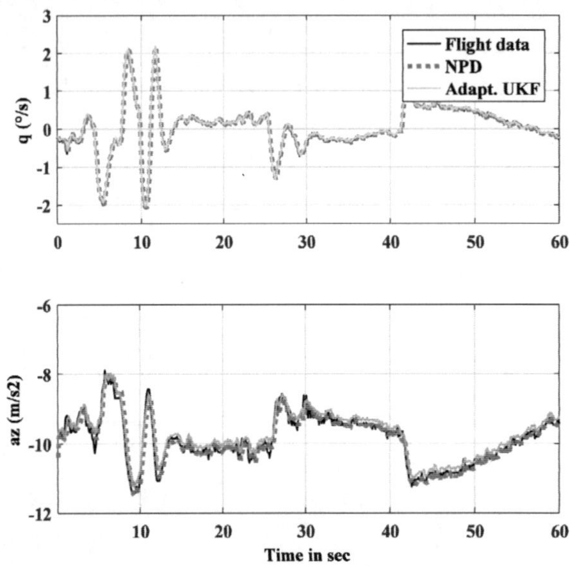

Figure 9. Validation of estimated time histories of NPD with flight data

We have established the neural model of an aircraft with a suitable selection of appropriate inputs and output of the network. Moreover, a separate consideration of neural network is made for the estimation of parameters from the expression of $\dot{q}$ and a_z . Longitudinal derivatives of Aircraft are estimated from flight data using the NPD method, and comparable with estimates obtained from the adaptive UKF. This shows that the NPD method has become the preferred approach to estimate parameters as it does not require the initial value of estimates and less computation. As the initial values of parameters are not available in a practical situation as well as filtering approach requires these initial parameters with high computational cost, application of the NPD approach gives an advantage over adaptive UKF to estimate aerodynamic parameters from flight data.

5. References

[1] J. Leitner, A. Calise, and J. V. R. Prasad, J. of GCD, **20**, pp. 972–979, (1997)

[2] J. Garcia-Velo, B. K. Walker, J. GCD, **20** , pp. 1257–1260, (1997).

[3] B. Lei, G. Xu, M. Feng, F. Van der Heijden, Y. Zou, D. de Ridder, and D. M. Tax,. *John Wiley & Sons*, (2017)

[4] V. N. Yavuztu¨rk, E. Topbas, U¨ . Kutluay, and Y. Yazicioglu, AIAA Atm. Flight Mech. Conf., pp. 1865, (2017)
[5] M. Majeed and I. N Kar, Aircraft Engg. & Aerospace Tech., **85**, pp. 267–279, (2013)
[6] B. Wang, D. Liu, W. Wang, and X. Peng, Microelects. Rel., **84**, pp. 253–262, (2018)
[7] B. Gao, S. Gao, G. Hu, Y. Zhong, and C. Gu, Aero. S& T, **73**, pp. 184–196, (2018)
[8] A. Ghosh, S. Raisinghani, and S. Khubchandani, J. of Aircraft, **35**, pp. 876–881, (1998)
[9] J. O. Pedro and P. Kantue,., 2011 8th Asian. IEEE Contl .Conf, pp. 1170–1175, (2011).
[10] U. J. Pesonen, J. E. Steck, K. Rokhsaz, H. S. Bruner, and N. Duerksen, J. of G.C.D, vol. 27, pp. 434–443, (2004)
[11] S. Saderla, R. Dhayalan, and A. Ghosh, Defence Sci. J., **67**, (2017)
[12] R. Ganguli, Adv. UAV Aero. Flight Stab. and Contrl, **283**, (2017)
[13] S. Raisinghani, A. Ghosh, and P. Kalra, The Aero. J., **102**, pp. 25–30, (1998)
[14] S. Raisinghani and A. Ghosh, *Sadhana*, **25**, pp. 181–191, (2000)
[15] M. Mohamed and V. Dongare, **8** (2018).
[16] M. Sinha, R. Kuttieri, and S. Chatterjee, J. of G.C.D, **36**, pp. 1162–1176, (2013)
[17] V. Dongare and M. Mohamed, Inter.Conf. on. IEEE, pp. 1–6, (2015)
[18] M. Partovibakhsh and G. Liu, IEEE Trans. on Cont. Sys. Tech., **23**, pp. 357–363, (2015)
[19] R. V. Jategaonkar, AIAA Inc., (2015)
[20] M. Majeed and I. N Kar, Inter. J. of Aerospace Innov., **2**, (2010)

IOP Conf. Series: Journal of Physics: Conf. Series **1215** (2019) 012026 doi:10.1088/1742-6596/1215/1/012026

Research on the Whole Station's Interior Orientation Cues Oriented to Navigation Enhancement

Zhu Lei[1,3], **Bai Xianglin**[1,3], **Tang Xiaohan**[2,3] **and Zhang Yanfei**[3*]

[1] Harbin Institute of Technology, Harbin, China
[2] Beijing Union University, Beijing, China
[3] Harbin Institute of Technology Robotics group, Harbin, China

zhangyf@hitrobotgroup.com

Abstract. The large and complex interior space of multi-cabin space station makes the astronauts easily happened to navigational disorder during the travel.With the need to enhance the navigation ability of the astronauts,based on a method study on space station intra-vehicular navigation simulation including both VR visual simulation and physiological HDBR simulation,multiple different sets of sample scenes are designed for the orientation cues numbers reflected by the whole visual orientation cues that has similar form,and the specific form of the whole visual orientation cues that reflects same orientation numbers.Through the experimental data analysis, the influence law of the whole station's interior orientation cues on the navigation task performance and associated physiological effects in the cabin are obtained.

1. Introduction

The multi-cabin space station has a complex three-dimensional axial space structure, and the astronauts are prone to navigation problems in the cabins, the astronauts mainly use visual cues to conduct space navigation[1], one of the visual environmental factors affecting intra-vehicular navigation is the whole station's interior orientation cues.

In this article, the whole station's interior orientation cues refers to the orientation information which is provided by the various elements in the environment, especially the elements that are composed and implied in the space station interior, one of which is the whole station's interior orientation cues.

In 2005, Aoki et al. conducted a comparative test on three sets of color schemes of the space station interior bulkhead, the experimental results showed the visual information were effective for improving the target object orientation ability and spatial memory ability of the virtual weightless environment[2]. The research results of Aoki et al. have provided some reference for the design of visual orientation cues in multi-cabin space stations, but there is no further research on whether there are other effective methods for enhancing navigation capability.

Therefore, based on a simulation system of the space station intra-vehicular navigation including both VR visual simulation and physiological HDBR simulation ,this paper studies the influence law of the orientation cues numbers reflected by the whole visual orientation cues which have similar form on the intra-vehicular navigation task performance and associated physiological effects, and the influence law of the specific form of the whole visual orientation cues which reflect same orientation cues numbers on the intra-vehicular navigation task performance and associated physiological effects, thus to provide an effective reference for enhancing the intra-vehicular navigation capability of astronauts.

2. The space station intra-vehicular navigation simulation system

IOP Conf. Series: Journal of Physics: Conf. Series **1215** (2019) 012026 doi:10.1088/1742-6596/1215/1/012026

In related studies, the VR simulation method is used to simulate the astronauts' visual navigation in multi-cabin space station 's interior by Aoki et al[2-5],and the physiological HDBR simulation is a mature method for simulating physiological effects of weightlessness[6,7],therefore, the VR visual simulation and physiological HDBR simulation are combined as the space station intra-vehicular navigation simulation system, as shown in figure 1. The simulation system is composed of the graphics workstation computer 1,the wide-field head-mounted display 2,the three-dimensional mouse 3, and the angle adjustable experimental bed 4.
The simulation system's scenario of the virtual space station refers to the structure of the international space station and is slightly modified, as shown in figure 2.The virtual space station consists of two cubic node cabins (Node1, Node2) which are 2×2×2 m3 in size and seven rectangular non-node cabins which are 2×2 ×6 m3in size. The interior surface of the cabin is mapped with pictures of the international space station interior, each cabin is distinguished by different visual signs and given unique name.

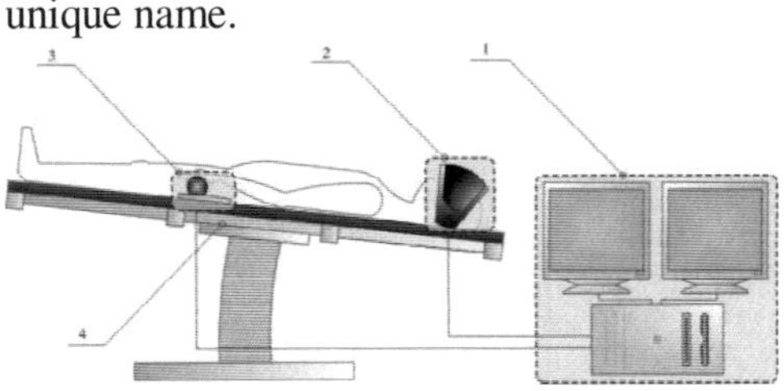

Figure 1. Schematic diagram of the space station intra-vehicular navigation simulation system.

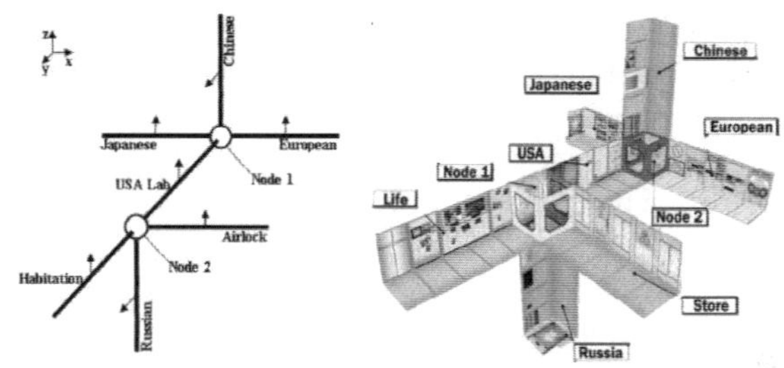

Figure 2. Virtual space station scenario.

The layout and visual signs of the space station cabin remain unchanged but the internal visual design is different. The whole orientation cues in the interior decoration of each cabin are arranged in the gap position of the wall cabin.

3. The experiment

3.1. The experimental subjects

A total of 110 (18-28 years old) subjects participated, all without visual or auditory impairment. The subjects were asked to participate in the mental rotation test and visual rotation ability test before the experiment. All the subjects were divided into 11 groups, with 10 people in each group. The distribution of gender and test scores in each group of subjects was balanced.

3.2. The experimental process

*3.2.1. The preparation stage of experiment.*The subjects were placed on a supine position with the -6 °head-down posture and weared a wide-view helmet-mounted display, kept this pose for a period of time; then, the simulation program of the virtual multi-cabin space station scene in the graphics workstation was started. The subjects operated the three-dimensional mouse to realize the roaming in the multi-cabin space station and to check the equipment availability.

*3.2.2. The navigation training stage of experiment.*In the training part,the subjects firstly followed the simulated astronauts walking along nine preset path in the space station,they went through three rectangular cabins and two node cabins that required the observation orientation of the subjects be consistent with the simulated astronaut.The subjects secondly passed through each cabin of the virtual space station,they must used the keyboard or handle to switch the orientation to complete the corresponding task.The subjects were asked to select one of all the paths on the screen and to go to the cabin on their own under the guidance of the simulated astronaut. After reaching the destination, the subjects needed to return to the starting point,if it was wrong,they were asked to try it again. The subjects visited each cabin three times (except for the American laboratory in the middle of the space station—all paths past through it).

3.2.3. The navigation testing stage of experiment. In the testing part, the subjects were asked to perform orientational and path-finding task,they were asked to indicate the orientation of their destination firstly, then to move as fast as possible to complete the path-finding task, and then to indicate the location of the starting point.There were two parts in this testing part,and each part had 9 tests.In each test,the wrong type r indicating the orientation, the response time indicating the destination and the operation time of completing the evacuation task were recorded.In the orientational task, if the orientational was correct, r=r0,or else r=r1.
In addition,the subjects were asked to dictate oral the space station cabin layout after the completion of all tasks,the staff drew a virtual space station layout map according to the description of the subject and confirmed with the subjects whether it matched the description, the error type e in this layout was recorded.When it was exactly true,e=e0; if the cabin sequence (structure) described by the subjects was correct, but the cabin interior orientation was wrong,e=e1;if the cabin sequence (structure) was wrong, e=e2.
In the process of simulated navigation and autonomous navigation, the head of the subjects received EEG continuous real-time monitoring on the head. Before and after the experiment and related nodes, they also received the head resting FMRI scan to obtain the objective information of the physiological state of the brain. Finally, the staff also asked and recorded the symptoms of motion sickness such as dizziness during the experiment.

3.3. The design of experimental sample scenes

3.3.1.Sample scenes design based on different numbers of the whole orientation cues. Six sets of sample scenes were designed,as shown in figure 3. It showed different orientation numbers of the whole orientation cues (n), where A1 scheme n=0, A2 scheme n=2, A3 scheme n =4, A4 n=6, A5 scheme n=6, A6 scheme n=6. The subjects were informed of the meaning of the orientation each color represented before the experiment began.

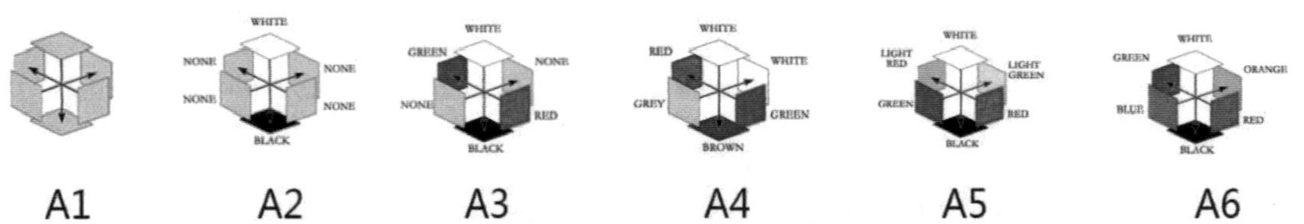

Figure 3. A sample scheme showing the numbers of whole orientation cues.

3.3.2.Sample scenes design based on different visual form of the whole orientation cues. Five sets of sample scenes were designed,as shown in figure 4. It reflected different visual form of the whole orientation cues. The B1 scheme reflected different color contrasts, the B2 scheme reflected the color brightness gradient, the B3 scheme reflected the graph arrangement,the B4 scheme reflected the text marking mode, and the B5 scheme reflected the graphic marking mode. Before the experiment began, the subjects were informed of the meaning of the orientation represented by each visual form.

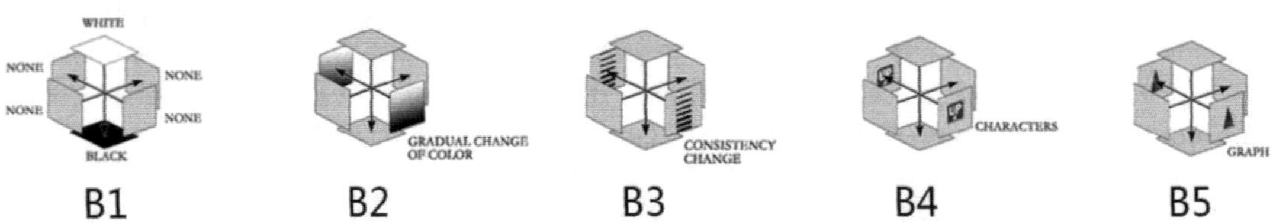

Figure 4. A sample scheme showing the visual form of the whole orientation cues.

3.4. The experimental results and analysis

3.4.1.The experimental results of the sample scenes which reflect the orientation numbers of the whole orientation cues. Through the statistics of the obtained experimental data, we can obtain the comparison chart of the accuracy rate of orientational task,the average time of the path-finding task,the accuracy rate of psychological map task and the proportion distribution of symptoms of motion sickness for different sample scenes which reflect different orientation numbers of the whole orientation cues.

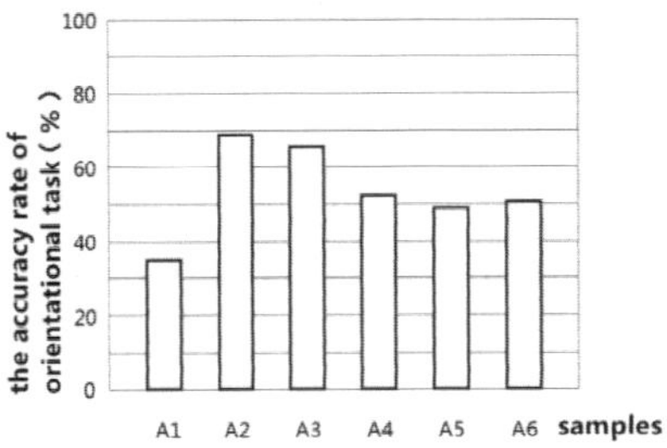

Figure 5. Comparison of the accuracy rate of orientational task for different samples.

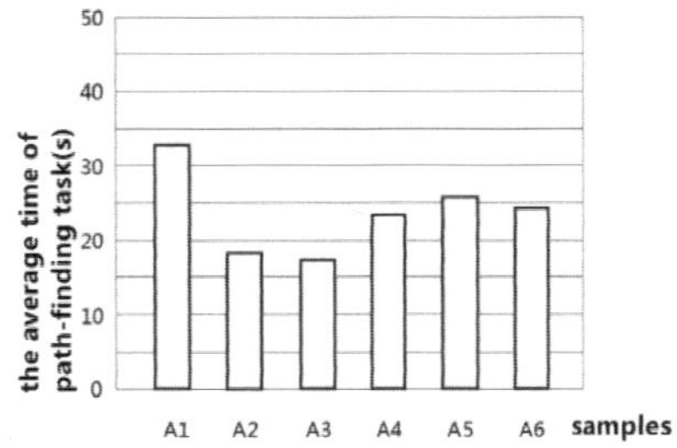

Figure 6. Comparison of the average time of path-finding task for different samples.

In figure 5,the accuracy rate of orientational task corresponding to A1(n=0) was significantly lower than others(P<0.05); the accuracy rate of the orientational task corresponding to A2(n=2) and A3(n=4) were significantly higher than A4(n=6)、A5(n=6) and A6(n=6)(P<0.05).

In figure 6,the average time of path-finding task corresponding to A1(n=0) was significantly higher than others(P<0.05);the average time of path-finding task corresponding to A2(n=2) and A3(n=4) were significantly lower than A4(n=6)、A5(n=6) and A6(n=6)(P<0.05).

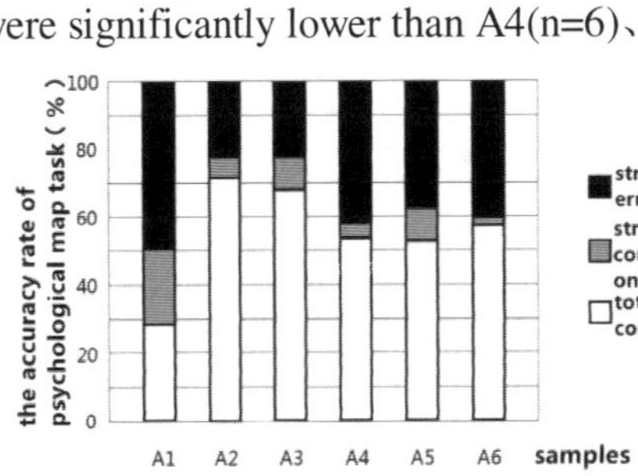

Figure 7. Comparison of the accuracy rate of psychological map task for different samples.

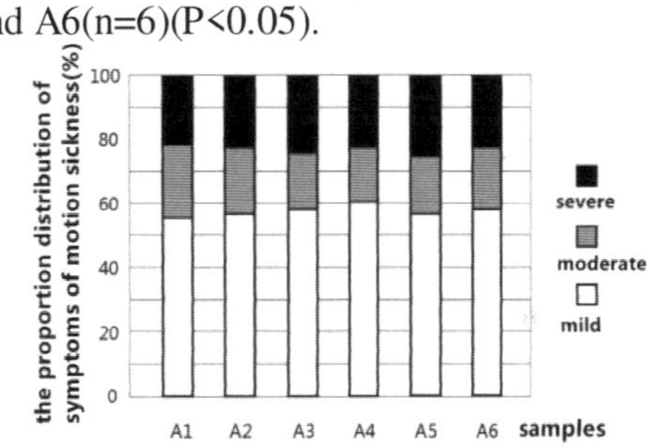

Figure 8. Comparison of the proportion distribution of symptoms of motion sickness for different samples.

In figure 7, the accuracy rate of psychological map task corresponding to A1 (n=0) was significantly lower than others (P<0.05); the accuracy rate of psychological map task corresponding to A2 (n=2) and A3 (n=4) were significantly higher than A4(n=6), A5(n=6) and A6 (n=6) (P<0.05).

In figure 8, there was no significant difference in the proportion of mild, moderate and severe motion sickness symptoms in all samples (P>0.05).

Conclusion and analysis: ompared with the situation without the whole orientation cues, the setting of the whole orientation cues can improve the navigation task performance to some extent, but the symptoms of motion diseases such as dizziness have not been significantly improved. As the orientation number n ($0 \le n \le 6$) of the whole orientation cues increases, the navigation task performance does not appear to improve continuously; when the orientation number of the whole orientation cues is n=2 or n=4, the navigation task performance is relatively high.

3.4.2.The experimental results of the sample scenes which reflect visual forms of the whole orientation cues. Through the statistics of the obtained experimental data, we can obtain the comparison chart of the accuracy rate of orientational task, the average time of the path-finding task, the accuracy rate of psychological map task and the proportion distribution of symptoms of motion sickness for different sample scenes which reflect different visual forms of the whole orientation cues.

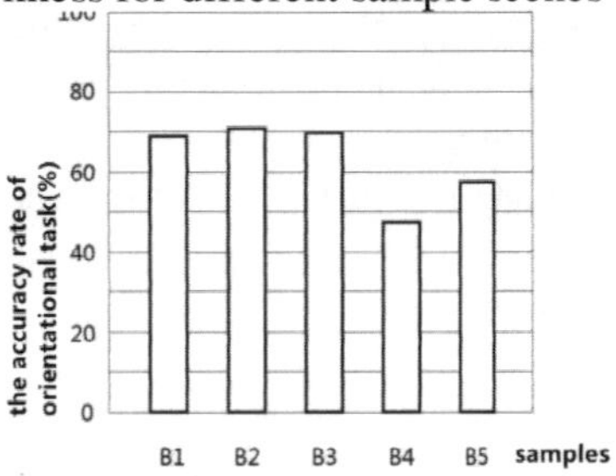

Figure 9. Comparison of the accuracy rate of orientational task for different samples.

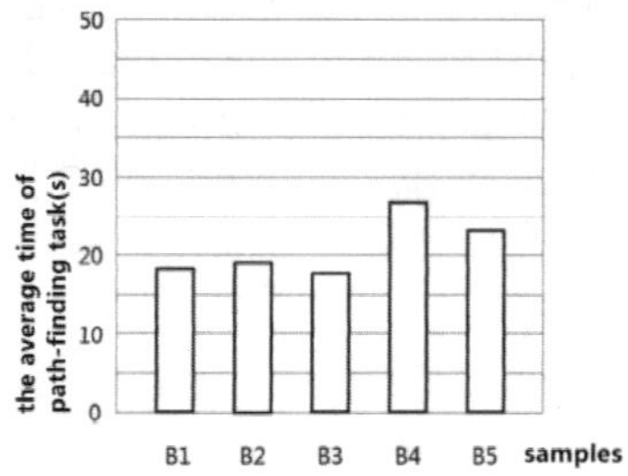

Figure 10. Comparison of the average time of path-finding task for different samples.

In figure 9, the accuracy rate of orientational task corresponding to B4(text mark) was significantly lower than others(P<0.05); the accuracy rate of the orientational to task corresponding to B1(color contrast), B2(brightness gradient) and B3(graphic arrangement) were significantly higher than B5(graphic identification)(P<0.05).

In figure 10, the average time of path-finding task corresponding to B4(text mark) was significantly higher than others(P<0.05);the average time of path-finding task corresponding to B1(color contrast). B2(brightness gradient) and B3(graphic arrangement) were significantly lower than B5(graphic identification).

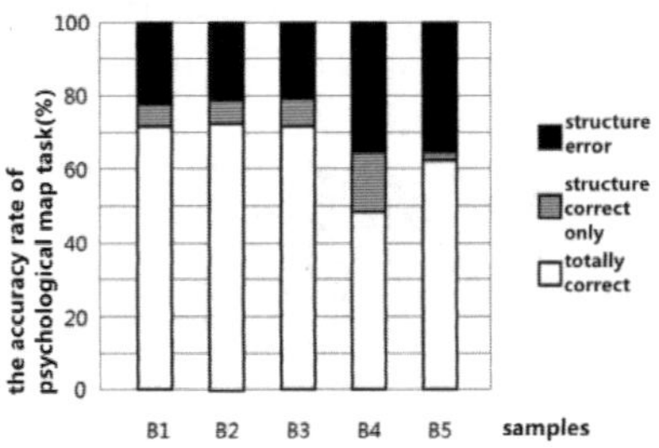

Figure 11. Comparison of the accuracy rate of psychological map task for different samples.

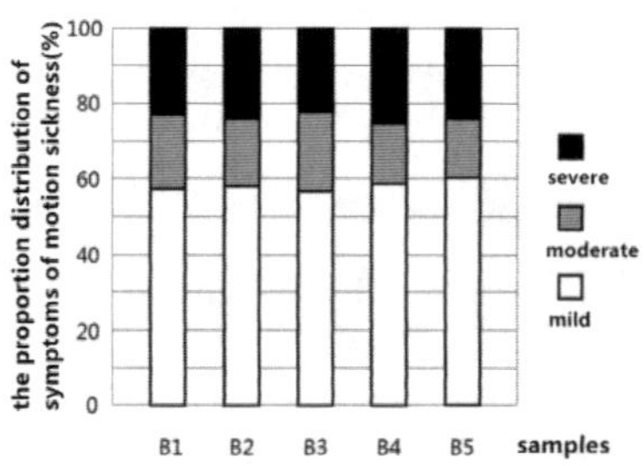

Figure 12. Comparison of the proportion distribution of symptoms of motion sickness for different samples.

In figure 11, the accuracy rate of psychological map task corresponding to B4(text mark) was significantly lower than others(P<0.05);the accuracy rate of psychological map task corresponding to B1(color contrast)、B2(brightness gradient) and B3(graphic arrangement) were significantly higher than B5(graphic identification) (P<0.05).

In figure 12, there was no significant difference in the proportion of symptoms of mild、 moderate and severe motor diseases in all samples.

Conclusion and analysis: compared with different visual forms of the whole orientation cues, the visual form of color and graphics are more effective in improving navigation task performance than text; large area distribution is more effective in improving navigation task performance than small area distribution; but the symptoms of motion diseases such as dizziness have not been significantly improved.

4. Conclusion

With the need to enhance the navigation ability of the astronauts, based on a method study on space station intra-vehicular navigation simulation including both VR visual simulation and physiological HDBR simulation, multiple different sets of sample scenes are designed for the orientation cues numbers reflected by the whole visual orientation cues that has similar form, and the specific form of the whole visual orientation cues that reflects same orientation numbers. Through the experimental data analysis, the influence law of the whole station's interior orientation cues on the navigation task performance and associated physiological effect in the cabin are obtained. The research could provide reference for the space station interior design engineering in the orientation of the cabin visual navigation capability.

5. References

[1] NASA 2001 *International Space Station Interior Color Scheme*
[2] Hirofumi Aoki, Ryuzo Ohno and TakaoYamaguchi 2005 *The Effect of The Configuration and The Interior Design of a Virtual type less Space Station on Human Spatial Orientation Conf.* p 1005-1016
[3] Manuel Vidal, Mark Lipshits, Joseph McIntyre and Alain Berthoz 2003 Gravity and spatial orientation in virtual 3d-mazes *Journal of Vestibular Research* 13:273-286
[4] M.Vidal, M.Amorim and A.Berthoz 2004 Navigating in A virtual three-dimensional maze: how do egocentric And allocentric reference frames interact? *Cognitive Brain Research* 19:244-258
[5] Wouldiam B. Lathrop and Mary K. Kaiser 2005 Acquisition Spatial Knowledge While Traveling Simple and Complex Paths with Immersive and Nonimmersive Interfaces *Presence* 14(3): 249-263
[6] Yan Xiaoxia 1990 Long-term ground simulation of weightlessness - bed rest and its physiological research *Advances in biochemistry and biophysics* 17(3): 181-184
[7] Wang Linjie and Zhang Dan 2003 Studies on motion diseases in space before, during and after spaceflight *Aerospace medicine and medical engineering* 16(5): 382-386

Acknowledgments
Project 51305099 supported by National Natural Science Foundation of China.

IOP Conf. Series: Journal of Physics: Conf. Series **1215** (2019) 012027 doi:10.1088/1742-6596/1215/1/012027

Research on Binding Memory of Icon Features Based on Event-related Potential

Y F Niu[1,2,*], B Z Shi[3], L C Qiu[3], B Li[2], Q Y Shen[1], C Q Xue[1]

[1] School of Mechanical Engineering, Southeast University, Suyuan Road, Nanjing 211189, China
[2] Science and Technology on Electro-optic Control Laboratory, Kaixuan West Road, Luoyang 471023, China
[3] Shanghai Academy of Spaceflight Technology, Zhichun Road, Shanghai 201109, China

nyf@seu.edu.cn

Abstract. The visual characteristics are important factors influencing the cognition and memory of the icon. In order to study the process of memory coding of feature binding in working memory of icon, the experiment requires 20 subjects to remember whether the icon features or feature binding are consistent and response quickly by pressing buttons, and the behaviour data and ERP data of the subjects are recorded. The experiment results show that compared with single feature memory of an icon, the accuracy under the icon-color feature binding memory condition is lower and the reaction time is longer while the P300 brain electric component with larger amplitude can be triggered. Studies have shown that the binding of visual features in icon memory requires additional cognitive resources. Compared with the other three memory conditions, the latency of the P300 under icon semantic-color feature binding is more significant, indicating that the task of semantic coding of icon is more difficult to complete and the cognitive load is greater.

1. Introduction

Icons are important ties between human-computer interaction in digital interfaces [1]. Design and use of icons could optimize the design and layout of the digital interface. The rich design expression of the icon can beautify the interface and improve the user's pleasure when interacting with the interface. Icon also has its unique cognitive advantage in the digital interface [2] which can improve the user's cognitive efficiency, the universality of the interface, and enhance the user's memory effect [3]. Therefore, it is important to study the impact of icon design on working memory in interface design and evaluation stage.

The visual features of object are important factors affecting working memory [4]. The visual features of icon have multiple dimensions, including shape, color and texture. According to the feature binding theory, these features are stored in the working memory in a bundled form. After the visual information of icon is input through the visual pathway, users' brain processes those icon in two different stages. The first phase is pre-attention phase, the features of each dimension of the icon are roughly bundled to form an object with an overall image. Then, in the process of visual perception, the shape, color and other features of the icon are bundled into an individual with a rough outline. The second stage is to after-attention binding stage. The detailed features of the icons, such as the details

[*] Niu Yafeng, Ph.D, lecturer, School of Mechanical Engineering, Southeast University, Email: nyf@seu.edu.cn

on the icon graphics, color matching, texture will be further bundled at this stage with the participation of attention resources. Combined with the contour image that has been formed in the first stage, the final complete image of the object is processed [5].
At present, researchers at home and abroad mainly use physiological measurements [6], eye tracking [7-9] and EEG analysis [10] to study the evaluation of the availability of digital interface elements. The multi-dimensionality of the icon itself has a significant impact on digital interface. Domestic and international scholars have done research based on visual cognitive model of the icon [11], visual search efficiency [12], and human-computer interaction efficiency [13]. There are few studies on icon feature bundling, and there are always divergence and controversy such as if the characteristics of each dimension of the object are stored in different brain region and do not affect each other, or will they compete for resources among each other in the buffer zone [14]? In this paper, event-related potential (ERP) technology is used to study the influence of color characteristics of icons on user's working memory, in order to explore the influence of color features on working memory of icon features.

2. Experiment Methods

2.1. Subjects
The subjects of this experiment were 20 students in the Department of Industrial Design of Southeast University, including 10 males and 10 females. All subjects were between 25 and 27 years old with an average age of 26 years old. None of the 20 subjects were color-blind, with a corrected visual acuity of 5.0, and were all right-handed. None of the subjects had a history of mental illness or trauma to the brain.

2.2. Experiment materials
Before the experiment, using Photoshop design 200 icons, the matching semantic word length is mainly two words, using the standard meaning of the icon. Then redesign these 200 icons in a linear icon design style to keep them in a same design style. The familiarity, specificity, visual complexity, and semantic distance of the 20 subjects were scored by 5-level Richter scale.
As shown in Figure 1, we select 16 concrete icons with high familiarity, semantic distance, and specificity as sample icons. Six of these 16 icons are taken as experimental memory items in each memory task. There are 6 colors of icons, red (RGB: 255, 0, 0), orange (RGB: 255, 153, 51), purple (RGB: 204, 102, 255).

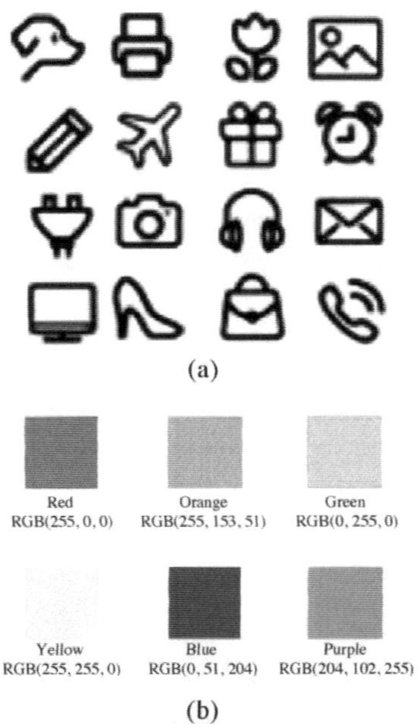

Figure 1. Icon and color used in memory experiment.

2.3. Procedure

The experiment uses a change-aware paradigm of single-stimulus detection. The experiment is a single factor design, which is divided into four conditions according to the type of memory: single graphic feature memory, single color feature memory, binding memory of color and icon, and binding memory of color and icon semantics. In the formal experiment, memory experiments under four memory conditions were performed in sequence. For each condition, 40 trials will be performed, 50% of which will not in the memory sequence, and 50% of the detection items are in the memory item. Under the condition of binding memory, there are three possible cases that the detection items are not in the memory sequence: the picture is in the memory column, the color is not; the color is in the memory column, the icon is not; the color and the icon are in the memory column but match wrongly.

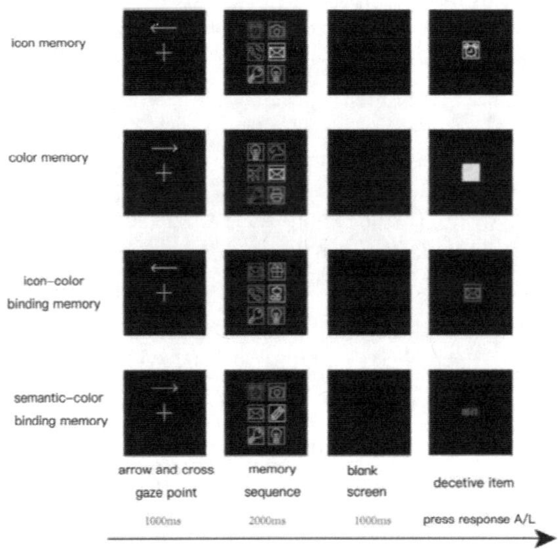

Figure 2. Experiment flow.

The experimental flow is shown in Figure 2. Before the start of each experiment, center of the screen will show a cross gaze point and a left or right indicator arrow above it. The presentation time is 1000 ms, and the subject is asked to remember the direction of the arrow. After that, 3×2 color icons are presented, and the presentation time is 2000ms. The subject is required to remember the characteristics or binding relationship of the three icon samples in the column indicated by the arrow according to the specific experimental task, and then the blank screen of 1000 ms appears, and finally the detection is performed. The user needs to determine whether the binding relationship of the feature or feature in the probe item appears in the three memory items of the memory sequence. Press A key if it appears, or L key if it does not appear. After the user makes a button response, the probe disappears on the screen and enters the next round after 1000ms. The formal experiment consisted of 4 sets of experiments, 40 memory tasks per group, a total of 160 memory tasks, and the subjects were allowed to rest between each group of experiments, which lasted about 20 minutes.

2.4. ERP data and process

Behavior data such as accuracy and reaction time was recorded using E-prime2.0. EEG data were collected and using ANT brain electrical equipment, and then the original EEG data was analyzed and processed using ASA4.1 software. Finally, SPSS was used to analyze the behavioral data and EEG data of the subjects in each group.

3. Experiment results

3.1. Behavioral data

As a result, it was found that the memory condition has a significant main effect, p = 0.000. Memory accuracy under graphic-color binding condition (0.789) is significantly lower than the accuracy under a single graphic memory condition (0.925), p=0.000. Memory accuracy under graphic-color binding condition (0.789) is significantly lower than the accuracy under a single-color memory condition (0.938), p=0. 000.The memory accuracy (0.795) under icon semantics - color binding conditions is significantly lower than the accuracy under a single graphics memory condition (0.925), p = 0.000. The memory accuracy (0.795) under icon semantics - color binding conditions is also significantly lower than the accuracy under single color memory conditions (0.938), p = 0.000. There was no significant difference between the single-color memory condition (0.938) and the accuracy under a single pattern memory condition (0.925), p=0.904. There was also no significant difference between the graphic-color binding condition (0.789) and the icon semantic-color binding condition (0.795), p=1.000.

There was no significant difference between icon memory task and color memory task in reaction time, p=0.475. The icon-color binding memory task reacts significantly longer than a single icon memory task, p=0.000, and is significantly longer than a single-color feature memory task, p=0.000. The semantic-color binding memory task is also significantly longer than a single icon memory task, p=0.000, which is significantly longer than a single-color feature memory task, p=0.000. The semantic-color binding memory task is significantly longer than the graphics-color memory task, p=0.000.

Table 1. Accuracy rate and Reaction time.

		ACC		RT	
Detective item	Sample	Mean	Standard deviation	Mean	Standard deviation
Icon	20	0.925	0.264	779.820	201.368
Color	20	0.938	0.242	728.440	229.291
Icon color binding	20	0.789	0.408	879.170	327.920
Semantic color binding	20	0.795	0.404	1023.080	408.573

3.2. ERP data

According to the EEG waveform, there are obvious p300 EEG components in the brain regions such as the central region and the top region in the memory experiment. At the same time, according to the existing research on the P300 neurogenic source [15-16], the six central electrodes (CP1, CP2) and the top region (POz, Pz, P3, P4) were selected for P300 EEG component analysis.

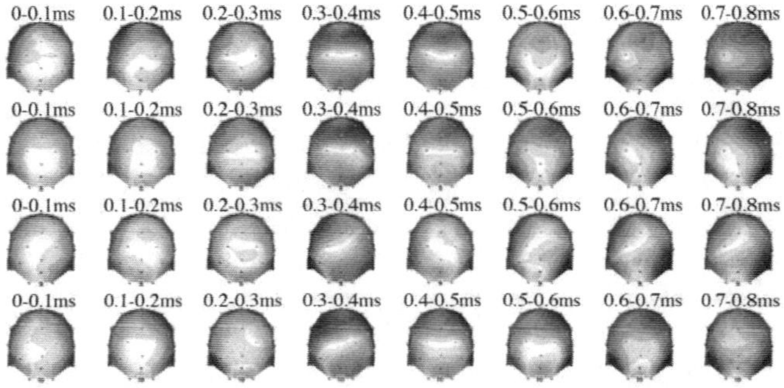

Figure 3. Icon feature binding memory experiment map.

First, a 4 (memory feature type: graphics, color, graphics-color bundling, semantic-color bundling) × 6 (electrodes: CP1, P3, CP2, P4, POz, Pz) repeated measurement analysis of variance are performed in the memory experiment under three characteristic memory conditions. It can be found that the electrode main effect is significant. F = 2.915, p = 0.015 < 0.05. There were also significant differences between memory feature types, F=6.449, p=0.004<0.05, there was no significant difference between electrodes and feature types, F=0.548, p=0.854.

In order to study whether there are significant differences in the four groups of characteristic memory tasks on these six electrodes, one-way ANOVA was performed on the average amplitude of the P300 components on each electrode. It can be seen from Table 5-16 that the amplitudes on the CP1, P3, P4, and POZ electrodes are significantly different among the memory tasks of each group (CP1: F=4.591, p=0.006<0.05; P3:F=4.504, p= 0.007<0.05; P4: F=4.833, p=0.005<0.05; POz: F=9.863, p=0.000<0.05). There was no significant difference in P300 amplitude between groups CP2 and Pz (CP2: F=1.429, p=0.245>0.05; Pz: F=1.507, p=0.223>0.05).

Table 2. Significant effect of 4 feature memory task p300 electrode amplitude

Electrode	Sum of square	df	Mean square	F	Sig.
CP1	14.097	3	4.699	4.591	.006
P3	11.993	3	3.998	4.504	.007
CP2	2.857	3	.952	1.429	.245
P4	10.995	3	3.665	4.833	.005
POz	23.557	3	7.852	9.863	.000
Pz	3.660	3	1.220	1.507	.223

Figure 4 is a graph of ERP brain electrical waveforms formed by four sets of characteristic memory tasks at the P4 electrode position. It could be observed from the figure that the graphic-color binding and icon semantic-color binding have more volatility than the P300 EEG component generated in the single feature memory task. It can also be seen from the P4 electrode waveform diagram that there is a significant difference in the appearance time of the P300 between the graphic-color binding and icon semantic-color binding memory task. The latency of P300 component of the color semantic- color binding memory task is significantly later than the other three memory tasks.

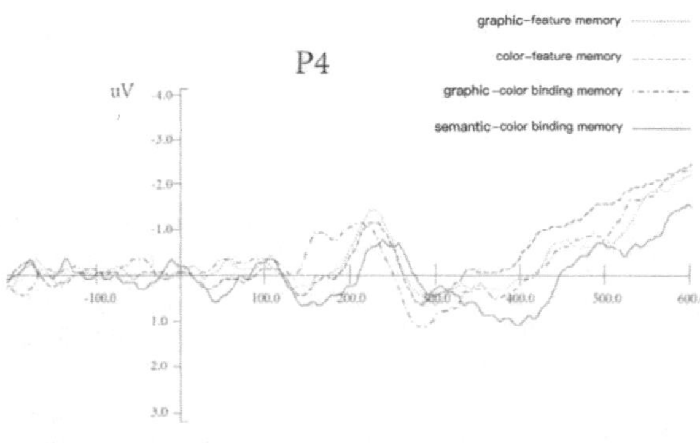

Figure 4. P4 electrode amplitude in 4 memory experiment.

A 2 (bundle type: graphic-color bundling, semantic-color bundling) × 6 (electrodes: CP1, P3, CP2, P4, POz, Pz) repeated measures analysis of variance was performed on the P300 latency data of the two different methods of the binding memory experimental group. It can be found that the main effect of the electrode is not significant, p=0.160>0.05. The main effect between the bundled memory types presented a significant edge, p=0.057>0.05, the interaction between the electrode and the feature type was not significant, p=0.191>0.05.

The six electrodes are further divided into three brain regions: left region (CP1, P3), right region (CP2, P4), and midline (P0z, Pz). After the two sets of bundled memory experiment EEG data 2 (bundle type: graphics - color bundling, semantic - color bundling) × 3 (brain area: left zone, right zone, midline) repeated measurement analysis of variance can be found, the interaction between brain zone and the group was not significant, F=2.280, p=0.123. The significance of the main effect of the brain area was significantly improved, F=4.122, p=0.028<0.05. A paired sample T test under the two bundled memory conditions was performed on the left, right, and midline brain regions. It was found that there was a significant difference between the two bundled memory tasks only in the midline of the brain region, p=0.035<0.05. The paired sample T test under two bundled memory conditions was continued for the P0z and Pz electrodes in the midline brain region. Only on the P0z electrode the latency of the graphics-color binding memory task is significantly earlier than the semantic-color binding memory task, p=0.030<0.05. P0z electrodes can be used as a focus for further analysis.

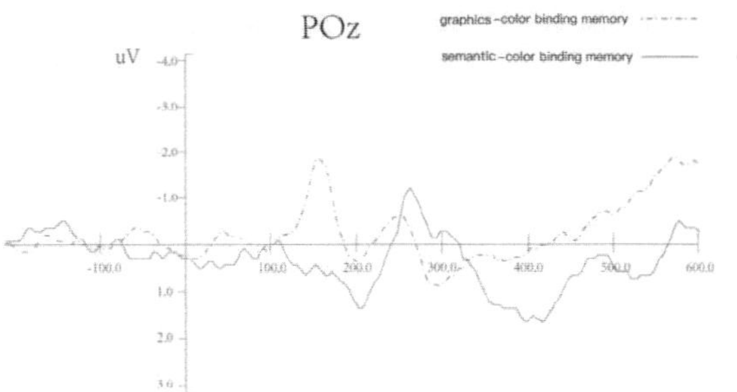

Figure 5. POz electrode amplitude in 2 memory experiments.

Figure 5 is an EEG waveform graph of the POz electrode in two binding memory experiments. The latency of the P300 component in the graphics-color bundle memory task is significantly earlier than the semantic-color bundle memory, indicating that the cognitive load generated by the participants in the semantic-color bundle memory task is greater. This is consistent with the analysis of behavioral data.

4. Discussions

4.1. Behavioral data results analysis

In this experiment, it can be obtained from the behavior data that the feature binding memory task is longer than the single-feature memory task with lower accuracy, indicating that the feature binding memory task is more difficult than the single-feature memory task. In the weak object storage theory, the concept of subsystems is no longer used, but that there is only one shared memory storage system, and all features and objects in working memory need to compete for cognitive resources in this shared storage system. [17]. The behavioral data results in this experiment verify the correctness of the weak object storage theory in the icon feature binding memory.

Icon semantics-color binding memory reacts longer than graphic-color binding memory experiment. According to the synchronous multi-encoding theory proposed by Wickens, in working memory, the object information is simultaneously encoded by both speech and visual encoding [18]. The significant difference between Icon semantics - color binding memory and graphic – color binding in response in react time is caused by differences in user's memory coding. In the graphic-color binding memory task, the double-coded memory was used by the participants, while the semantics-coded memory was mainly used in the icon semantic-color binding memory, indicating that the task of semantically encoded memory is more difficult and the cognitive load is greater.

4.2. P300 results analysis

P300 is currently the most widely studied ERP component in the field of working memory, and its incubation period is around 250-600ms. P300 is affected by many types of tasks, such as type of task, the importance of stimulation, attention, working memory, decision-making, and emotion. From the data of the P300 brain electrical component, it can be found that P300 mainly appears in the top and central regions of the brain. The P300 waveforms of each electrode in the single-color feature memory, the single-color feature memory, and the graphic-color binding memory test are roughly similar. And the P300 generated by graphics-color bundling and icon semantic-color binding has a larger amplitude than it in the single feature memory task, indicating that the subject has a larger cognitive load in the feature-bound memory task.

Comparing with the behavioral data of the subjects, it can be found that the graphics-color bundle memory and the icon semantic-color bundle memory are significantly lower than single graphics feature memory and the single-color feature memory in accuracy. At the time of the reaction, the graphics-color bundling and icon semantic-color bundling memory tasks are significantly lower than the single graphical feature memory and the single-color feature memory. At the same time, the P300 latency in the semantic-color-bound memory experiment is significantly later than the latency in the other three tasks, indicating that the feature-bound memory task is more difficult than the single-feature memory. Behavioral data is consistent with the analysis of EEG data which support the conclusion that with larger cognitive load comes later P300 volatility [19].

5. Conclusions

In this paper, ERP is used to study the influence of the color characteristics of icons on the working memory of icons from the perspective of feature binding memory. The research shows that there are significant differences in the response time, accuracy rate and P300 composition of the icon under the single feature memory and the feature binding memory condition of the icon. The feature binding memory of the icon generates a larger cognitive load than the single feature memory. Bundling of features requires additional cognitive resources. The research results could be the scientific guidance and has reference value for revealing the characteristics of icon binding memory rules, enhancing the usability of icons, and exploring more effective icon design methods. In following research, the feature binding relationship in same dimension of multicolor icons and its impact on icon memory will be explored.

6. References

[1] King L E 1971 *Traffic Engineering & Control* **12** 518-520
[2] Horton W K 1994 The ICON Book: Visual Symbols for Computer Systems and Documentation. (John Wiley & Sons, Inc)
[3] Baddeley A 2000 *Trends in Cognitive Sciences* **4** 417
[4] Treisman A M, Gelade G 1980 *Cognitive Psychology* **12** 97-136
[5] Wheeler M E, Treisman A M **2002** *Journal of Experimental Psychology General* **131** 48-64
[6] Proaps A B, James ETC 2014 *Proceedings of the Human Factors and Ergonomics Society 58th Annual Meeting* 1476-1480
[7] Zhou X, Xue C, Zhou L and Niu Y 2018 *International Journal of Pattern Recognition & Artificial Intelligence* **32** 1850016
[8] Kovesdi C, Spielman Z, Leblanc K 2017 *Nuclear Technology* **4** 1-10
[9] Chen X, Ran J 2017 *Applied Ergonomics* **65** 551-561
[10] Giraudet L, Imbert J, Bérenger M 2015 *Behav Brain Res* **294** 246-253
[11] Wang H, Huang Y, Chen M 2016 *Journal of Computer-Aided Design & Computer Graphics* **28** 1740-1749
[12] Gong Y, Zhang S, Shen F 2016 *Journal of Computer-Aided Design & Computer Graphics* **28** 1115-1120
[13] Gong Y, Yang Y, Zhang S 2012 *Journal of Computer-Aided Design & Computer Graphics* **24** 001145-1150
[14] Chen Q 2013 The detection process and development of simple-visual feature figure feature bindings.(Shenyang: Shenyang Normal University)
[15] Polich J, Kokb A 1995 Biol Psychol **41** 103-146
[16] Barcelo F, Cooper P S 2018 *Psychophysiology* **55** 12814
[17] Olson I R, Jiang Y 2002 *Perception & Psychophysics* **64** 1055-67
[18] Wickens D D 1973 *Memory & Cognition* **1** 485-490
[19] Wei J 2010 Principle and technique of event-related brain potentials. (Beijing: Science Press)

Acknowledgment
This work was supported jointly by National Natural Science Foundation of China (No. 71801037, 71871056, 71471037), Science and Technology on Electro-optic Control Laboratory and Aerospace Science Foundation of China (No. 20165169017), SAST Foundation of China (SAST No. 2016010) and Equipment Pre research & Ministry of education of China Joint fund.

IOP Conf. Series: Journal of Physics: Conf. Series **1215** (2019) 012028 doi:10.1088/1742-6596/1215/1/012028

Research on Influence of Microburst on Ballistic Characteristics of Rockets

Chen Jianwei[1], Wang Liangming[1] and Li Zijie[1]

1 Nanjing University of Science and Technology, 210094 Nanjing, People's Republic of China

Abstract. In order to study the influence of microburst on the ballistic characteristics of rockets, an engineering model was established by using the vortex ring method based on the basic principle of hydrodynamics and the theory of the exterior ballistic. The wind field model is combined with the six-degree-of-freedom rigid-body ballistic model of rockets. This study examined a kind of empennage-rocket whose flight process was influenced by a microburst. The simulation results show that the microburst model based on the vortex method is simple and the three-dimensional characteristics are good, which can describe the main wind field characteristics of the microburst in a certain extent. Meanwhile, the microburst affects the ballistic characteristics of rockets such as the trajectory, velocity and angular motion. The extent of the effect is concerned with the vertical wind velocity along the central axis and the vortex ring radius of the microburst model.

1. Introduction

Wind shear is the vector difference between the two points of the wind of space. It is known as low-level wind shear when occurred at 600m below. Low-level wind shear is an atmospheric phenomenon associated with strong convective weather, frontal weather, geography and environmental factors [1]. A large number of researches indicate that the microburst is one of the most dangerous and most probable type of wind shear. It is a small, strong, downward impulse that rushes to the ground to produce a horizontal airflow in all directions [2]. When the airplane and other aircrafts such as rockets moving through the microburst wind shear region, the rapid changes in wind speed will change the state of the aircraft to affect its flight stability and trajectory. In the initial stage of the launch, especially when the rocket is out of the deflector, the speed of the rocket is low and the anti-disturbance capability is weak. The change of the wind speed can cause a large distance and direction deviation on the trajectory of the rocket [3]. In the rocket range test and the actual battles, the complex and varied surrounding climate and the geographical environment are very favourable for the microburst wind shear to form. Therefore, how to consider and reflect the influence of the microburst on the rocket flight and the ballistic characteristics is extremely important to the launching and control of rockets.

The effect of the change in wind field on the process of the rocket flight has always been an important issue in the ballistic research. A simplified wind field model for the trajectory calculation was established according to the statistical characteristics of the wind field, and the wind shear was described by a linear function [4-5]. The model of the atmospheric wind field at 100 km below was developed, and the analysis for the movement of a long-range rocket under the influence of high-altitude winds was studied[6]. A numerical simulation was conducted to analyze the impact of different wind fields on the trajectory of the projectile based on the external ballistics and the meteorological theory [7]. Simplified models of two kinds of low-level wind shear were established and their effect on the fly-by-ground flight of the aircraft is studied [8]. The main structure and the evolution process of a microburst formed in Beijing area in 1997 was simulated by a non-hydrostatic and elastic mesoscale model [9]. A model of vortex ring induced microburst is developed [10]. On the basis of previous studies [10], a multi-vortex ring model and an inclined vortex ring model for microburst were designed and its influence on the take-off and the landing process of the aircraft was explored [11]. The multiple and inclined vortex ring model of the microburst was combined with the atmospheric turbulence velocity field [12].

It can be seen from the above research that the studies on the microburst wind shear is mainly focuses on the modeling method and the influence on the navigation or the flight of the airplane, but little is known about the research on the influence of the microburst on the flight and the ballistic characteristics of the projectile. This paper gives the engineering model of the microburst wind shear according to the flow features combined with the rocket's six-degree-of-freedom (6-DOF) trajectory model. We analyzed the ballistic characteristics of the rocket under the influence of the microburst, meanwhile, the influence of the characteristic parameters of the microburst on the ballistic characteristics of the rocket is studied. The results would provide some reference for the research of the ballistic trajectory and the launch control of the rocket.

2. Microburst model based on vortex ring method

Microburst is one of the most common convective storm wind shear, which behave as a localized subsided airflow. When the airflow reaches the ground, it collides with the ground and spreads to the surrounding. The formation process of a microburst is shown in Figure.1. There are several engineering modeling methods of a microburst, such as the dipole-based panel method, and the vortex ring method. The former method requires a large number of mathematical integral operations and the solving process are complex. This paper used the vortex ring method for the microburst wind field modeling.

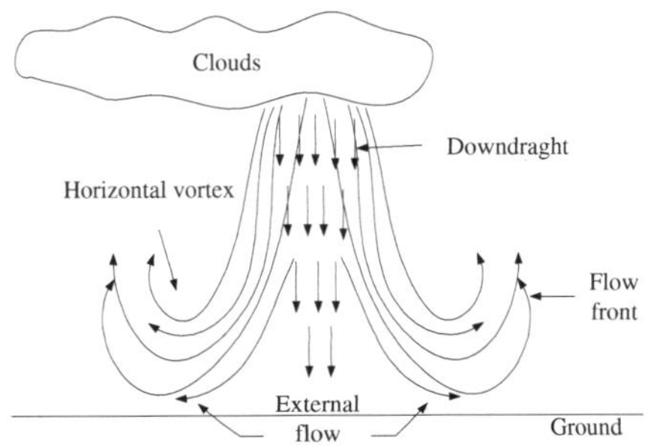

Figure 1. Forming process of microburst wind field.

2.1. Modeling process

Since the velocity field induced by the vortex ring is very similar to the wind field of the microburst, the wind field model of the microburst can be constructed by using the vortex ring principle in fluid mechanics. The vortex ring model in the ground coordinates system (O-xyz) is shown in Figure.2, in which the center of the main vortex ring is O_P and the radius of the main vortex ring is R.

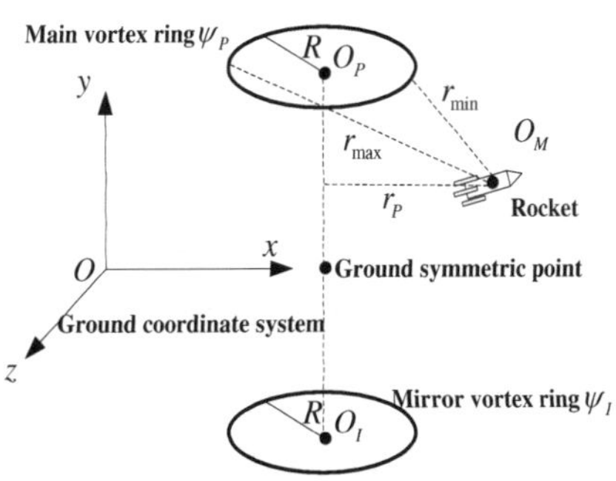

Figure 2. Schematic diagram of vortex ring model.

The main vortex curvilinear equation can be written as

$$\begin{cases}(x_c - x_p)^2 + (y_c - y_p)^2 = R \\ z_c = z_p\end{cases} \quad (1)$$

where (x_c, y_c, z_c) are the coordinates of the points on the curve and (x_c, y_c, z_c) is coordinate of O_p. The main vortex circulation equation is

$$\psi_P = \frac{\Gamma}{2\pi}(r_{\max} + r_{\min})F(k) \quad (2)$$

where $r_{\max}$ and $r_{\min}$ are the maximum distance and the minimum distance from any point in space $O_M(x_M, y_M, z_M)$ to the vortex ring curve, $F(k)$ is the elliptic integral function, and Γ is the vortex strength ,which is determined by the vertical velocity in the vortex center $V_{wz}(0)$ and the vortex ring radius R.

$$\Gamma = 2RV_{wz}(0) \quad (3)$$

The variable is given by

$$k = \left|\frac{r_{\max} - r_{\min}}{r_{\max} + r_{\min}}\right| \quad (4)$$

If $0 \le k \le 1$, $F(k)$ can be approximated as

$$F(k) \approx \frac{0.788k^2}{0.25 + 0.75\sqrt{1-k^2}} \quad (5)$$

Based on Eq. (2), the radial (parallel to the horizontal plane) induced velocity and the axial (parallel to axis oz) induced velocity can be obtained by

$$\begin{cases} v_r^P = \dfrac{1}{r_P}\dfrac{\partial \psi_P}{\partial z_R} \\ v_z^P = -\dfrac{1}{r_P}\dfrac{\partial \psi_P}{\partial r_P}\end{cases} \quad (6)$$

where r_p is distance from O_M to the central axis of the vortex ring ($r_P = \sqrt{(x_M - x_P)^2 + (y_M - y_P)^2}$).The velocity components along axis ox and axis oz calculated from Eq. (6) are

$$\begin{cases} v_x^P = \dfrac{x_M - x_P}{r_P} v_r^P \\ v_z^P = \dfrac{y_M - y_P}{r_P} v_r^P\end{cases} \quad (7)$$

In order to make the vertical component of the wind vector on the ground zero, a mirror vortex ring should be configured at the symmetry point $O_I(x_p, y_p, -z_p)$ of the center of the main vortex ring with respect to the horizontal plane. The mirror vortex circulation equation is $\psi_I = -\psi_p$ .So the wind vectors induced by the main vortex and the mirror vortex are equal in magnitude and symmetric with respect to the ground. Following the same steps with Eq. (6) and Eq. (7), the induced velocity v_x^I, v_y^I and v_z^I of the mirror vortex ring at the point O_M can be obtained, and the resultant velocity of the two vortex rings at the point are expressed as

$$\begin{pmatrix} V_{wx} \\ V_{wy} \\ V_{wz} \end{pmatrix} = \begin{pmatrix} v_x^P \\ v_Y^P \\ v_Z^P \end{pmatrix} + \begin{pmatrix} v_x^I \\ v_Y^I \\ v_Z^I \end{pmatrix} \tag{8}$$

Thus, the streamline equation of point is given by

$$\psi = \psi_P + \psi_I = \frac{\Gamma}{2\pi}\left[\frac{0.788k^2(r_{\max} + r_{\min})}{0.25 + 0.75\sqrt{1-k^2}} - \frac{0.788k^{'2}(r_{\max}^{'} + r_{\min}^{'})}{0.25 + 0.75\sqrt{1-k^{'2}}}\right] \tag{9}$$

where ψ_I is the streamline of the mirror vortex ring, $r_{\max}^{'}$ and $r_{\min}^{'}$ are the maximum distance and the minimum distance from point O_M to the mirror vortex ring curve.

There are two special cases of the above-mentioned vortex ring model:

- The induced velocity along the central axis of the vortex ring ($r_p = 0$) calculated from Eq.(6) is close to infinity. The potential function of the vortex ring can be introduced here to complete the derivation of the induced velocity along the central axis.

$$\begin{cases} V_{wx} = V_{wy} = 0 \\ V_{wz} = \dfrac{\Gamma}{2R}\dfrac{1}{(1+(\frac{z_M - z_P}{R})^2)^{1.5}} \end{cases} \tag{10}$$

- The induced velocity at any points along the vortex filament ($r_p = R, z_M = z_P$) calculated from Eq.(6) is also close to infinity. Thus, a method based on Rankine compound vortex is used here [10]. It takes the vortex core as an annular cylinder with a radius of r as shown in Figure.3.The vorticity inside the vortex core is evenly distributed. The velocity along the vortex filament is kept at 0, while the flow outside the vortex core is still subject to the streamline equation ψ_p.From the vortex core to the vortex radius, the induced velocity is linearly distributed.

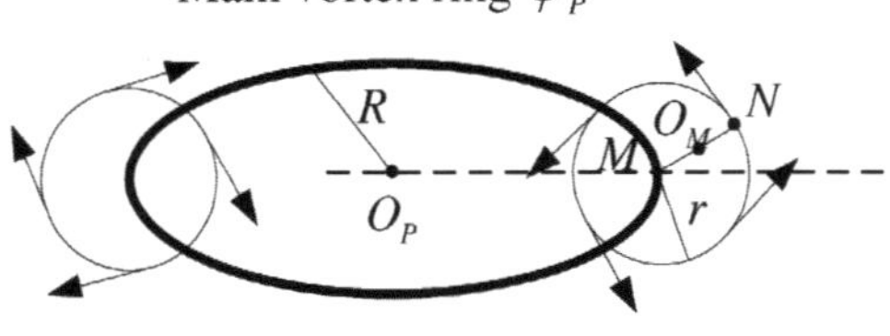

Figure 3. Schematic diagram of wind vector in vortex core of compound vortex.

When point O_M is in the vortex core, the coordinate of point M can be obtained by combing the linear equation of the projectile on the horizontal plane of line O_PO_M and $(x-x_P)^2+(y-y_P)^2 = R^2$, thus, the coordinate of point N is

$$N = M + \frac{r}{|MO_M|} \bullet MO_M \tag{11}$$

Based on Eq. (6), Eq. (7) and the coordinate of point N obtained from Eq. (11), the wind vector V_w^N of point N can be calculated, and the wind vector of point O_M is expressed as

$$V_w^M = \frac{|MO_M|}{r} V_w^N \tag{12}$$

According to the above-mentioned derivation, the wind speed vector at any point in the ground coordinate system space can be calculated.

IOP Conf. Series: Journal of Physics: Conf. Series **1215** (2019) 012028 doi:10.1088/1742-6596/1215/1/012028

2.2. Numerical simulation verification

In order to verify the rationality and effectiveness of the microburst model based on the vortex ring method, the numerical simulation method is used here to obtain the wind field model. The simulation results are compared with the measured wind field distribution from the relevant data.

According to the statistical data of the intensity and the spatial scale of the microburst wind field, it usually occurs below 600m and the horizontal outflow range is within 4km. For the moderate intensity microburst, the vertical velocity of the central axis is about 10m/s. According to the above information, the basic parameters of the vortex ring model in the ground coordinate system are shown in Table 1. Based on these parameters, the wind vector diagram of vertical section of vortex ring is shown in Figure.4(a) and the wind vector diagram of horizontal section(y=100m) is shown in Figure.4(b).

Table 1. Basic parameters of vortex ring.

Parameters	Value
Coordinate of center of main vortex ring (m)	23.56
Radius of vortex ring (m)	34.64
Radius of vortex core (m)	23.76
Vertical flow velocity of the vortex center (ms^{-1})	27.9

Figure.4(a) shows that the vertical velocity near the central axis of vortex ring is higher, and along the horizontal direction, the farther away from the central axis of the vortex, the smaller the wind speed is. Figure.4(b) shows that the maximum wind velocity appears at the center of the vortex ring in the horizontal section, and the wind velocity is distributed around the center of the vortex ring and gradually decreases.

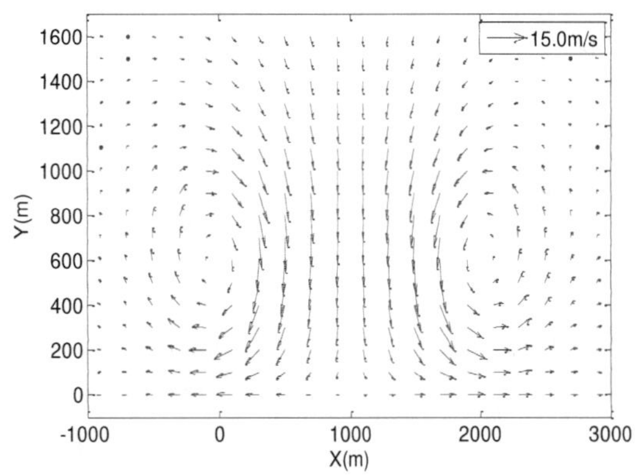

Figure 4(a). Wind vector diagram of vertical section

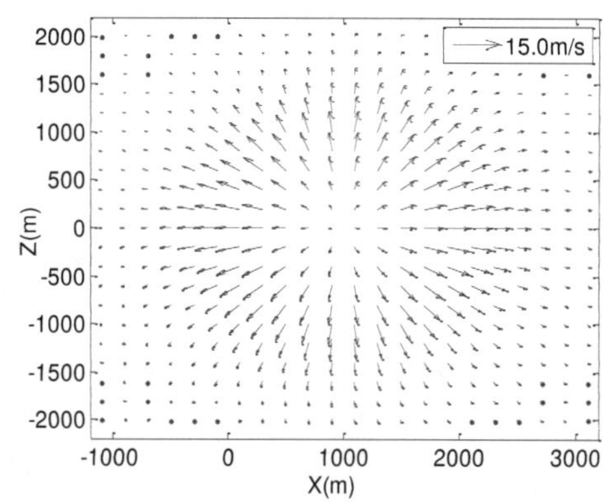

Figure 4(b). Wind vector diagram of horizontal section(y=100m)

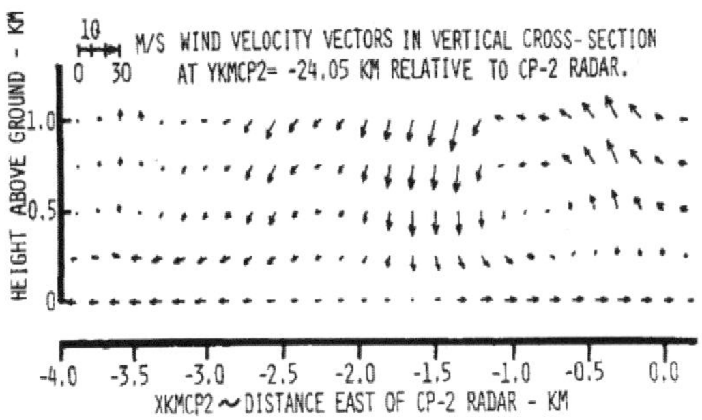

Figure 5(a). JAWS 1982 wind-shear data showing a downburst flow pattern

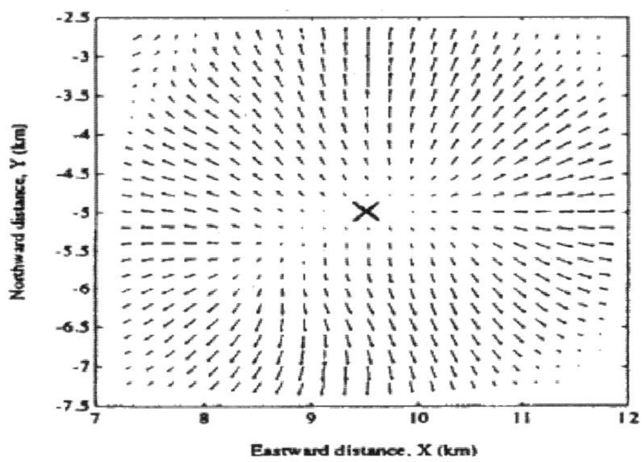

Figure 5(b). Wind field of 1988 microburst event of DEN by NASA

From the comparison of the simulation results in Figure.4(a), Figure.4(b) and the measured wind field data in Figure.5(a), Figure.5(b), it can be seen that the wind field model based on the vortex ring method is very similar to the actual microburst wind distribution, and it also has a good three-dimensional characteristic.

3. 6-DOF rigid-body trajectory model for rockets

Rocket is regarded as a free rigid body. The movement of the rocket is composed of centroid movement and the movement encircling the centroid of rockets. Based on the theory of the external ballistics[14], we select a reasonable reference coordinate system for rocket force analysis. when we set up the rocket equation of motion, the rocket is treated as a constant mass object. Meanwhile, we ignore the gas inertia force and its torque, and take no account of the influence of gunpowder combustion caused by the acceleration of mass center of gravity and the rate of change in inertia.

The motion vector equation of rocket mass center is decomposed into the ballistic coordinate system ignoring the minor factors such as the earth rotation, and the equations of the scalar form are obtained by projecting the motion vector equation around the center of mass to the projectile axis coordinate system [14]. The motion of the center of mass of rocket relative to the inertial coordinate system follows the theorem of center of mass motion

$$m\frac{\mathrm{d}\boldsymbol{v}}{\mathrm{d}t}=\mathbf{F} \tag{13}$$

Where t is the flight time, m is the mass of rocket, $\boldsymbol{v}$ is the velocity vector on the mass center of rocket. By projecting Eq.(13) onto the ballistic coordinate system, the components of the net force $\mathbf{F}$ on the rocket in the ballistic coordinate system are F_{x_2}, F_{y_2} and F_{z_2} are obtained. Thus, the scalar equation of mass center of the rocket in the ballistic coordinate system is given by

$$\begin{cases}\dfrac{\mathrm{d}v}{\mathrm{d}t}=\dfrac{1}{m}F_{x_2}\\[2mm] \dfrac{\mathrm{d}\theta_1}{\mathrm{d}t}=\dfrac{1}{mv\cos\psi_2}F_{y_2}\\[2mm] \dfrac{\mathrm{d}\psi_2}{\mathrm{d}t}=\dfrac{1}{mv}F_{z_2}\end{cases} \tag{14}$$

Where θ_1 is the velocity pitch angle and ψ_2 is the velocity yaw angle. Eq. (14) determines the relationship between the center-of-mass velocity of the rocket and the direction change and the force acting on the rocket, and the equation that describing the change of the coordinates of the mass center of the rocket is as follows

$$\begin{cases} \dfrac{\mathrm{d}x}{\mathrm{d}t} = v\cos\psi_2 \cos\theta_1 \\ \dfrac{\mathrm{d}y}{\mathrm{d}t} = v\cos\psi_2 \sin\theta_1 \\ \dfrac{\mathrm{d}z}{\mathrm{d}t} = v\sin\psi_2 \end{cases} \tag{15}$$

Where x, y and z are the coordinates of the mass center of the rocket in the ground coordinate system, and the motion of the rocket around the mass center depends on the theorem of moment of momentum.

$$\frac{\mathrm{d}\boldsymbol{G}}{\mathrm{d}t} = \boldsymbol{M} \tag{16}$$

Where $\boldsymbol{G}$ is the moment of momentum acting on the mass center of rocket and $\boldsymbol{M}$ is the external torque acting on the mass center of rocket. Projecting Eq.(15) to the projectile axis coordinate system and omit some minor factors, the dynamic equation and kinematics equation for the motion around the mass center of rocket are obtained as

$$\begin{cases} \dfrac{\mathrm{d}\omega_\xi}{\mathrm{d}t} = \dfrac{1}{C} M_\xi \\ \dfrac{\mathrm{d}\omega_\eta}{\mathrm{d}t} = \dfrac{1}{A} M_\eta - \dfrac{C}{A}\omega_\xi\omega_\zeta + \omega_\zeta^2 \tan\varphi_2 \\ \dfrac{\mathrm{d}\omega_\zeta}{\mathrm{d}t} = \dfrac{1}{A} M_\zeta + \dfrac{C}{A}\omega_\xi\omega_\eta - \omega_\zeta\omega_\eta \tan\varphi_2 \end{cases} \tag{17}$$

$$\begin{cases} \dfrac{\mathrm{d}\varphi_a}{\mathrm{d}t} = \dfrac{1}{\cos\varphi_2}\omega_\zeta \\ \dfrac{\mathrm{d}\varphi_2}{\mathrm{d}t} = -\omega_\eta \\ \dfrac{\mathrm{d}\gamma}{\mathrm{d}t} = \omega_\xi - \omega_\zeta \tan\varphi_2 \end{cases} \tag{18}$$

Where ω_ξ, ω_η and ω_ζ are the components of the rocket rotational angular velocity in the projectile axis coordinate system, C and A are the polar moment of the inertia and the equator moment of the inertia of the rocket respectively, M_ξ ,M_η and M_ζ are the components of the external torque acting on the rocket in the projectile axis coordinate system, φ_a and φ_2 are the pitch angle and the yaw angle of rocket, γ is the angle of the rotation of the rocket.

The 6D rigid-body trajectory differential equations of the rocket can be obtained by associating Eq.(14), Eq.(15) and Eq.(17), Eq.(18). When the structural parameters and aerodynamic parameters of the rocket are given and the firing conditions, weather conditions and other starting conditions are determined, the trajectory differential can be solved by using the ballistic trajectory integral method.

4. Simulation and analysis

4.1. Simulation conditions

As an example, a kind of 122mmempennage-rocket is studied, and the basic parameters and initial launching conditions of projectile are given by table.2 and table.3 respectively. During the boost phase of the rocket projectile, the mass of the rocket is reduced by a uniform rate, and the moment of inertia and the position of the center of mass change with time.

Table 2. Basic parameters of vortex ring.

Parameters	Value
Diameter of projectile(m)	0.122
Length of projectile(m)	2.9
Specific impulse(s)	250
Engine's working time(s)	3.0
Initial mass of rocket in boost phase(kg)	70.01
Initial equator moment of inertia (kg•m^2)	40.05
Initial polar moment of inertia (kg•m^2)	0.147
Initial position of center of mass(m)	1.533
Mass change rate of projectile(kg/s)	-8.0

Table 3. Initial launching condition of rocket.

Parameters	Value
Initial velocity , m/s	40
Elevation angle , deg	50
Direction angle , deg	0
Rotational angular velocity, rad/s	0
Launching position in ground coordinate system	(0,0,0)

4.2. Influences on trajectory and impact point

Figure.6 shows the trajectory of the rocket under the influence of the microburst wind field with a central vertical induced velocity of $V_0 = 10\text{m/s}$, $V_0 = 15\text{m/s}$ and $V_0 = 20\text{m/s}$ respectively. figure 7 shows the trajectory of the rocket under a vortex ring radius of $R = 800\text{m}$, $R = 900\text{m}$, $R = 1000\text{m}$ and $R = 1100\text{m}$ respectively. In addition, the trajectory under windless condition is also shown in figure.6 and figure.7 for comparison. Table.4 summarizes the data of ballistic performance under different central vertical induced velocity and vortex ring radius.

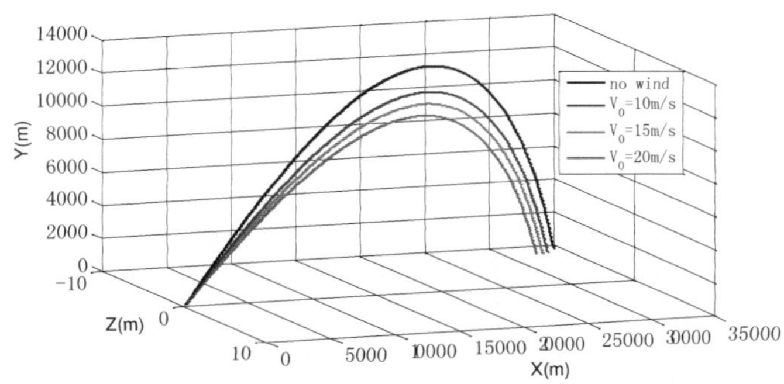

Figure 6. Trajectory of rocket under different central vertical wind velocity

As shown in figure.6 and the data in group 1 and group 2 of table.4, due to the impact of the vertical airflow in the boost phase of the rocket, the lower dynamic pressure caused a corresponding decrease in lift and a static stabilization moment acting on the rocket. The curvature of the trajectory is reduced because of the decrease of the trajectory inclination angle. So the maximum trajectory height, the longitudinal range and transverse range of the projectile are less than that under the non-wind condition. It can be seen from the data in group 2, 5, 7 of table.4 and figure.6 that the relative velocity between the projectile and the air decreases with the decrease of the aerodynamic force in the pitch direction due to the increase of the vertical induced velocity of the microburst center. Thus, the maximum trajectory height and the projectile's longitudinal range and transverse range are reduced correspondingly. The flight time becomes shorter.

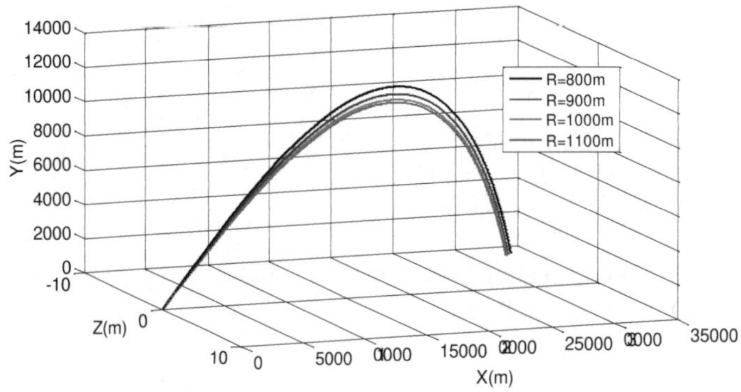

Figure 7. Trajectory of rocket under different vortex ring radius

As shown in Figure.7 and the data in group 3, 4, 5, 6 of Table.4, when the radius of the vortex ring increases, the effective intensity of the microburst on the trajectory increases correspondingly. Thus, the ballistic parameters such as the flight time, the maximum trajectory height and the range of the rocket are reduced. However, as the radius of the vortex ring increases, the reduction of the trajectory characteristic data decreases. This is because the rocket will pass through the affected area of the microburst under the action of the thrust in a short time. On the other hand, the influence of the radius of the vortex ring on the wind velocity is weaker, which means the effect on the dynamic pressure on the projectile is small. Therefore, the increase in the radius of the vortex ring representing the horizontal spatial scale of the microburst wind field makes a limited influence on the rocket trajectory.

Table 4. Ballistic performance under different model conditions.

Group number	Vertical wind velocity(m/s)	Vortex ring radius(m)	Flight time(s)	Maximum trajectory height(m)	Impact point velocity(m/s)	longitudinal range(m)	transverse range(m)
1	0	1000	105.8	13126.3	367.6	34478.8	-9.3
2	10	1000	99.2	11590.8	357.4	33505.2	-8.5
3	15	800	99.3	11605.7	357.5	33514.4	-8.6
4	15	900	97.3	11164.2	354.5	33185.2	-8.4
5	15	1000	95.9	10851.4	352.6	32949.9	-8.2
6	15	1100	95.1	10690.4	351.6	32826.5	-8.1
7	20	1000	92.5	10133.1	348.3	32369.8	-8.0

4.3. Influence on Velocity Characteristics

IOP Conf. Series: Journal of Physics: Conf. Series **1215** (2019) 012028 doi:10.1088/1742-6596/1215/1/012028

The most direct effect of the microburst wind field on a rocket is the change of its velocity. Figure.8 shows the curve of the rocket's velocity during the flight varying with time. Under action of the thrust of the engine, the rocket reaches at a speed of nearly 1000m/s in a short time. The wind speed of the microburst wind field is smaller than that of the high speed of the projectile so that the speed change is not obvious when the rocket crosses the area

Due to the long flight time of the passive free flight, the velocity and acceleration of the projectile are gradually accumulated by the influence of the wind speed, resulting in the decrease of the impact point velocity and longitudinal range compared to that of the windless condition. In addition, the higher the vertical induced wind speed, the smaller the projectile's impact point velocity. From Figure.8(b), it can be seen that the increase of the vortex ring radius also decreases the impact point velocity of the projectile, but the velocity decrease is not linearly related to the increase of vortex ring radius.

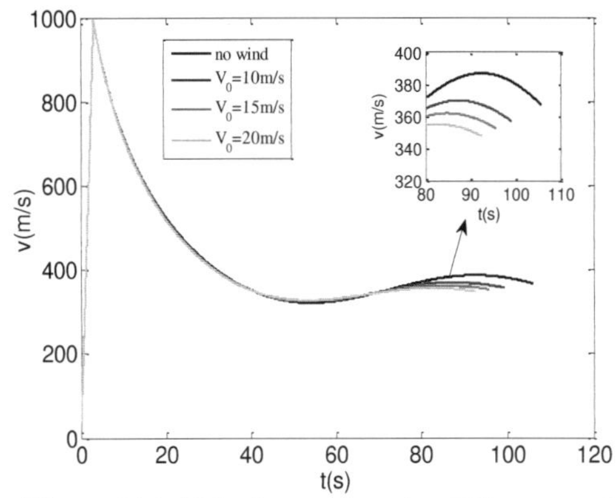

Figure 8(a). Velocity curve under different vertical wind velocity

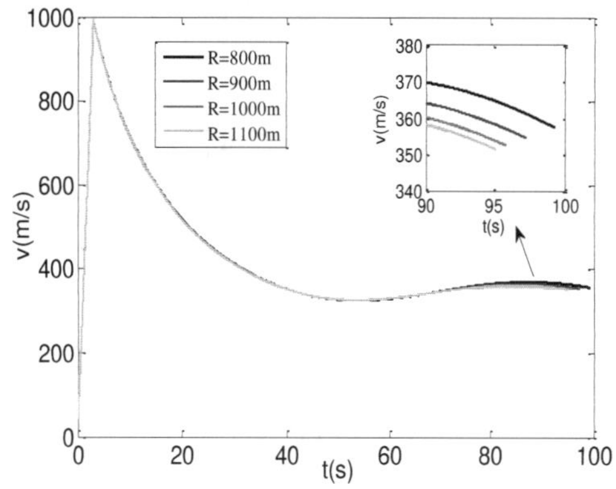

Figure 8(b). Velocity curve under different vortex ring radius

Due to the long flight time of the passive free flight, the velocity and acceleration of the projectile are gradually accumulated by the influence of the wind speed, resulting in the decrease of the impact point velocity and longitudinal range compared to that of the windless condition. In addition, the higher the vertical induced wind speed, the smaller the projectile's impact point velocity. From Figure.8(b), it can be seen that the increase of the vortex ring radius also decreases the impact point velocity of the projectile, but the velocity decrease is not linearly related to the increase of vortex ring radius.

4.4. Influences on angular motion

The Figure.9 shows the curve of the attack angle of the rocket under the microburst with different wind field characteristics parameters.

It can be seen from Figure.9(a) and Figure.9(b) that the lift force acting on the rocket is obviously reduced, as well as the pitch moment compared to the windless condition. The axis of the projectile moves along with direction of the airflow of microburst and the pitch attack angle is changed from positive to negative. The amplitude of attack angle increases with the vertical induced wind speed. However, due to the stabilizing moment of the tail-fin projectile, the rocket gradually returns to steady state, and the attack angle amplitude decreases. Compared to the windless condition, the recovery time of the attack angle of rocket increases. The lateral wind velocity component is small because the central axis of the microburst is in the launching plane of the rocket. Therefore, the microburst wind shear has little influence on the yaw attack angle and its restore time of returning to steady.

Similar conclusions can be obtained from Figure.(9c) and Figure.(9d) that, the influence of the microburst on the pitch attack angle of the projectile is greater than that on the yaw attack angle. The larger the radius of the vortex ring, the greater the impact of the microburst on the attack angle of the rocket.

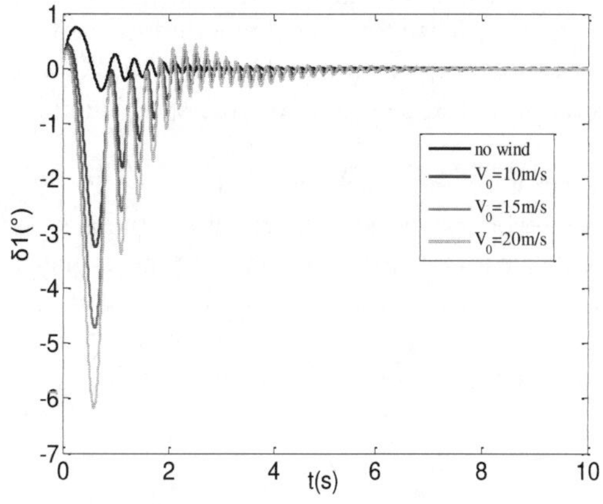

Figure 9(a). Pitch attack angle curve

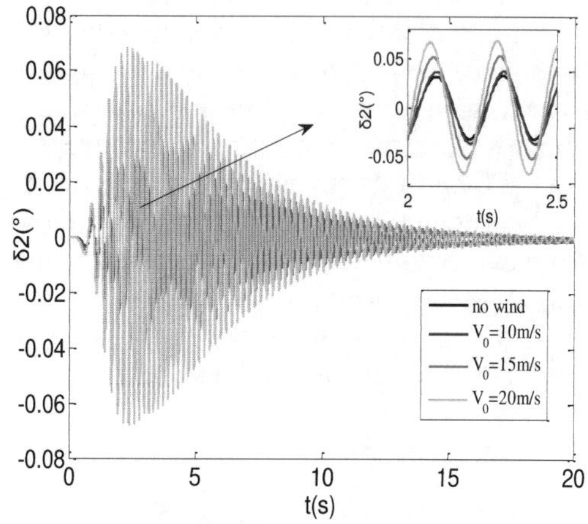

Figure 9(b). Yaw attack angle curve

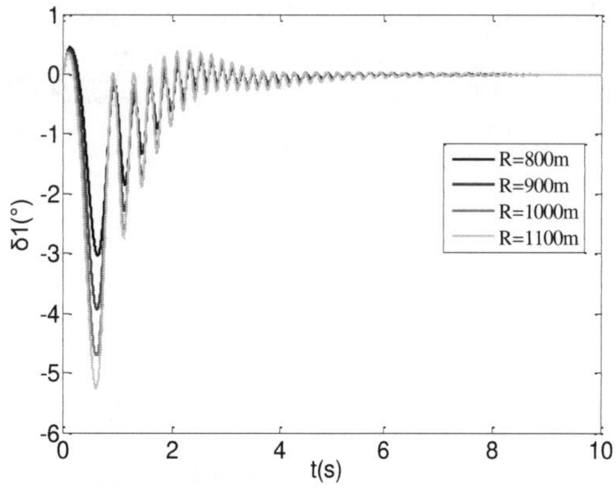

Figure 9(c). Pitch attack angle curve

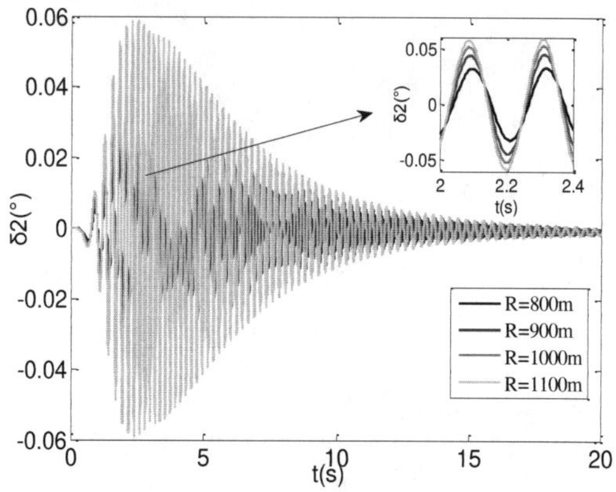

Figure 9(d). Yaw attack angle curve

5. Conclusions

Footnotes should be avoided whenever possible. If required they should be used only for brief notes that do not There are three main conclusions of this research:

- The microburst wind field model based on the vortex ring method is simple and has good three-dimensional characteristics. It can reflect the wind field characteristics of the microburst to a certain extent;
- The microburst has certain influence on the trajectory parameters such as the flight time, the range, the lateral deviation, the maximum trajectory height and the impact point velocity of the rocket. The greater the vertical induced wind velocity or the radius of vortex ring, the higher the degree of the influence of the microburst;
- Under the influence of the microburst, the pitch attack angle of the rocket is reduced to a negative value and the time of the rocket returning to steady state increases. When the central axis of the microburst is in the vertical launching plane of the projectile, the effect of the wind field on the pitch attack angle is greater than that of the yaw attack angle of rockets;

6. References

[1] Ran Z, Gang X and Yue X 2012 *Electronics Optics and Control* vol 19 pp 51-56
[2] Renbiao W, Yanfei H and Hai L 2013 *Jouranl of Electronics & Information Technology* vol 10 pp 121-125
[3] Jianming W, Honghua P and Tao X 2004 Journal of Ballistics vol 16 pp 11-15
[4] Songlan Z and Xiaoli L 2006 *Journal of Projectiles, Rockets, Missiles and Guidance* vol 26 pp 550-552

[5] Xin W, Wei Z and Zipeng H 2007 *Acta Armamentarii* vol 28
[6] Chenming L and Xiaoli L 2006 *Journal of Projectiles, Rockets, Missiles and Guidance* vol 26 pp 550-552
[7] Hanzhou W, Weidong S and Lei Z 2015 *Journal of Ordnance Engineering College* vol 4 pp 38-42
[8] Huang Z, Li X and Deng Y 2006 *World Aviation Conference* vol 2 pp 5384-5390
[9] Hongen L 2001 *Acta MeteorologicaSinica* vol 259 pp 183-195
[10] Ivan M 1986 *Journal of Aircraft* vol 23 pp 232-236
[11] Zhengxin G and Hongbin G 2008 *Journal of System Simulation* vol 3 pp 6524-6528
[12] Shijie L, Hongjun X and Yuan X 2014 *FLIGHT DYNAMICS* vol 32 pp 303-307
[13] A Data Fusion Algorithm for Multi-sensor Microburst Hazard Assessment i, AIAA(1992)
[14] Zipeng H 2008 *Exterior Ballistics of Shells and Rockets* (Beijing: Beijing University of Science & Technology Press) p 76
[15] Mingyou X 2004 *Exterior Ballistics of Rockets* (Harbin : Harbin Institute of Technology Press) p112

Acknowledgments
This work was supported by the Project of National Natural Science Foundation of China (grants 61603191, 51305204).

IOP Conf. Series: Journal of Physics: Conf. Series **1215** (2019) 012029 doi:10.1088/1742-6596/1215/1/012029

Assurance and Evaluation Method for Reliability Requirements Completion of Aviation Product Based on CMM

Yuan Li[1], Wei Wang[1], Weijia Feng[1], Qi Zhang[1] and Huichao Liu[1]

[1]School of Reliability and Systems Engineering, Beihang University, Beijing, China

Abstract. Modern aviation product needs to meet the reliability requirements during the development process. In this paper, a comprehensive assurance system of reliability requirements completion is established and a method to evaluate the assurance system capabilities of reliability requirements completion based on CMM (Capability Maturity Model) is proposed. The evaluate index system in each level is described and AHP method is applied to determine the weight of each index factor in maturity. The evaluation results can comprehensively and effectively direct the aviation product reliability work during the development process and ultimately ensure the reliability of aviation product meets the requirements.

1. Introduction

As the complexity of modern aviation product increases, reliability work needs to be implemented. The goal of reliability work is to ensure that the aviation product meets the specified reliability requirements and maintains and improves its reliability level. Further to ensure the system can meet the requirements of operational readiness and mission success and the requirements for guaranteed resources can reduce [1]. The reliability requirements refer to a series of reliability qualitative and quantitative requirements and specific reliability work items requirements required by the ordering party in the contract. Capability maturity model related reliability have been partially studied. Researchers at the University of Maryland have used CMM to evaluate the reliability capabilities of electronics manufacturers [2-4]. The concept of reliability-centered system engineering capability was putting forward and the weapon equipment reliability-centered system engineering capability maturity model framework was established in the paper of researcher Kang Rui [5]. Li Guosheng added the maintainability and supportability to the reliability engineering capability maturity model [6]. This paper studies the factors that ensure the reliability requirements completed and builds the capability maturity model and proposes the evaluation method.

2. Assurance system of reliability requirements completion

The assurance system of reliability requirements completion refers to a series of engineering organization activities implemented in the development process which to ensure the completion of reliability requirements of aviation product. From the perspectives of organization, technology, and process, the hierarchy model of the assurance system of reliability requirements completion is shown as figure 1. The overall assurance work of reliability requirements completion of aviation product is divided into two categories which are the work of the implementation process and the work of the management process. The work of the management process is to serve the work of the implementation process. In order to establish the assurance system, the reliability requirements of product and the reliability work specified in national military standards are analyzed firstly. Then the factors affecting the completion of reliability requirements are summarized. Finally, the correlation among the factors is considered and the structure of the assurance system is determined. With further practice and deeper learning, the assurance system will be further refined.

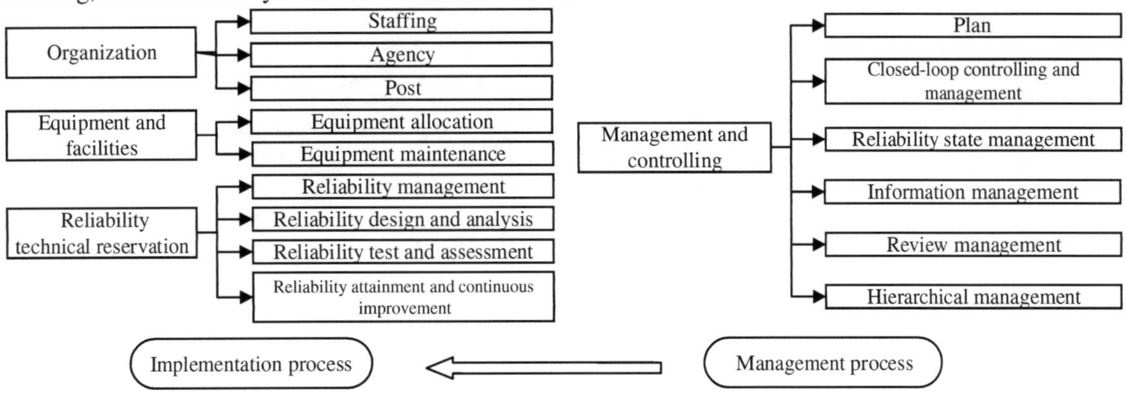

Figure 1. Assurance system of reliability requirements completion.

3. Assurance capability of reliability requirements completion evaluation

This paper defines the assurance capability of reliability requirements completion as the ability to develop the aviation product to meet the specified reliability requirements under the constraints on various factors of the assurance system of reliability requirements completion. A which means assurance capability maturity of reliability requirements completion is used to measure the system capability and A is a number within [0, 5]. In this paper, fuzzy comprehensive evaluation method and analytic hierarchy process (AHP) method are used to evaluate assurance capability maturity of reliability requirements completion. The evaluation system construction idea is shown in figure 2.

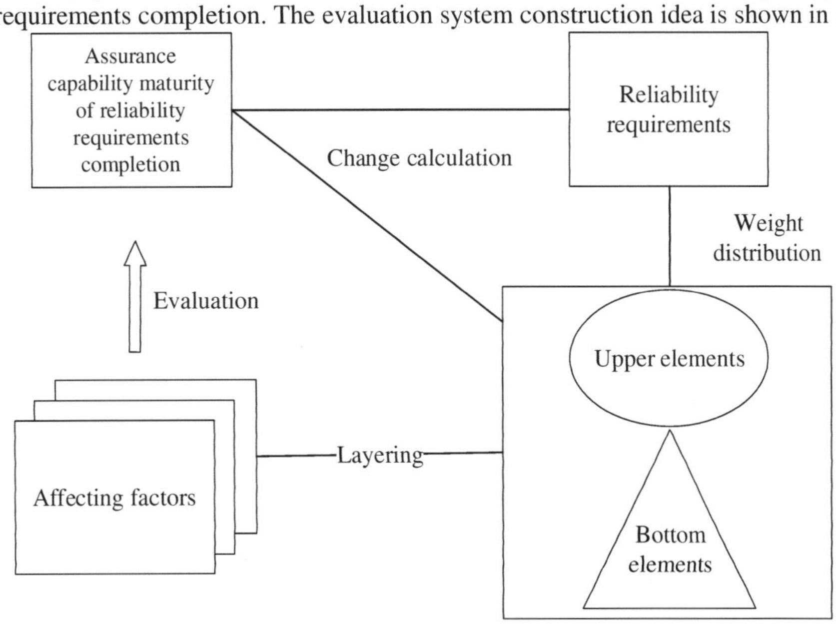

Figure 2. Assurance capability of reliability requirements completion evaluation system construction idea.

3.1. Maturity level

Maturity level is the core content of the capability maturity mode [7]. According to the principle of traditional capability maturity models and referring to the CMMI continuous representation model, ISO/IEC 15504 and other models [8-9], the hierarchical framework is established as shown in figure 3. Maturity level classification and level description:

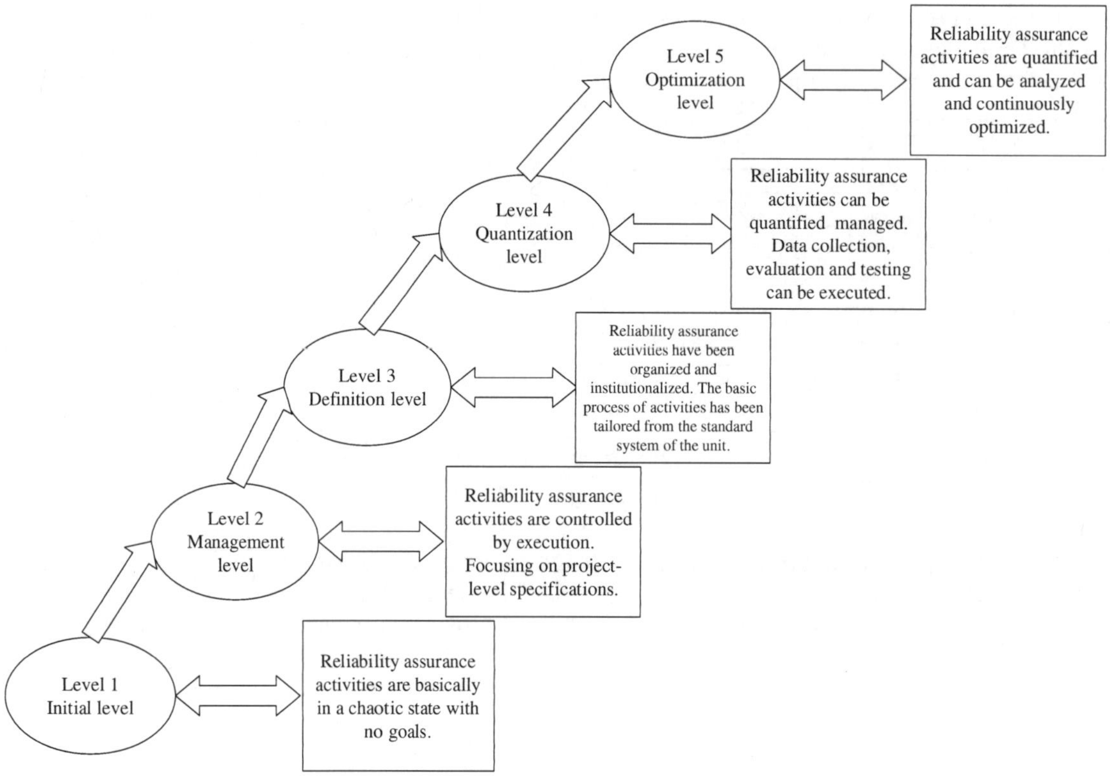

Figure 3. Hierarchical framework of maturity level.

(1) Initial level. Reliability assurance activities are executed under the chaotic management state or not executed. The organization has the reliability assurance activities without corresponding norms and mature goals. Work is not performed in accordance with the principles and ideas of reliability assurance. Activities such as complete planning, tracking, analysis, improvement and experience dissemination are not implemented. Work steps are not systematic. Reliability assurance activities and the outputs obtained are partially valid. Many problems exist, including unidentified information on the collaborative process across units, geographies, and interdisciplinary, invalid of complex product development and management, and uncertainty project success.

(2) Management level. The typical project is the core at most of the time and project-level specifications are focused. The reliability assurance activity program is developed specifically including objectives, plans, and processes. The requirements for reliability assurance activities are clearly defined and activities are planned, tracked, evaluated and accepted. Reliability assurance organizations are established and organized with relevant guidelines to support and encourage. Activity resources are allocated. Appropriate corrective actions exist when there is a deviation from the operational effect and the plan or the process is not executed. The successful experience of similar projects in the past can be reproduced.

(3) Definition level. The standards-suitable reliability assurance activity process is fully established. Project-level activities are absorbed, summarized and improved. The processes related to reliability assurance activities are clearly defined at the organizational level. The project reliability assurance activities are obtained through reductions of organizational standard activities and reasonable adjustment according to the actual technical status of the project. The tailored activity contributes activity improvement information to the organization's active asset library. The biggest feature of this level is standardization.

(4) Quantization level. Reliability assurance activities are measurable and quantifiable. Statistics and other quantitative methods are used as management means and quantitative quality, process and performance indicators are used as the standard for management. Existing defects can be identified

IOP Conf. Series: Journal of Physics: Conf. Series **1215** (2019) 012029 doi:10.1088/1742-6596/1215/1/012029

and trends in work processes can be predicted. Measures can be implemented before problems arise to reduce costs. Good software tools are used as support generally.

(5) Optimization level. Reliability Assurance activities are able to meet current and anticipated goals and accommodate changes. Innovative methods and techniques are used centrally. It is possible to achieve continuous improvement activities by actively collecting data and screening and analyzing the data. Activity defects can be identified and defect patterns can be analyzed. The root cause of defects can be discovered and corresponding improvements can be taken. The defect prevention plan was developed and tracked. Reliability assurance staff are continuously trained and upgraded. The activity process is the result of continuous optimization.

The quantification level and the optimization level are based on the first three levels and the processes are further quantitatively predicted and controlled and analyzed and optimized according to the cost, efficiency and quality requirements of the organization's activities. Quantization level and optimization level are relatively difficult to achieve, and they are stages to realize potential by actively discovering and solving problems in an open and positive environment.

3.2. Evaluation index

The evaluation indexes are established based on the factors of the assurance system of reliability requirements completion. The evaluation indexes are shown in table 1.

Table 1. Evaluation index

Objective	The first-level assessment index	The second-level assessment index	The third-level assessment index
Assurance capability of reliability requirements completion	Basic capability	Organization	Set up independent reliability organizations and work management agencies or positions
			Set up dedicated reliability information management agencies or positions
			Equipped with a complete reliability team
			Establish a complete reliability training system
			Establish a sound job evaluation system for reliability professionals
		Equipment and facilities	Configure the required device
			Guaranteed number of devices
			Guarantee the advanced nature of the equipment
			Guarantee equipment reliability
	Management and controlling capability	Plan	Top-level documents and requirements related to overall planning reliability work
			Develop a work plan that meets product reliability requirements and characteristics
			Establish a reliability assurance specification for applicability
		Closed-loop controlling and management	Reliability requirements work management requirements and control plan development
			Ability to carry out management control work according to management requirements and control plan system
			Apply feedback adjustment mechanism to control plan during work development
		Reliability state management	Reliability state identifier
			Reliability state control
			Reliability state review
			Reliability state documentary
		Information management	Establish and operate a complete reliability information system
			Establish a standardized information system operation

		requirements
		Timely and accurate collection of information data
		Closed-loop management of information
	Review management	Reliability review plan setup in coordination with the development plan
		Organize reliability review activities according to the review plan
		Develop review criteria and detailed review checklists for each reliability work
		Closed-loop control of review results
	Hierarchical management	Selection of transferee and supplier
		Contracts and agreements between the transferee and the supplier
		Develop a monitoring plan to manage the reliability of the contractor and supplier
	Reliability management	Develop a reliability plan
		Develop a reliability work plan
		Reliability growth management
	Reliability design and analysis	Timeliness of design analysis work
		Normative design analysis work
		The correctness of the design analysis method
		Coordination between design analysis work
		Design analysis uses the credibility of the data
		Iterative design analysis work
		Design credibility of analytical work results
Reliability technology capability	Reliability test and assessment	Develop appropriate test and evaluation plans as required
		Develop the right experimental protocol and outline
		Responsible for the determination of the test unit and its responsibilities
		Integrity of data collection and processing during testing
		Scientific and credibility of assessment methods and results
		Closed-loop solution to exposed problems in the test
	Reliability attainment and continuous improvement	Reliability requirements indicators met the standard
		Supervised measures to correct and improve external field problems

3.3. Weight calculation of evaluation indexes

This paper uses the AHP method to establish a weight set, which classifies various related factors in a complex problem into an interrelated and orderly hierarchy. Based on certain objective judgments, the first-level, the second-level and the third-level assessment indexes are compared and analyzed. In the weight calculation of the three layers of assessment indexes, some experts are invited to construct the discriminant matrix by comparing the index among each other. 1~9 are used as comparison scales among evaluation factors. By solving these discriminant matrices, it can be calculated that the three weights of the first-level assessment index which are denoted as $\omega_i (i=1,2,3)$. The weights of the second-level assessment index for the first-level assessment indexes are denoted as ω_{ij} $(i=1,2,3; j=1,2\cdots)$. The weights of the third-level assessment index for the second-level assessment indexes are denoted as ω_{ijk} $(i=1,2,3; j=1,2\cdots; k=1,2\cdots)$. The weight set of index is denoted as $W_{ij}=\{\omega_{ij1}, \omega_{ij2}, \ldots, \omega_{ijk}\}(i=1,2,3; j=1,2\cdots)$.

3.4. Scoring criteria of index

In this paper, fuzzy comprehensive evaluation method is used to determine the scoring criteria shown as table 2.

Table 2. Scoring criteria of index

The second-level assessment index	The third-level assessment index	Maturity level				
		Initial level	Management level	Definition level	Quantization level	Optimization level
Plan	Top-level documents and requirements related to overall planning reliability work	None or disorder	Project level management	Standardization of organization and systematization	Quantitatively management	Continuous optimization
…	…	…	…	…	…	…
Hierarchical management	Contracts and agreements between the transferee and the supplier	Very Bad	Bad	Average	Good	Excellent
…	…	…	…	…	…	…

3.5. Maturity calculating

(1) A membership function whose span is [0,1] to quantify index and experts scores are used to get fuzzy evaluating matrix with the third-level assessment index. Degree of membership x_{ijk}^{n} indicates the membership degree of the maturity level v_n of factor X_{ijk}.

$$x_{ijk}^{n} = \frac{d_{ijk}^{n}}{d} \tag{1}$$

where d is the number of experts, d_{ijk}^{n} is the number of experts whose score is v_n.

(2) The fuzzy evaluating matrix is shown as equation (2):

$$X_{ij} = \begin{bmatrix} x_{ij1}^{1} & x_{ij1}^{2} & x_{ij1}^{3} & x_{ij1}^{4} & x_{ij1}^{5} \\ \vdots & \vdots & \vdots & \vdots & \vdots \\ x_{ijk}^{1} & x_{ijk}^{2} & x_{ijk}^{3} & x_{ijk}^{4} & x_{ijk}^{5} \end{bmatrix} \tag{2}$$

(3) Maturity evaluation vector of the second-level assessment index is shown as equation (3):

$$A_{ij} = W_{\mathrm{ij}} \cdot X_{ij} \tag{3}$$

(4) Maturity evaluation vector of the first-level assessment index is shown as equation (4):

$$A_i = \mathrm{W_i} \cdot A_{ij} \tag{4}$$

(5) System evaluation value is shown as equation (5):

$$A = \mathrm{W_i} \cdot A_i \tag{5}$$

3.6. Result analysis

The system evaluation value A can reflect the level of assurance capability maturity of reliability requirements completion. The corresponding description can be found in figure 3.

4. Conclusion

This paper aims to help the aviation product meet the reliability requirements. An assurance system of reliability requirements completion and a method to evaluate the assurance system capabilities of reliability requirements completion based on CMM of aviation product are proposed. The result can not only reflect the level of assurance system of reliability requirements completion, but also expanded the application of capability maturity model in reliability. Further study will enrich the assurance system of reliability requirements completion.

References

[1] Yang Jing 2014 Review of software reliability and safety work requirements. *Quality & reliability* 5: 47-52

[2] Bate, Roger, et al 1995 A systems engineering capability maturity model, Version 1.1. No. CMU/SEI-95-MM-003. Carnegie-Mellon Univ Pittsburgh PA Software Engineering Inst

[3] Tiku, Sanjay 2005 Reliability capability evaluation for electronics manufacturers

[4] Tiku, Sanjay, Michael Azarian, and Michael Pecht 2007 Using a reliability capability maturity model to benchmark electronics companies. *International Journal of Quality & Reliability Management* 24.5: 547-563

[5] Kang Rui 2008 Research of reliability-centered system engineering capability maturity model. *Engineering & technology* 3: 63-66

[6] Li Guosheng 2012 RMS engineering capability evaluation based on proximity. *Ship electronic engineering* 32.11: 84-86

[7] Duncan Nancy Bogucki 1995 Capturing Flexibility of In-formation Technology Infrastructure: A Study of Resource Characteristics and Their Measure. *Journal of Management Information Systems* 12: 21-37

[8] Team, Cmmi Product 2001 CMMI for Systems Engineering/Software Engineering, Version 1.1, Continuous Representation. CMMI-SE/SW, V1.1, Continuous

[9] ISO IEC, ISO/IEC TR 15504:1998 Information technology — Software process assessment Part 3:Performing an assessment

IOP Conf. Series: Journal of Physics: Conf. Series **1215** (2019) 012030 doi:10.1088/1742-6596/1215/1/012030

Approach and Application of Semi-Blind Source Separation for Aero-Engine Vibration Signals Using ICA-R

Lujie Shi, Yankai Wang*, Mingfu Liao, and Yunfan Jiang

Northwestern Polytechnical University, School of Power and Energy, 710129, Xi'an, China

ykwang@nwpu.edu.cn

Abstract. In this paper, the approach and application of semi-blind source separation (SBSS) in aero-engine vibration signal is studied. Firstly, the features of aero-engine vibration signal and difficulties for blind source separation (BSS) are summarized, and the SBSS incorporated the available prior knowledge is match to the goal of signal processing. Then, the ICA with reference (ICA-R) algorithm based on classical FastICA is introduced, with Newton iteration and gradient descent iteration approach to obtain optimal solution. The unique parameters in ICA-R for aero-engine vibration signal are also provided. Finally, the efficacy and the accuracy of the ICA-R algorithm are verified by numerical simulations and real engine vibration signals. The approach of SBSS in this paper perfectly suited to handle aero-engine vibration source separation and it lead to efficient implementation in fault diagnosis.

1. Introduction

Vibration signal offers important information for aero-engine condition monitoring and fault diagnosis. However, for the vibration signal of the whole engine, how to judge the position of the specific vibration is imperative to fault diagnosis. Blind source separation (BSS) is the issue of recovering the various independent vibration sources, which is of great significance for locating faulty part accurately. Independent Component Analysis (ICA)[1-2] is a statistical technique that are widely used to detect source or independent component for solving BSS problem.

The semi-blind source separation (SBSS) that incorporates the available prior knowledge into the separation process has gradually become a more suitable method for processing complicate signals. Specifically, ICA with References (ICA-R) [3] impose prior knowledge into mathematical process, forces the optimization process converge to the target component by formulating a rough template as reference signals.

In term of SBSS, Lin[4] introduced the pre-processing and separation vector standardization methods to improve the iteration speed of one-unit ICA-R algorithm. Zhang[5] studied reference signal construction method based on the second-order cumulant. De Vos[6] provided spatially constrained ICA(scICA) to deals with the situation when one mixing vector is exactly known. Wang[7] proposed an original CICA (OCICA) algorithm based on fixed-point iterative to simplify the learning rate parameters in the iterative process. Wang[8] extended the CICA method to the complex domain, and proposed cyclic stationary constraint (ICA-CC) and spatial constraint (ICA-SC) algorithms. Yang[9] used the rotation speed information to separate signals in the wavelet domain, and effectively extracted the information of main vibration sources such as high and low-pressure rotor, combustion chambers and accessory casings. Such methods provided good result but suffer from complexity in algorithm and application in aero-engine.

Combined with the engineering background of aero-engine vibration signals, this paper presents an SBSS approach based on ICA-R. Section2 discusses the exclusive features of aero-engine vibration signal and describes the basic SBSS model. The ICA-R algorithm is presented in Section3. Finally, the methodology is verified by number simulation and a convincing case in application of engine in Section4.

2. Blind source separation model for aero-engine vibration signals

2.1. Basic BSS model

Generally, it can be considered as an instantaneous mixing model for independent signals[10]. The basic mixing-demixing model without noise is:

$$\boldsymbol{x} = \boldsymbol{A}\boldsymbol{s} \tag{1}$$

$$\boldsymbol{y} = \boldsymbol{W}\boldsymbol{x} \tag{2}$$

Where $\boldsymbol{A}$ is mixing matrix, represents the force transmission path; $\boldsymbol{W}$ is demixing matrix, and $\boldsymbol{W} = [\boldsymbol{w}_1^{\mathrm{T}}, \boldsymbol{w}_2^{\mathrm{T}}, \cdots, \boldsymbol{w}_m^{\mathrm{T}}]$, while $\boldsymbol{w}$ is demixing vector. And different components during mixing-demixing process are systematically displayed as following:

(1) the source signal denoted by $\boldsymbol{s}$.
(2) the raw observed signal denoted by $\boldsymbol{x}$.
(3) the observed signal after preprossing denoted by z.
(4) the reference signal denoted by $\boldsymbol{r}$.
(5) the demixing signal (output estimation) denoted by $\boldsymbol{y}$.

2.2. BSS of aero-engine: a list of exclusive features

Due to the complicated structure of the aero-engine, there are difficulties BSS is faced with when dealing with aero-engine vibration signals. These conflicts have never been explicitly recognised before.

2.2.1. Coupling of vibration sources. There are physical relevance of different parts in aero-engine such as inter-shaft bearing and common bearing cavity, whereas the basic assumption of BSS is the sources are irrelevant from each other. It is general that the dynamics relationship between excitation and response for different parts coupled, which caused vibration sources are not independent. Taking dynamic equations of dual-rotor with the inter-shaft bearing as an example:

$$m_l \ddot{x}_l + d\dot{x}_l + k_2 x_l = F_{bearing} + F_{fault} + m_l g \tag{3}$$

$$m_h \ddot{x}_h + d\dot{x}_h + k_2 (x_h - x_l) = F_{bearing} + F_{fault} + m_h g \tag{4}$$

The fault force in any rotor can transmits to the other rotor via the inter-shaft bearing, which means the vibration signals in low and high-pressure rotor are relevant in statistics.

2.2.2.Nonlinear mixing. In can be seen from equation (1) and (2) that the BSS adopt linear equation to demix signals, while the fact that strong nonlinear mixing in aero-engine are rarely described by simple linear algebraic equations. For instance, a rigid rotor on anisotropic flexible supports including gyroscopic effects[11] is expressed as:

$$\begin{bmatrix} \Omega \boldsymbol{G} & \boldsymbol{M} \\ \boldsymbol{M} & \boldsymbol{0} \end{bmatrix} \frac{\mathrm{d}}{\mathrm{d}t} \begin{Bmatrix} x \\ \dot{x} \end{Bmatrix} + \begin{bmatrix} \boldsymbol{K} & \boldsymbol{0} \\ \boldsymbol{0} & \boldsymbol{-M} \end{bmatrix} \begin{Bmatrix} x \\ \dot{x} \end{Bmatrix} = \begin{Bmatrix} F \\ F \end{Bmatrix} \tag{5}$$

Ignoring the nonlinear mixing will reduce the accuracy of BSS, or even lead to wrong result.

2.2.3.Overmuch number of sources. BSS algorithm require the number of observed signals to be equal or greater than the number of source, in other words, that $\boldsymbol{s}$ and $\boldsymbol{y}$ must have same number of rows. Yet sensibly the vibration sources cannot be all observable from limited sensors (six in test-rig normally). Figure 1 illustrates the infinite vibration source of aero-engine. It is obvious that deficient measurement unable to meet the requirement of BSS model.

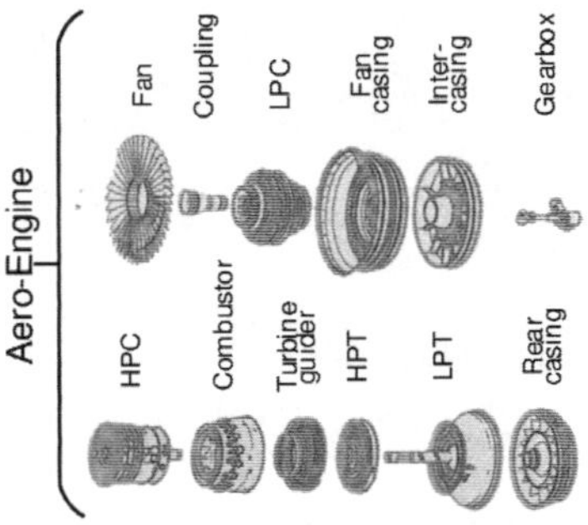

Figure 1. Typical example where the number of vibration source can hardly be inferred

2.3. Prior knowledge and SBSS

Actually, the application of BSS wish to extract the desired subset of source, like low and high-pressure rotor and bearing, and automatically discard the rest of uninteresting sources from the observed mixtures.

Also, the engine vibration test is not blind totally. The vibration of aero-engine is not a completely stochastic process while only the statistical properties of the data are utilized in BSS.

Some of source signals can be identified by traditional signal processing (such as Fourier Transform). As shown in figure 2, some parameters are already known: (1) the amplitude and frequency of the low and high-pressure rotor, including fundamental frequency and the multiplier frequency; (2) the characteristic frequency main parts like bearing, driving gear, etc.

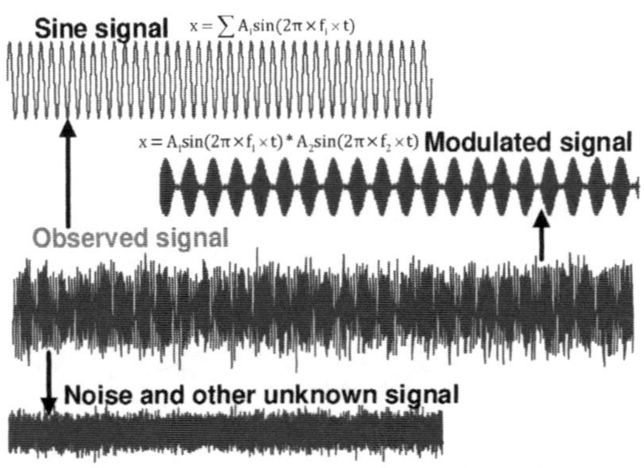

Figure 2. Some prior knowledge about the source signals are available in aero-engine test.

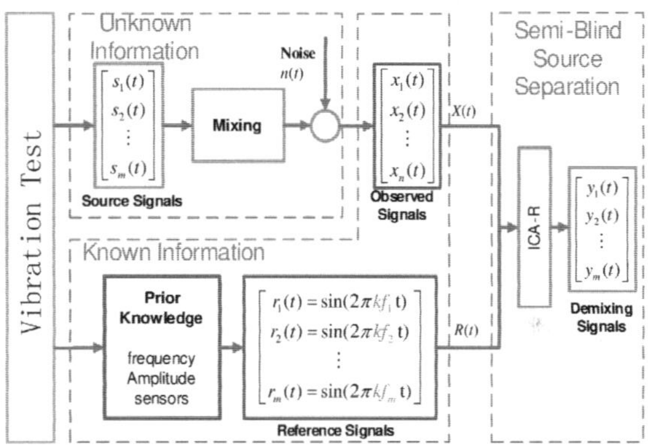

Figure 3. Framework of SBSS of aero-engine vibration signals

Figure 3 gives the framework of SBSS, the prior knowledge needs to be fully used to avoid three difficulies and improve its feasibility. Based on this goal, the ICA with reference (ICA-R) that incorporates reference signals can extract the interesting source signal accurately.

3. Approach of semi-blind source separation

3.1. Classical FastICA algorithm

FastICA based on maximization of negentropy [12] is classical algorithm to solve BSS problem, which are introduced firstly here.

Before the FastICA, the raw observed signal should be preprossed including centering step and whitening step to decorrelate observed signals and orthogonalize the demixing vector. The preprossing phase is introduced detailed in reference [1,2,13]. In this context, all observed signal are assumed been preprossed and donated as z.

An reliable approximation of negentropy is defined as:

$$J(\boldsymbol{y}) \propto \{\mathrm{E}[\mathrm{G}(\boldsymbol{y})] - \mathrm{E}[\mathrm{G}(\boldsymbol{v})]\}^2 \tag{6}$$

Where J is negentropy; $\boldsymbol{y} = \boldsymbol{w}^{\mathrm{T}}\boldsymbol{z}$; $\boldsymbol{v}$ is a Gaussian variable with zero mean and unity variance. There is an inherent constraint condition $\mathrm{E}[(\boldsymbol{w}^T\boldsymbol{z})^2] = \|\boldsymbol{w}\|^2 = 1$ in deducing negentropy. So the equation (6) transfer to:

$$\begin{aligned} &\max\ \mathrm{E}[\mathrm{G}(\boldsymbol{w}^T\boldsymbol{z})] \\ &\text{s.t.}\ \ \mathrm{E}[(\boldsymbol{w}^T\boldsymbol{z})^2] = \|\boldsymbol{w}\|^2 = 1 \end{aligned} \tag{7}$$

The equation (7) can be solved by Lagrange multipliers method, and the Lagrange function and its gradient is given by equation (8) and equation (9).

$$\mathrm{L}(\boldsymbol{w}) = \mathrm{E}[\boldsymbol{z}g(\boldsymbol{w}^T\boldsymbol{z})] + \beta\boldsymbol{w} = 0 \tag{8}$$

$$L'[\boldsymbol{w}] = \frac{\partial L}{\partial \boldsymbol{w}} = \mathrm{E}[\boldsymbol{z}\boldsymbol{z}^T g'(\boldsymbol{w}^T\boldsymbol{z})] + \beta\boldsymbol{I} \tag{9}$$

Where $g(\boldsymbol{y}) = \mathrm{G}'(\boldsymbol{y})$. The optimal solution of Lagrange function is searched by Newton iteration scheme as equation (10) and equation (11).

$$\boldsymbol{w}_{k+1} = \boldsymbol{w}_k - \frac{\mathrm{L}(\boldsymbol{w}_k)}{\mathrm{L}'(\boldsymbol{w}_k)} \tag{10}$$

$$\boldsymbol{w}_{k+1} = \boldsymbol{w}_k - \frac{\mathrm{E}\{\boldsymbol{z}g[\boldsymbol{w}_k{}^T\boldsymbol{z}]\} + \beta\boldsymbol{w}_k}{\mathrm{E}\{\boldsymbol{z}\boldsymbol{z}^T g'[\boldsymbol{w}_k{}^T\boldsymbol{z}]\} + \beta} \tag{11}$$

The approximate equation (12) is correct because of $\boldsymbol{z}$ is whitened, by which the equation (11) is simplified to equation (13).

$$\begin{aligned} \mathrm{E}[\boldsymbol{z}\boldsymbol{z}^T g'(\boldsymbol{w}^T\boldsymbol{z})] &\approx \mathrm{E}[\boldsymbol{z}\boldsymbol{z}^T]\mathrm{E}[g'(\boldsymbol{w}^T\boldsymbol{z})] \\ &= \mathrm{E}[g'(\boldsymbol{w}^T\boldsymbol{z})]\boldsymbol{I} \end{aligned} \tag{12}$$

$$\boldsymbol{w}_{k+1} = \boldsymbol{w}_k - \frac{\mathrm{E}\{\boldsymbol{z}g[\boldsymbol{w}_k{}^T\boldsymbol{z}]\} + \beta\boldsymbol{w}_k}{\mathrm{E}\{g'[\boldsymbol{w}_k{}^T\boldsymbol{z}]\} + \beta} \tag{13}$$

The both sides of equation (13) multiply the $\mathrm{E}\{g'[\boldsymbol{w}_k^{\mathrm{T}}\boldsymbol{z}]\} + \beta$, and the final FastICA iterative formula is obtained as:

$$\begin{cases} \boldsymbol{w}_{k+1} = \mathrm{E}\{\boldsymbol{z}g[\boldsymbol{w}_k{}^T\boldsymbol{z}]\} - \mathrm{E}\{g'[\boldsymbol{w}_k{}^T]\}\boldsymbol{w}_k \\ \boldsymbol{w}_{k+1} = \boldsymbol{w}_{k+1} / \|\boldsymbol{w}_{k+1}\| \end{cases} \tag{14}$$

In addition, $g(\bullet)$ is the nonlinear function. The vibration signal of aero-engine, like sine and cosine signal, is sub-Guassian signal. The references[14-15] tell a better choice of $g(\bullet)$ and $g'(\bullet)$ for aero-engine vibration separation is as follows:

$$\begin{aligned} g(\boldsymbol{y}) &= -\boldsymbol{y} - \tanh(\boldsymbol{y}) \\ g'(\boldsymbol{y}) &= -1 - (1 - \tanh^2(\boldsymbol{y})) \end{aligned} \tag{15}$$

FastICA provide decent solution to BSS problems due to clear iteration and cubic convergence speed.

3.2. ICA-R algorithm

The ICA-R take into account the prior knowledge as the reference signal into basic ICA paradigm. The ICA-R algorithm based on FastICA in term of aero-engine vibration signal is introduced here.

On one hand, maximization of negentropy can be adopted as the objective function. On the other hand, the reference signals can be created to represent rough templates of specific component vibration, and minimization of closeness measure function[16] can be adopted as the other objective function:

$$\begin{cases} \max\ J(\boldsymbol{w}) = \sum E[G(\boldsymbol{w}^T z)] \\ \min\ f(\boldsymbol{w}) = \varepsilon(\boldsymbol{y}, \boldsymbol{r}) \end{cases} \tag{16}$$

Where $\varepsilon(\boldsymbol{y},\boldsymbol{r})$ is closeness measure function to scale distance between the demixing signal $\boldsymbol{y}$ and the corresponding reference signal $\boldsymbol{r}$.

For the first objective function with maximizing negentropy, it can be iterated by equation (14). For the second objective function with minimizing closeness measure function, it can be updated by gradient descent method:

$$\boldsymbol{w}_{k+1} = \boldsymbol{w}_{k+1} + \nabla f(\boldsymbol{w}) \tag{17}$$

Where $\nabla f(\boldsymbol{w})$ is the gradient of the closeness measure function with respect to the demixing vector $\boldsymbol{w}$. And the demixing vector must be standardized after each iteration step as equation (14). The final iteration formulation of ICA-R is given as:

$$\begin{cases} \boldsymbol{w}_{k+1} = \mathrm{E}\{zg[\boldsymbol{w}_k{}^T z]\} - \mathrm{E}\{g'[\boldsymbol{w}_k{}^T]\}\boldsymbol{w}_k \\ \boldsymbol{w}_{k+1} = \boldsymbol{w}_{k+1} + \nabla f(\boldsymbol{w}) \\ \boldsymbol{w}_{k+1} = \boldsymbol{w}_{k+1} / \|\boldsymbol{w}_{k+1}\| \end{cases} \tag{18}$$

In summary, the steps to achieve the ICA-R algorithm are shown as figure 4.

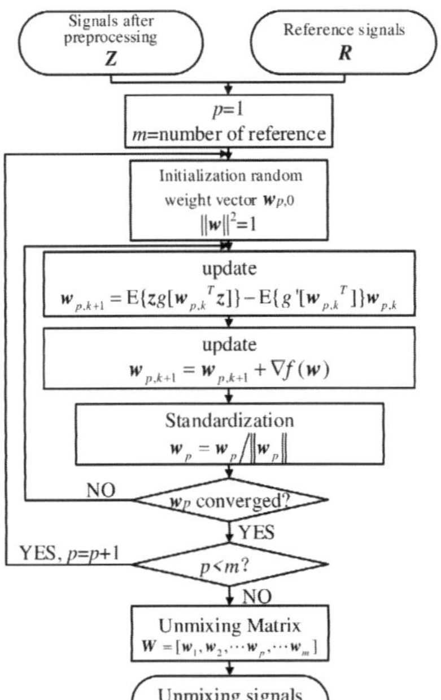

Figure 4. Flow chart of ICA-R algorithm

3.3. Analyses of parameter selection

Due to the complicated structure of the aero-engine, there are difficulties BSS is faced with when dealing with aero-engine vibration signals. These conflicts have never been explicitly recognised before.

3.3.1.Closeness measure function. The mean square error (MSE) is used as the similarity measure in previous studies because of simple format, easy derivative and rapid iteration. Mean square error is shown in equation (19) and the derivative function to demixing vector is shown in equation (20).

$$\varepsilon(\boldsymbol{y},\boldsymbol{r}) = \mathrm{E}[(\boldsymbol{y}-\boldsymbol{r})^2] \tag{19}$$

$$\begin{aligned} f(\boldsymbol{w}_k) &= \mathrm{E}[(\boldsymbol{w}_k^T\boldsymbol{Z}-\boldsymbol{r})^2] \\ f'(\boldsymbol{w}_k) &= \mathrm{E}[(\boldsymbol{w}_k^T\boldsymbol{Z}-\boldsymbol{r})^2]' = \mathrm{E}[2\times z(\boldsymbol{w}_k^T\boldsymbol{Z}-\boldsymbol{r})] \end{aligned} \tag{20}$$

3.3.2. The reference signal for aero-engine. Low and high-pressure rotor, whose vibration is close to sine signal, are main aero-engine vibration sources in form of rotor fundamental frequency and its multiple frequency. So standard sine or sine superimpose signal can be selected as reference signal.

$$r = \sum A_i \sin(2\pi \times f_i \times t + \varphi_i) \tag{21}$$

Where A is amplitude, f is frequency (Hz), and φ is phase (rad). Subscript i denote different order of fundamental frequency. The fundamental frequency (rotation speed) of the high and low-pressure rotor is known in aero-engine vibration test, and amplitude and other frequency can be obtained easily through the FFT spectrum.

4. Verification and application

4.1. Numerical simulation on synthetic signals

The vibration of the aero-engine includes periodic vibration, modulated vibration, shock vibration, random vibration and so on. In order to verify the effectiveness of the ICA-R algorithm in aero-engine, the sine signal, modulated signal and the pulsed signal are selected as the three source signals for experiment:

$$\begin{cases} s_1 = 10\sin(2\pi \times 10 \times \mathrm{t}) \\ s_2 = 8\cos(2\pi \times 75 \times \mathrm{t}) \times 2\sin(2\pi \times 0.8 \times \mathrm{t}) \\ s_3 = 15\mathrm{sawtooth}(2\pi \times 30 \times \mathrm{t}) \end{cases} \tag{22}$$

The mixing matrix is generated as non-linear formation (equation(23)), and the waveform of source signals and mixing signals are as shown in figure 5 and 6.

$$\begin{bmatrix} \sin & 0.22 & 0.16 \\ 0.22 & \cos & 0.33 \\ 0.25 & 0.15 & \sin \\ 2\sin & 2\sin & 2\sin \end{bmatrix} \tag{23}$$

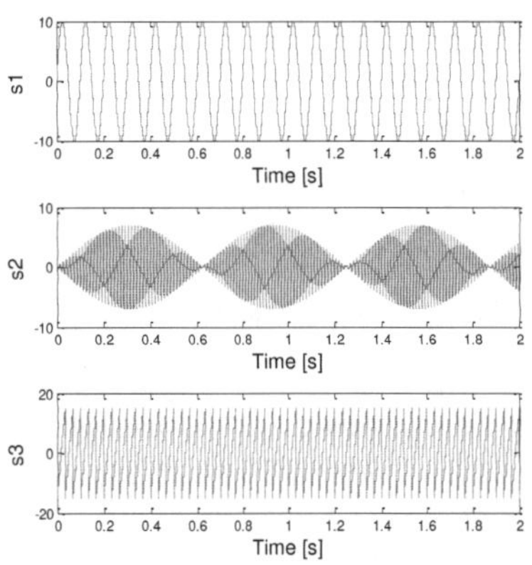

Figure 5. Synthetic source signals

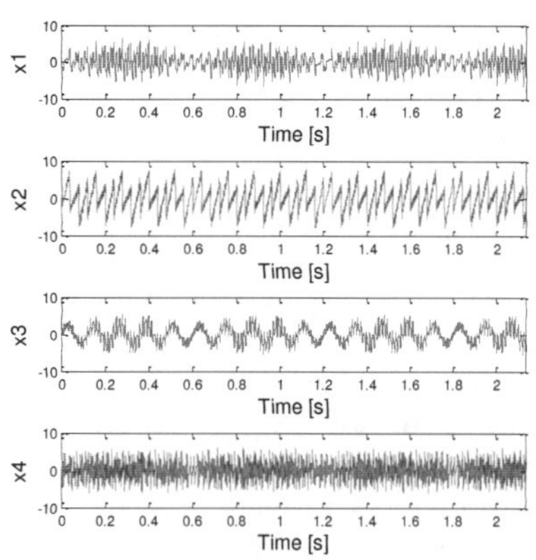

Figure 6. Mixing signals

The characteristics of the source signals are substantially invisible from the mixing signals. For *s1*, it can be assumed that amplitude and frequency are known; For *s2*, it can be assumed that higher frequency is known; For *s3*, the reference are formulated by square signal with same amplitude and frequency. The three reference signals are shown in equation (24). The demixing signals calculated by FastICA and ICA-R are shown in figure. 7 and 8.

$$\begin{cases} r_1 = 10\sin(2\pi \times 10 \times \mathrm{t}) \\ r_2 = 8\cos(2\pi \times 75 \times \mathrm{t}) \\ r_3 = 15\,\mathrm{square}(2\pi \times 30 \times \mathrm{t}) \end{cases} \tag{24}$$

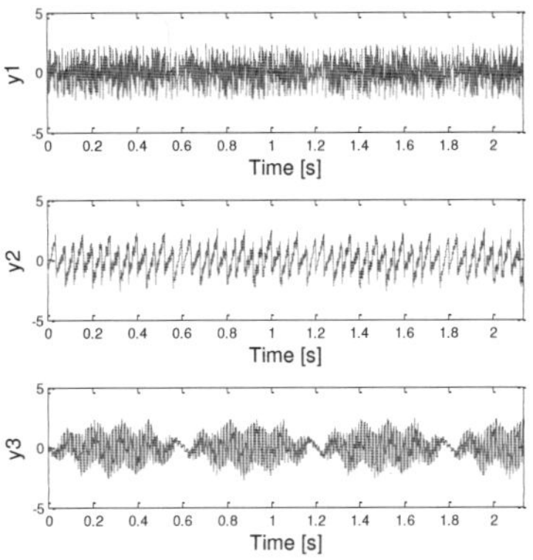

Figure 7. Demixing signals by FastICA

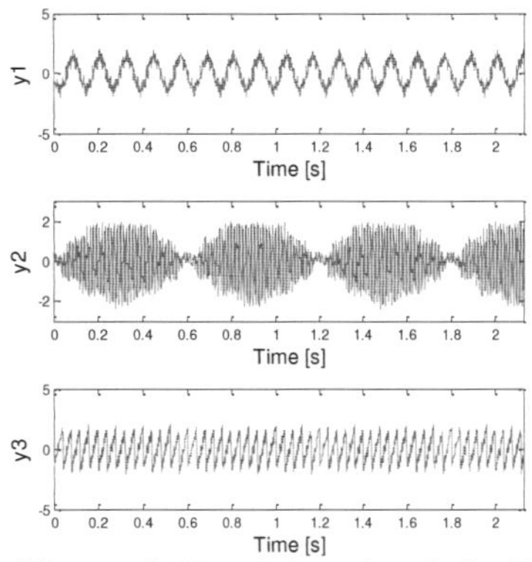

Figure 8. Demixing signals by ICA-R

The correlation distance between the source signals and demixing signals can be calculated to evaluate the quality of two algorithms, which is defined as:

$$d_{ij} = \frac{\left|\sum_{i=1}^{n}(s_i - \bar{s}_i)(y_j - \bar{y}_j)\right|}{\sqrt{\sum_{i=1}^{n}(s_i - \bar{s}_i)^2 \times \sum_{i=1}^{n}(y_j - \bar{y}_j)^2}} \tag{25}$$

And correlation distance is[0.1629, 0.8843, 0.8777] in FastICA, while [0.9390, 0.9678, 0.9621]in ICA-R. That means the ICA-R can extract source signals yet the FastICA cannot process non-linear mixing problem.

4.2. Case study with real-world turbine engine vibration signals

The application is on an actual dual-rotor turbine test rig wherein the amplitude of vibration was abnormally large during all test. Six acceleration sensors were measured on the front air intake casing, intermediate casing and rear casing, and in vertical and horizontal direction respectively. The sampling method is uniform period sampling tracking rotation speed of high-pressure rotor. A 16 periods raw observed in stationary state are displayed in frequency domain in figure 9. At the time, fundamental frequency of low and high-pressure rotor is f_1=149.7Hz and f_2=203.9Hz respectively.

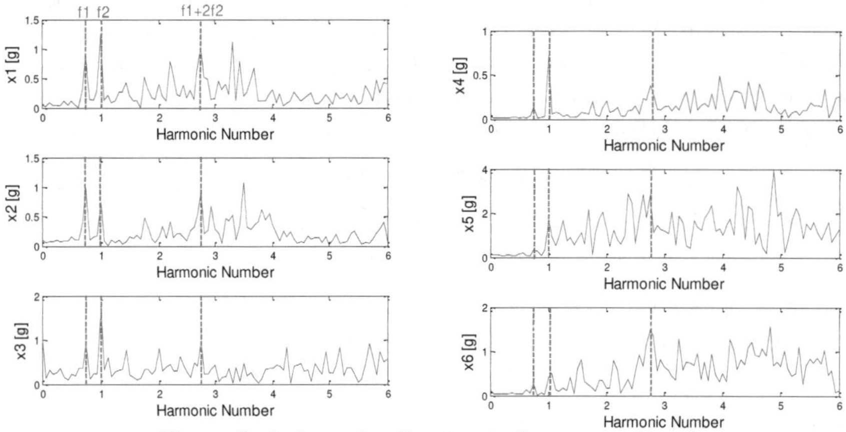

Figure 9. 6 channels vibration in frequency domain

A firstly FFT analysis revealed that combined frequency of f_1+2f_2 existed throughout the test. The FastICA is unable to separate rotor signals no matter how many source numbers are assumed, and SBSS is applied to further investigate which part produced the combined frequency.

According to the figure 9, there are f_1, $2f_1$, f_2, $2f_2$, f_1+2f_2, and the amplitude of each frequency are also known. The reference signals of the low and high-pressure rotors are formulated respectively:

$$\begin{cases} r_1 = 1\sin(2\pi \times f_1 \times \mathrm{t} + \pi) + 0.5\sin(2\pi \times 2f_1 \times \mathrm{t} + 9\pi/5) \\ r_2 = 1\sin(2\pi \times f_2 \times \mathrm{t} + 7\pi/5) + 0.5\sin(2\pi \times 2f_2 \times \mathrm{t} + 2\pi/3) \end{cases} \tag{26}$$

The separation results are shown in time domain and frequency domain respectively in figure 10.

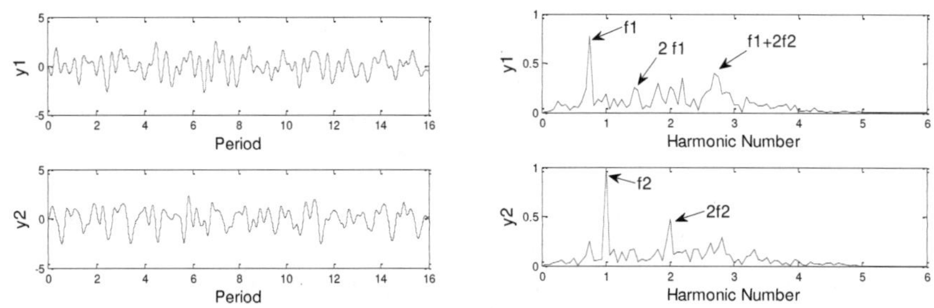

(a) Demixing signals in time domain (b) Demixing signals in frequency domain

Figure 10. Demixing signals by ICA-R

It can be seen from the figure 10 that, there are f_1, $2f_1$, f_1+2f_2 in signal $y1$, which is vibration signal of low-pressure rotor clearly; there are f_2, $2f_2$ in signal $y2$, which is vibration signal of high-pressure rotor clearly. Hence, it can conclude the combined frequency comes from the low-pressure rotor, and can be check to originate from rubbing between low-pressure rotor and stator[17].

Apart from separating rotor signals, the approach is general enough to apply in other mechanical circumstances, such as bearing, gear, stator, any part that have characteristic frequency.

5. Conclusions

In this work, the approach and application of ICA-R algorithm in aero-engine vibration signals are ascertained, and the SBSS provides promising method for fault diagnosis and condition monitoring of aero-engine.

This paper followed the approach of not only explaining the algorithm, but also presenting the engineering background. The paper listed three difficulties related with aero-engine vibration signal when it comes to be processed by BSS: (1)un-independent, (2) nonlinear mixing and (3) overmuch sources. Because of effective utilization of the prior knowledge the SBSS perfectly suited to demix aero-engine vibration signal.

The ICA-R algorithm based on FastICA is introduced with detailed explanations. The maximizing negentropy and minimizing closeness between the reference signal and the demixing signal are selected as two objective functions. And the unconstrained optimization problem is solved by Newton method and gradient descent method. This work also contributes to provide the proper closeness measure function and reference parameters for aero-engine vibration signal. The advantage and superiority of this ICA-R algorithm compared to the earlier methods is that simple update formulation and without manual intervention except reference signal.
Moreover, the experiment in synthetic signals and real-world signals have illustrated the use of the ICA-R approach, and the approach provide a efficacious way to diagnose fault for aero-engine.
A difficulty in ICA-R is phase varies randomly in different acquisition channels and different frequency and cannot be judged in advance for creating reference signal. Unfortunately, the phase affects the similarity measure extremely, and the phase deviation is likely to cause failure of SBSS. Further exploration of how to decide phase in ICA-R is worth pursuing.

6. References

[1] Tharwat A 2019 Independent component analysis: An introduction. *Appl. Comput. Inf.*, (*Preprint*)
[2] Choi S, Cichocki A, and Park H 2005 Blind source separation and independent component analysis: A review. *Neural Inf. Process.* **6** pp.1-57
[3] Lu W, and Rajapakse J C 2006 ICA with reference *Neurocomputing* **69** pp.2244-2257
[4] Lin Q, Zhang Y, Yin F, Liang H, and Calhoun V C 2007 A fast algorithm for one-unit ICA-R. *Inf. Sci.* **177** pp.1265-1275
[5] Zhang Z 2008 Morphologically constrained ICA for extracting weak temporally correlated signals. *Neurocomputing* **71** pp.1669-1679
[6] De Vos M, Lathauwer L De and Van Huffel S 2011 Spatially constrained ICA algorithm with an application in EEG processing. *Signal Process.* **91** pp.1963-1972
[7] Wang Z 2011 Fixed-point algorithms for constrained ICA and their applications in fMRI data analysis. *Magn. Reson. Imaging* **29** pp.1288-1303
[8] Wang X, Huang Z, Zhou Y and Ren X 2013 Approaches and applications of semi-blind signal extraction for communication signals based on constrained independent component analysis: The complex case. *Neurocomputing* **101** pp.204-216
[9] Yang G, Jing J, Ming Y, Yan Y and Chen C 2017 Blind vibration source separation method based on signal feature of aircraft engine. *J. Xi'an Jiaotong University* **51** pp.20-27
[10] Antoni J 2004 Blind separation of vibration components: Principles and demonstrations. *Mech. Syst. Sig. Process.* **19** pp.1166-1180
[11] Friswell M I,. Penny J E T, Garvey S D and Lees A W *Dynamics of Rotating Machines* Cambridge university press 2010
[12] Hyvarinen A, Karhunen J and Oja E. Independent component analysis (John Wiley & Son, 2001)
[13] Hyvarinen A 1999 Fast and robust fixed-point algorithms for independent component analysis. *IEEE Trans. Neural Networks.* **10** pp.626 -634
[14] Dermoune A and Wei T 2013 FastICA algorithm: five criteria for the optimal choice of the nonlinearity function. *IEEE Trans. Signal Process.* **61** pp.2078-2087
[15] Song X and Liao M 2006 Vibration signal separation technical for double-rotor aero-engines. *Mech. Sci.* **25** pp.487-496
[16] Li C, Liao G and Shen Y 2010 An improved method for independent component analysis with reference. *Digital Signal Process.* **20** pp.575-580
[17] Wang Y, Wang L, Liao M and Ding X 2014 Exploring vibration characteristics of dual-rotor engine's rotor-to-case rub-impact. *Mech. Sci. Tech. Aero. Eng.* **33** pp.614-620

Acknowledgment
This work was supported by the National Natural Science Foundation of China (Grant No. 51775436).

Study of the Possibility of Using the Mechanics of Inflatable Braking Devices to Remove the Spacecraft

Vsevolod V. Koryanov[1 *], Alexey G. Toporkov[2], Victor P. Kazakovtsev[3] and Anton A. Nedogarok[2]

[1] Assoc. Prof., Cand. of Sc. (Eng.), First deputy head of Department, Bauman Moscow State Technical University, Moscow, Russian Federation
[2] Assistant, Bauman Moscow State Technical University, Moscow, Russian Federation
[3] Professor, Doctor of Sc. (Eng.), Bauman Moscow State Technical University, Moscow, Russian Federation

vkoryanov@mail.ru

Abstract. The problem of accumulation of space debris in near-Earth space is very relevant now. More than 15,000 debris have been recorded, and this represents a danger for current and future missions to space. A solution to this issue is to remove these spacecrafts from orbit. One of the ways to do it is to use inflatable device. These inflatable braking device can be installed on the landing vehicles on Earth, before their launch, or after they have been labelled as space debris. Studies have already been conducted about inflatable device. Using inflatable device instead of rigid device has advantages, such as a small mass and a great compacity, which is convenient to respect the required launching volume. This paper is dedicated to use of various inflatable brake devices and they are compared with each other.

1. Introduction

The braking device is designed to reduce the lifetime of a spacecraft in a near-earth orbit after its active operation has ceased for its intended purpose. The removal of the spacecraft (SC) from orbit is expected to be carried out by natural aerodynamic deceleration in the upper rarefied atmosphere of the Earth [1, 2, 3]. During the operation of the spacecraft in orbit, the device is in a folded compact state.

After the decision is made to terminate the target use of the spacecraft or by command of the long-term timer, the device is activated. At the same time, the compactly laid thin-film shell of the device under the action of pressurization takes a volumetric shape, which leads to an increase in the area of the projection of the spacecraft and its ballistic coefficient by tens to hundreds of times.

In contrast to mechanical opening, when using gas pressurization, it is necessary to solve the problem of gas leakage and pressure loss due to the porosity of the film, damage to the shell by particles of space debris. Maintaining the shape of the shell in these conditions is assumed by equipping the device with a gas reserve for multiple pressurization.

In addition, the device must ensure reliable issuance and execution of the activation command after several years of the spacecraft operation, and even after its possible partial or complete failure. To meet these requirements, the device must contain its own built-in separate control unit and a power source that are independent of the onboard network of the spacecraft and whose reliability is higher than the reliability of the spacecraft's control system and power supply.

Finally, in the device during the entire period of the active life of the spacecraft, a supply of compressed gas of boost or reagent stock for the production of cold gas due to chemical transformations should be stored.

IOP Conf. Series: Journal of Physics: Conf. Series **1215** (2019) 012031 doi:10.1088/1742-6596/1215/1/012031

2. Mass calculation

The mass of the inflatable braking device (IBD) m_{IBD} includes the masses

1) Mass of film m_{film},
2) The mass of pressurized gas with a balloon and an opening system (when stored in cylinders in a compressed state) or a reactant with a reactor (using a cold gas chemical generator) m_{pg},
3) Body mass m_{body} – mass of control system, elements of a separate power system for an inflatable braking device and other parts not dependent on the area of disclosure.

In this way,

$$m_{IBD} = m_{film} + m_{pg} + m_{body}. \tag{1}$$

Mass of film

$$m_{film} = S_{film} \cdot \delta_{film} \cdot \rho_{film}, \tag{2}$$

where:
S_{film}, m^2 – film area,
δ_{film}, mcm – film thickness,
$\rho_{film}, \frac{kg}{m^3}$ – density of the film material.
The mass of the substance and pressurization devices m_{pg} is selected on the basis of a range of standard sizes of gas generators or cylinders.
Estimation of the mass of the body, components and assemblies in the first approximation is taken depending on the area of the film in proportion to

$$m_{body} = k_{body} \cdot S_{film}, \tag{3}$$

where:
k_{body} – coefficient of proportionality between the area of the film and the mass of the body. This coefficient is calculated separately for each spacecraft. For example, for the spacecraft Mayak [4].

3. Estimate of time from de-orbit

The reduction of spacecraft from low near-earth orbital orbits is considered. The moment of the termination of the ballistic existence will conditionally be considered the moment when the height of the orbit decreases to 100 km.
For an approximate estimate of the lifetime, the differential equation for the focal parameter is solved [5]:

$$\frac{dp}{dt} = 2r\sqrt{\frac{p}{\mu}}N, \tag{4}$$

or for the major axis of the orbit [6]

$$\frac{da}{dt} = \frac{2a^2}{\sqrt{\mu p}}N. \tag{5}$$

where r, m is the orbit radius, p, m is the orbit parameter, $\mu, \frac{kg^3}{s^2}$ is the gravitational parameter, $N, \frac{m}{s^2}$ is the transversal projection of the disturbing acceleration.
The initial conditions are values $h_k = 400\ km, 600\ km$ и $800\ km$.
To estimate the time of spacecraft descent, the initial near-circular orbits are с $e \approx 0$, for which $a \approx p \approx r$. These ratios are also valid in the course of further descent in the upper atmosphere along a spiral path.
Perturbing transversal acceleration N in this case is equal to the acceleration of the resistance force of the atmosphere

$$N = \sigma\rho(h)v^2, \tag{6}$$

where:

$\sigma = \frac{c_{xa} S}{2m}, \frac{\text{m}^2}{\text{kg}}$ – ballistic coefficient,
$\rho(h), \frac{kg}{m^3}$ – dependence of the density of the Earth's upper atmosphere on height,
$v, \frac{m}{s}$ – the flow velocity of the gas.
Presumably the spacecraft and the inflatable braking device are coaxial to the velocity vector of the spacecraft dilution relative to the moving atmosphere; the period of stabilization to this position is not considered in this work.
A more accurate estimate can be obtained by the joint solution of the complete system from the differential equations of motion of the center of mass of the spacecraft, as well as the angular motion of the spacecraft around the center of mass, including other typical perturbing factors of low orbits.

4. The estimated volume of the film in the folded state

Experiments conducted, in particular, in the development of the student satellite "Mayak" revealed that the volume of the folded film V_{film}, l is approximately

$$V_{film} = k_{film} \cdot S_{film} \cdot \delta_{film}, \tag{7}$$

where:
k_{Vfilm} – coefficient of proportionality between the volume of the film material and the volume of the film in the folded pressed state, according to the results of work on the Mayak spacecraft $k_{Vfilm} \approx 0{,}0128$ [4],
S_{film}, m^2 – shell area,
$\delta_{film}, \text{mcm}$ – film thickness.
Industry offers a huge amount of film materials with different composition, thickness, purpose. Film materials were selected according to the following criteria:
1) Preservation of mechanical, chemical and electromagnetic properties in the temperature range of the stay of a spacecraft on the Earth's orbit (-150 ... 250°) during the entire lifetime of the spacecraft and its subsequent de-orbit.
2) Flexibility and strength sufficient to preserve the integrity and tightness of the shell after installation, pressing into the compartment, storage, subsequent disclosure and pressurization.
3) Low porosity.
4) Manufacturability cutting and fasteners. For example, weldability for the possibility of soldering hermetic joints, or ease of cutting.
5) Mass production and reasonable cost.
Currently, the development of promising materials with improved properties. Currently, they are either too expensive or have technological limitations, but in some perspective they can enter the mass market, and they deserve mention in this review, such as, for example, graphene film.

5. Pressurization system

In the framework of this work, braking devices of inflatable type are considered. Charging involves the use of gas to open and maintain the shape of an aerodynamic device. In view of the similarity of the task, will consider the overpressure of the boosting device as equal to the excess internal pressure for stratostats in the Earth's atmosphere. This will simplify the assessment of the possibility of using these or other materials for an inflatable braking device: if in the conditions of a ground-based experiment, the shells maintain their shape and integrity at this pressure, then they will be able to maintain the shape and withstand similar breaking stress in space conditions. According to the publications, the overpressure of atmospheric balloons is about 150 Pa according to the publications of the Laboratory of Autonomous Systems at Stanford University [7] or 180 Pa [8]. It is also assumed that in space conditions even less excessive pressure will be required, since in orbit the gravity will not counteract the opening of inflatable structures.
The mass of gas required for a single deployment of the cylinder, according to the equation of state of an ideal gas

$$pV = \frac{m_{pg}}{M} RT, \tag{8}$$

makes up

$$m_{pg} = \frac{pVM}{RT}, \tag{9}$$

where:
$m_{нд}, kg$ – boost gas mass,
p, Pa – boost pressure (in this case approximately equal to 150 Pa),
V, m^3 – the volume of the expanded shell,
$R = 8{,}315 \frac{J}{mol \cdot K}$ – universal gas constant,
T, K – temperature of gas filling device.
Pressurized structures are vulnerable to pressure leaks, both due to the leakage of the original components of the structure (film porosity) or breakdowns obtained, and due to changes in pressurized gas of a chemical or physical nature (for example, temperature heating or compression). In this paper, only the problem of leaks is considered. To assess the time interval between repeated aspirations, a review of the operation of aerostatic systems, in particular, super-pressure stratostats, was conducted. Currently, Google and NASA are developing sealed (super pressure) stratostats with a flight duration of 100 days [8]. Considering the similarity of sealed balloons and the space inflatable braking device, will consider 100 days as a typical interval with which it is necessary to produce boost (during this time the gas in the aerodynamic device is completely replaced by repeated activation of the charge device).

6. Comparison of various configurations of inflatable braking devices

In principle, the inflatable device can be positioned closely adjacent to the spacecraft (see Figure 1), or remotely on a cable/rod. In the absence of protruding parts of the spacecraft from the deployment of a cylinder that can damage it, the most effective is to be placed closely. This placement prevents the spinning of the spacecraft relative to the balloon and possible impacts on it. If there are protruding parts on the body of the spacecraft from the side of the disclosure of the braking device, the cylinder can be placed on a rigid rod. However, with such a scheme, the cylinder is less protected from damage when it strikes the hull, and the overall rigidity of the structure is reduced. In this paper, being considered only the layout closely.
Calculations were made for the masses and volumes of the film and the pressurization system of three spherical shells with diameters of 5 m, 10 m and 15 m, as well as pyramidal shells of equal diameter (4 faces, angle of inclination to the axis 60°) (and conical 60°), pressurized with nitrogen (N_2). The maximum and minimum values of the mass of the gas were taken for temperatures of, respectively, -120° and +120°, taking into account possible daily temperature differences [9], the level of illumination, the greenhouse effect inside the film. The maximum and minimum masses of the film m_{film} and the volumes in the laid state V_{film} are given for films 50 and 100 microns thick, respectively. The corresponding tables with the results of calculations are given below for each section.

6.1. Spherical braking device

Figure 1. Possible view of a spacecraft with a spherical inflatable braking device.

Spherical braking device [10, 11] has the simplest design, the scheme of pressurization and disclosure. However, this brake device has the largest amount of gas and film area, which must be used to manufacture such a brake device. In addition, this design, in comparison with other forms, has not the highest coefficient of resistance $C_x = 1,0$.

The lifetime of the bundle of spacecraft and activated inflatable braking device with the three ballistic coefficients considered in the table is shown in Figures 2 and 3.

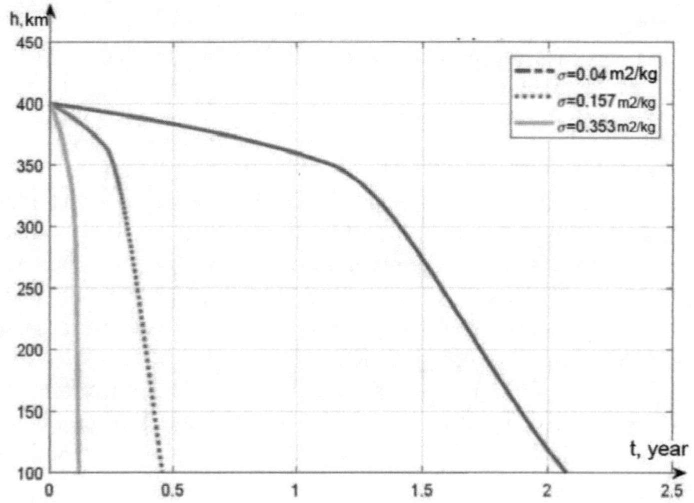

Figure 2. Estimation of the lifetime of a spacecraft with three different ballistic coefficients at an initial altitude of 400 km.

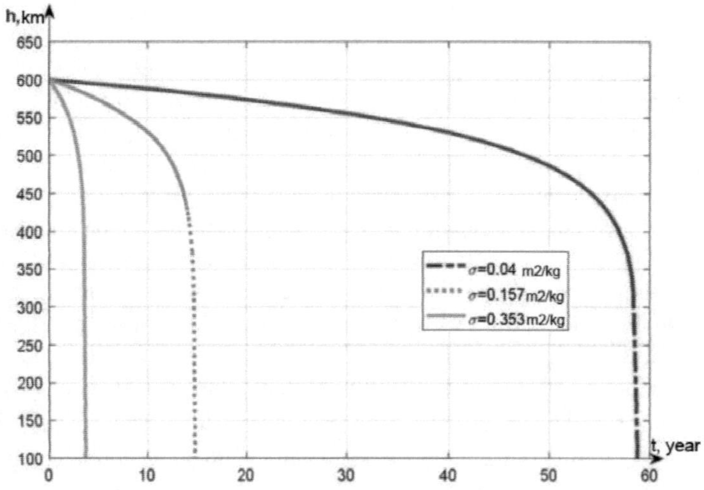

Figure 3. Estimation of the lifetime of a spacecraft with three different ballistic coefficients at an initial altitude of 600 km.

In the case of a spherical device, the area of the shell film is calculated as the surface area of the sphere using the formula

$$S_{film} = 4\pi R_{sph}^2. \tag{10}$$

R_{sph}, m – radius of the spherical shell.

The volume of gas filling the device is defined as

$$V = \frac{4}{3}\pi R_{sph}^3, \tag{11}$$

the area of the middel of the spherical shell

$$S_m = \pi R^2. \tag{12}$$

S_m, m^2 – middel area.

6.2. Pyramidal braking device

The pyramidal device [12, 13] is a regular pyramid, the edges of which are formed by inflated aerobals, with a film stretched between them. The base of the pyramid, which after stabilization of the tandem "spacecraft + inflatable braking device" will be leeward (not blown), is not supposed to be covered with film (see Figure 4a).

Figure 4. A possible view of an apparatus equipped with a braking device with a pyramidal inflatable braking device a) without air rails at the base; b) with aerobals at the base for rigidity

To give the brake device greater rigidity and stability, the base of the pyramid can be additionally reinforced with inflatable beams around the perimeter (see Figure 5b).

Both options compared to the shell of a spherical shape have several important advantages:

- to maintain the volume, a smaller supply of pressurized gas or reagent is required,
- increased survivability;
- less probability of penetration of the aerobals due to the fact that from any direction of impact they occupy only a small part of the projection area of the apparatus, and not every film penetration will fall on aerobals ;
- aerobals can be pneumatically separated by bypass valves, which even in case of penetration and descent, if one beam is damaged, prevent leakage from the other beams ;
- when the number of sides is more than 4, the functionality of the device is preserved when one or several air racks are lowered, due to the fact that the adjacent girders support the volume of the structure.

A rough estimate of the mass of gas and film is made on the basis of the total length of the air racks, without taking into account the gas in the adapters, corners and folds. The shape of the beam in the first approximation is cylindrical:

$$V = n \cdot L \cdot \frac{1}{4}\pi d_{\mathrm{a}}^2, \tag{13}$$

n – the number of faces of the aerodynamic device,,
d_{a}, m – diameter of airbals,
L, m – the length of the edge of the pyramid (aerobals).
The midel area pyramidal shell

$$S_m = \frac{\left(\frac{V}{n \cdot \frac{1}{4}\pi d_{\mathrm{a}}^2}\right)^2 - h_{\mathrm{p}}^2}{2} \sin\frac{360°}{n}. \tag{14}$$

h_p, m – the height of the pyramid.

Thus, the area of the mid-section grows in proportion to the square of the volume, $S_m \sim V^2$, so this form of fairing is the most economical of the proposed.

The coefficient of resistance of the pyramid when blowing along the axis (at zero angles of attack and slip) is calculated by the expression

$$C_x = 2\cos^3 \alpha_{bf}. \tag{15}$$

α_{bf} – the angle between the face and the base of the pyramid.

Estimated film area required for assembly

$$S_{film} = n\left(\underbrace{L\pi d_a}_{sleeve} + \underbrace{\frac{1}{2}\sqrt{L^2 - h_p^2}\sin\frac{360}{2n}\sqrt{L^2 - \left(L^2 - h_p^2\right)\left(\sin\frac{360}{2n}\right)^2}}_{side} \right) \quad (16)$$

7. Results

The following Table 1 shows the final qualitative comparison of the two types of shells, from which it can be concluded that pyramidal-type devices have a significant advantage.

Table 1. Comparison of the shells of various shapes.

Criterion	Type	
	Sphere	Pyramid
Dependence of projection area on volume	$S_m \sim V^{2/3}$	$S_m \sim V^2$
The probability of penetration of the beam during the passage of the fragment through the projection of an inflatable braking device	100%	$\ll 100\%$ (for 4-sided pyramid with base 10 m – 2%)
Potential survivability when breaking through	Low	High
Mechanical stability	High	Determined by beam diameter

Comparing the approximate masses of an inflatable braking device, which should be mounted on a 250-kg spacecraft for removal from orbits of 400 km and 600 km, it becomes even more obvious that pyramidal devices are the only types of devices considered that provide a rational ratio of the mass of an inflatable braking device to the mass of the reconstructed spacecraft (up to 20% for any variant of the initial conditions and duration of the data). Spherical inflatable devices, in view of vulnerability, small C_x and large required volumes are realizable only for heights of about 400 km.

8. Conclusion

1. Taking into account the results obtained, it can be said that the use of passive brake devices in the form of inflatable structures can significantly help in the fight against comic debris in orbit.
2. The main advantage of inflatable structures before rigid ones is a small mass and the possibility of compact stacking in the required volume when putting into orbit.
3. At the same time, there is a probability of breakdown of the inflatable structure by the fragments of the space debris and the occurrence of leaks of the supercharged gas, which can significantly reduce the efficiency of such a system. The results section shows the results of comparing two forms of inflatable braking devices.

References

[1] V.V. Koryanov. Method of calculating the descent of the spacecraft in the atmosphere using technology adaptation landing in different environmental conditions, IAC-14-C2.3.4, 65th Astronautical Congress, Toronto, Canada, 2014, 29 September – 03 October.
[2] Michele Iacovazzo, Valerio Carandente, Raffaele Savino, Gennaro Zuppardi. Longitudinal stability analysis of a suborbital re-entry demonstrator for a deployable capsule. *Acta Astronautica. Volume 106*, January–February 2015, pp. 101–110.
[3] V.V. Koryanov, V.P. Kazakovtsev. Dynamics of angular motion of landing vehicle in martian atmosphere with allowance for small asymmetries. *International Journal of Mechanical Engineering and Robotics Research*, Volume 7, Issue 4, 1 July 2018, Pages 385-391.
[4] Scientific significance. [Electronic resource]: Project "Mayak". - Access mode: http://cosmomayak.ru/about/science.

[5] M.D. Koptev. Dynamics and control of a swarm of nanosatellites in the task of studying the earth's magnetosphere [Electronic resource]. - Access mode: http://www.keldysh.ru/microsatellites/ Bachelor_Thesis_Koptev.pdf. (15.05.2018)
[6] M. Macdonald. Handbook of Space Technology / Malcolm Macdonald, Viorel Badescu. // [Electronic resource]. - Access mode: URL: https://books.google.ru/books?id=w_65BQAAQBAJ. (15.05.2018)
[7] Low Cost, High Endurance, Altitude-Controlled Latex Balloon for Near-Space Research (ValBal) [Electronic resource]. - Access mode: URL: https://asl.stanford.edu/wp-content/papercite-data/pdf/Suskho.Tedjarati.ea.AERO2017.pdf. (15.05.2018)
[8] The Super Pressure Balloon (SPB) [Electronic resource]. - Access mode: URL: https://sites.wff.nasa.gov/code820/docs/outreach/The Super Pressure Balloon.pdf. (15.05.2018)
[9] M. Finckenor. A Researcher's Guide to International Space Station. Space Environmental Effects [Electronic resource]. - Access mode: URL: https://www.nasa.gov/sites/default/files/files/NP-2015-03-015-JSC_Space_Environment-ISS-Mini-Book-2015-508.pdf. (15.05.2018)
[10] V.S. Zarubin, V.N. Zimin, G.N. Kuvyrkin. Temperature Distribution in the Spherical Shell of a Gauge-Alignment Spacecraft. *Journal of Applied Mechanics and Technical Physics* (2017) 58. pp. 1083-1090. https://doi.org/10.1134/S0021894417060141.
[11] V.V. Leonov. The design features of inflatable large-scale mirror concentrators for space high-temperature solar power plant. Proceedings of the international astronautical congress, IAC 67, Making Space Accessible and Affordable to All Countries. 2016.
[12] V.I. Mayorova, N.A. Nerovnyy, D.A. Rachkin, S.M. Tenenbaum, O.S. Kotsur, A.S Popov. Current status of BMSTU-Sail space experiment. 66 International Astronautical Congress 2015, Volume 14. pp. 10679-10683.
[13] N.A. Nerovny, I.E. Lapina, A.S. Grigorjev. Light radiation pressure upon an optically orthotropic surface. *Journal of Quantitative Spectroscopy and Radiative Transfer*. Volume 202, November 2017, pp. 64-73. https://doi.org/10.1016/j.jqsrt.2017.07.016.

Acknowledgment
The research was performed at Bauman Moscow State Technical University with the financial support of the Ministry of Education and Science of the Russian Federation under the Federal Target Program "Research and development on priority directions of scientific and technological complex of Russia for 2014-2020". Agreement # 14.577.21.0247 (unique identifier RFMEFI57717X0247).

IOP Conf. Series: Journal of Physics: Conf. Series **1215** (2019) 012032 doi:10.1088/1742-6596/1215/1/012032

A Cognitive Study of Multicolour Coding for the Head-up Display (HUD) of Fighter Aircraft in Multiple Flight Environments

Bai Ling[1, 2], Li Bo[2], Shi Bingzheng[3], Qiu Lingcun[3], Xue Chengqi[1] and Niu Yafeng[1, 2 *]

[1]School of Mechanical Engineering, Southeast University, 211189 Nanjing, China
[2]Science and Technology on Electro-optic Control Laboratory, 471023 Luoyang, China
[3]Shanghai Academy of Spaceflight Technology, 201109 Shanghai, China

nyf@seu.edu.cn

Abstract. By analyzing flight environments and HUD key information of fighter combat mission, based on the theory of color psychology and the regulation of the avionics system, this study simplified 10 kinds of typical flight environments and combined with 7 types of key information colors into 70 color schemes. Taking reaction accuracy and reaction time as performance indicators and testing participants' subjective feelings, to evaluate pilots' recognition ability of key information and general information with different color schemes in different flight environments, so as to provide the scientific basis for information coding design of flight HUD. The experimental results showed that the cognitive performance of the bicolor coding is significantly better than that of the monochrome coding by comparing the mean reaction time in various environments, and the cognitive performance of color schemes with the hue of 320 or 340 was better than that of other color schemes in most environments.

1. Introduction

The fighter is mainly aimed at shooting down and expelling the aircraft of the other side. In modern combat mode, fighter aircraft also begin to undertake integrated missions such as reconnaissance, ground attack and electromagnetic suppression, which depends not only on the pilot's first-class combat capability but also on the update and upgrade of modern military equipment. Head-up display (HUD) was created in the aviation sector in the 1950s, is the main flight status display widely used in modern fighter now. HUD mainly projects flight key information and combat control information on the transparent screen directly in front of the pilot, so that the pilot can observe the external environment and internal flight information at the same time. 90% of the flight information obtained by the pilot is through the visual channel, and the color as one of the fastest human visual perception elements, compared with other elements such as shapes, it improves recognizability of flight information [1], so the color-coding is an important encoding way in HUD visual information display, it helps pilots to achieve a variety of flight information cognition and integration, and improves the pilot's cognitive performance. By detecting the performance of military pilots in simulated combat, Derefeldt [2] et al. found that when using a color-coded symbol system, pilots can fire missiles faster, reducing the possibility of being attacked, proving the practical effect of color-coding.

The HUD display system of many countries mainly adopts the green monochrome coding method (with the wavelength of 555nm). Under the condition of high or low brightness, human eyes respond best to green about the comfort level in long-term use, but at the same time, the recognizability and saliency of green are not ideal in some cases [3]. In the actual research, pilots generally reflected that the monochrome display was not conducive to the most effective acquisition of information in many cases. For example, under the condition of high-altitude flight in cloudy weather, the contrast between green HUD information and white flight environment is poor, and the information is less recognizable [4]. The monochrome coded HUD has poor situational awareness, which cannot guarantee the best dynamic and static display effect of information in various complicated flight environments, especially is not conducive to prompt warning, abnormal and key information. So by color-coding to distinguish the key information and general information in the HUD, can help the pilot to keep high efficient and rapid cognitive level, avoid the additional cognitive load in the process of scanning head-up display (HUD) and head-down display (HDD), reduce the error rates in visual information process and fighting process, thereby enhance the overall combat capability.

At present, some countries have applied the multi-color coding method to helmet display (HMD) with a significant effect [5]. At the present stage, the research on multi-color interface coding mainly focuses on civil aircraft and mostly is the color-coding of HDD. There are relatively few studies on the multi-color coding of fighter HUD. Xiong Duanqin et al. [4] used the dynamic simulation flight mode to study the monochrome and multi-color HUD information and found that the combination of green and magenta has the best visual effect. Zhang Lei et al. [6] proposed that the matching of environment -- target color which should be selected according to different mental load levels and task requirements, has a significant impact on the identification of the target.

Accordingly, because of fighter special transparent material of HUD, overlapping characters color and environments color is the key factor affecting pilot's cognitive performance and comfort in a realistic environment. However, few studies involve the matching of multi-color coding and environments color. On this basis, this study designed and conducted cognitive experiment of HUD interface with the bicolor coding to investigate pilots' recognition ability of key information and general information in different flight environments, compared the cognitive performance of different color schemes to propose the best color scheme and to provide scientific basis for the design of the fighter HUD interface.

2. Description of experimental design

2.1. Experimental equipment and environment setting

In this experiment, E-prime2.0 software was used to program the experiment and record the data. After the experiment, EXCEL was used to sort and summarize the experimental data. Finally, by using SPSS to process and analyse the experimental data. The materials of the experiment will be shown in a 15.6 inches Lenovo screen, and all of the experimental images are in 1024 * 768 pixels. The brightness of the control display was 120 cd/ m2 and the indoor illumination was 500 Lux. The whole process of the experiment was completed in the ergonomic laboratory of Southeast University.

2.2. Selection of experimental subjects

The experimental subjects are 20 undergraduates, graduate students and doctoral students (male 16, female 4) from Southeast University who have relevant mechanical knowledge or understanding of aircraft structure, aged between 22 and 27, with normal vision or corrected vision (all above 1.0) and no color blindness or color weakness.

2.3. Design of experimental materials

2.3.1. Selection of HUD flight environment and extraction of environment color. To select HSB Color Space as the color-coding tool for this experiment. First of all, the experiment selected 10 representative fighter environment pictures, followed by (1) mountain scene (high flight altitude) (2) mountain scene (low flight altitude) (3) snow mountain scene (low flight altitude) (4) city scene at night (low flight altitude) (5) desert scene (low flight altitude) (6) plain scene (low flight altitude) (7) sea scene (mid-low flight altitude) (8) cloud scene (high flight altitude) (9) purple sunset scene (low

IOP Conf. Series: Journal of Physics: Conf. Series **1215** (2019) 012032 doi:10.1088/1742-6596/1215/1/012032

flight altitude) (10) orange sunset scene (low flight altitude), which covering the current predictable world environments of high and low altitude flight. In order to reduce the distraction of various colors in the flight environments and ensure the reliability of the experiment, these 10 flight environments are purified by ITTI model [7]. Extracting three image features (color, brightness, direction) from images, and then using the center-surround operator to calculate the difference in different scales and normalizing it. The visual significance measurement results of 3 features were combined into a saliency map, and then obtained primary color of the saliency map through mathematical expectation calculation, and 10 pure color images of the environments were labeled as Hn (n=1, 2, 3... 10), as shown in figure 1.

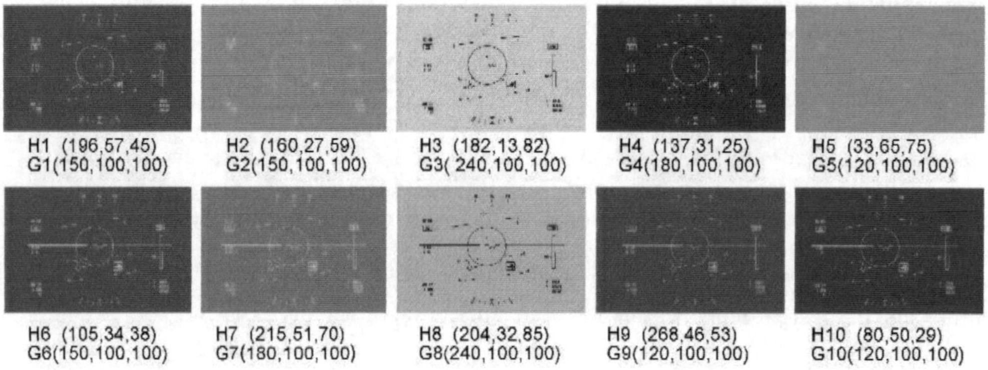

Figure 1. Flight environments (H) and HUD primary colors (G) with HSB values.

2.3.2. Extraction of HUD primary color. This paper aims at the pilots' recognition ability of key information's color, so it is necessary to ensure high recognition of HUD primary color in flight environments, which can improve the operability and Scientificity of the whole experiment, and eliminate the possibility that poor readability of the primary color affects the reading of key information. The selection of HUD primary color range is within 120 to 240 degrees and take the integer in a unit of 30 degrees [8]. After the visual acuity measurement [9] and screening, 10 types of HUD primary colors were selected corresponding to 10 types of environment colors, which are labeled as Gn (n=1, 2, 3... 10), as shown in figure 1. The experiment mainly points at the research of hue value in HSB color system, and values of brightness and saturation both are 100%.

2.3.3. Selection of key information colors. Li Weimin [10] proposed that the colors usually selected on the display screen include red, yellow and green, and blue and purple should be used as few as possible because the visual acuity of blue and purple are poor compared with other colors, according to their color psychological characteristics, they are not suitable to be used as colors to provide highlighting. Based on the theory of color design and color psychology, and combining with provisions of color selection in avionics system [11], the paper summarized the color selection basis of different display functions in fighter display system, as shown in table 1.

The table 1 shows that warm colors (centered on orange) stimulate the nerve to excite it and make the person into a state of mental concentration. When multiple colors occur at the same time, the visual perception of warm color is closer than the actual space distance. Hence, it can affect the pilot's vision in the first time enabling to quickly capture and confirm the information. Cool colors (centered on blue) inhibit people's sense of excitement and presents a more contractive vision effect. When multiple colors occur at the same time, the visual perception of cool color is farther than the actual space distance, which is not conducive to the acquisition of visual information. Therefore, the experiment select the warm colors within the range of 300 to 60 degrees on Munsell 12 color ring as the objects of study. Taking 20 degrees as the unit (the brightness and saturation both are 100%) to extract into 7 kinds of colors and labeled as Kn. The HSB values are K1 (300,100,100), K2 (320,100,100), K3 (340,100,100), K4 (0,100,100), K5 (20,100,100), K6 (40,100,100), K7 (60,100,100).

IOP Conf. Series: Journal of Physics: Conf. Series **1215** (2019) 012032 doi:10.1088/1742-6596/1215/1/012032

2.3.4. Selection of flight mission. By the analysis of the complexity and difficulty of flight missions, the Air to Air with Long Range Missile interface was finally selected as the stimulation interface. According

Table 1. Optional colors with display function in avionics system.

Color	Display Function	Emotional Response	Physiological Response	Sense of Advance and Recede
Red	Warning, prohibition, activation, limit state	Excited, nervous	Sense of impending, visual fatigue; Stimulate nervous excitement, improve work efficiency	Advancing color
Orange	Prompt, caution, alarm	Excited, restless	Blood circulation increases; Relieve fatigue	Advancing color
Yellow	Abnormal/critical alarm, attention, status notification	Optimism, inspiration	Activate memory and active thinking	Advancing color
Green	Start up, normal operation, standard mode	Harmony, self-discipline, peace	Facilitates concentration of thought; Lower blood pressure	Normal Color
Blue	Sky background, notification	Calm, trust, fresh, profound	Lower blood pressure; Balance metabolism	Receding Color
Mauves	Less use, radioactive, corrosive	Mysterious and dreary	Depresses motor nerve, lymphatic and cardiac pulse system; hypnosis	Receding Color
Magenta	caution	Relaxing, soothing, gentle, romantic	Decreases adrenalin production; Promotes action and relaxes muscles; Stable mood	Advancing color
White	Scale and related figures	Stability, comfort and monotony	Keep blood pressure normal, keep actions simple and clear	Advancing color

to the research on general display information and priority of the fighter cockpit [12], the target box and aiming circle as the key information are the color-coding objects of this study.

In the experiment, experimental samples total 10 (HUD primary colors) × 7 (HUD key information colors) + 10 (monochrome HUD samples) = 80. The monochrome HUD interface is set as a control group to verify that the bicolor coding is superior to the monochrome one. The specific experimental materials as shown in figure 2.

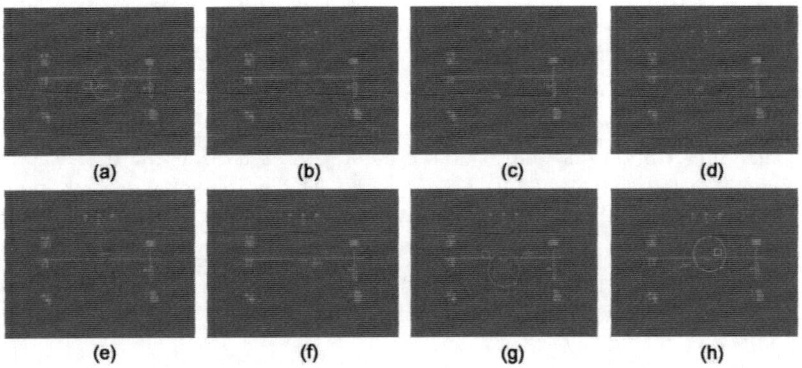

Figure 2. HUD Color schemes in H1 flight environment (a) G1 (b) K1+G1 (c) K2+G1 (d) K3+G1 (e) K4+G1 (f) K5+G1 (g) K6+G1 (h) K7+G1.

3. Experimental methods and process

The whole experiment is divided into two parts. The first part is the objective experiment, including the practice and the formal experiment. The formal experiment includes 2 tasks, first, subjects are required to judge whether the target enters the aiming circle, Press the keyboard "1" for "yes" (entered the aiming circle) and "2" for "no" (not entered the aiming circle), then determine whether the target designator box is on the same side of waterline as velocity vector, press the keyboard "1" for "yes" (on the same side) and "2" for "no" (not on the same side). The second part is the subjective evaluation, which requires the subjects to evaluate the information accessibility and color harmony of different color schemes.

First, the subject reads the experimental instructions and enters the practice phase by pressing the space bar. The practice operation is consistent with the formal experiment so that the subject can be familiar with the experiment. After the practice phase, the formal experiment is conducted, and the experimental program records the response time and accuracy of 2 tasks. The experimental samples appear randomly in order to obtain the most natural response data of the subjects. The experiment is repeated twice. In order to prevent visual fatigue from affecting the experimental results, subject is allowed to rest three times (30s/Time) during the formal experiment, and it will take about 25min for each subject to complete the experiment. The experimental process is shown in figure 3.

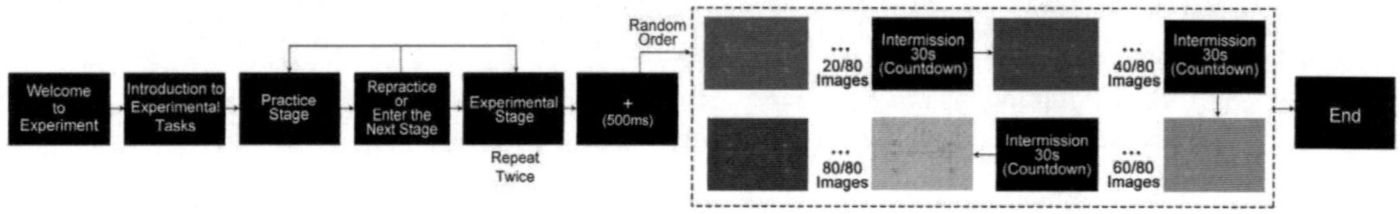

Figure 3. Experimental flow chart.

4. Experiment results and analysis

4.1. Cognitive performance results and analysis

In 10 kinds of flight environments, the distribution of average reaction time about subjects' recognition of HUD information under 8 types of color-coding is shown in figure 4. The accuracy of this experiment is not in high reference to reveal the relationship between the variables, however, reaction time can quantitatively reflect the relations and significant differences among various variables. Therefore, the analysis of this experimental conclusion mainly based on the reaction time and supplemented by the accuracy. First, it can be seen from figure 4 that reaction time intervals in some environments are significantly lower than that in other environments. For example, the intervals in H5 and H8 environments are at a low level, indicating that the difference in environment does have an impact on information identifiability. Second, under some conditions, the reaction time intervals present a minor differences. For example, the intervals in H1, H4 and H10 environments are small, which means the flight environment with low Brightness, such as mountain scene (high flight altitude), city scene at night and sunset scene, which are close to the dark environment, have less interference to identifying information, so the differences of reaction time is not large.

In figure 4, the point at the lowest position in the ordinate represents this point with the minimum average reaction time, which indicates that the pilot has the highest cognitive performance at this point. Thus, color scheme Gn + Kn, corresponding to the point, can be used as the best color scheme in the environment represented by the abscissa of that point. The color scheme with the best reaction accuracy and reaction time are: G1+K2 in H1; G2+K2 in H2; G3+K1 in H3; G4+K3 in H4; G5+K2 in H5; G6+K3 in H6; G7+K6 in H7; G8+K3 in H8; G9+K7 in H9 and G10+K3 in H10. Based on the above objective experimental results, it can be seen that the bicolor coding has relative advantages in accuracy and reaction time of identifying key, abnormal and attention information compared with the monochrome coding.

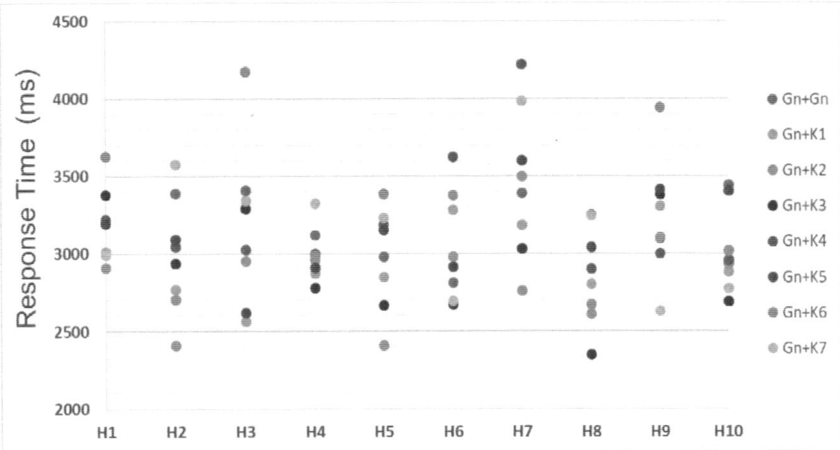

Figure 4. Average response time under 80 color schemes in 10 types of flight environments.

In order to further discuss the double-effect results of color schemes and flight environments, the experimental data were imported into IBM SPSS software for processing, and significance test of difference was performed by using one-way ANOVA. The analysis result is shown in table 2.

F represents difference level, P represents the significance level (significant difference: $P < 0.05$). The table 2 shows that different color schemes have a significant influence on reaction time ($F = 2.449$, $P = 0.028 < 0.05$). Different flight environments also have a significant effect on reaction time ($F = 2.062$, $P = 0.047 < 0.05$), and P values is near to the critical value 0.05, suggesting that the influence of environment on the reaction time is significantly lower than that of color schemes. Thus it has statistical significance to study the color-coding of HUD in different flight environments.

Table 2. One-way ANOVA (dependent variable: reaction time).

Source	Type III sum of squares	df	Mean square	F	Sig.
Corrected Model	3053763.916a	16	190860.245	2.231	0.013
Intercept	763932784.221	1	763932784.221	8930.339	0.000
Environments	1587335.490	9	176370.610	2.062	0.047
Color schemes	1466428.425	7	209489.775	2.449	0.028
Error	5389242.633	63	85543.534		
Total	772375790.770	80			
Corrected Total	8443006.549	79			

4.2. Subjective evaluation results and analysis

After the experiment, subjects were asked to select the optimal HUD interface in terms of information availability and color harmony, and the population distribution is shown in figure 5 and figure 6. In this experiment, information availability refers to the identifiability or visual salience. In figure 5, Gn+K7 is rated as the color scheme with the highest information availability in 70% of the flight environments. Locally, there is an obvious tendency of population dominated by Gn+K7 in H4 and H9, with the proportion remaining at around 75%. Color harmony in this experiment refers to the color comfort of the display interface. In figure 6, Gn+K7 is evaluated as the best color scheme with high color harmony in 60% of flight environments. Locally, the proportion of Gn+K7 is slightly higher than that of other color schemes in H1, H2, H5, H7, H9, and H10, accounting for about 45%, especially up to 60% in the H1.

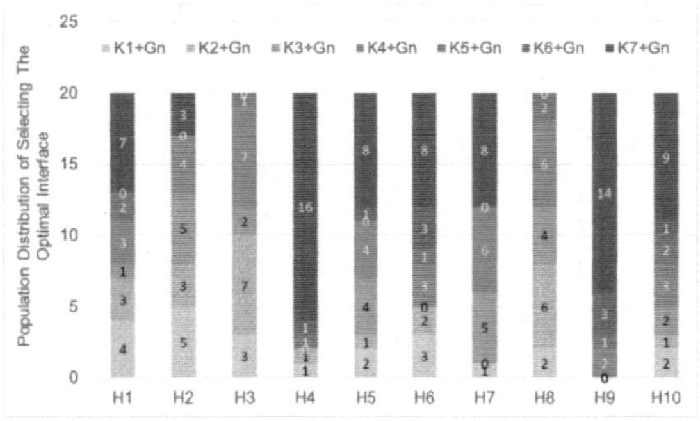

Figure 5. Population distribution of information availability.

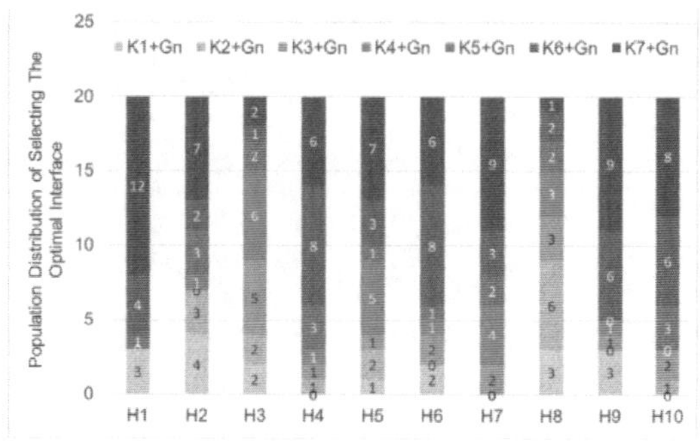

Figure 6. Population distribution of color harmony evaluation.

5. Discussion

10 kinds of flight environments, including different samples of topography, climate, height and time, were used as the dividing basis to judge the effectiveness and differences of HUD color-coding in the actual flight environment at present. Overall, there are significant differences in the identifiability of HUD information under the same color-coding in different environments, and there are also significant differences in the identifiability of HUD information under the different color-coding in the same environment. HUD interfaces with any color schemes in dark environment have obvious advantages compared with those in bright environment, which is because the dark environment has less interference with the pilot's recognition ability. The result shows that the cognitive performance of bicolor coding is better than that of monochrome coding. According to the average reaction time, the cognitive performance of color schemes with the hue of 320 or 340 is better than that of other color schemes in various environments.

5.1. Topographic conditions

According to topography, in the environment, such as a mountain scene or a plain scene, where the main color is green and the brightness is low (about 30%~40%), the display effect of key information color with the hue of 320 or 340 is better than that of other colors; in a snow mountain scene or a cloud scene where the main color is white and the brightness is high (about 70%~100%), the display effect of key information color with the hue of 300 or 340 is better than that of other colors; in the environment with cool color and moderate brightness (about 40%~70%),such as a sea or purple sunset scene, the display effect of color with the hue of 40 or 60 is better than that of other colors. In a desert scene, the display effect of color with the hue of 320 is better than that of other colors.

5.2. Climatic conditions

According to climate, in the polar climate, the color with the hue of 300 is better than other colors. In the desert climate, the color with the hue of 320 is better than other colors. In the continental climate and rainforest climate, the display effect of the color with the hue of 320 and 340 are better than that of other colors. In the marine climate, the display effect of the color with the hue of 40 is superior to that of other colors.

5.3. Altitude conditions

According to the flight height, in the high-altitude flight condition, the environment is mainly land, marine or cloud, and it can be obtained from the above analysis that the display effect of color with the hue of 320 or 340 is better than that of other colors. Environment colors are more complicated when flying at low altitude, but in most cases, the color with the hue of 320 or 340 is better than other colors. From the overall analysis, the flight height has little impact on display effect. Regardless of the high, medium and low altitude conditions, the color schemes with the best overall effect are Gn+K2 and Gn+K3.

5.4. Day-night time conditions

According to the time, the cognitive performance of flying at night is relatively high. In the near-night environment with low brightness (about 20%~40%), such as nightfall and dusk in the city, the display effect of color with the hue of 340 is better than that of other colors. During the daytime, cognitive performance has quite differences from the change of the environments, the cognitive performance in the environment with high brightness (about 70%~100%) is significantly higher than that in other conditions, and the color scheme has a strong contrast with environment colors, therefore the hue value between 300 and 0 is significantly better than that between 20 and 60. While in the environment with 40%~70% brightness and saturation, the contrast between the color schemes and environment colors is not strong enough to enable human eyes to achieve the best cognitive state, so the hue value between 20 and 60 is significantly better than that between 300 and 0.

5.5. Subjective evaluation

From the subjective analysis, in the environment with high brightness (about 70%~100%), the highest evaluation score of information availability is the color with the hue of 320 and 20, and in other environments, the color with the hue of 60 gets the highest score. In the environment with low brightness (about 20%~40%), the highest evaluation score of color harmony is the color with the hue of 40, and in other environments, the color with the hue of 60 gets the highest score. On the whole, the color with the hue of 60 gets the highest score both of information availability and color harmony in most of the environments. However, the materials presented in the experiment are images instead of real-time flight pictures. Considering the physiological and psychological effects of long-term flight on human beings, it is necessary to design a further experiment to judge the reliability of such subjective evaluation.

6 Conclusion

Some conclusions about ergonomic design of HUD color-coding are drawn: the cognitive performance of bicolor coding is significantly better than that of monochrome coding, and the cognitive performance of color schemes with the hue of 320 and 340 were better than that of other color schemes in different environments. In the subjective evaluation, subjects prefer yellow (the hue of 60) to be the color of key information in most environments. The results not only further develop HUD design of color-coding about the key, warning, and abnormal information, but also have a preliminary study on the classification of flight environment. It provides a preliminary experimental basis for the information coding design of HUD based on different environments. In order to obtain reliable and usable experimental basis, dynamic HUD display system should be adopted for experimental verification in the future.

7. References

[1] Wei L I and TangY C 2000 Visual performance of color-coding in character displays. *Journal of Beijing Institute of Technology* **20(4)** 485—488

[2] Derefeldt G et al. 1999 Improvement of tactical situation awareness with color-coded horizontal-situation displays in combat aircraft. *Displays* **20(4)** 171-184

[3] Xiao X, Wanyan X, Zhuang D and Wei Z 2013 Ergonomic design and evaluation of visual coding for aircraft head-up display. *International Conference on Biomedical Engineering and Informatics IEEE* pp748-752

[4] Xiong D et al. 2016 Effects of multicolor display on HUD information in dynamic simulated flight. *Space Medicine & Medical Engineering* **29(2)** 127-132

[5] Post D L, Geiselman E E and Goodyear C D 1999 Benefits of color coding weapons symbology for an airborne helmet-mounted display. *Human Factors* **41(4)** 515

[6] Zhang L, Zhuang D M and Xiao-Ru W Y 2009 The color coding of information based on different mental workload and task type. *Acta Armamentarii* **30(11)** 1522-1526

[7] Itti L, Koch C and Niebur E 1998 A model of saliency-based visual attention for rapid scene analysis. *IEEE Transactions on Pattern Analysis and Machine Intelligence* **20(11)** 1254-59

[8] Jing L I and Xue C 2016 Color encoding research of digital display interface based on the visual perceptual layering. *Journal of Mechanical Engineering* **52(24)** 201-208
[9] Bai T 2015 The cognitive ergonomical study of color character on helmet-mounted displays. (Doctoral dissertation, Nanjing: Southeast University)
[10] Li W, Guan J and Tao Z 2011 Study on display design airworthiness of SVS for civil aviation aircraft based on rough set. *IEEE 2011 10th International Conference on Electronic Measurement & Instruments* Chengdu pp 244-249
[11] Kafurke P 1981 An evaluation of the effectiveness of color coded tactical symbology applied to military maps R 88
[12] MIL.STD-1787B (USAF) 1996 Aircraft display symbology S

Acknowledgments

This work was supported jointly by National Natural Science Foundation of China (No. 71801037, 71871056, 71471037), Science and Technology on Electro-optic Control Laboratory and Aerospace Science Foundation of China (No. 20165169017), SAST Foundation of China (SAST No. 2016010) and Equipment Pre research & Ministry of education of China Joint fund.

IOP Conf. Series: Journal of Physics: Conf. Series **1215** (2019) 012033 doi:10.1088/1742-6596/1215/1/012033

Numerical Study of Fillet Surface Effect on the Ducted Performance Using OpenFOAM

Aldias Bahatmaka[1,*], Dong Joon kim[2] and Hadi Pranggono[3]

[1]Interdisciplinary program of Marine Convergence Design, Pukyong National University, 48513 Busan, South Korea
[2]Naval Architecture and Marine Systems Engineering Department, Pukyong National University, 48513 Busan, South Korea
[3]Aquaculture Department, Pekalongan University, 51115 Pekalongan, Indonesia

aldias@pukyong.ac.kr

Abstract. In the present study, the effect of fillet surface on the performance of the ducted propeller using Reynolds-Average Navier Stokes (RANS) method is applied. The numerical simulations compare the propeller model with fillet on the surface and without fillet. OpenFOAM is used to simulate the performances and solve the problems in the open water propeller simulation. MRF approach is applied for this simulation. Kaplan model with the 19A nozzle is investigated by comparing the results with experimental data. The effect of fillet surface on the ducted propeller are presented and discussed in this study.

1. Introduction

The recent years, lot of numerical simulation to improve the performances of the ducted propeller. The nozzle of ducted propeller is used to improve the propeller efficiency on the ship. The ducted provides the extra thrust and minimize the cavitation. Other reason to use the ducted propeller to reduce underwater noise or protect the propeller damage. Generally, there are 2 types of ducted propeller. They are an acceleration duct or Kort nozzle and decelaration duct. The efficiency of the duct propeller is therefore greater than the open propeller [1].

The computational fluid dynamics (CFD) method is already growing up to solve the engineering problem and compared with experimental data. Several study in CFD for propellers have been conducted [2-5]. In other studies, Optimization for propeller have been investigated, such as using genetic algorimth in the propeller optimization [6], propeller shape optimization is also investigated [7]. Other references are conducting the improving propeller performances for the ducted propeller [8-12].

In the present study, OpenFOAM is used to simulate the performance of the ducted propeller. OpenFOAM is open-source CFD libraries which has ability to do numerical simulation. Several references have been validated the OpenFOAM simulation and the results were estimated and close to the experimental data [13-15]. Therefore, proving OpenFOAM is a powerfull solution to solve the engineer problems, especially for the ducted propeller simulation.

In this research, MRF (Moving Reference Frame) method is applied to do the simulation. MRF is one of the meshing strategies in the numerical simulation. For the propeller simulation, MRF is having robust and closer to the experimental results than the AMI (Arbitrary Mesh interface), and already conducted in the OpenFOAM [16].

IOP Conf. Series: Journal of Physics: Conf. Series **1215** (2019) 012033 doi:10.1088/1742-6596/1215/1/012033

The fillet surface is one of important part to make the 3D model for engineer. The propeller design is complex model and should be investigate the effect of fillet surface between the blade propeller and the hub. This research shows the difference between model between fillet and without fillet on the ducted propeller. The result also compared the experimental data, to validate the approach method.

2. Geometry and conditions

The present study used the Kaplan Ka4-70 with the 19 nozzle. the main dimensions of the ducted propeller are listed in table 1.

Table 1. Main dimensions of the ducted propeller

Parameters	**Value**
Diameter (D)	300.0 cm
Number of blades	4
Pitch of ratio	1.2
Expanded area ratio	0.7
Length of duct (LD)	0.5D
Location of propeller	0.5LD
Clearance between duct and propeller	3 mm

19A is used for the duct type. 19A is mostly used with Kaplan series. The duct section can be seen in figure 1.

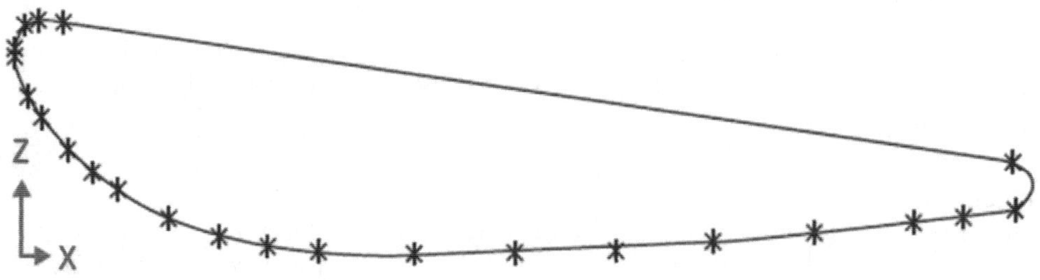

Figure 1. The 19A duct section

Figure 2 shows the three dimensional of the ducted propeller. The model is created in Rhinoceros 3D program.

Figure 2. The 3D model of ducted propeller

3. Numerical approach

The numerical method predicts the flow around the propeller. These conservation equations are generally known as the Navier-Stokes equation and they are used in their incompressible form. Turbulence is modelled with the two-equation k-ω SST model. The simulations were conducted in two different methods, i.e., the steady-state computation that is based on the Semi-Implicit Method for Pressure Linked Equations (SIMPLE) algorithm and PimpleDyMFoam method, which is an implementation of the PimpleFoam solver that allows the implementation of dynamic meshes.

3.1. Governing equations

In the Reynold-Average Navier-Stokes equation, a large number of problems are considered in the open water test. Its main objective is to solve the average velocity field (thrust and torque) that depends on the mean flow characteristics. Velocity and pressure field can be expressed as:

$$u = \bar{u} + u' \tag{1}$$

$$p = \bar{p} + p' \tag{2}$$

Where $\bar{u}$ is value of velocity ; $\bar{p}$ is value of pressure. Then u' is the fluctuating part of velocity and p' is the fluctuating part of pressure, it can be seen in Eq. (1) and (2). The time-average continuity equation is substitution from velocity and pressure equation into the Navier-Stokes equation. It can be expressed as:

$$\frac{\partial \bar{u}}{\partial t} + \nabla.(\bar{u}.\bar{u}) - \nabla.(v\nabla\bar{u}) = -\nabla\bar{p} + \nabla.(\overline{u'u'}) \tag{3}$$

Where $\overline{u'u'}$ is the Reynold stress tensor in Eq. (3), it depends on the fluctuating variables of turbulence flow. The open water propeller characteristics were then conventionally presented in the form of the thrust and torque coefficient KT detailed out in Eq. (4), KQ in Eq. (5) denotes the term of the advanced coefficient, and J is elaborated through Eq. (6).

$$K_T = \frac{T}{\rho n^2 D^4} \tag{4}$$

$$K_Q = \frac{Q}{\rho n^2 D^5} \tag{5}$$

$$J = \frac{V_A}{nD} \tag{6}$$

The propeller efficient is expressed by:

$$\eta_0 = \frac{K_T}{K_Q}\frac{J}{2\pi} \tag{7}$$

Where η0 and π shown in Eq. (7) denotes the efficiency of the propeller and motion in circle, which value is generally 3.14.

3.2. Boundary condition and method

Moving Reference Frame (MRF) is the steady-state solver employing the SimpleFoam as solver of the flow follows the blade rotation according to the Moving Reference Frame (MRF) method. The MRF method can be seen in figure 3.

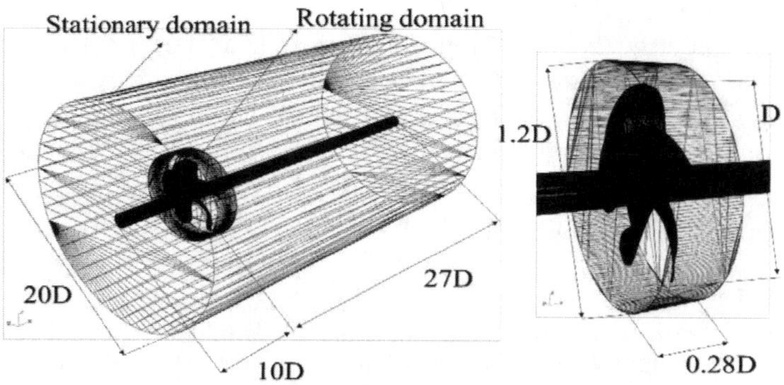

Figure 3. The MRF method for propeller simulation [16]

The domain consists of an internal rotating cylinder enclosing the propeller and an external stationary cylinder with a radius of 10D. In addition, the inlet uniform boundary condition is located at 10D upstream of the propeller plane, and the outlet that is a constant pressure condition was located at 27D downstream of the propeller.

3.3. Benchmark Study

Validation of the numerical is needed for CFD. Therefore, in this study we compare the numerical simulation using MRF method with the experimental data. The comparison results are listed in table 2.

Table 2. Comparison with experimental data

	KT duct	KT duct Exp.	KT prop Exp.	KT prop Exp.	10Kq Exp.	10Kq Exp.	Efficiency	Efficiency Exp.
0.200	0.169	0.162	0.531	0.562	0.619	0.655	0.275	0.275
0.300	0.127	0.127	0.468	0.504	0.600	0.636	0.372	0.378
0.400	0.088	0.084	0.404	0.445	0.576	0.606	0.450	0.471
0.500	0.052	0.052	0.346	0.389	0.545	0.572	0.505	0.540
0.600	0.020	0.028	0.287	0.325	0.504	0.526	0.544	0.588
0.700	-0.011	-0.015	0.229	0.258	0.453	0.474	0.561	0.607
0.800	-0.047	-0.050	0.170	0.179	0.394	0.408	0.545	0.558

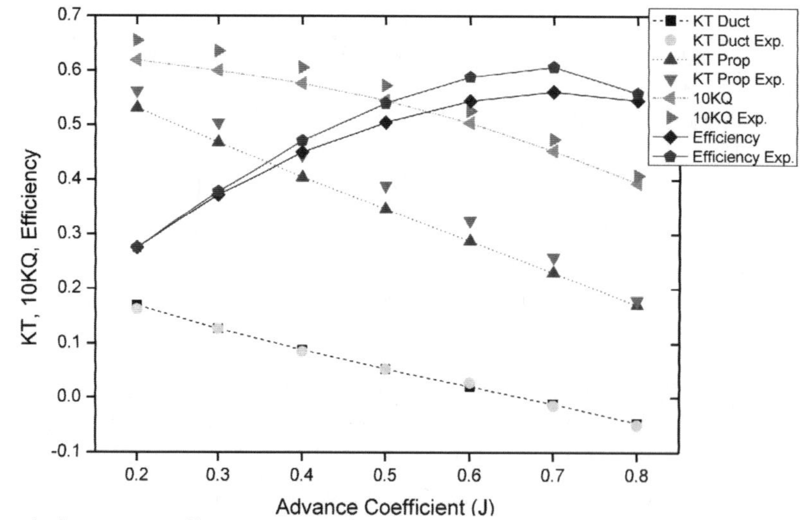

Figure 4. Open water diagram, numerical simulation is compared to experimental data

From the Figure 4, it shows the numerical simulation is having close results to experimental data. It means, this method can be applied to produce the propeller performances.

4. Fillet surface on the propeller model

The Fillet surface is one of the method to make the three dimensional model. Mostly, fillet surface is using to complex model, such as propeller model. To make the smooth model is needed the technique and some parameters to produce the best model. In this study, we considered the fillet surface, and investigate the effect of flow around the blade and the hub of the propeller. Then, we compared the 2 model, between model with fillet surface and without fillet surface. For this research, the radius of fillet surface is constant, 1 mm. The detail model can be seen in figure 5 for the original model and figure 6 for modified with fillet surface.

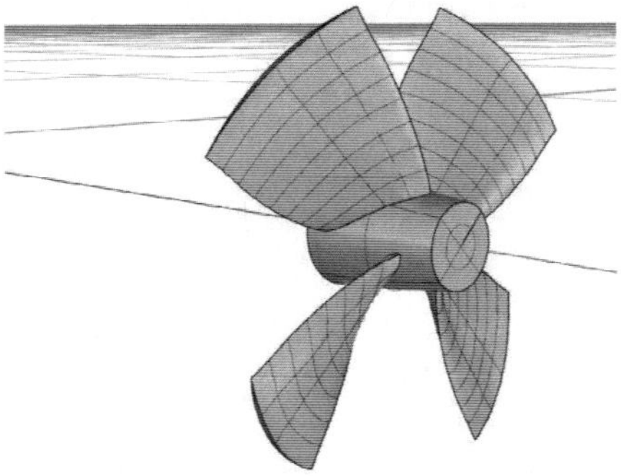

Figure 5. Original model (without fillet surface)

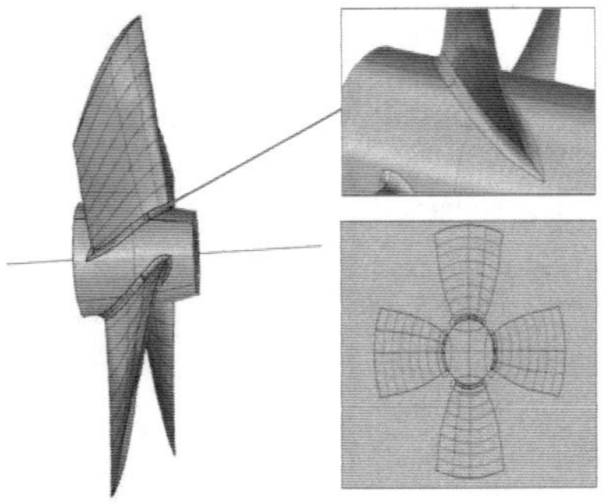

Figure 6. Modified model (with fillet surface)

5. Mesh convergence study

After conducted the modelling, starting the simulation with decide the mesh strategies. In this research, meshing is divided to be 3, e.g. Coarse, Medium and Fine Mesh. The difference of the meshes are listed in table 3.

Table 3. Meshing strategies

Results	Coarse (0.5M)	Medium (2M)	Fine (5M)
T (*N*)	130.870	217.070	219.922
Q (*Nm*)	6.585	9.942	14.942

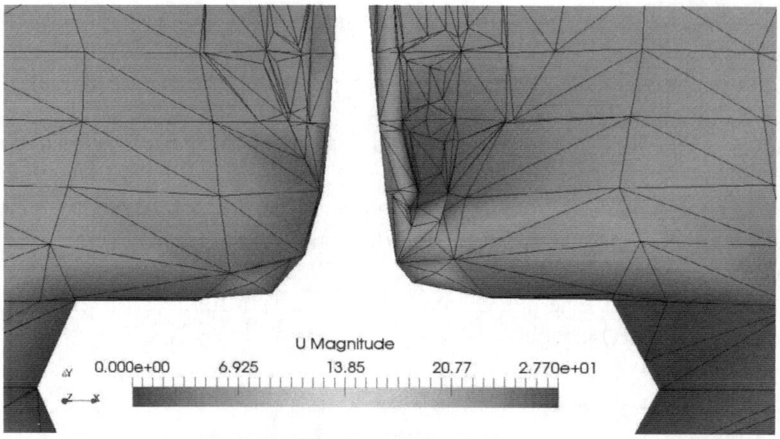

(a). Coarse mesh

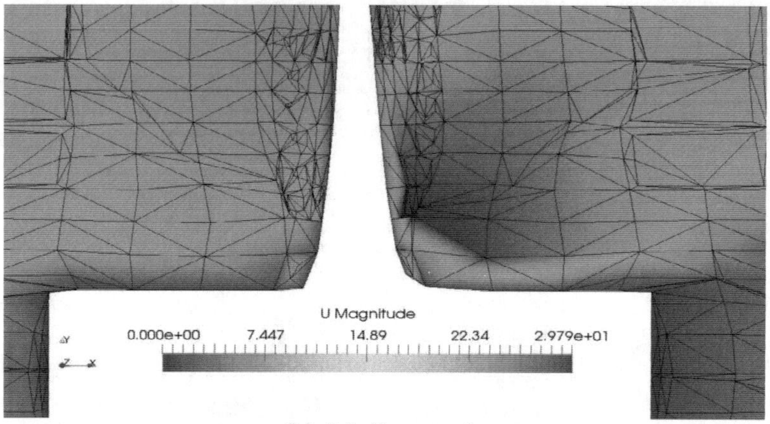

(b). Medium mesh

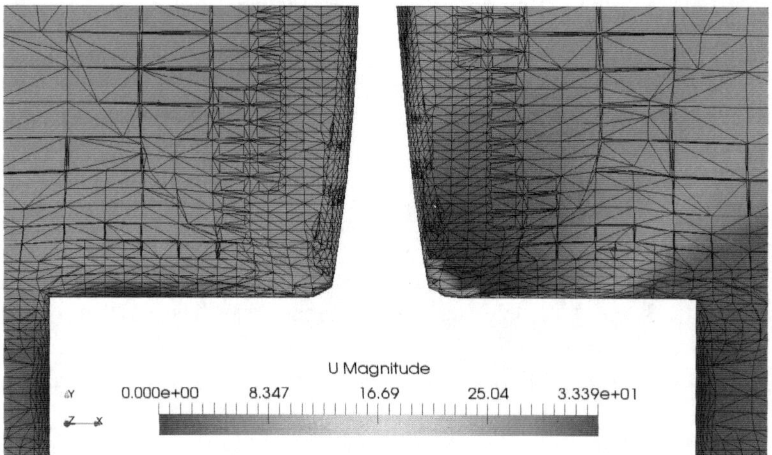

(c). Fine mesh

Figure 7. Comparison of the meshing strategies

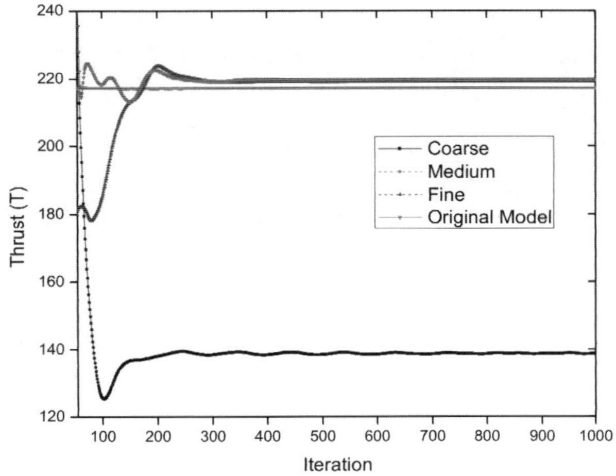

Figure 8. Comparison of thrust for all meshes

Figure 7 shows the contour distribution for all meshes and figure 8 presents the thrust differences for all meshes. From the Figures, it can be seen that Fine Mesh is relatively close to the original model and it is proved by the value of the thrust.

6. Results and discussions

The numerical result shows the difference between both models. The simulation is comparing the force and the torque. From the result, there is the little difference for the performances. In this stage, we calculated on the velocity of J=0.8. The detail results are presented in table 4.

Table 4. Results of fillet surface effect analysis

Results	Original model	Fillet surface	Differences (%)
T (*N*)	219.818	219.222	0.271
Q (*Nm*)	14.988	14.942	0.315

From the table, it can be seen the difference between original model and modified model is just little bit, less than 1% for the force and torque. It does not have much effect to the ducted propeller performance.

The detail velocity distribution can be seen in figure 9 and figure 10 for pressure distribution. From all the figures comparison we can see that there are no striking differences between the original and fillet model for both velocity and pressure distributions.

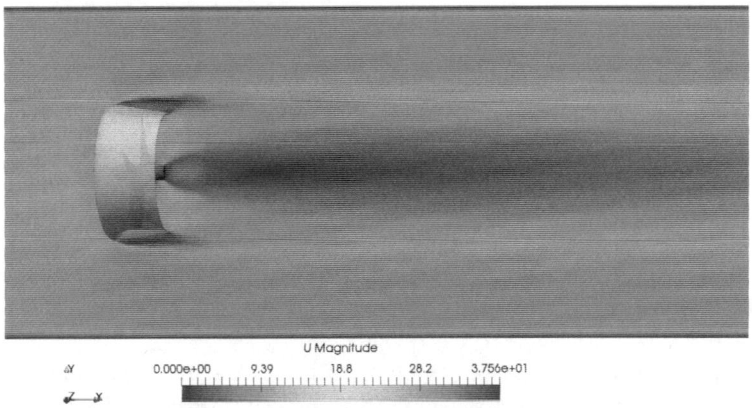

(a). Velocity distribution of original model

IOP Conf. Series: Journal of Physics: Conf. Series **1215** (2019) 012033 doi:10.1088/1742-6596/1215/1/012033

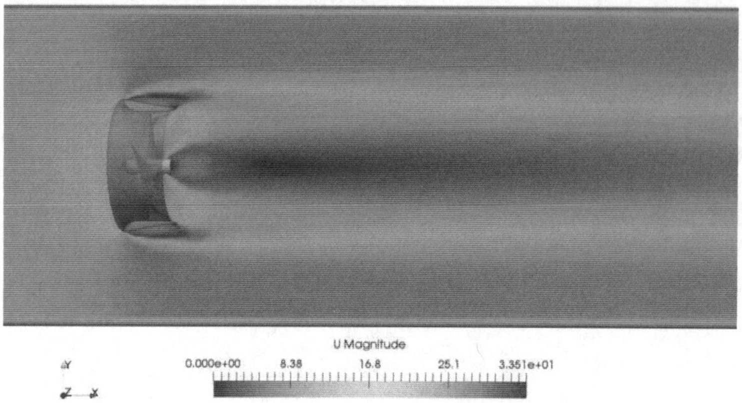

(b). Velocity distribution of modified model

Figure 9. Comparison of the velocity distribution

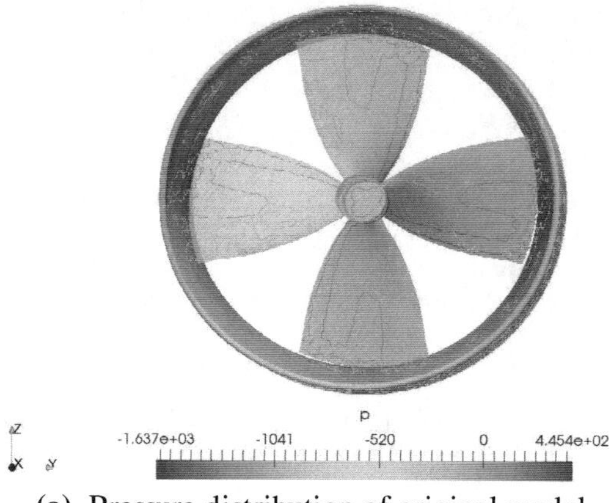

(a). Pressure distribution of original model

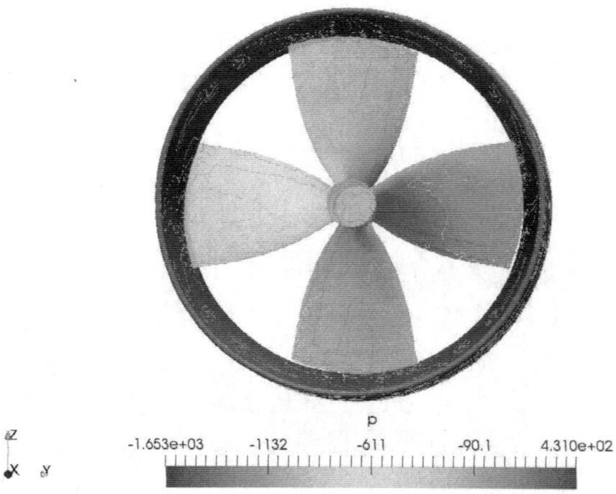

(b). Pressure distribution of modified model

Figure 10. Comparison of the pressure distribution

7. Conclusions

The study has applied the RANS solver in OpenFOAM to analysis the effect of fillet surface for the Kaplan propeller on the ducted propeller. OpenFOAM can performed the numerical simulation for the ducted propeller. MRF approach is applied for this simulation and the method can be used and the result is getting close to the experimental data. The simulations present the comparison between original model and modified model in CFD simulation. There is little bit difference effect from both model. The difference is not effecting to performance of the performance of the ducted propeller, because it just less than 1% of the differences. We assumed the fillet surface will just effect to cavitation performance, it's because of the force and torque did not increase too much. Then, this research should be continued for the effect of fillet surface on the cavitation performance of the ducted propeller.

8. Acknowledgments

This paper was supported by the BK21 plus MADEC human resource development group of Pukyong National University, Busan, South Korea.

9. References

[1]. Ghose, J. P. and Gokarn, R 2004 Basic Ship Propulsion. (Allied Publishers Pvt. Limited).
[2]. Bahatmaka, A., Kim, D.-J., Chrismianto, D., J.D, S. & Prabowo, A. R 2017 Numerical investigation on the performance of ducted propeller. MATEC Web Conf. **138**, 1–8
[3]. Santhosh Babu, K. P. and Padmanabhan, S 2017 Design of marine propeller blade with different blade sequences analyse the hydro formation under pressure hydrodynamic fill. ARPN J. Eng. Appl. Sci. **12**, 4996–5002
[4]. Shen, Z., Wan, D. and Carrica, P. M 2014 RANS Simulations of Free Maneuvers with Moving Rudders and Propellers Using Overset Grids in OpenFOAM
[5]. Kukudzhanov, V. N 2013 Numerical continuum mechanics. 1, xviii + 429
[6]. Bahatmaka, A., Kim, D.-J., Chrismianto, D., Hai, N. & Prabowo, A. R 2017 Optimization of thrust propeller design for an ROV (Remotely Operated Vehicle) consideration by Genetic Algorithms. MATEC Web Conf. **138,** 07003
[7]. Joung, T.-H., Sammut, K., He, F. and Lee, S.-K 2012 Shape optimization of an autonomous underwater vehicle with a ducted propeller using computational fluid dynamics analysis. Int. J. Nav. Archit. Ocean Eng. **4**, 44–56
[8]. Willemsen, J. C 2013 Improving Potential Flow Predictions for Ducted Propellers. (UNIVERSITY TWENTE).
[9]. Chamanara, M. and Ghassemi 2016 H. Hydrodynamic Characteristics of the Kort-Nozzle Propeller by Different Turbulence Models. **4**, 169–172
[10]. Ii, N. F 2014 The Nonlinear Actuator Disk Method as Applied to Open and Ducted Rotors
[11]. Elik, F. C., G Uner, M. and Ekinci, S 2010 An Approach to the Design of Ducted Propeller. Trans. B Mech. Eng. **17**, 406–417
[12]. Yongle, D., Baowei, S. and Peng, W. 2015 Numerical investigation of tip clearance effects on the performance of ducted propeller. Int. J. Nav. Archit. Ocean Eng. **7**, 795–804
[13]. Lloyd, T. P. P., Turnock, S. R. and Humphrey, V. F. F 2011 Unsteady CFD of a marine current turbine using OpenFOAM with generalised grid interface. Proc. 14th Numer. Towing Tank Symp. 1–6
[14]. Turunen, T., Siikonen, T., Lundberg, J. and Bensow, R 2014 Open-Water Computations of a Marine Propeller Using Openfoam. 11Th World Congr. Comput. Mech. 5Th Eur. Conf. Comput. Mech. 6Th Eur. Conf. Comput. Fluid Dyn. Vols **Ii - Iv** 1123–1134
[15]. Bensow, R. E 2013 Simulation of unsteady propeller blade loads using OpenFOAM. 16th Numer. Towing Tank Symposium
[16]. Bahatmaka, A. and Kim, D 2018 The Influence of Meshing Strategies on the Propeller Simulation by CFD. **4**, 78–85

IOP Conf. Series: Journal of Physics: Conf. Series **1215** (2019) 012034 doi:10.1088/1742-6596/1215/1/012034

Design of Online Monitoring System for Pulse Detonation Engine

Z Y Qian[1], J H Li[2], W L Zhang[3] and B K Wang[4]

[1]Master, Power and Engine College, Northwestern Polytechnical University, Xi'an China

[2]Associate Professor, Power and Engine College, Northwestern Polytechnical University, Xi'an China

[3]Doctor, Power and Engine College, Northwestern Polytechnical University, Xi'an China

[4]Master, Power and Engine College, Northwestern Polytechnical University, Xi'an China

Abstract. The PDE monitoring equipment being applied is large in size, harsh in working conditions, unable to carry out outdoor experiments, limiting the flexibility of PDE's original rational experiment. An online monitoring system was designed to meet needs of mobility, real-time, portability, none effect on PDE thrust test and so on. The paper introduces the whole scheme of the online monitoring system, the hardware design of each function module and the software design of the whole scheme. The experimental results show that the on-line monitoring system can realize the measurement of working parameters in the principle experiment of pulse detonation engine in time and accurately and improve the flexibility of operation

1. Introduction

With the advantages of high thermal cycle efficiency and simple structure, pulse detonation engine (PDE) is recognized as the new aerospace vehicle power device of the 21st century[1-2]. Currently the basic laws of the PDE and various factors that affect its performance are mainly explored through principle tests and state monitoring is an important part of the principle test[3].

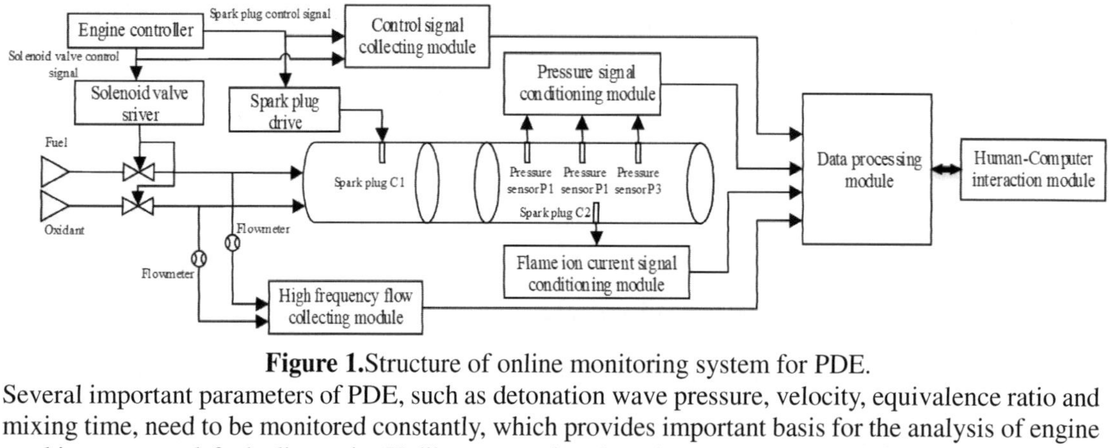

Figure 1. Structure of online monitoring system for PDE.

Several important parameters of PDE, such as detonation wave pressure, velocity, equivalence ratio and mixing time, need to be monitored constantly, which provides important basis for the analysis of engine working state and fault diagnosis. Unlike conventional engines, parameters of PDE vary with time periodically, which also raises higher requirements for the performance of the real-time monitoring system[4-5]. During the process of the principle test, pressure signal is collected by the pressure sensors mounted on the detonation chamber and is transmitted to the signal collector through the charge amplifier. The ion current signal of the detonation cavity is passed through the ion converter to the signal acquisition instrument. The testers must analyze the data collected by the acquisition instrument manually. However, due to the large size of the charge amplifier and ion converter, the long transmission distance between device and the PDE and the power supply voltage of the charge amplifier and the ion converter which needs to be AC220V, the flexibility of the principle tests have been greatly restricted. To simplify the outdoor experimental conditions, implement the real-time monitoring function in outdoor experiments and reduce the error caused by the outdoor test on the thrust test, this paper presents a high integration real-time monitoring system. This real-time device reduces the volume and weight of the device, greatly simplifies the system structure and improves experimental flexibility so that the PDE principle experiment can be carried out in a more complex and changeable environment. It also implements the function of monitoring parameters of PDE in each working cycle and improves the accuracy of the monitoring results, which is of great significance to PDE work mechanism research, fault diagnosis analysis and boundary condition exploration.

2. Design and performance requirements of PDE on-line monitoring system

2.1 Design scheme of PDE on-line monitoring system

The structure of online monitoring system for PDE is shown in Figure 1. It is mainly composed of working state monitoring unit, mixing time and equivalence ratio measuring unit and main control unit. During the experiment the engine controller controls the opening of fuel valve and oxidizer valve and ignition of spark plug during one ignition cycle to form detonation wave. During the ignition process the timing of ignition must be matched well with the filling time of the gas or it will cause fuel waste or reduce the intensity of detonation wave, and even lead to the failure of ignition[6]. The on-line monitoring system monitors the flow of fuel and oxidizing agent, timing of valve opening and spark plug ignition, changes of detonation chamber pressure and state of combustion in PDE working cycle, calculating PDE equivalent ratio and mixing time and obtaining the time sequence characteristic analysis of key parameters in PDE working cycle, which provides reference for the fault diagnosis and engine working mechanism exploration.

The main components and functions of each unit of the online monitoring system

(1) working state monitoring unit is composed of a detonation chamber pressure signal conditioning module and a flame ion current signal conditioning module to monitor the pressure change of the detonation chamber and calculate a flame propagation speed.

(2) mixing time and equivalence ratio measuring unit is composed of a high frequency flow collecting module and a control signal collecting module, monitoring the flow of fuel and oxidant and the engine controller ignition signals.

(3) main control unit is composed of a data processing module and a human-computer interaction module. The data processing module optimizes and calculates the data collected by the pressure signal conditioning module and the ion current signal conditioning module so that the system could calculate the exact equivalence ratio and mixing time through the mixing time and the equivalence ratio measuring unit and display the results on top of the human-computer interaction module.

2.2 Performance requirements for on-line monitoring systems

(1) Monitoring the detonation chamber pressure, ion current, fuel flow and oxidant flow state in each ignition cycle.

(2) Monitoring the change of valve control signal and spark plug control signal in each ignition cycle.

(3) Monitoring the equivalence ratio and mixing time in each ignition cycle.

(4) Monitoring the timing sequence of the ignition signal and the opening signal of the fuel and oxidizer valves in each ignition cycle.

(5) The Man-machine interface shows the engine ignition times and ignition success rate and the change of signals in each ignition time.

3. Hardware design of PDE on-line monitoring system

3.1 Hardware principle of on-line monitoring system

According to the performance requirement of PDE on-line monitoring system, the hardware principle of the monitoring system is designed as shown in Figure 2. It mainly includes six function modules, namely the control signal acquisition module, the high frequency flow collection module, the pressure signal conditioning module, the ion current conditioning module, the data processing module, the man-machine interaction module.

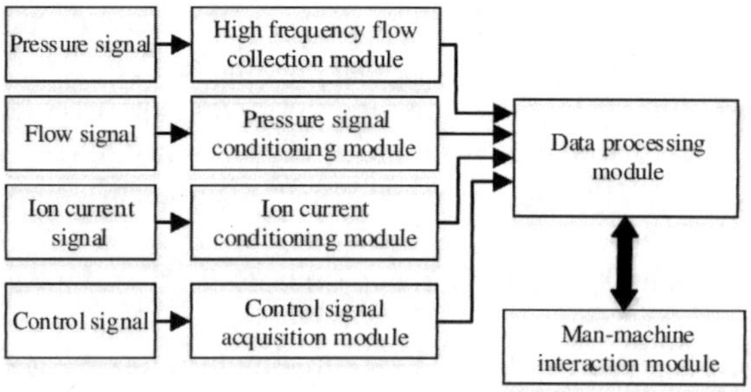

Figure 2. Hardware principle of the monitoring system.

3.2 Design of the data processing module and the man-machine interaction module

The on-line monitoring hardware system uses the STM32F4 of the ST Microelectronics company as the core processor, it takes the cortex M4 as the kernel, its main frequency reaches 168MHz, it has the 4-way UART to satisfy the communication demand, it has 12-bit ADC sampling channels to achieve high sampling accuracy. The data processing module includes STM32F4 microcontroller and its minimum system. STM32 minimum system is the smallest circuit which can make STM32 work normally, including STM32 microcontroller, Jtag download interface, clock circuit, starting circuit, power supply circuit and reset circuit. The working voltage of the STM32 is 5V, which can be supplied by the battery [7-9]. The Man-machine interaction module adopts RS232 Configuration screen, the screen has the following advantages: (1) Adopting CORTEX-M3+FPGA dual-core structure makes refresh speed fast (2) Excellent color difference performance, rich content of control, support touch operation, (3) ease of development, easy to debug[10].The data processing module communicates with the Man-machine interaction module through the UART, the design is shown in Figure 3.

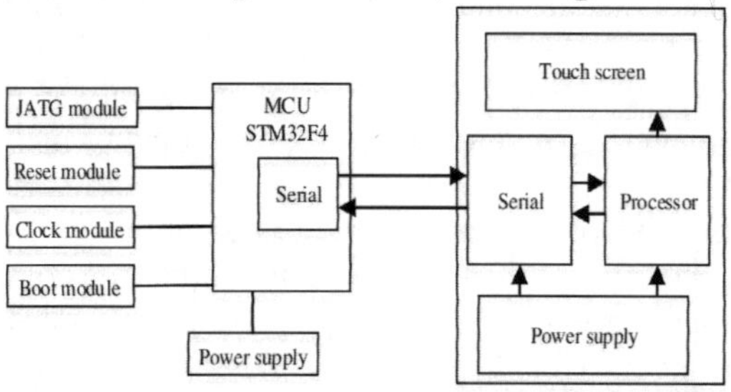

Figure 3. Design of the data processing module and the man-machine interaction module

3.3 Design of the ion current conditioning module

The ion current signal conditioning module adopts the spark plug as the ion current sensor. By adding biased voltage on the detonation chamber the ions produced by combustion are directed to move and produce ion current.

The system figures out specific information of combustion by the intensity change of ion current. Ion current signal conditioning module principle as shown in Figure 4, the module uses high voltage silicon reactor as signal isolating element, bias voltage positive pole connects the Spark plug Center electrode while the negative pole connects the detonation chamber. The ion current signal is converted to the voltage signal by the resistor element, the adjusting sliding resistance causes the voltage signal to change within the appropriate voltage range. the voltage signal is amplified by the amplifier and sent to the data processing module. The electrical components are connected by shielded cables to reduce the influence of the external interference on the ion current signal.

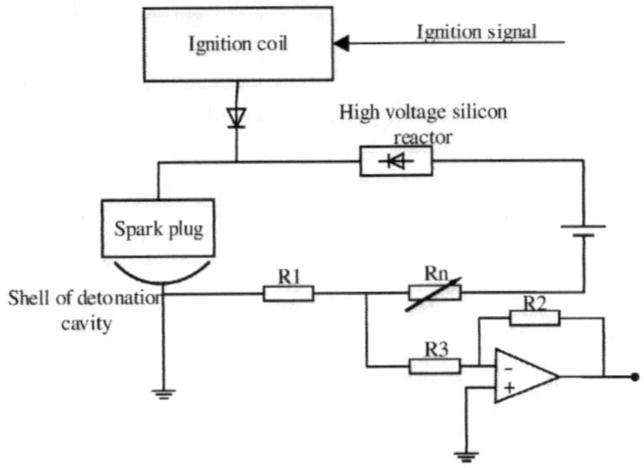

Figure 4. Design of the ion current conditioning module.

3.4 Design of the high frequency flow collection module and the pressure signal conditioning module
The pressure sensor is directly measured by pressure sensor, and the flow of fuel and oxidizer is measured directly by flowmeter. The pressure transducer and the flowmeter output signal are 4~20ma current signals. The current signal is converted to a voltage signal by a resistor element and transmitted to a data processing module through an ADC conversion module.

3.5 Design of the control signal acquisition module
The control signal should be of medium speed or high speed optocoupler to ensure that the signal is not delayed or deformed after optocoupler. In this paper, high speed optical coupling 6N137 is selected to control signal acquisition. The principle of the control signal acquisition module is shown in Figure 5. After the control signal is sent to the 6N137, the control signal is transmitted to the data processing module through the phase logic. To improve the system signal-to-noise ratio the control signal is isolated from the digital circuit and the power of the control signal is isolated from the power supply of the digital circuit.

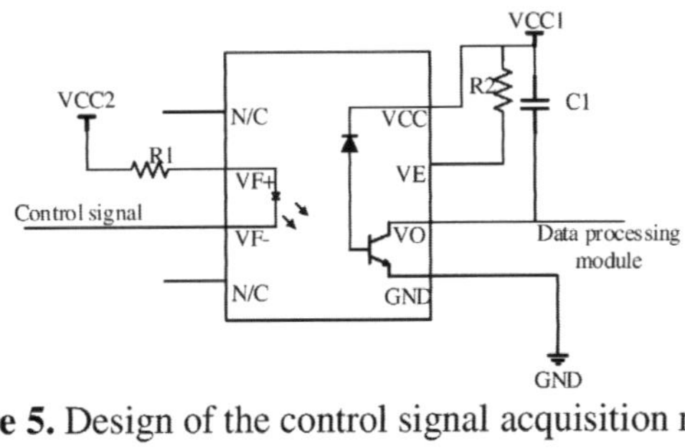

Figure 5. Design of the control signal acquisition module.

4. Software design and implementation of PDE on-line monitoring system

Based on C language and firmware library that is provided by ST company and that is used for application development, the system uses Keil uVision integrated development environment to develop, which greatly improve the development efficiency. According to the concept of modular software design, the software system is divided into hardware initialization module, data acquisition module, business processing module and human-machine communication module. The hardware initialization module includes UART, Timer, AD, data initialization. data acquisition module including acquisition of AD output and control signal. The Man-machine interaction module is responsible for user setting parameters, obtaining status, etc. Combining the finite state machine model, the system divides the task into the editing mode and the view mode. System state parameters can be modified in edit mode and state parameters in view mode can only be displayed and cannot be modified. In this paper, the instruction issued by human-computer interaction module is used as the signal of triggering state switching, triggering the system state change. The program flowchart is shown in Figure 6.

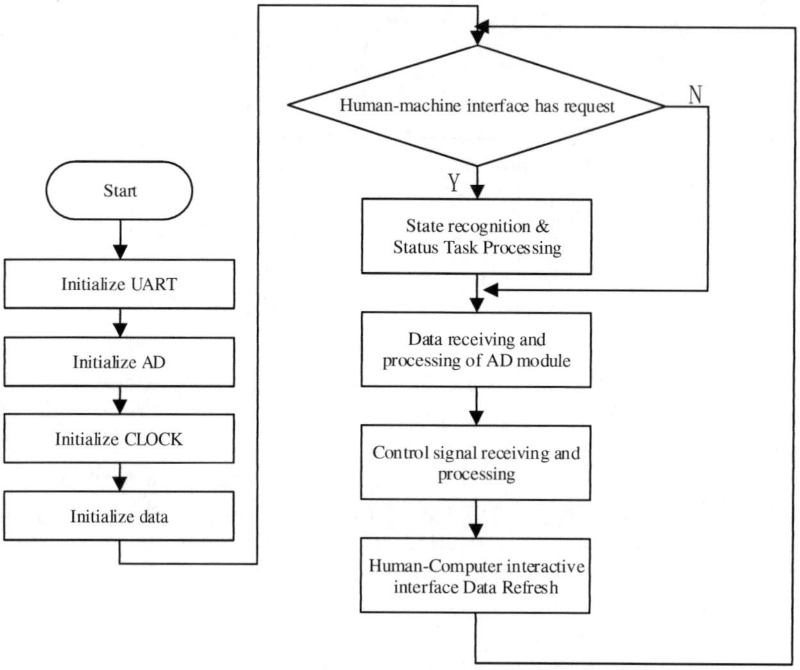

Figure 6. Software design for online monitoring system.

After the system running, the related hardware module is initialized firstly, then the cycle of work is entered. In a work cycle, the system first processes the instructions issued by the human-computer interaction module, setting the system state and executing state tasks. After that, data acquisition and data processing are carried out, and finally the parameters are updated on the human-machine interaction interface.

5. Conclusion

The on-line monitoring system for PDE realizes real-time monitoring of detonation engine working condition in each ignition cycle and improves the accuracy and real-time of detection. It also has small size, light weight, portability, high integration, easy to operate, simple structure and other advantages, reducing the error caused by the device quality to the thrust measurement and simplifying the requirements for working environment, which makes the principle experiment can be carried out outdoors. The system accurately reflects the working state of the engine, which provides an important basis for the fault diagnosis and engine working principle exploration.

6. References

[1] Yan C J ,Fan W and Wang J.Y 2005 *Pulse detonation engine technology*(China: Northwestern Polytechnical University Press).

[2] Meyers M. (2007). Detonation. In Dynamic Behavior of Materials (pp. 244-270). Hoboken, NJ, USA: John Wiley & Sons.

[3] Dai Y and Qu W D. 2009. Control strategy analysis for pulse detonation engine performance model. *Microcomputer Information*, 25(1), 3-5.

[4] Panicker, P. K. 2008. The development and testing of pulsed detonation engine ground demonstrators. *Dissertations & Theses* - Gradworks.

[5] Zhang W L, C M Pei and J H li. 2018."Design of Monitoring and Control System for Multi - Tube Ignition Pulse Detonation Engine". *J. NPU* 36(1):162-168.

[6] Xiong C, Qiu H. 2014.Influencing Factors on the Performance of Pulse Detonation Engine at

Test Bench. *International Journal of Turbo & Jet-Engines* **31(1)** 19-27.

[7] Li Y J. 2009.Long-Range Multi-Channel Data Acquisition and Control System. Taiyuan North University of China

[8] Hu X B. 2012.Design and Realization of Multi-Channel Real-Time Data Acquisition and Storage System. Taiyuan North University of China

[9] Wang Z Y, Wang M.S and Gu W R. 2013."A High-Linearity Pipelined ADC with Opamp Split-Sharing in a Combined Front-End of SlH and MDAG. *IEEE Trans on Circuits and Systems I*, 60(11) 2834-2844.

[10] Liu X G, Ling C H, Lv F Y, Liu Z M and Wu M. 2017. Design of High Precision Analog Signal Isolation Circuit Based on HCNR201. *Instrument Technique and Sensor* **01**

A Static Friction Model for Elastic-plastic Contacting Surfaces Using Statistically Homogenized Technique

Xiaoya ZHENG[1, *], yuanfeng YU[1]

[1]School of Astronautics, Northwestern Polytechnical University, Xi'an 710072, China

zhengxy_8042@nwpu.edu.cn

Abstract. An improved static friction model is developed for elastic-plastic contacting surfaces. Two random variables of asperity height and curvature for Gaussian isotropic contacting surfaces are assumed to govern asperities distribution and they satisfy a joint probability distribution function. Based on the contact parametric representation of the KE model proposed by Kogut and Etsion, the expected macroscopic values of normal stress and shear stress are obtained by using statistically homogenized technique. Then the effect of roughness, the energy of adhesion and material properties on static friction coefficients are studied and the results show similar trend with literatures. A comparison of the present model with the SV friction model proposed by Sista and Vemaganti shows that the adhesion force has a more significant effect for smoother surface and SV model is more suitable for heavily loaded contacts. The static friction coefficient is related to the ratio of hardness to elastic modulus H/E. At the same roughness and normal load, lower H/E leads to lower critical interference ω_c, higher plasticity index ψ, and smaller static friction coefficient.

1. Introduction

Static friction force exists generally in applications of mechanical engineering and is defined as the force counteracting relative motion of two contacting surfaces[1]. According to the widely employed Coulomb friction model, the static friction force is calculated by multiplying the normal force by a constant static friction coefficient. In the past several decades, numerous experimental studies indicated that many factors would affect the static friction coefficient, including contacting materials, surface topography, the value of normal load and environmental conditions. To large extent it depends on the normal load and changes nonlinearly[2-4].

Bowden and Tabor[5] provided a scientific understanding of friction mechanics of two rough contacting surfaces. Tabor[6] summarized three important components of static friction without lubrication, namely, (1) real contact area, (2) nature and strength of junctions formed at the contact interface and (3) the way in which the material of the contact asperities is sheared and ruptured.

Chang et al.[7] incorporated surface roughness and adhesion force into a comprehensive friction model for unlubricated metallic surfaces. The above three components can be easily understood from the definition of the static friction coefficient, μ:

$$\mu = \frac{Q_{\max}}{F} = \frac{Q_{\max}}{P - F_S} \tag{1}$$

where $Q_{\max}$ is the tangential load (the static friction force) related to the force breaking all the junctions to initiate sliding. F is the normal load. In order to keep the force balance (see Figure 1), the normal load equals to the adhesion force F_s subtracted from the actual contact force P. The actual contact force P is related to real contact area and the adhesion force F_s is related to the strength of the junctions formed at the interface. Note that the adhesion effect is taken into account compared to the classical Coulomb model.

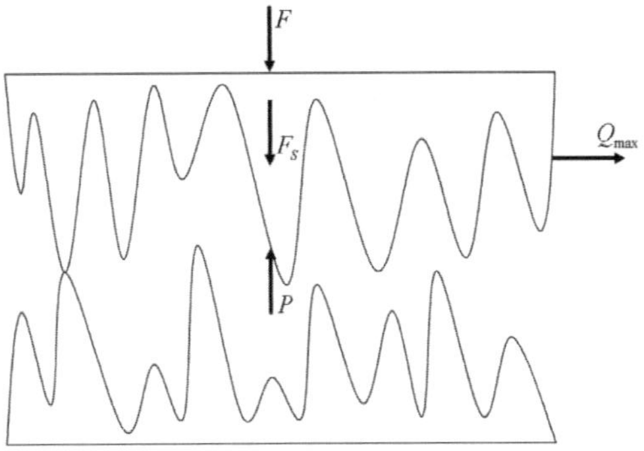

Figure 1. The forces acting on two rough contacting surfaces

Besides the experimental studies, many researchers have been focused on the estimation of static friction coefficient by the asperity-based model. An important asperity-based model was proposed by Greenwood and Williamson[8] (GW model). The basic assumptions in GW model are (1) the rough surface is isotropic, (2) asperities are spherical and (3) they have uniform radius, but heights vary randomly satisfying Gaussian distribution. Chang, Etsion and Bogy[7] (CEB model) used these assumptions to estimate the static friction coefficient. Based on the previous work, Kogut and Etsion [9-12] carried on much investigation. They employed the finite element analysis to simulate the elastic-plastic contact between a deformable sphere and a rigid plane and obtained a parametric expression of the actual contact load, adhesion force and shear coefficient of a single asperity. They extended their work to a static friction model named as the KE model[9]. But, both the CEB model and the KE model have limitations due to their basic assumptions. The rough surface of constant radius asperity may be unrealistic in nature.

The constant radius assumption was further relaxed by Nayak[13] and Francis[14].They defined the rough surface by using two random variables represent asperity height and radius, which satisfying a joint probability distribution function (PDF). Tworzydlo et al.[15] used this approach to develop a new asperity-based model (TCOY model), and combined statistically homogenized technique with a nonlinear finite element analysis of surface asperities. Sista and Vemaganti[16](SV model) brought together the advantages of the TCOY model and the KE model. The SV model employed the contact parametric representation of a single asperity from the KE model and the statistically homogenized technique from the TCOY model. However, the simple adhesion model adopted in SV model, which is proposed by Derjaguin, Muller and Toporov[17-18](DMT model) is unable to describe the adhesion force in different deformation stages. The adhesion force can be neglected as the contact becomes more and more plastic. But in the elastic stage and in the early elastic-plastic stage, the adhesion force may be significant.

To resolve this issue and provide a more accurate static friction model, we integrate the SV model with an accurate general solution for elastic-plastic adhesion proposed by Kogut and Etsion[11] for parametric representation of adhesion force. Then we use the proposed model to study the static friction coefficient for metallic surfaces by varying roughness, the energy of adhesion and materials properties.

2. Modeling and analysis of rough surfaces

Rough surfaces are assumed as isotropic, Gaussian random distribution with asperity height and curvature as the random variables. The joint PDF of the two random variables is given by[14-16]

$$f(\xi,\eta)=\frac{\sqrt{3}}{\pi\sqrt{1-\beta^2}}\{\eta^2-1+\exp(-\eta^2)\}$$
$$\times\exp\left[-\frac{1}{2(1-\beta^2)}\{\xi^2-2\beta\xi\eta+\eta^2\}\right] \tag{2}$$

in where ξ — Non-dimensional form of asperity height,
η — Non-dimensional form of asperity curvature,
β — Wavelength spectrum parameter.
The above non-dimensional parameters are defined as

$$\xi=\frac{Z}{\sigma} \qquad \eta=\frac{\sqrt{1.5}k}{\ddot{\sigma}} \qquad \beta=\sqrt{1.5}\frac{\dot{\sigma}^2}{\sigma\ddot{\sigma}} \tag{3}$$

where Z is asperity height and k is asperity curvature, σ, $\dot{\sigma}$ and $\ddot{\sigma}$ are standard deviations of asperity height, slopes and curvature, respectively. They can be represented by the profile spectral moments

$$m_0=\sigma^2 \qquad m_2=\dot{\sigma}^2 \qquad m_4=\ddot{\sigma}^2 \tag{4}$$

Thus the non-dimensional form of asperity height, asperity curvature and wavelength spectrum parameter are given as

$$\xi=\frac{Z}{\sqrt{m_0}} \qquad \eta=\sqrt{\frac{1.5}{m_4}}k \qquad \beta=\sqrt{\frac{1.5m_2^{\ 2}}{m_0m_4}} \tag{5}$$

The asperity density can be calculated from the profile spectral moments

$$D_p=\frac{m_4}{6\sqrt{3}\pi m_2} \tag{6}$$

In this work, the contact between two rough surfaces is modeled by a perfectly smooth rigid plane in contact with a rough surface which is called sum surface (see Figure 2). Let R and d denote the asperity radius and separation distance from rigid plane to sum surface. This equivalent method of two contacting rough surfaces has generally been used in literatures.
The interference is defined as

$$\omega=Z-d \tag{7}$$

The interference is positive when asperity is in contact.

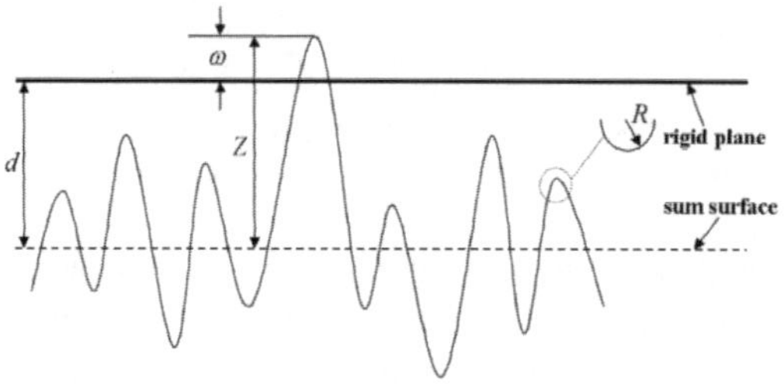

Figure 2. The equivalent model of two rough contacting surfaces

Note that the contact load $\overline{P}$, adhesion force $\overline{F}_S$ and the maximum static friction force $\overline{Q}_{max}$ of a single asperity depend on its own interference. Let the separation distance d be normalized by σ, namely $h=d/\sigma$. Then the contact load $\overline{P}$, adhesion force $\overline{F}_S$ and the static friction force $\overline{Q}_{max}$ are functions of random variables ξ and η.
The normal load of a single asperity is given by

$$\overline{P}=\overline{F}+\overline{F}_S \tag{8}$$

For a given separation distance, we can obtain the expected macroscopic normal stress using statistically homogenized technique[14-15]

$$\langle \tau_N \rangle = D_p \int_{\eta=0}^{\infty} \int_{\xi=h}^{\infty} \bar{F}(\xi,\eta) f(\xi,\eta) d\xi d\eta \tag{9}$$

Similarly, for a given separation distance, the expected macroscopic shear stress at the inception of sliding is given by

$$\langle \tau_S \rangle = D_p \int_{\eta=0}^{\infty} \int_{\xi=h}^{\infty} \bar{Q}_{\max}(\xi,\eta) f(\xi,\eta) d\xi d\eta \tag{10}$$

Note that the adhesion force exists in non-contacting asperities. Eq. (9) can be expressed in another form

$$\langle \tau_N \rangle = D_p \Big(\int_{\eta=0}^{\infty} \int_{\xi=h}^{\infty} \bar{P}(\xi,\eta) f(\xi,\eta) d\xi d\eta - \int_{\eta=0}^{\infty} \int_{\xi=-\infty}^{\infty} \bar{F}_S(\xi,\eta) f(\xi,\eta) d\xi d\eta \Big) \tag{11}$$

The representation of $\bar{P}(\xi,\eta)$, $\bar{F}_s(\xi,\eta)$ and $\bar{Q}_{\max}(\xi,\eta)$ are given in next section. Eqs. (10) and (11) can be calculated by numerical integrals. Then the static friction coefficient is obtained by

$$\mu = \frac{\langle \tau_N \rangle}{\langle \tau_S \rangle} \tag{12}$$

3. KE Elastic-plastic Contact parameters of a Single Asperity

The evolution of the elastic-plastic contact can be divided into three distinct stages. The first one is fully elastic contact stage for $\omega/\omega_c<1$. The second one is elastic-plastic contact stage which is further divided into two regime for $1\leq\omega/\omega_c\leq6$ and $6\leq\omega/\omega_c\leq110$. And the third one is fully plastic contact for $110<\omega/\omega_c$. ω is the interference and ω_c is the critical interference. The parametric representation for the normal load of a single asperity is taken from the finite element analysis[12]

$$\frac{\bar{P}}{\bar{P}_c} = \begin{cases} \left(\dfrac{\omega}{\omega_c}\right)^{1.5} & \omega/\omega_c < 1 \\ 1.03\left(\dfrac{\omega}{\omega_c}\right)^{1.425} & 1 \leq \omega/\omega_c \leq 6 \\ 1.40\left(\dfrac{\omega}{\omega_c}\right)^{1.263} & 6 \leq \omega/\omega_c \leq 110 \\ \dfrac{3}{K}\left(\dfrac{\omega}{\omega_c}\right)^{1.0} & 110 < \omega/\omega_c \end{cases} \tag{13}$$

where $\bar{P}_c=(2/3)KH\pi\omega_c R$ is the critical contact load at yielding inception($\omega=\omega_c$). The critical interference ω_c is given by

$$\omega_c = \left(\frac{\pi KH}{2E}\right)^2 R \tag{14}$$

in where

$$K = 0.454 + 0.42\upsilon_1$$
$$\frac{1}{E} = \frac{1-\upsilon_1^2}{E_1} + \frac{1-\upsilon_2^2}{E_2} \tag{15}$$

In Eq.(15), E and v denote elastic modulus and Poisson's ratio. Subscript 1 and 2 represent the materials of the two contacting surfaces and material 1 is the softer one.
The parametric representation for the adhesion force is given by[11,17]

$$\frac{\bar{F}_S}{\bar{F}_{S0}} = \begin{cases} 0.979(\frac{\omega}{\omega_c})^{0.298}\left(\frac{\varepsilon}{\omega_c}\right)^{-0.290} & \omega/\omega_c < 1 \\ 0.792(\frac{\omega}{\omega_c})^{0.356}\left(\frac{\varepsilon}{\omega_c}\right)^{-0.321} & 1 \le \omega/\omega_c \le 6 \\ 1.193(\frac{\omega}{\omega_c})^{0.093}\left(\frac{\varepsilon}{\omega_c}\right)^{-0.332} & 6 \le \omega/\omega_c \le 110 \end{cases} \tag{16}$$

$$\frac{\bar{F}_{Snc}}{\bar{F}_{S0}} = \frac{4}{3}\left[\left(\frac{\varepsilon}{\varepsilon-\omega}\right)^2 - 0.25\left(\frac{\varepsilon}{\varepsilon-\omega}\right)^8\right] \quad \omega/\omega_c < 0 \tag{17}$$

where $\bar{F}_S$ and $\bar{F}_{Snc}$ are the adhesion force for contacting asperities and non-contacting asperities, respectively. $F_{S0} = 2\pi R\Delta\gamma$ is the adhesion force at point contact (ω=0) and $\Delta\gamma$ is the energy of adhesion. ε is the equilibrium distance and the value is usually assumed between 3 to 5 Å. In this work, the equilibrium distances are given as 1.58 Å and 1.82 Å for steel surfaces and aluminum surfaces respectively[15].

The parametric representation for static friction force is given by[10]

$$\frac{\bar{Q}_{max}}{\bar{P}_c} = \begin{cases} 0.516(\frac{\omega}{\omega_c})^{0.982} & \omega/\omega_c < 1 \\ -0.01(\frac{\omega}{\omega_c})^{4.425} + 0.09(\frac{\omega}{\omega_c})^{3.425} \\ -0.40(\frac{\omega}{\omega_c})^{2.425} + 0.85(\frac{\omega}{\omega_c})^{1.425} & 1 \le \omega/\omega_c \le 6 \\ 0 & 6 < \omega/\omega_c \end{cases} \tag{18}$$

Note that the critical interference ω_c, the normal load $\bar{P}$, the adhesion force $\bar{F}_S$ and the static friction force $\bar{Q}_{max}$ are functions of the asperity radius of curvature. This indicates that asperities have different critical interference ω_c if they have different radius of curvature. At the same interference ω, the sharper and the broader asperities are in different deformation stages. And the normal stress and shear stress depend on which stage of deformation the asperity is in. Therefore, it is necessary and crucial to consider the asperity curvature as a random variable.

In the implementation, Eqs. (13), (16), (17) and (18) are first transformed to the functions of the asperity height and curvature. By substituting the transformed equations into Eqs. (10) and (11), we can calculate the expected macroscopic normal stress and shear stress. The forms of the expected macroscopic normal stress and shear stress are expressed by summation of integrals. The integrals and their corresponding limits of integration represent the contribution of the asperities in elastic, elastic-plastic and fully plastic contact, respectively. Then the static friction coefficient can be easily obtained. The non-dimensional separation distance d, is $0 \le d \le 3$ based on the basic assumptions of no interaction between neighboring asperities[9]. For a given non-dimensional separation distance, the relations of the static friction coefficient and the expected macroscopic normal stress can be plotted.

4. Results and Discussion

The contact for steel on steel is chosen as an example. The material properties of steel are given as elastic modulus E = 197 GPa, hardness H = 7100 MPa, and Poisson's ratio of v = 0.2920.Two different pairs corresponding to smooth and rough surfaces are selected to study the effect of surface roughness on the static friction coefficient.

Let Z_1 and Z_2 be asperity height of two rough surfaces. And Z_S defines the sum surface. If Z_1 and Z_2 are Gaussian and isotropic, Z_S will also be Gaussian and isotropic. Furthermore, when Z_1 and Z_2 are uncorrelated, the spectral moments of Z_S will be the sum of Z_1 and Z_2[13,20], namely

IOP Conf. Series: Journal of Physics: Conf. Series **1215** (2019) 012035 doi:10.1088/1742-6596/1215/1/012035

$$\begin{aligned}(m_0)_S &= (m_0)_1 + (m_0)_2 \\ (m_2)_S &= (m_2)_1 + (m_2)_2 \\ (m_4)_S &= (m_4)_1 + (m_4)_2\end{aligned} \qquad (19)$$

Then the statistical parameters m_0, m_2, and m_4 for sum surfaces can be derived. The statistical parameters for individual surface are listed in Table 1.

Table 1. Statistical parameters for rough surfaces

Case	Individual surface parameters		
	$(m_0)_1/(m_0)_2$ (μm^2)	$(m_2)_1/(m_2)_2$	$(m_4)_1/(m_4)_2$ (μm^{-2})
1 smooth	6.25/6.25	5.0E-5/5.0E-5	1.2E-9/1.2E-9
	6.25/6.25	5.0E-5/5.0E-5	1.2E-9/1.2E-9
2 rough	100.0/100.0	2.0E-4/2.0E-4	1.2E-9/1.2E-9
	100.0/100.0	2.0E-4/2.0E-4	1.2E-9/1.2E-9

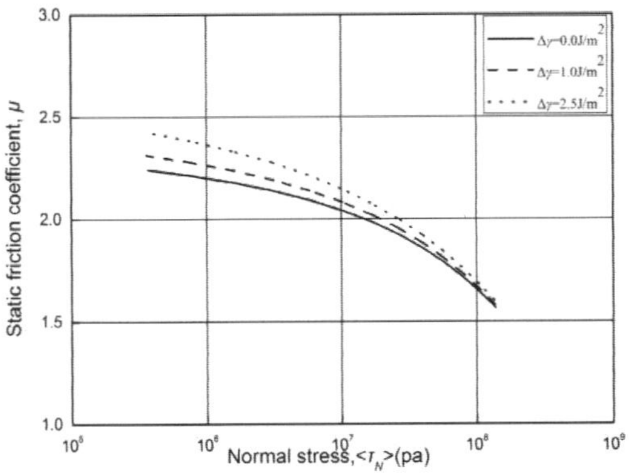

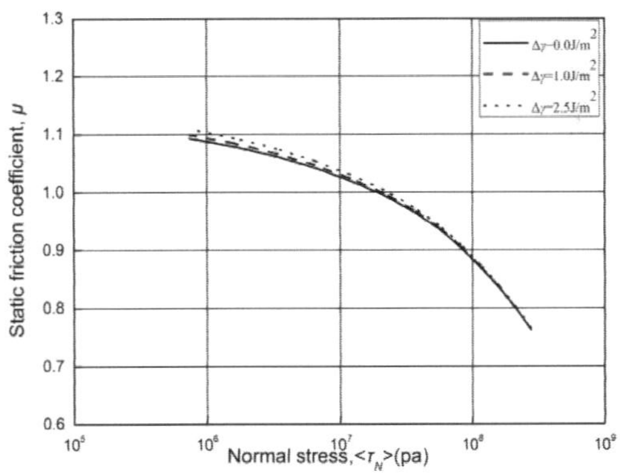

Figure 3. Static friction coefficient vs. normal stress

Figure 4. Static friction coefficient vs. normal stress

The energy of adhesion $\Delta\gamma$ are 0.0 J/m^2, 1.0 J/m^2, 2.5 J/m^2. Figure 3 depicts the trend of static friction coefficient with respect to the normal stress for three different energy of adhesion for the roughness of case 1. And Figure 4 is for the roughness of case 2.

As illustrated in Figure 3 and Figure 4, with the increase of normal load, the separation distance decreases, and more asperities come into the stage of plastic deformation. According to the finding in reference 9, when the interference $\omega \geq 6\omega_c$, asperities cannot support any tangential load. The contact force increases faster than the static friction force and the adhesion force becomes negligible compared to the normal load. Hence, the static friction coefficient reduces.

It also can be found that the static friction coefficient increases with the increase of the energy of adhesion over thc lower normal stress range and the effect diminishes as the normal stress increases. High energy of adhesion decreases the separation distance, and more asperities come into contacting for a given normal stress, thus can support larger tangential load, therefore, the static friction coefficient increase. On the other hand, for the high normal stress, the effect of adhesion can be neglected and the curves of different $\Delta\gamma$ are superimposed.

Figure 3 and Figure 4 show that the static friction coefficient is lower for rougher surface. This agrees with the experimental observations[21].

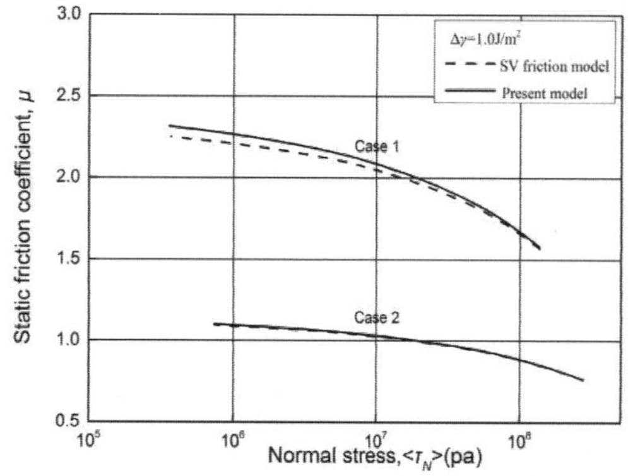

Figure 5. Comparisons of two static friction models

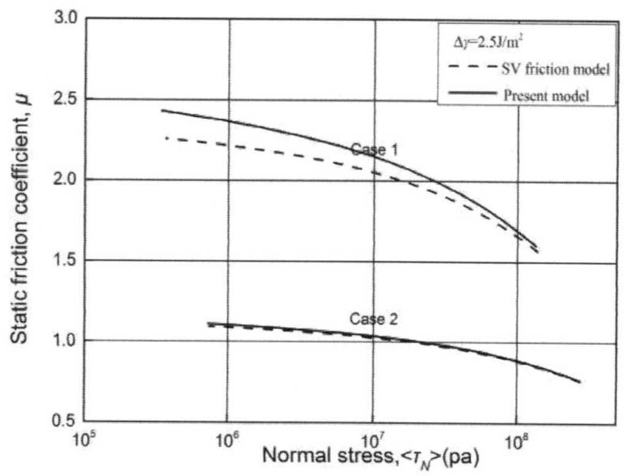

Figure 6. Comparisons of two static friction models

Figure 5 and Figure 6 depict a comparison of two friction models when the energy of adhesion $\Delta\gamma$ takes 1.0 J/m^2 and 2.5 J/m^2 for two roughness cases. Because the adhesion is stronger for smoother contacts, the effect of case 1 is more significant than case 2 and the difference becomes larger as the decreasing normal stress. It shows clearly that the adhesion force has a more significant effect on the static friction coefficient, especially for clean surfaces in contact. Note that the DMT adhesion model is a simplified model and thus the SV friction model is more acceptable for the heavily loaded contacts.
To study the effect of material properties on static friction coefficient, we calculate the contact for aluminum on aluminum case. The statistical parameters of Table 1 are still used. The material properties of aluminum are elastic modulus $E = 68.9$ GPa, hardness $H = 545$ MPa, and Poisson's ratio $v = 0.3320$. The energy of adhesion $\Delta\gamma$ takes 1.0 J/m^2.

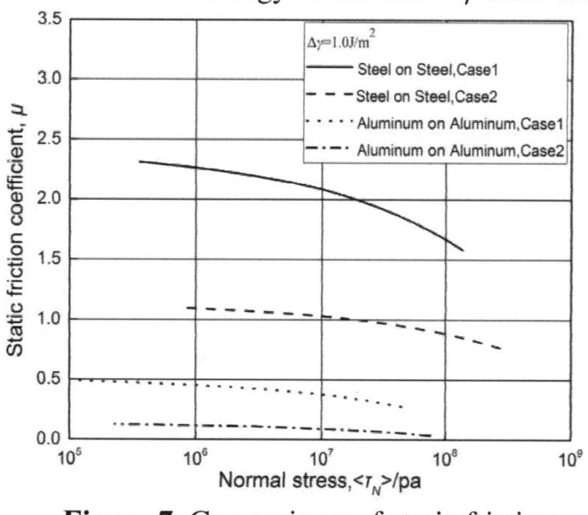

Figure 7. Comparisons of static friction coefficient for two materials

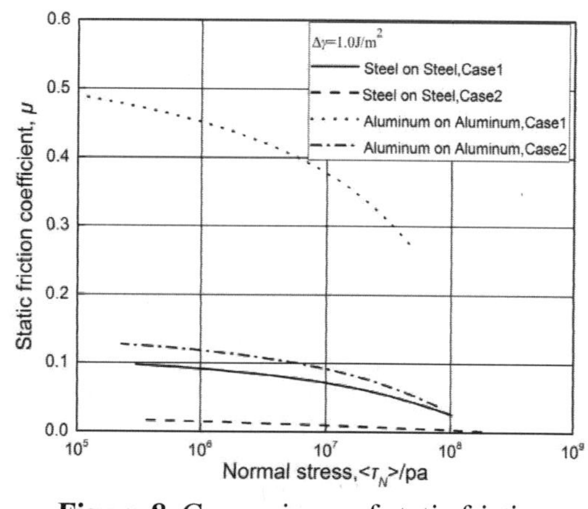

Figure 8. Comparisons of static friction coefficient for two materials

As shown in Figure 7, the static friction coefficient for the contact of aluminum on aluminum is considerably lower than the contact of steel on steel. Comparing the material properties, the H/E ratio of steel is 0.036, whereas it is 0.0079 for aluminum. From Eq. (14), the critical interference ω_c is related to the H/E ratio. According to the definition of plasticity index $\psi(=(\omega_c/\sigma)$-0.5) proposed by Greenwood and Willamson , the lower H/E ratio is corresponding to the lower critical interference ω_c and higher plasticity index ψ. Soft materials have high ψ values and the contact is mostly plastic[19]. Therefore, at a given normal stress, the contact for aluminum includes more fully yielded asperities compared to the contact for steel. As mentioned above, the yielded asperities cannot support any tangential load at the second stage of elastic-plastic deformation and have no contribution to the static friction force. Thus the static friction coefficient for the contact for steel on steel is larger.
In order to further study the effect of H/E ratio, the hardness of steel is changed to 700MPa[16], the other parameters remain the same. In Figure 8, the static friction coefficient of steel is considerably lower than that of aluminum. Comparing the material properties, the H/E ratio of steel is 0.0036,

whereas it is 0.0079 for aluminum. That is to say, at the same roughness, the surfaces with larger values of *H*/*E* have the higher the static friction coefficient. This conclusion agrees with the analytical results by Sista and Vemaganti[16]. In addition, the values of static friction coefficient are closely related to the statistical parameters of surfaces topography, especially they decrease with an increase in the statistical parameter m_2[16].Because the above statistical parameters are hypothetical and the order of statistical parameter m_2 of roughness surfaces are lower than the values given in other literatures[15,21], the values of static friction coefficient can change largely.

5. Conclusions

An improved static friction model is developed to estimate the static friction coefficient for elastic-plastic contacting surfaces. The asperity heights and radius of curvature are assumed as the random variables. A joint PDF is adopted to describe the rough surface. Based on the parametric representation of the response of a single asperity and statistically homogenized technique, the static friction coefficient is obtained. The present friction model uses the KE adhesion model instead of the DMT adhesion model.

The proposed model shows that static friction coefficient is lower for rougher surface. For a given roughness, it decreases with the increase of normal load and the energy of adhesion $\Delta\gamma$. Through the comparison of SV friction model and the present model, it shows that the smooth surface has a significantly larger adhesion effect. In the low and medium load range, the difference between two models is significant. The KE adhesion model is a more accurate description of the adhesion force compared to the DMT adhesion model, in my view, the SV friction model is more acceptable for the heavily loaded contacts.

Plasticity index ψ is inversely proportional to the critical interference ω_c which is proportional to the *H*/*E* ratio. Under the same roughness and normal stress, the plastic deformation occurs at larger plasticity index for material with lower the ratio *H*/*E* and hence the static friction coefficient is smaller.

Conflict of Interest

None declared.

References

[1] M.Stembalski, P.Preś and W.Skoczyński, “Determination of the friction coefficient as a function of sliding speed and normal pressure for steel C45 and steel 40HM,” Archives of Civil & Mechanical Engineering,vol.13,no.4,pp.444–448,2013.

[2] M.Tajdari and M.Javadi, “A new experimental procedure of evaluating the friction coefficient in elastic and plastic regions,” Journal of Materials Processing Technology, vol.177,no.1-3,pp.247–250, 2006.

[3] Q.Zhao and F.E.Talke, “Effect of environmental conditions on the static behavior of laser textured hard disk media,” Tribology International, vol.33,no.3-4,pp.281–287, 2000.

[4] I.Etsion and M.Amit, “The effect of small normal loads on the static friction coefficient for very smooth surfaces,” Journal of Tribology, vol.115,no.3,pp.406–410, 1993.

[5] F.P.Bowden and D.Tabor, “The friction and lubrication of solids,” Oxford, Oxford University Press,1950.

[6] D.Tabor, “Friction—The present state of our understanding,” Journal of LubricationTechnology, vol.103,no.2,pp.169–179,1981.

[7] W.R.Chang, I.Etsion and D.B.Bogy, “An elastic-plastic model for the contact of rough surfaces,” Journal of Tribology, vol.109,no.2,pp.257–263, 1987.

[8] J.A.Greenwood and J.B.P.Williamson, “Contact of nominally flat surface,” Proceedings of the Royal Society A, vol.295, pp.300–319, 1966.

[9] L.Kogut and I.Etsion, “A static friction model for elastic-plastic contacting rough surfaces,” Journal of Tribology, vol.126, no.1,pp.34–40, 2004.

[10] L.Kogut and I.Etsion, “A semi-analytical solution for the sliding inception of a spherical contact,” Journal of Tribology, vol.125, no.3, pp.499–506, 2003.

[11] L.Kogut and I.Etsion, "Adhesion in elastic-plastic spherical microcontact," Journal of Colloid & Interface Science, vol.261,no.2, pp.372–378, 2003.

[12] L.Kogut and I.Etsion, "Elastic-plastic contact analysis of a sphere and a rigid flat," Journal of Applied Mechnics, vol.69, no.5, pp.657–662, 2002.

[13] P.R.Nayak, "Random process model of rough surfaces in plastic contact, "Journal of Lubrication Technology, vol.26, no.3, pp.305–333, 1973.

[14] H.Francis, "Application of spherical indentation mechanics to reversible and irreversible contact between rough surfaces," Wear, vol.45, no.2, pp. 221–269, 1977.

[15] W.W.Tworzydlo, W.Cecot, J.T.Oden and C.H.Yew, "Computational micro- and macroscopic models of contact and friction:formulation,approach and applications,"Wear,vol.220,no.2, pp.113–140, 1998.

[16] B.Sista and K.Vemaganti, "A computational study of dry static friction between elastoplastic surfaces using a statistically homogenized microasperity model," Journal of Tribology, vol.137, no.2, pp.021601–021601–11, 2015.

[17] Ning Yu and A.A.Polycarpou, "Adhesive contact based on the Lennard-Jones potential: a correction to the value of the equilibrium distance as used in the potential," Journal of Colloid and Interface Science,vol.278,no.2,pp.428–435,2004.

[18] G.A.George,Sinan Müftü and Nazif Mohd Azhar, "A nan-scale multi-asperity model for contact and friction," in Proceeding of 2002 ASME/STLE Joint International Tribology Conference,Cancum, Mexico, 27-30 October 2002.

[19] W.R.Chang, I.Etsion and D.B.Bogy, "Static friction coefficient model for metallic rough surfaces," Journal of Tribology, vol.110,no.1,pp.57–63,1988.

[20] I.Etsion and M.Amit, "The effect of small normal loads on the static friction coefficient for very smooth surfaces," Journal of Tribology, vol.115,no.3,pp.406–410,1993.

[21] Chul-Hee Lee and A.A.Polycarpou, "Static friction experiments and verification of an improved elastic-plastic model including roughness effects," Journal of Tribology,vol.129,no.4,pp.754–760, 2007.

Acknowledgment
This work is supported by National Natural Science Foundation of China under Grant No. 11502207.

IOP Conf. Series: Journal of Physics: Conf. Series **1215** (2019) 012036 doi:10.1088/1742-6596/1215/1/012036

Error Compensation for 3D Digital Laser Scanning Data Based on the Surface of Aluminum Coated F-12 Aramid Plain Weave Fabric and a Test of Configuration for Large Balloon

Yong Zheng Shen[1], Guo Chang Lin[1] and Hui Feng Tan[1 *]

[1] National Key Laboratory of Science and Technology for National Defence on Advanced Composites in Special Environments, Harbin Institute of Technology, Harbin 150001, China

tanhf@hit.edu.cn

Abstract. The measurement of the configuration parameters for the large balloon is an important content, including volume, surface area and specific size under a certain internal pressure. Non-contact measurement techniques using laser scanning have the advantage of fast acquiring large numbers of points. However, compared to their contact-based counterparts, these techniques are known to be less accurate. Furthermore, this accuracy depends on different parameters such as materials properties, surface texture, product color, etc. The large balloon is made by aluminum coated F-12 aramid plain weave fabric pieces. The digitizing errors for high-speed 3D digital laser scanning data based on the surface of F12 fabric material are analyzed and characterized in this paper. With error characterisation, an empirical model is obtained relating the errors to the influencing factor. Establishing a reasonable test scheme and using the technology of laser scanning, the spatial information of the surface for large balloon can be obtained quickly. The empirical formula is used to compensate for 3D laser scanning data of large balloon. The problem of compensation for 3D laser scanning data for improved test accuracy is addressed. The more accurate parameters of the configuration for large balloon are quickly obtained. This provides a method for the rapid, high-efficiency and high-precision test of configuration for large balloon.

1. Introduction

The configuration parameters of large balloon in an inflated state are important data for evaluating the performance, such as size, volume, and surface area. Cathey H M et al.[1,2] used the photogrammetry technique to obtain the configuration of the super pressure balloon. Then the overall shape of the balloon was estimated by the local shape of the balloon. Welch J V et al.[3] conducted a test study on the deformation and strain of the surface of the balloon by photogrammetry technology. The three-dimensional local shape of the balloon was obtained. When using digital photography to measure the configuration parameters of the balloon, the spatial coordinates of few reflective points on the surface can be obtained. And these data of reflective points is used to evaluate the configuration of the balloon. Although the precision of data of the spatial coordinates of reflective points is high, the number of reflector points is small, which will lead to deviation when used to evaluate the overall configuration parameters of large balloon. At the same time, the method takes long time, large workload, and inefficiency for large balloon. For the structure of inflatable reducer, NASA[4] proposed a measurement method for the configuration by the three-dimensional laser scanning. Non-contact measurement techniques using laser scanning have the advantage of fast acquiring large numbers of points. However, the compensation problem of scanning data is not considered. Furthermore, this accuracy depends on different parameters such as materials properties, surface texture, product color, etc. The influence of the surface characteristics of the material of the balloon for the accuracy of scanning data is not considered.

Some scholars have studied the error compensation problem of 3D line laser scanning. The laser sensors currently on the market do not have the accuracy of the classical dynamic contact sensors. Taking into account and correcting these induced errors can be considered in two manners: either by developing a theoretical model that allows taking them into account during the calibration phase, or by developing an experimental model of correction of the systematic errors after the measurement. Developing a theoretical model to cope with these drawbacks is rather complex. Determining its parameters through calibration is all the more difficult. Moreover the parametrical correction models do not automatically take into account the local defects. The use of a non-parametrical model can prove more accurate and easier to implement when calibrating an imager with a flecked test pattern freely moving in the field of view[5]. Therefore, in order to solve this problem, the approach of a correction after measurement from an experimental model has been chosen usually. Xie J et al.[6] have used a look-up-table (LUT) combined with a complex interpolation algorithm to correct or adjust rectify the non-linearity between the scanning speed of the laser focus on the image field and the angular velocity. Prieto F[7] has determined a model of errors based on the repeatability of the positioning and orientation parameters. However, his study was carried out with static measurement positions that do not correspond to real functioning conditions. Feng H Y and Liu Y[8,9] have studied the effect of the distance d and the ortho-incidence angleβon the measurement of the distance between a plane and the center of a sphere. To carry out the design of the model, Feng and Liu's experimental setup, including the influence of the angleα, has been used. The setup consists of a ceramic plate (gauge block) as well as a reference sphere also in ceramic. VanGestel N et al.[10] have studied the influence of the three parameters (d, α, β) on the distribution of measurement points of a reference plane, when taking into account the thermal effects. Isheil A et al.[11] presented aims at improving the accuracy of the line laser scanning techniques through an error correction procedure based on an experimental process that concerns mechanical parts. And the influence of the three parameters, defining the relative position and the orientation between the sensor and the surface, is studied.

The large balloon is made by aluminum coated F-12 aramid plain weave fabric pieces. The digitizing errors for high-speed 3D digital laser scanning data based on the surface of F12 fabric material are analyzed and characterized in this paper. With error characterisation, an empirical formula is obtained relating the errors to the influencing factor. The problem of compensation of 3D laser scanning data for improved inspection accuracy is addressed. The empirical formula is used to compensate for 3D laser scanning data of a large balloon. The more accurate parameters of the configuration for large balloon are quickly obtained.

2. Error characterisation

2.1. Definition of error influencing factors

As shown in figure 1, a typical 3D digital laser scanning process can be described by principle of laser ranging. The principle of point cloud coordinate acquisition is to measure the horizontal and vertical angles, as well as the distance between the pulsed laser emission and recovery time difference. The $X_P Y_P$ as the plane, and the Z_P is the vertical direction. The space coordinates of the target point P relative to the origin of the coordinate are obtained by the distance L and the angle between the point and the axis, which can be described as

$$\begin{cases} x_p = L\cos\alpha\cos\beta \\ y_p = L\cos\alpha\sin\beta \\ z_P = L\sin\alpha \\ L = \dfrac{c \bullet t}{2} \end{cases} \tag{1}$$

where the c is the speed of light, t is the time between the pulsed laser emission and recovery. Where the t, α and β are obtained by the internal modules of the scanner. The distance L is calculated by the t. And the distance L is determined by the actual distance d. The distance d is the distance from the scanner to the measured surface. Considering the working mode of the scanner, the parameter d is considered as the influencing factor investigated in this paper.

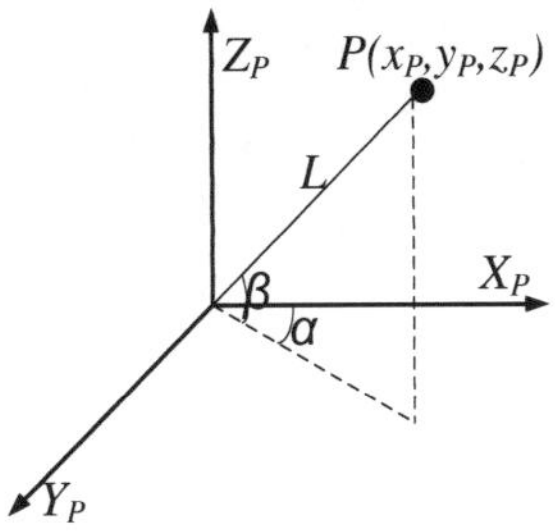

Figure 1. The principle of process for typical 3D digital laser scanning.

2.2. Error evaluation

The test set-up for error evaluation is shown in figure 2. The testing system consists of laser scanner, a reference plane and a reference ball. The laser scanner is a high-speed 3D digital laser scanner named the Trimbe TX5. The scanner can measure with 976,000 points per second and an optimum 120 meters. The reference plane is made up of F12 fabric material with a length and width of 0.5m. The front of F12 fabric material is aluminized. The back of F12 fabric material is scanned in this paper. The reference ball is rigidly mounted with a distance of *Do* onto the reference plane. The diameter of the reference ball is 100.025mm. The spherical deviation of the ball is 6μm. The error evaluation covers a range of the distance d from 1m to 7.5m. The distance increment is 0.5m, and, hence, there are 14 distance set-ups. And five tests is chose for each distance setup. In total, there are 70 tests.

Each test corresponds to one distance. The reference ball and reference plane are both scanned. The distance between the ball and the plane, as shown in figure 2, is determined from the scan data and compared against the value measured by 3D line laser scanner that is considered as the benchmark. So, the error is defined as

$$e = (D_0 - D) \cdot \frac{d}{D_0} \tag{2}$$

where e denotes the error, D and *Do* denote the measured distance by laser scanner and that by the 3D line laser scanner, respectively. The selection of D for the tests takes into consideration the spherical nature of the ball. Scanning of the ball would be the same for different projected angle set-ups.

The scene diagram of test is shown in figure 3. The results of the 3D laser scanning and the line laser scanning is shown in figure 4.

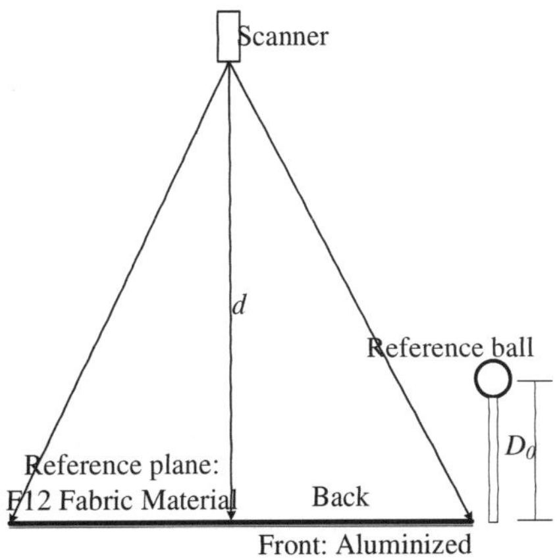

Figure 2. The test set-up for error evaluation.

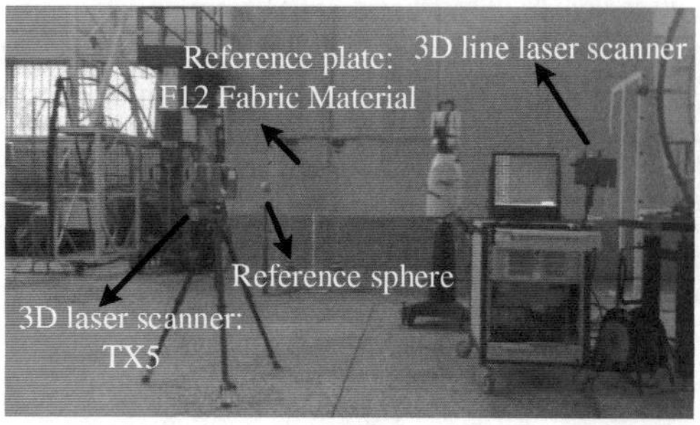

Figure 3. The scene diagram of test.

(a) (b)

Figure 4. The results of the laser scanning: (a) 3D laser scanning, (b) line laser scanning.

2.3. Empirical formula

In each test conducted, the error is averaged over five scans. The figure 5 shows the error map based on the test data. The least-squares method is applied for data fitting, considering the one variable, the distance *d*. This suggests that the systematic error can be effectively modeled by a linear relationship with the distance *d*. The empirical formula obtained from the least-squares fitting is given as

$$e = -0.27338d + 0.31439 \tag{3}$$

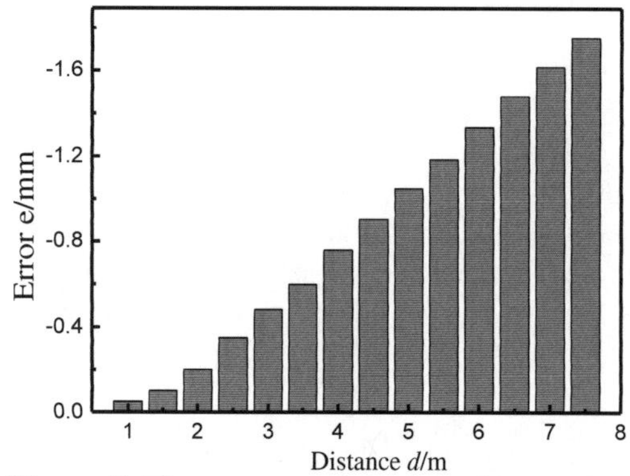

Figure 5. The error map based on the test data.

3. Error compensation

To carry out error compensation, first, the concerned parameters are computed, based on the original scan data set. Secondly, the distance d of each scan point is computed, and then the scan point is compensated based on equation (3).

3.1. Case study

A case study is conducted to compensate for the scan data of a large balloon. The large balloon is made by F12 fabric pieces. The size of the balloon is larger. The laser scanner TX5 is placed inside the balloon. The laser scanner TX5 is located at the center of the bottom of the balloon. And the bottom of the balloon is in contact with the horizontal ground. The cloud of scanning points for the large balloon under a certain internal pressure is shown in figure 6.

Under each state of the balloon, only one station is needed to scan the entire surface of balloon. And each station takes less than 3 minutes. Therefore, with this method, the complete configuration information of the surface for the large balloon can be quickly obtained.

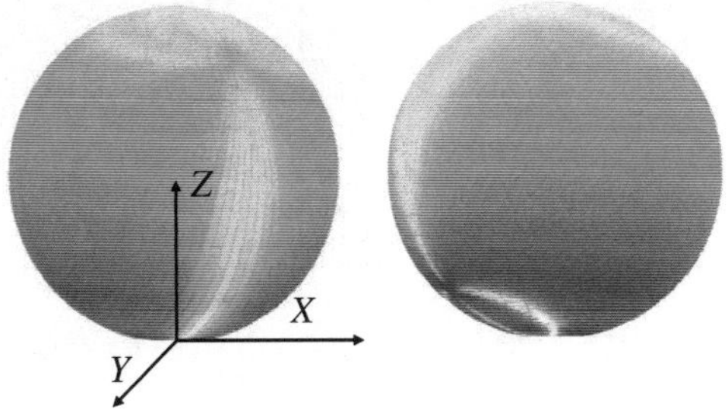

(a) (b)

Figure 6. The cloud of scanning points for the large balloon under a certain internal pressure: (a) front view, (b) side view.

3.2. Estimation of the distance d

In light of equation (3), compensation is carried out based on the distance *d*, and it's values may be estimated based on the scan data set. A spatial coordinate system is established with the coordinate origin close to the scanner position and the direction perpendicular to the horizontal ground as the *z*-axis direction. The spatial coordinate system *XYZ* is shown in figure 6. So the distance *d* of the *i*th scan point may be estimated as below

$$d_i = \sqrt{{x_i^j}^2 + {y_i^j}^2 + {z_i^j}^2} \tag{4}$$

where the d_i denotes the estimated distance, the x_i^j, y_i^j and z_i^j represent the coordinates of the *i*th scan point.

3.3. Error compensation

Based on the estimated distance *d*, the compensation value is computed using equation (3). The method of the compensation is based on point adjustment. The compensation follows the principle of azimuth invariance, that is, the point after compensation is located on the connecting line between the point before compensation and the origin. α_i and β_i represent the azimuth of the *i*th scan point in the coordinate system. As is shown in figure 7. So the error compensation is applied to the components in the *x*-, *y*- and *z*- directions, and may be given as

$$\begin{cases} x_i^{j+1} = x_i^j - |e_i| \cdot \cos\beta_i \cos\alpha_i \\ y_i^{j+1} = y_i^j - |e_i| \cdot \cos\beta_i \sin\alpha_i \\ z_i^{j+1} = z_i^j - |e_i| \cdot \sin\beta_i \end{cases} \tag{5}$$

where x_i^{j+1}, y_i^{j+1} and z_i^{j+1} represent the coordinates of the *i*th scan point after compensation. The e_i represents the the error of the *i*th scan point. The α_i and β_i can be calculated by the values of the coordinates for the *i*th scan point.

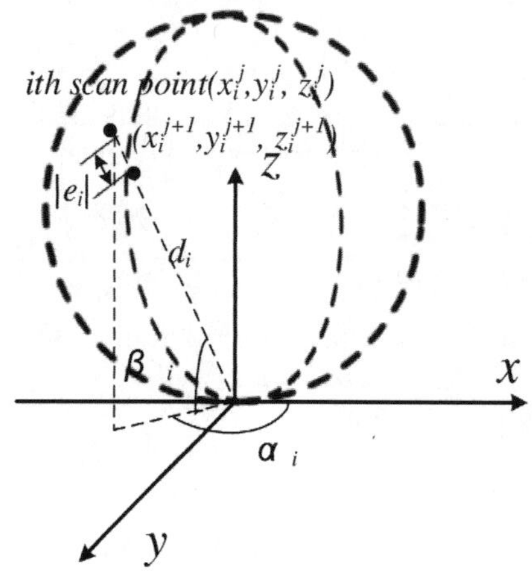

Figure 7. The schematic diagram of error compensation.

4. Results and discussions

The method of the compensation proposed in this paper is used to compensate the scanning data of surface for the large balloon. And it can improve the accuracy of the scanning data. So the more accurate values of volume, surface area and space size for configuration parameters can be obtained under internal pressures. This provides an important data reference for the flight tests of balloon and so on.

The volume and surface area of configuration parameters calculated before and after the compensation of the scanning data are shown in the figure 8 and figure 9. The V_B is expressed as the inside volume of the balloon. The S_B is the surface area of the balloon. The ΔV_B denotes the difference between the volume of the balloon before and after compensation. The ΔS_B denote the difference between the surface area of the balloon before and after compensation. It can be seen from the diagram that the values of volume and surface area which calculated after the compensation are smaller than those calculated before compensation under the same internal pressure. The values of the volume and surface area of the balloon gradually increase with the increase of the internal pressure, and increase almost linearly. With the change of the internal pressure of balloon, the difference of the volume and surface area is little fluctuating.

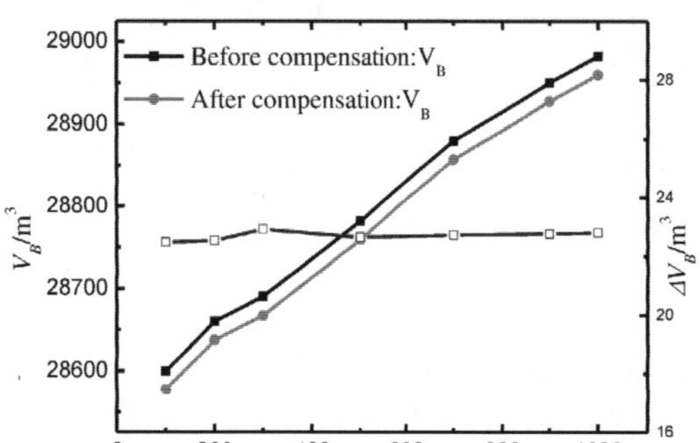

Figure 8. The volume of configuration parameters calculated before and after the compensation of the scanning data for the large balloon.

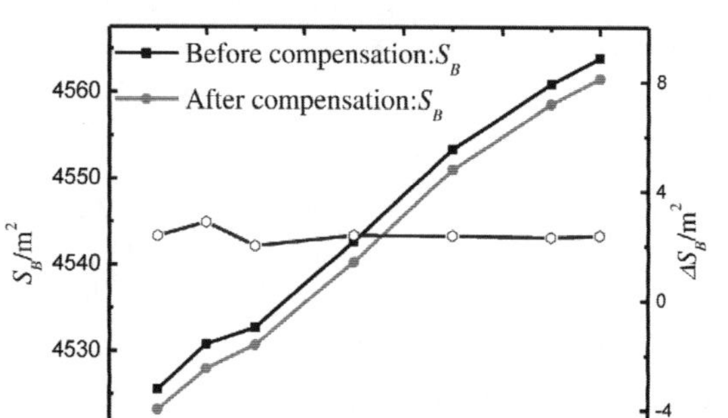

Figure 9. The surface area of configuration parameters calculated before and after the compensation of the scanning data for the large balloon.

Therefore, according to the method of surface configuration test for large balloon and the compensation of the scan data, the more accurate configuration parameters of the large balloon can be quickly obtained. This provides a more efficient, fast, and accurate method for surface configuration testing of the large balloon.

After that, the balloon can be further analyzed. The scanned data of the flap of the balloon can be intercepted after compensation, and the model of flap can be reconstructed. Then the finite element analysis method is used to analyze the strain field of the surface of the balloon. The finite element numerical results of the balloon is shown in figure 10.

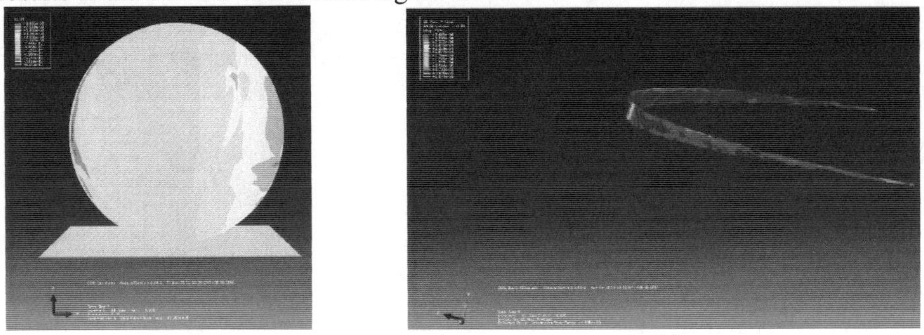

(a) (b)

Figure 10. The finite element numerical results of the balloon:

(a) displacement U1, whole, (b) maximum principal strain, flap.

5. Conclusions

The large balloon is made by aluminum coated F-12 aramid plain weave fabric pieces. In this paper, a methodology is put forward for compensating the error of the 3D laser scanning data. This includes error characterisation and compensation. The digitizing errors for high-speed 3D digital laser scanning data based on the surface of F12 fabric material are analyzed and characterized.

The laser scanner TX5 is placed inside the balloon. Under each state of the balloon, only one station is needed to scan the entire surface of balloon. And each station takes less than 3 minutes. The complete configuration information of the surface for the large balloon can be quickly obtained. Therefore, according to the method of surface configuration test for the large balloon and the compensation of the scan data, the more accurate configuration parameters of the large balloon can be quickly obtained. This provides a new method for the rapid, high-efficiency and high-precision test of configuration for the large balloon.

6. References

[1] Cathey H M 2009 The NASA super pressure balloon – A path to flight *Advances in Space Research* **44** pp 23–38

[2] Rabha M B, Boujmil M F, M and Cathey H M 2007 Evolution of the NASA Ultra Long Duration Balloon *AIAA Balloon Systems Conf.* (Williamsburg VA) p 2615

[3] Welch J V, Wang S R, Joseph R and Blandino K M 2005 Super pressure balloon non-linear structural analysis and correlation using photogrammetric measurements *AIAA 5th Aviation, Technology Integration and Operations Conf.* p 7447

[4] Leo L, Christopher K, Ben T, David J, Glen B, Brian G, Dennis B, Robert D and Charles P 2012 Design and testing of the inflatable aeroshell for the IRVE-3 flight experiment *53rd AIAA/ASME//AHS/ASC Structures, Structural Dynamics and Materials Conf.* (Honolulu, Hawaii) p 1515

[5] Orteu JJ, Cornille N, Garcia D, Sutton, S A and Mcneill S R 2005 Calibrage d'imageurs avec prise en compte des distorsions *Instrumentation Mesure and Métrologie* (Lavoisier) **4** p 105-24

[6] Xie J, Huang S, Duan Z, Shi Y and Wen S 2005 Correction of the image distortion for laser galvanometric scanning system *Optics and Laser Technology* **37** p 305–11

[7] Prieto F 1999 Me'trologie assiste'e par ordinateur: apport des capteurs 3D sans contact PhD Thesis (INSA Lyon, France)

[8] Xi F, Liu Y and Feng H Y 2001 Error compensation for three-dimensional line laser scanning data *International J. of Advanced Manufacturing Technol* **18** p 211–6
[9] Feng H Y, Liu Y and Xi F 2001 Analysis of digitizing errors of a laser scanning system. *J.* of the International Societies for Precision Engineering and Nanotechnology **25** p 185–91
[10] VanGestel N, Cuypers S, Bleys P and Kruth J P 2009 A performance evaluation test for laser line scanners on CMMs *Optics and Lasers in Engineering* **47** p 336–42
[11] Isheil A, Gonnet J P, Joannic D and Fontaine J F 2011 Systematic error correction of a 3D laser scanning measurement device *Optics and Lasers in Engineering* **49** p 16–24

IOP Conf. Series: Journal of Physics: Conf. Series **1215** (2019) 012037 doi:10.1088/1742-6596/1215/1/012037

Two Problems for a Strip with a Transverse Crack: Exact Solutions

Guangming Yu [1,2]**, M D Kovalenko**[3,4]**, I V Menshova**[3,5] **and A P Kerzhaev**[3*]

[1] School of Civil Engineering, Qingdao University of Technology, Qingdao 266033, China

[2] College of Civil Engineering and Architecture Xinjiang University, Urumqi 830047, China

[3] Institute of Earthquake Prediction Theory and Mathematical Geophysics, Russian Academy of Sciences, Moscow 117997, Russia

[4] Institute of Applied Mechanics, Russian Academy of Sciences, Moscow 125040, Russia

[5] Bauman Moscow State Technical University, Moscow 105005, Russia

Abstract. In this paper for the first time we have constructed the exact solutions of two boundary value problems of the theory of elasticity for an infinite strip with a central transverse crack on which a constant normal stress is given (even-symmetric deformation). In the first problem the sides of the strip are free, while in the second they are rigidly clamped. The solution is represented in the form of series in Papkovich–Fadle eigenfunctions. The expansion coefficients (Lagrange coefficients) have the form of simple Fourier integrals. The final formulas are simple and can easily be used in engineering.

1. Introduction

The boundary value problem of the theory of elasticity for an infinite strip (when its sides are free or rigidly clamped) with a central transverse crack on which a constant normal stress is given has been the subject of many studies (e.g. see [1]), but no exact solutions have been obtained. A review of publications on this topic, including recent works, can be found in the paper [2]. In this paper the problem for a strip with a transverse crack is considered as a boundary value problem on the junction of two semi-strips on which a longitudinal displacement discontinuity is given. In the general case, a great variety of boundary conditions can be specified at the junction of the semi-strips: solution continuity, discontinuities of displacements or stresses, etc. Boundary conditions on the long sides of the right and left semi-strips can also differ (a mixed boundary value problem for a strip with a crack). In the most general case, the solutions on the right and left of the junction line will be described by series in Papkovich–Fadle eigenfunctions, the theory of which can be found in [3–6]. However, two complete minimal systems of basis functions will always be involved in the boundary conditions at the junction of the semi-strips. Their union will not be minimal, therefore it is impossible to construct a system of functions biorthogonal to this union and, consequently, it is impossible to find closed formulas for the unknown coefficients of expansions. The essence of the proposed method is to extract a basis system of functions, to which we can construct biorthogonal systems of functions, from the non-minimal one and then to find explicit expressions for the unknown coefficients of expansions. It is achieved through the introduction of two analytical functions. They are analogues of the Kolosov–Muskhelishvili potentials.

2. Formulation of boundary value problems

Let us consider the infinite strip $\{|y|\le 1, |x|<\infty\}$ with the central transverse crack $\{\gamma : x=0, |y|\le \alpha <1\}$ on which the constant normal stress $\sigma_x(0,y)=p$ is given (even-symmetric deformation). We consider two cases: 1) the sides of the strip are free (figure 1); 2) the sides of the strip are rigidly clamped (figure 2).

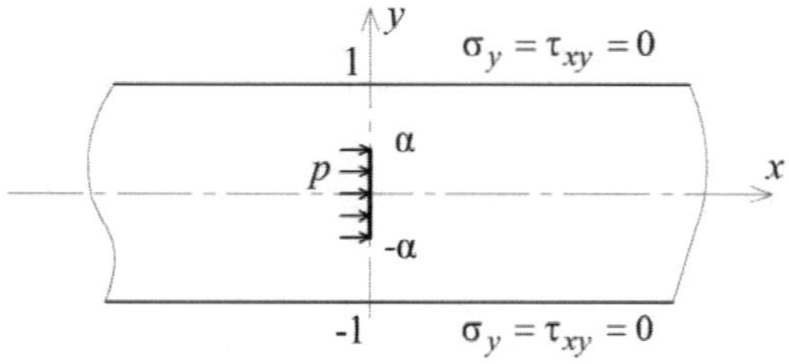

Figure 1. Scheme of the problem in the first case.

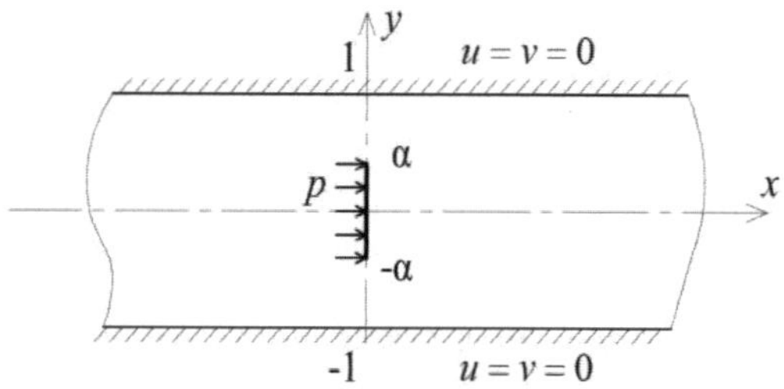

Figure 2. Scheme of the problem in the second case.

The solutions to both problems are constructed in the form of explicit expansions in Papkovich–Fadle eigenfunctions. If we do not take elementary solutions into account, then, regardless of the type of boundary conditions for $y=\pm 1$, the solutions to the problems under consideration can be represented as follows $(x\ge 0,\ \mathrm{Re}\lambda_k<0)$:

$$
\begin{aligned}
U(x,y)&=\sum_{k=1}^{\infty} a_k\xi(\lambda_k,y)e^{\lambda_k x}+\overline{a_k}\xi(\overline{\lambda_k},y)e^{\overline{\lambda_k}x},\\
V(x,y)&=\sum_{k=1}^{\infty} a_k\chi(\lambda_k,y)e^{\lambda_k x}+\overline{a_k}\chi(\overline{\lambda_k},y)e^{\overline{\lambda_k}x},\\
\sigma_x(x,y)&=\sum_{k=1}^{\infty} a_k s_x(\lambda_k,y)e^{\lambda_k x}+\overline{a_k}s_x(\overline{\lambda_k},y)e^{\overline{\lambda_k}x},\\
\sigma_y(x,y)&=\sum_{k=1}^{\infty} a_k s_y(\lambda_k,y)e^{\lambda_k x}+\overline{a_k}s_y(\overline{\lambda_k},y)e^{\overline{\lambda_k}x},\\
\tau_{xy}(x,y)&=\sum_{k=1}^{\infty} a_k t_{xy}(\lambda_k,y)e^{\lambda_k x}+\overline{a_k}t_{xy}(\overline{\lambda_k},y)e^{\overline{\lambda_k}x},
\end{aligned}
\tag{1}
$$

where $U(x,y)=Gu(x,y)$ and $V(x,y)=Gv(x,y)$ are the displacements along the x and y axes respectively, G is the shear modulus, ν is Poisson's ratio.

In the first problem the eigenvalues λ_k are the zeros of the entire function of exponential type $L(\lambda)=\lambda+\sin\lambda\cos\lambda$, in the second they are those of the entire function of exponential type $L(\lambda)=\frac{(3-\nu)\sin 2\lambda}{8\lambda}-\frac{1+\nu}{4}$. The numbers $M_k=L'(\lambda_k)/2\lambda_k$ (the prime denotes the derivative).

The Papkovich–Fadle eigenfunctions for the free strip have the form:

$$
\begin{aligned}
\xi(\lambda_k,y)&=\left(\frac{1-\nu}{2}\sin\lambda_k-\frac{1+\nu}{2}\lambda_k\cos\lambda_k\right)\cos\lambda_k y-\frac{1+\nu}{2}\lambda_k y\sin\lambda_k\sin\lambda_k y,\\
\chi(\lambda_k,y)&=\left(\frac{1+\nu}{2}\lambda_k\cos\lambda_k+\sin\lambda_k\right)\sin\lambda_k y-\frac{1+\nu}{2}\lambda_k y\sin\lambda_k\cos\lambda_k y,\\
s_x(\lambda_k,y)&=(1+\nu)\lambda_k\{(\sin\lambda_k-\lambda_k\cos\lambda_k)\cos\lambda_k y-\lambda_k y\sin\lambda_k\sin\lambda_k y\},\\
s_y(\lambda_k,y)&=(1+\nu)\lambda_k\{(\sin\lambda_k+\lambda_k\cos\lambda_k)\cos\lambda_k y+\lambda_k y\sin\lambda_k\sin\lambda_k y\},\\
t_{xy}(\lambda_k,y)&=(1+\nu)\lambda_k^2\{\cos\lambda_k\sin\lambda_k y-y\sin\lambda_k\cos\lambda_k y\},
\end{aligned}
\tag{2}
$$

while for the clamped strip they are as follows:

$$\xi(\lambda_k, y) = -\frac{1}{4}(1+\nu)(\sin\lambda_k \cos\lambda_k y - y\cos\lambda_k \sin\lambda_k y),$$

$$\chi(\lambda_k, y) = \left(\frac{\nu+1}{4}\sin\lambda_k + \frac{\nu-3}{4\lambda_k}\cos\lambda_k\right)\sin\lambda_k y + \frac{\nu+1}{4}y\cos\lambda_k\cos\lambda_k y,$$

$$s_x(\lambda_k, y) = \left(-\nu\cos\lambda_k - \frac{1+\nu}{2}\lambda_k\sin\lambda_k\right)\cos\lambda_k y + \frac{1+\nu}{2}\lambda_k y\cos\lambda_k\sin\lambda_k y, \tag{3}$$

$$s_y(\lambda_k, y) = \left(\frac{1+\nu}{2}\lambda_k\sin\lambda_k - \cos\lambda_k\right)\cos\lambda_k y - \frac{1+\nu}{2}\lambda_k y\cos\lambda_k\sin\lambda_k y,$$

$$t_{xy}(\lambda_k, y) = \left(\frac{\nu-1}{2}\cos\lambda_k + \frac{1+\nu}{2}\lambda_k\sin\lambda_k\right)\sin\lambda_k y + \frac{1+\nu}{2}\lambda_k y\cos\lambda_k\cos\lambda_k y.$$

3. Determination of the expansion coefficients

The expansion coefficients a_k in formulas (1) are determined from the conditions on the crack as a solution to the problem of conjugation of two functions analytical in the right and left semi-strips. The meaning of their introduction is to extract minimal systems of functions, to which it is possible to construct biorthogonal systems of functions, from the right and left complete systems of the Papkovich–Fadle eigenfunctions involved in the boundary conditions at the junction of the semi-strips, and then, with the help of them, to find the required coefficients of expansions.
With the help of formulas (1), we construct the following two expressions:

$$-\tau_{xy}(x,y) + \frac{i}{2}[\sigma_y(x,y) - \sigma_x(x,y)] = \sum_{k=1}^{\infty} 2\,\mathrm{Re}\left[a_k\Phi(\lambda_k, y)e^{\lambda_k x}\right],$$

$$-\left[(1+\nu)\frac{dU(x,y)}{dy} + \frac{1-\nu}{2}\tau_{xy}(x,y)\right] - i\sigma_x(x,y) = \sum_{k=1}^{\infty} 2\,\mathrm{Re}\left[a_k F(\lambda_k, y)e^{\lambda_k x}\right]. \tag{4}$$

Here

$$\Phi(\lambda_k, y) = (1+\nu)\lambda_k^2(i\cos\lambda_k + y\sin\lambda_k)e^{i\lambda_k y},$$

$$F(\lambda_k, y) = (1+\nu)\lambda_k[i(\lambda_k\cos\lambda_k - \sin\lambda_k) + \lambda_k y\sin\lambda_k]e^{i\lambda_k y}. \tag{5}$$

We consider the functions of the complex variable $z = x+iy$ $(x>0,\ \mathrm{Re}\,\lambda_k<0)$ analytical in the right semi-strip and similar functions analytical in the left semi-strip for $(x<0,\ \mathrm{Re}\,\lambda_k>0)$:

$$\Phi(\lambda_k, z) = (1+\nu)\lambda_k^2(i\cos\lambda_k - iz\sin\lambda_k)e^{\lambda_k z},$$

$$F(\lambda_k, z) = (1+\nu)\lambda_k[i(\lambda_k\cos\lambda_k - \sin\lambda_k) - iz\lambda_k\sin\lambda_k]e^{\lambda_k z}. \tag{6}$$

Denote

$$\omega_1 = \lambda_1,\ \omega_2 = -\lambda_1,\ \omega_3 = \lambda_2,\ \omega_4 = -\lambda_2, \ldots\ (\mathrm{Re}\,\lambda_k < 0),$$

$$a_k^+ = A_k,\ a_k^- = -A_k, \tag{7}$$

where a_k^+, a_k^- are the unknown expansion coefficients for the right and left semi-strips respectively.
From the conjugation conditions for $x=0$ we obtain the following two equations:

$$\sum_{k=1}^{\infty} 2\,\mathrm{Re}(A_k\Phi(\omega_k, y)) = 0,\ \sum_{k=1}^{\infty} 2\,\mathrm{Re}(A_k F(\omega_k, y)) = 2\frac{dU(y)}{dy}. \tag{8}$$

Let the jump of longitudinal displacements be given at the junction of the semi-strips:

$$2U(y) = \begin{cases} 2\sqrt{\alpha^2 - y^2} & (|y| < \alpha < 1), \\ 0 & (1 \geq |y| > \alpha). \end{cases} \tag{9}$$

The solution to the system of equations (8) is constructed in the same way as in [5], namely with the help of the functions $\Phi_k(y)$ and $F_k(y)$ biorthogonal to the functions $\Phi(\lambda_k, y)$ and $F(\lambda_k, y)$ respectively. The biorthogonal functions are determined from the equations (see [5]):

$$\int_{-\infty}^{\infty} \Phi(\lambda, y)\Phi_k(y)dy = \frac{\lambda L(\lambda)}{\lambda - \lambda_k}, \int_{-\infty}^{\infty} F(\lambda, y)F_k(y)dy = \frac{L(\lambda)}{\lambda - \lambda_k}. \quad (10)$$

The biorthogonal functions have the following structure: they consist of finite parts equal to zero outside the segment $|y| \le 1$ and nonfinite parts defined on an infinite spiral lying on the Riemann surface of the logarithm, which can be replaced by an infinite straight line in calculating the integrals connected with determining the expansion coefficients. However, only the finite parts of biorthogonal functions are used in solving boundary value problems.

The final solution to the problem is given by the formulas

$$U(x,y) = -\sum_{k=1}^{\infty} 2\mathrm{Re}\left\{u_k \frac{\xi(\lambda_k, y)}{\lambda_k M_k} \frac{\mathrm{Im}(\overline{\lambda_k} e^{\lambda_k x})}{\mathrm{Im}\lambda_k}\right\},$$

$$V(x,y) = -\sum_{k=1}^{\infty} 2\mathrm{Re}\left\{\frac{u_k}{\lambda_k} \frac{\chi(\lambda_k, y)}{M_k} \frac{\mathrm{Im}(\overline{\lambda_k} e^{\lambda_k x})}{\mathrm{Im}\lambda_k}\right\},$$

$$\sigma_x(x,y) = -\sum_{k=1}^{\infty} 2\mathrm{Re}\left\{\frac{u_k}{\lambda_k} \frac{s_x(\lambda_k, y)}{M_k} \frac{\mathrm{Im}(\overline{\lambda_k} e^{\lambda_k x})}{\mathrm{Im}\lambda_k}\right\}, \quad (11)$$

$$\sigma_y(x,y) = -\sum_{k=1}^{\infty} 2\mathrm{Re}\left\{\frac{u_k}{\lambda_k} \frac{s_y(\lambda_k, y)\lambda_k \overline{\lambda_k}}{M_k \lambda_k^2} \frac{\mathrm{Im}(\lambda_k e^{\lambda_k x})}{\mathrm{Im}\lambda_k}\right\},$$

$$\tau_{xy}(x,y) = -\sum_{k=1}^{\infty} 2\mathrm{Re}\left\{u_k \overline{\lambda_k} \frac{t_{xy}(\lambda_k, y)}{\lambda_k M_k} \frac{\mathrm{Im}(e^{\lambda_k x})}{\mathrm{Im}\lambda_k}\right\},$$

in which the numbers

$$u_k = \int_{-\alpha}^{\alpha} U(y)u_k(y)dy \quad (12)$$

and the functions (δ is the Dirac delta function)

$$u_k(y) = \frac{1}{1+\nu}\left\{\frac{\lambda_k \cos\lambda_k y}{\sin\lambda_k} - \left[\delta(y-1) + \delta(y+1)\right]\right\}. \quad (13)$$

We consider the Lagrange series (the expansion of one function into a series in a single system of Papkovich–Fadle eigenfunctions)

$$-\frac{y}{\sqrt{\alpha^2 - y^2}} = \sum_{k=1}^{\infty}\left(\frac{u_k}{\lambda_k}\frac{F(\lambda_k, y)}{M_k} + \frac{\overline{u}_k}{\overline{\lambda}_k}\frac{F(\overline{\lambda}_k, y)}{\overline{M}_k}\right). \quad (14)$$

The numbers u_k are determined by the formula (see [4])

$$u_k = -2\lambda_k \int_{-\alpha}^{\alpha} U(y)\frac{-\cos(\lambda_k y)}{2(1+\nu)\sin\lambda_k}dy = \frac{\pi\alpha J_1(\alpha\lambda_k)}{2(1+\nu)\sin\lambda_k}, \quad (15)$$

where $J_1(\alpha\lambda_k)$ is Bessel's function.

With the help of a Cauchy-type integral (see [7]), we continue the functions $F(\lambda_k, y)$ to the right semi-strip. Then the series (14) can be considered as the limit value as $x \to +0$ of the series

$$\sum_{k=1}^{\infty}\left(\frac{u_k}{\lambda_k}\frac{F(\lambda_k, z)}{M_k} + \frac{\overline{u}_k}{\overline{\lambda}_k}\frac{F(\overline{\lambda}_k, z)}{\overline{M}_k}\right) \quad (z = x + iy) \quad (16)$$

converging as $x \to +0$ to the function $-y/\sqrt{\alpha^2 - y^2} + i0$ – the limit value of the function $iz/\sqrt{\alpha^2 + z^2}$ analytical everywhere outside the crack γ in the complex plane z. Therefore, according to the second expression of (4), the series

$$\sigma_x(+0,y)=\sum_{k=1}^{\infty}\left(\frac{u_k}{\lambda_k}\frac{s_x(\lambda_k,y)}{M_k}+\frac{\bar{u}_k}{\bar{\lambda}_k}\frac{s_x(\bar{\lambda}_k,y)}{\bar{M}_k}\right)\quad (|\,y|<\alpha) \tag{17}$$

converges to zero for $|\,y|<\alpha$. However, since the Papkovich–Fadle eigenfunctions $s_x(\lambda_k,y)$ are self-balanced on the segment $[-1,1]$, the series (17) actually converges to a constant p that is determined from the self-equilibrium condition of the normal stress $\sigma_x(0,y)$ as

$$p=\frac{1}{\alpha}\int_{\alpha}^{1}\sigma_x(0,y)dy. \tag{18}$$

In the problem for the clamped strip the numbers u_k are determined by formula (12) in which

$$u_k(y)=\frac{1}{2}\left\{\frac{\lambda_k^2\cos\lambda_k y}{\cos\lambda_k}+\left[\delta(y-1)+\delta(y+1)\right]\right\}. \tag{19}$$

The value of the constant stress $\sigma_x(0,y)=p\quad(|\,y|<\alpha)$ is obtained from the equilibrium condition

$$\int_{-1}^{1}\sigma_x(0,y)dy+2\int_{0}^{\infty}\tau_{xy}(x,1)dx=0.$$

The solution to the problem for the clamped strip is given by the same formulas (11) into which it is necessary to substitute the numbers λ_k, M_k and Papkovich–Fadle eigenfunctions (3) corresponding to the problem considered.

4. Conclusions

1. The same formulas (11), but with other numbers u_k,λ_k,M_k and other Papkovich–Fadle eigenfunctions, give a solution to an even-symmetric problem for a strip in which a constant normal stress is given on a crack, and there are stiffeners along the sides. Herewith we can take into account either both rigidities of the stiffener or any of them.
2. If the boundary functions on the long sides of the left semi-strip are different from the boundary conditions on the right (a mixed boundary value problem), the numbers $\omega_{2k}=-\lambda_k$ must be replaced by the numbers μ_k $(\mathrm{Re}\,\mu_k>0)$ – the roots of the characteristic equation generated by the boundary conditions on the left. A convergent infinite product of the form (see [8])

$$\prod_{k=1}^{\infty}\left(1-\frac{\lambda}{\mu_k}\right)\left(1-\frac{\lambda}{\bar{\mu}_k}\right)\left(1-\frac{\lambda}{\lambda_k}\right)\left(1-\frac{\lambda}{\bar{\lambda}_k}\right)\quad(\mathrm{Re}\,\mu_k>0,\mathrm{Re}\,\lambda_k<0)$$

is assigned to the characteristic numbers μ_k and λ_k. The final solution on the right and left is still represented by formulas (11) with the corresponding numbers u_k,λ_k,μ_k,M_k and Papkovich–Fadle eigenfunctions.
3. On the basis of [4–6], it is not difficult to construct a solution similar to (11) for an infinite strip with a periodic system of vertical cracks. If we assume in the periodic solution for the clamped strip that the length of the discontinuity is equal to the width of the strip, then, cutting out of this strip a rectangle the ends of which coincide with the axes of the cracks, we will obtain an exact solution to the classical problem of the theory of elasticity for a rectangle two sides of which are rigidly clamped and to the other ones of which a constant normal stress is applied.
4. In the case where a constant tangential stress is given on a crack, the formulas for displacements and stresses will be different, but the scheme for solving the problem remains the same.
5. The functions (6) can be related to the analytical Kolosov–Muskhelishvili functions (see [9]) by simple linear transformations. However, the representation (6) is more general than the Kolosov–Muskhelishvili representation, since with its help it is possible to solve more complicated problems. The application of the functions (6) for solving the classical problems for a plane with cracks can be found in the paper [10].

IOP Conf. Series: Journal of Physics: Conf. Series **1215** (2019) 012037 doi:10.1088/1742-6596/1215/1/012037

5. References

[1] Goldstein R V, Ryskov I N and Salganik R L 1970 Central transverse crack in an infinite strip *Int. J. Fract.* **6**(1) pp 104–5

[2] Zhuravlova Z 2017 Stress analysis near the tips of a transverse crack in an elastic semi-strip *Appl. Math. Mech.-Engl. Ed.* **38**(7) pp 935–56

[3] Kerzhaev A P, Kovalenko M D and Menshova I V 2018 Borel transform in the class W of quasi-entire functions *Complex Anal. Oper. Theory* **12**(3) pp 571–87

[4] Kovalenko M D and Shulyakovskaya T D 2011 Expansions in Fadle–Papkovich functions in a strip. Theory foundations *Mech. Solids* **46**(5) pp 721–38

[5] Kovalenko M D, Menshova I V and Shulyakovskaya T D 2013 Expansions in Fadle–Papkovich functions: examples of solutions in a half-strip *Mech. Solids* **48**(5) pp 584–602

[6] Kovalenko M D, Menshova I V and Kerzhaev A P 2018 On the exact solutions of the biharmonic problem of the theory of elasticity in a half-strip *Z. Angew. Math. Phys.* **69** 121

[7] Bremermann H J 1965 *Distributions, Complex Variables and Fourier Transforms* (Reading: Addison-Wesley)

[8] Levin B Ja 1980 *Distribution of Zeros of Entire Functions* Translations of Mathematical Monographs vol. 5 (Providence: American Mathematical Society)

[9] Muskhelishvili N I 1953 *Some Basic Problems of Mathematical Theory of Elasticity* (Groningen: Noordhoff)

[10] Kerzhaev A P 2015 *Proc. All-Russian Scientific School-Conf. "Mechanics of Limit State and Related Questions" dedicated to the 85th anniversary of Prof. D D Ivlev* vol 2 (Cheboksary: Chuvash. gos. ped. un-t) pp 152–9 (in Russian)

Acknowledgments This work was partially supported by the Russian Foundation for Basic Research, Grant No. 16-31-60028 mol_a_dk, and National Natural Science Foundation of China, Grant No. 51674150.

IOP Conf. Series: Journal of Physics: Conf. Series **1215** (2019) 012038 doi:10.1088/1742-6596/1215/1/012038

Numerical Investigation on Flow Fields of SVC Nozzle with Bypass Injection

J W SHI[1], Z X WANG[1], L ZHOU[1]

[1]Shaanxi Key Laboratory of Internal Aerodynamics in Aero-engine, Collaborative Innovation Center for Advanced Aero-Engine, School of Engine and Energy, Northwestern Polytechnical University, China

Abstract. Shock vector control (SVC) is one typical fluidic thrust vectoring technology. It has simple working principle and wide working range, but not high enough vector efficiency. In the paper, a bypass injection concept was proposed on the SVC technology to improve the vector efficiency and vector angle. The flow mechanism of SVC nozzle with bypass injection was investigated numerically after the turbulence model was validated by experimental data. The enhanced performance of SVC nozzle was estimated and the influence of critical affecting parameters, bypass injection angle (θ .ad.) and bypass injection position (Xj.ad.) on vector performance were studied. Results show that, the vector efficiency was increased from 1.21 °/%-ω to 1.93 °/%-ω by adopting bypass injection, and the vector angle has an increase by about 58.9%. The vector performance increased with the increase of bypass injection angle when induced shock wave did not interacted with the opposite wall. The better vector performance was achieved with the configuration of the bypass injection slot was upstream of secondary injection slot when SPR is less than 1.0. In wide working ranges, a vector angle of 16°-20°, a vector efficiency larger than 2.0 °/%-ω were obtained while thrust coefficient of a SVC nozzle kept larger than 0.90.

1. Introduction

Thrust vectoring (TV) is a necessary technology for the 4th or later fighter plane. It provides extra force and moment in three dimensions besides thrust, and helps fighter plane to enhance agility and maneuverability at post-stall regime, to shorten taking off and landing distance and to enlarge the fight envelop *etc.* [1-2] Generally, there are two types of TV, the mechanical TV and the fluidic TV. The mechanical one is a fully developed technology, which was investigated since 1970s, and has different kinds of forms, including 2D mechanical TV used in F119, axis mechanical TV used in 31f and EJ-200, and three swivel bear nozzle used in F-135 *etc.* [3-5] However, the complex mechanical structures also bring out higher weight, larger lift cycle costs and gas leakage problem, which motivate researchers around the word to explore new concepts of TV with better performance. That is the fluidic TV. Compared with the mechanical TV, it has a weight reduction by 43%-80%, an improvement of engine thrust-to-weight ratio by 7%-12% and a reduction in nozzle procurement and life cycle costs by 37%-53% [6].

Different fluidic TVs were proposed based on active flow control. In these concepts, a secondary flow was adopted to control the primary flow, and methods including co-flow ejection, counter-flow suction, transverse injection were investigated [7]. Correspondingly, the concept of co-flow, counter-flow, throat shifting (TS), double throat nozzle (DTN) and shock vector control were formed. Among them, the SVC was emphasized due to its suitability on exhaust system with high nozzle pressure ratio. The working principle of SVC is based on transvers jet injected into supersonic primary flow at the divergent section of nozzle. In the supersonic flow, an induced shock wave forms and deflects the primary flow, therefore the thrust vectoring is achieved [8]. In SVC nozzle, the main side-force for primary flow deflection is the high pressure formed upstream of injection slot and the jet injection moment.

Nowadays, numerical and experimental investigations were conducted, and two aspects of work were focused on. One aspect is the study on the complex flow mechanism and the influence of aerodynamic and geometric parameters on vector performance, including nozzle pressure ratio (NPR), secondary pressure ratio (SPR), nozzle inlet total temperature, secondary inlet total temperature, free stream Ma number, secondary injection position, secondary injection area and secondary injection position [9-11]. In the previous study, critical conclusions were found that the lower NPR brings better vector efficiency; the optimal SPR for largest vector angle exists when the induced shock wave interacts with nozzle lip; better performance is obtained when the secondary injection position moves forwards; and the increase on vector angle can be achieved by 50% when the secondary injection angle is optimised etc. The other aspect is the study on the multi-function thrust vectoring and vector performance optimization. In the work of Chiarelli and Wing, the SVC and Co-flow method were combined and the thrust vectoring with pitch and yaw function was achieved [12-13]. In the work of Giuliano, the investigation on a SVC nozzle with different flow path cross-sections and trailing-edge shapes were conducted at jet exit test facility, and multi-axis thrust vectoring control was obtained [14]. In the work of Federspiel and Anderson, the SVC / throat skewing combined concept and mechanical / fluidic combined concept were proposed, which extended the usage of SVC method and showed that better performance can be achieved by the combinations of different fluidic TV methods [15].

In the previous work, the advantages of SVC nozzles were fully displayed, but the remaining problem, not high enough vector efficiency, also should be considered. Generally, the vector efficiency of SVC nozzle is about 0.8-1.4 °/%-ω, which means about a secondary flow of 14.3% - 25% of primary flow is needed. It has large influence on the working performance of an aero-engine. Therefore, in the paper a concept of SVC with bypass injection was studied to improve the vector performance and to reduce the secondary flow demand. The flow characteristics was investigated, the vector performance of SVC nozzle and SVC nozzle with bypass injection was compared, and the effect of bypass injection angle and bypass injection position on vector performance were conformed.

2. Numerical Approach

2.1. The geometric model

The basic geometric model of a SVC nozzle with bypass injection is a convergent-divergent SVC nozzle, and a bypass flow is introduced from the nozzle convergent section to the nozzle divergent section, seen in Figure 1. There three flows in the nozzle, the primary flow, secondary flow and bypass flow. Consequently, the geometric parameters consist of the three parts.

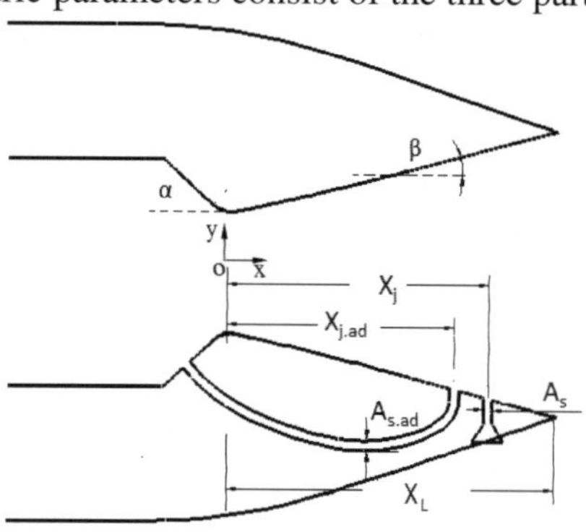

Figure 1 The schematic of the improved SVC nozzle

For the primary nozzle, a convergent-divergent configuration is adopted with a convergent angle (α) of 42°, a divergent angle (β) of 13.5 ° and an expansion ratio (A_9 / A_8) of 2.34 which results in a designed nozzle pressure ratio (NPR_D) of 13.88. The coordinate origin is at center of nozzle throat. The nozzle throat area (A_8) is 3890 mm^2. The nozzle aspect ratio at nozzle exit is 2.2. For the secondary flow, the injection position (X_j,) is 0.79, which is the dimensionless distance to the coordinate origin by the length of nozzle divergent section (L_D)). The secondary injection area ratio (As / $A8$) is 0.08, which is the dimensionless injection area by nozzle throat area. The secondary injection angle (θ) is 90º. For the bypass flow, the definitions of bypass injection position ($X_{j.ad}$) and bypass injection area ratio ($A_{s.ad}$./A_8) are similar with those of X_j and A_s/ A_8. In this study, the $X_{j.ad.}$ varies from 0.69 to 0.89, which means bypass injection is arranged upstream or downstream of the secondary flow. $A_{s.ad.}$ /A_8 is 0.08. The bypass injection angle ($\theta_{ad.}$) is from 90 ° to 120 °.

2.2. Governing equations

The flow characteristics of a SVC nozzle with bypass injection was obtained by solvig solving the compressible form of conservation equations continuity, momentum and energy equations in Cartesian coordinates, shown as follow. A commercial software of Fluent of ANSYS 18.2 was used to conducted the simulation work.

Continuity:

$$\frac{\partial \rho}{\partial t}+\frac{\partial}{\partial x_i}(\rho u_i) =0 \tag{1}$$

Momentum:

$$\frac{\partial}{\partial t}(\rho u_i)+\frac{\partial}{\partial x_i}(\rho u_i u_j) =-\frac{\partial p}{\partial x_i}+\frac{\partial \tau_{ij}}{\partial x_j}+\rho g_i + F_i \tag{2}$$

Energy:

$$\frac{\partial \rho E}{\partial t}+\frac{\partial}{\partial x_i}[u_i(\rho E+p)]=\frac{\partial}{\partial x_i}[(\alpha+\frac{C_p \mu_t}{P_{rt}})\frac{\partial T}{\partial x_i}+u_j(\tau_{ij})_{eff}] \tag{3}$$

Where τ_{ij} is the shear stress tensor and is defined as follow:

$$(\tau_{ij})_{eff}=\mu_{eff}(\frac{\partial u_i}{\partial x_j}+\frac{\partial u_j}{\partial x_i})-\frac{2}{3}\mu_{eff}\frac{\partial u_i}{\partial x_i}\delta_{ij} \tag{4}$$

The equations solved are the fully three-dimensional compressible Reynolds averaged Navier-Stokes equations, which are discretized in finite volume form on each of the hexahedral control volumes. The second implicit scheme adopted for the time discretization, and the second order upwind scheme is used in the spatial discretization. The implicit density-based algorithm is used to solve the equation, and Roe averaged flux difference splitting (Roe-FDS) is chosen for the flux type.

2.3. The computational grid

The computational grid of the SVC nozzle with bypass injection is shown in Figure 2. Due to the geometric symmetry of the SVC nozzle, half model was adopted for investigation. The hexahedral meshes were produced all over the computational zones. Meshes near nozzle wall and around secondary injection slot and bypass injection slot were refined, with the y$^+$ about 1.0. The final grid was generated with about 3.2 million cells.

2.4. The validation of turbulence model

The flow fields of the SVC nozzle with bypass injection are featured by jets interactions, flow separation, vortices, shock waves and shear layers *etc.* Before the numerical simulation, the turbulence model needs to be validated. Several turbulence models, including the *S-A* model, the *SST k-ω* model, the *SAS* model, the realizable *k-ε* model, and the *RNG k-ε* model, are chosen and validated by the experimental data from Spaid and Zukoshi[16]. The experiment was conducted at a free stream Ma of 2.61 with the ratio of injected jet total pressure to free stream total pressure of 1.15. Noticeably, the boundary condition is similar to that of the SVC nozzle with bypass injection, thus the best validated turbulence model will be used in the study. The numerical and experimental results of pressure distributions along the nozzle center line are shown in Figure 3.

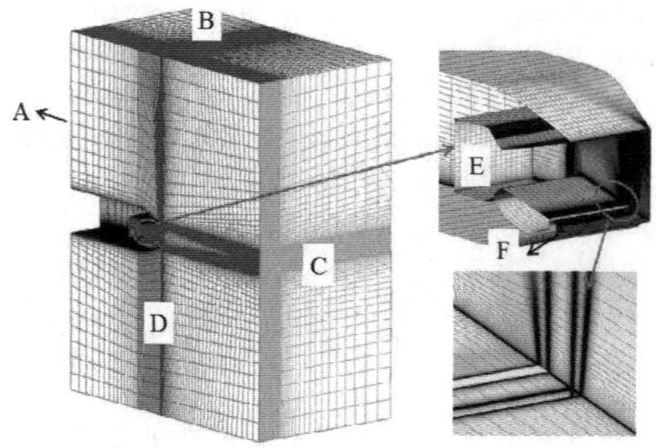

A: outer inlet, B: Pressure far-field, C: Outlet, D: Symmetric plane, E: nozzle inlet, F: Secondary inlet

Figure 2 The computational grid

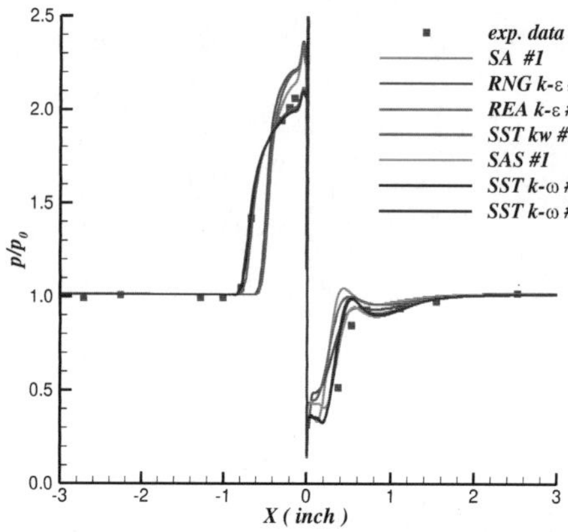

Figure 3 Turbulence model validated by experimental data

It is seen that the predicted separation position of boundary layer upstream of jet injection slot by the *S-A* model, the Realizable *k-ε* model, and the *RNG k-ε* model did not match the experimental data well, and the predicted peak pressure is about 10% larger than the experimental data while the predicted high pressure zone is about 20% smaller. The *SST k-ω* model and *SAS* model predicted separation position and pressure profiles very well, and in the high pressure zone the maximum pressure difference with the experimental data is within 2%. Additionally, the two turbulence models also performed better at predicting pressure distributions downstream of jet injection port. But in the comparison of calculation time, the simulation with the *SAS* turbulence model took time more than that of the *SST k-ω* model by about 50-80%. Therefore, The *SST k-ω* model was proven to be the best choice.

2.5. The Boundary conditions

Pressure inlet boundary is specified for nozzle inlet (E), secondary flow inlet (F) and outer flow inlet (A), shown in Figure 2. On these boundaries the total pressure (p_n^*, p_s^*, p_0^*), the total temperature (T_n^*, T_s^*, T_0^*) and flow angle are prescribed. Nozzle total pressure and secondary flow total pressure are calculated from *NPR* and *SPR*, which vary from 6 to 16 and from 0.6 to 1.5, respectively. The total temperature of nozzle inlet is 800 K, while secondary flow total temperature is obtained from the isentropic relationship between pressure and temperature, shown as follow:

$$T_s^* = T_0 \times \left(P_s^* / P_0\right)^{\frac{\gamma-1}{\gamma}} \tag{5}$$

At outlet boundary (C), static pressure (P_0) is imposed and the other variables were extrapolated from the interior parameters. Symmetry boundary condition is adopted at the center face (D). A free-stream Ma∞ and flow direction are provided for far-field pressure boundary (B). Impermeable, no-slip and adiabatic wall boundaries were applied on the solid wall to ensure zero normal mass, momentum and energy crossing the mesh face at the wall boundary.

IOP Conf. Series: Journal of Physics: Conf. Series **1215** (2019) 012038 doi:10.1088/1742-6596/1215/1/012038

3. Results and discussions

3.1. working principle and flow characteristics of a SVC nozzle with bypass injection

The sturcture of a SVC nozzle with bypass injection was shown in Figure 4. It is seen that the bypass flow is from the high pressure zone of the nozzle to the diverernt section. It is injected into the supersonic flow at nozzle divergent section and assists the secondary flow to control primary flow deflecting. With the bypass injection, the transverse jet penetration depth is enhanced, the strength of induced shock wave becomes stronger, the position of the induced shock wave moves forwards, and the separation zone upstream of injection enlarges, which results in larger high pressure zone on nozzle wall and larger side-force for vectoring. Meanwhile, the bypass injection has tiny influence on secondary flow. So the secondary flow mass keeps unchanged, and the vector angle increases, therofore, the vector efficiency is improved. When a larger vector angle is required, the demanded secondary flow mass of a sVC nozzle with bypass injection is smaller compared with a SVC nozzle, and it has less influence on the performance of an aero-engine.

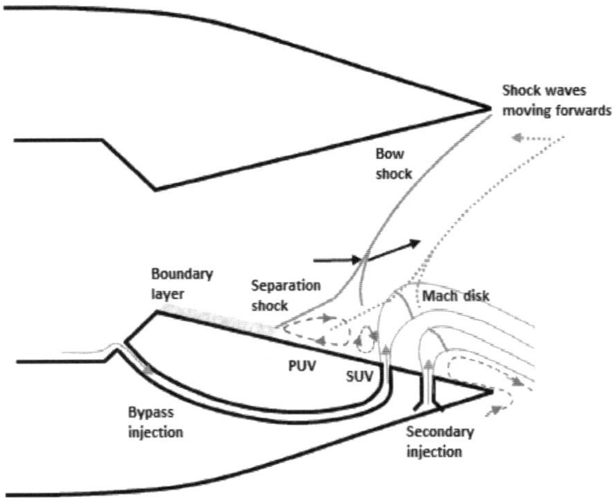

Figure 4 The sketch map of working principle of the improved SVC nozzle

The flow characteristics of a SVC nozzle with bypass injection are featured by transvers jets (secondary flow and bypass flow) into the supersonic at nozzle divergent section and the shock wave / boundary layer interactions in a confined space. In this section, the flow characteristics of a SVC nozzle with bypass injection were investigated numerically. The concerned parameters are as follow: the secondary injection area ratio (A_s / A_8) and the bypass injection area ratio ($A_{s.ad} / A_8$) are 0.08, respectively; the secondary injection angle (θ) is 90°, which is the same with the bypass injection angle ($\theta_{ad.}$); the nozzle pressure ratio (NPR) is 13.88, the nozzle inlet total temperature (T_n^*) is 800K; the secondary pressure ratio (SPR) is 1.0, and the secondary injection total temperature (T_s^*) is 622 K (according to Eq.5).

For the SVC nozzle with a bypass injection, two under-expanded jets (secondary flow and bypass flow) were injected into the supersonic primary flow. A strong obstacle was generated and an induced shock wave was caused. Correspondingly, there was a strong adverse pressure gradient, which forced the boundary layer upstream of injection slot to separate. Then another shock wave was induced, i.e. the separation shock wave. When the two shock waves interacted, they merged into a "λ" shock wave, seen in Figure 5. It has a stronger shock wave leg and a weaker one. The "λ" shock wave caused unbalanced pressure distributions on nozzle up and down walls, which plays an important role in deflecting primary flow.

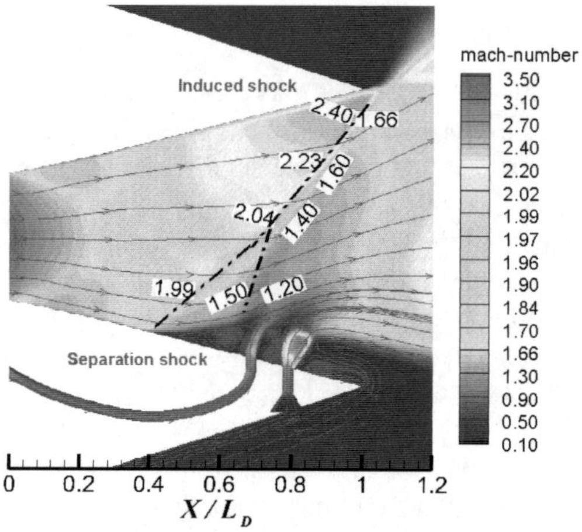

Figure 5 Ma number and streamlines distributions on the symmetric plane

The influence of the "λ" shock wave on the flow characteristics of a SVC nozzle with bypass injection is concerned with the acceleration of the primary flow. In one aspect, the acceleration of primary flow made the Ma number distributions upstream of shock waves non-uniform, which varied from 1.80 to 2.40. When primary flow passed through shock waves, the flow mechanical energy losses were different, and the Ma number downstream of shock wave, ranging 1.2 to 1.66, were obtained. In the other aspect, the acceleration and shock wave losses brought non-uniform flow characteristics at nozzle exit plane. And there are typically five different flow losses zones on the exit plane, seen in Figure 5, which is distinguished according to the total pressure recovery coefficient (σ) that is the ratio of local total pressure to the inlet total pressure of nozzle.

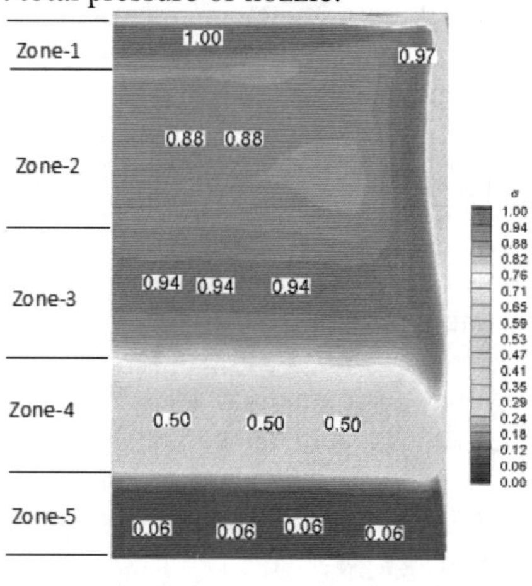

Figure 6 Total pressure recovery coefficient distributions on the exit plane

In zone-1, small flow loss exits, and σ is about 1.0. It is mainly because that primary flow in this zone did not affected by shock waves. In zone-2 and zone-3, primary flow passed through shock wave main branch and shock wave legs, respectively. The σ in zone-2 is about 0.88, and is a little smaller than σ (0.94) in zone-3. The primary reason is that the primary flow of zone-2 has larger Ma number upstream of shock wave, as is known, when shock angle is the same, the larger Ma number upstream of shock is, the more drastic flow loss there is. Another reason is that the flow of zone-3 passed through two weak shock waves, while the flow of zone-2 passed through a stronger one. Usually, the multi-shock waves can result in less flow loss than that of a stronger one. For the zone-4, the flows are from injected jets (secondary flow and bypass flow) and shear layers between jets and primary flow. The σ is about 0.50, which is comparatively low. The mainly reason is that there are Mach discs in under expanded jets, which are normal shock waves and bring out tremendous flow losses. In zone-5, ambient air flowed into low pressure zone downstream of injection slot, and those low powerful flows filled up the zone, which finally resulted in a much smaller σ of 0.07. Because the flow in zone-5 was not from primary flow, the mass averaged σ on nozzle exit plane was not suitable for estimating the overall performance of the SVC nozzle with bypass injection. So another parameter, the thrust coefficient (C_{fg}) was adopted, and in this case the C_{fg} of was about 0.907.

There are complicated vortices in the the SVC nozzle with bypass injection because of the transverse jets into accelerated supersonic primary flow. A pair of counter-rotating span-wise vortices were witnessed upstream of injection slots. The front vortex is formed by the boundary layer separation due to shock waves, named as primary upstream vortex (PUV), which is clockwise and bigger in size, seen in Figure 5. The position and size of it can be identified by limited streamlines distributed on nozzle down wall, seen in Figure 7 (a). The solid line denotes the convergent limited streamlines, which is the boundary separation line. It starts from a saddle point and ends with a nod point, and it is the front boundary of PUV. The following dashed line denotes the divergent limited streamlines, which represents the attachment position of flow separation, and it is the rear boundary of PUV. Additionally, the zone between solid and dashed lines is used to judge the stream-wise size of PUV. The other vortex is an anti-clockwise vortex, named as secondary upstream vortex (SUV), and it is a litter smaller. It is formed due to the entrainment of bypass jet and PUV, and in size it was confined by the attachment line and bypass injection slot. There is also another pair of counter-rotating vortices between bypass injection slot and secondary injection slot. As is analyzed, they were formed mainly by the entrainment of secondary flow and bypass flow. From the divergent limited streamlines (dashed line) between injection slots, it is found that the vortex near secondary jet is larger, because the secondary flow has higher speed and correspondingly the larger entrainment intensity. Besides, there is another divergent limited streamline on nozzle wall downstream of injection slots, marked by dashed line too, which was a sign of attachment of separation flow. In this zone, low pressure was formed due to injected jets, and ambient air flowed around nozzle lip into the zone, resulting in an anti-clockwise lip vortex, at the entrainment of which and secondary injection flow a small vortex was also induced, seen in Figure 5 and Figure 6.

The role of the bypass flow is to enhance the vector performance of a SVC nozzle, e.g. vector angle (δ_p), thrust vector efficiency (η). A comparison on the vector performance of a SVC nozzle with and without bypass flow was conducted. Results showed that an improved vector angle of 14.28° was achieved with assistant of the bypass flow, which was 58.9% larger than that of a SVC nozzle without bypass flow. Meanwhile a η of 1.93°/% -ω was achieved, which is also much larger than that of a SVC nozzle. Additionally, the bypass flow equivalently increased the flow flux of a nozzle, which changed the working character of an aero-engine, and it was benefit for net thrust. Based on the preliminary results above, it is inferred that the SVC with bypass injection method may offer more desirable option for fluidic thrust technology. Thus in the following sections, to get more understanding of the SVC nozzle with bypass injection, the influence of two critical geometric parameters on the flow characteristics and vector performance were investigated in wide working conditions.

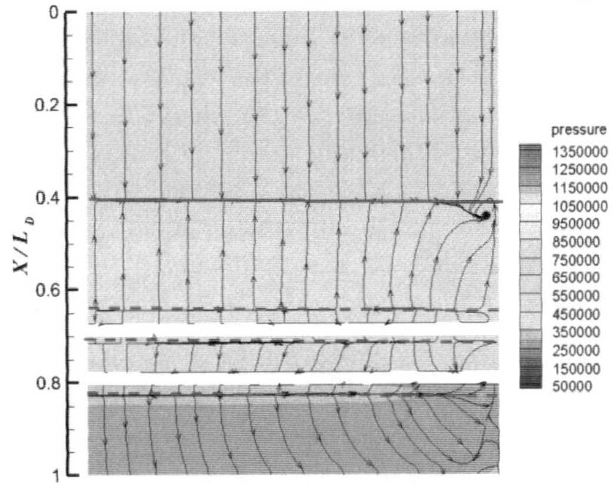

(a) Nozzle down wall

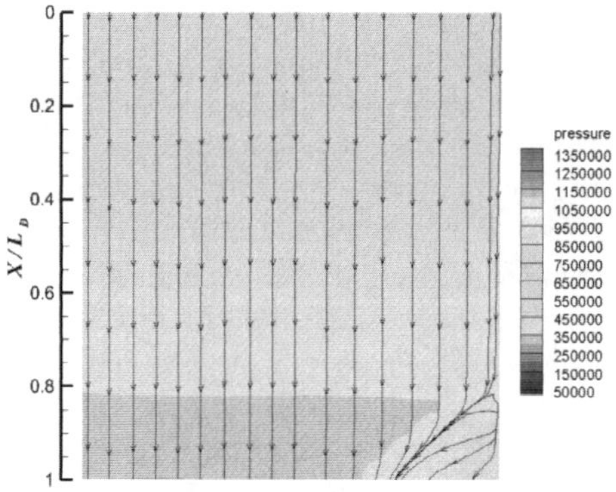

(b) Nozzle up wall

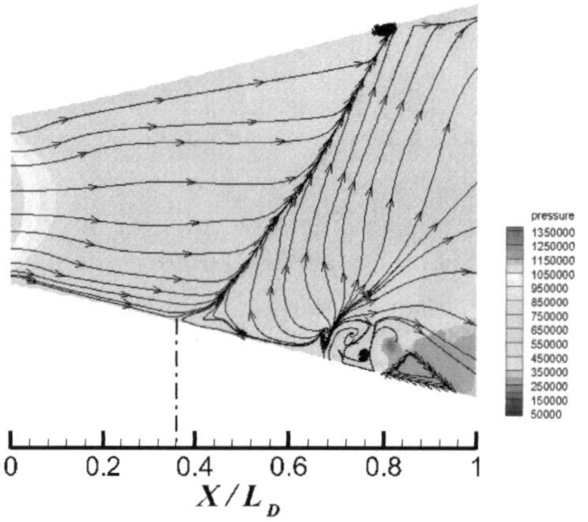

(a) Nozzle side wall

Figure 7 Limited streamlines on internal walls of the improved SVC nozzle

3.2. The influence of bypass injection angle on vector performance of a SVC nozzle with bypass injection

The influence of bypass injection angles ($\theta_{.ad.}$) on the performance of a SVC nozzle with bypass injection was investigated with $X_{j\cdot\ ad}$ of 0.89 and $A_{s.ad}$ of 0.08. $\theta_{.ad.}$ has a varies from 90 °, 100 °, 110 °, 120° to 130 °. The bypass injection angle affects the trajectory and jet penetration depth of secondary flow, and alters the separation length upstream of injection slot.

Figure 8 shows the flow characteristics of a SVC nozzle with bypass injection at *NPR* of 13.88 and *SPR* of 0.6, 1.0 and 1.5. With the increase of the bypass injection angles, the forward component of bypass flow moment increased, which has larger impact on secondary injection flow and made the peak point of jet trajectory promote, seen figure 8 (e). But it is found that the jet trajectory kept unchanged below the Mach dick. And the height of Mach dick is not suitable to analyse the jet penetration depth and vector performance. Instead, the jet trajectory can indicate the induced shock wave angle and the shock wave strength better. But it is difficult to obtain the relation expressions of the jet trajectory on other aerodynamic and geometric parameters.

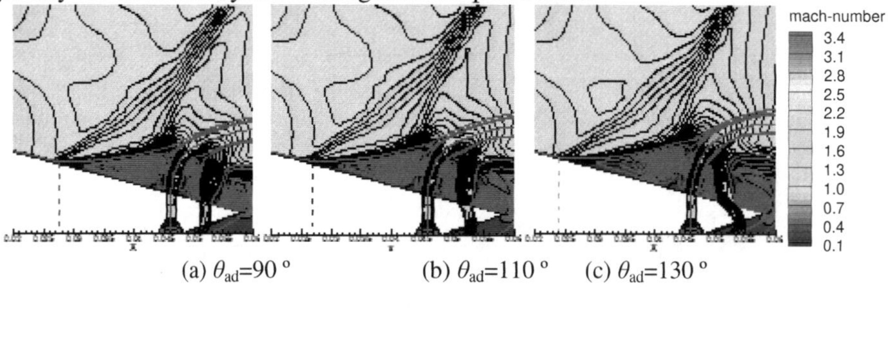

(a) θ_{ad}=90 ° (b) θ_{ad}=110 ° (c) θ_{ad}=130 °

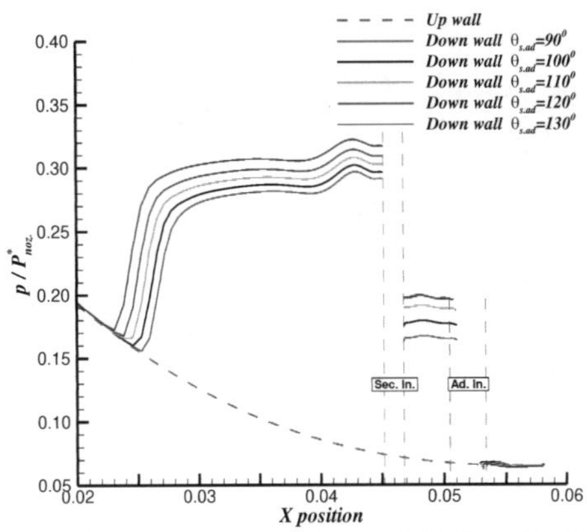

(d) Pressure distributions on nozzle wall (*NPR*=13.88, *SPR*=1.0)

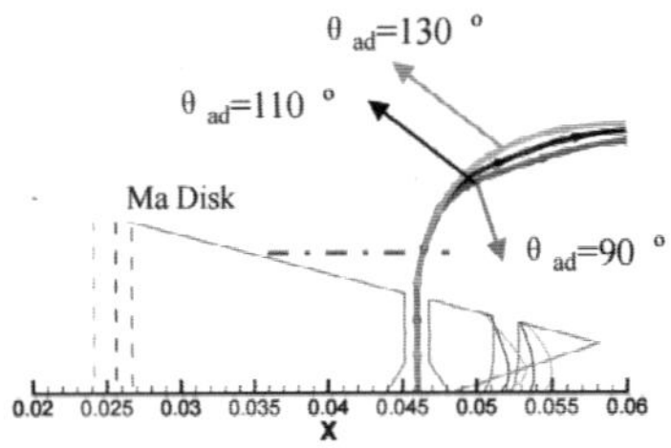

(e) The jet trajectory of secondary flow (*NPR*=13.88, *SPR*=1.0)

Figure 8 The flow characteristics of a SVC nozzle with bypass injection at different bypass injection angles

The pressure differences on nozzle upper and down walls and the vertical component of jets moment determine the side-force and vector angle. Figure 8 (d) shows the pressure distributions on nozzle internal walls at different bypass injection angles. With the increase of bypass injection angle, the influence of secondary flow on primary flow was enhanced, the strength of the induced shock wave increased and the induced shock wave moved forwards, which resulted in the separation point of boundary layer upstream of injection slot moved forwards too. Therefore, the high pressure zone enlarged. The pressure on the wall between secondary injection slot and bypass injection slot increased due to the interaction of the two jets until the θ_{ad} was 120 °. Additionally, the back pressure of bypass flow also increased with the increase of bypass injection angle, therefore the bypass flow mass had a decrease of about 3% -5%. On the other hand, the vertical component changed by （1- cos (θ_{ad}-90), due to the increase of bypass injection angle. These two aspects are negative to vector performance, but are minor factors.

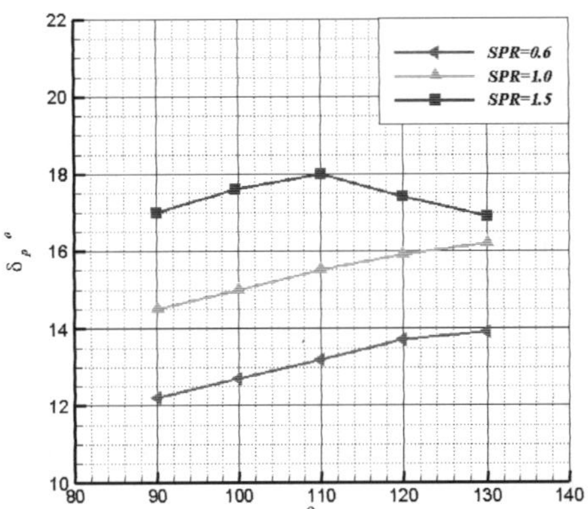

(a) The vector angle

IOP Conf. Series: Journal of Physics: Conf. Series **1215** (2019) 012038 doi:10.1088/1742-6596/1215/1/012038

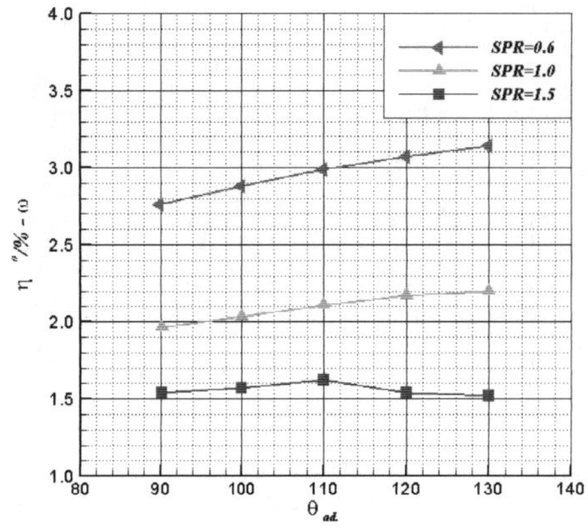

(b) The vector efficiency

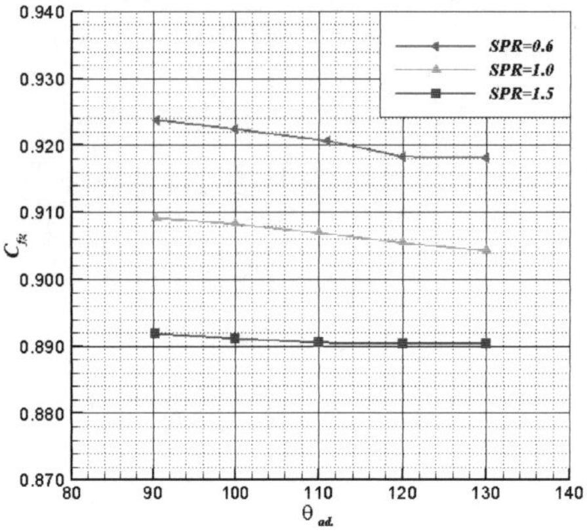

(c) The thrust coefficient

Figure 9 The vector performance of a SVC nozzle with bypass injection at different bypass injection angles

The vector angle of a SVC nozzle with bypass injection is shown in Figure 9 (a). At the condition of constant secondary flow mass and little changed bypass flow mass, the vector angle has a increase by 12% when bypass injection angle increased from 90 ° to 130 °. It is concluded that the bypass injection angle is a critical parameter. Under the geometric restrictions, the larger bypass injection is benefit to high vector and vector efficiency. But the design process, the bypass injection angle should be coupled with the secondary pressure ratio. Because at larger secondary pressure ratio (e.g. *SPR* of 1.5), the induced shock wave interacted with the opposite wall at larger bypass injection angle, which is bad for vector performance.

The secondary flow mass kept unchanged when nozzle pressure ratio and bypass injection angle altered. At *SPR* of 0.6, 1.0 and 1.5, the corrected secondary mass flow ratio is about 0.044、0.077、0.111. Therefore, the vector efficiency has the similar varying principle with vector angle versus the bypass injection angle at different *SPRs*, seen in Figure 9 (b). Noticeably, the larger *SPR* is, the larger the vector angle is, but the less vector efficiency. It is also seen that with the corrected secondary flow mass less than 10%, the vector angle of 16° - 20° and vector efficiency higher than 2.0 °/%-ω were achieved by modifying the *SPR* and bypass injection angle at large working ranges. It can satisfy the thrust vectoring demand of fighter planes and has less influence on the performance of an aero-engine.

The thrust coefficient varying with the bypass injection angle is seen in Figure 9 (c). With the increase of bypass injection angle, the induced shock wave angle increased and came closer to the nozzle lip. More primary flow went through the shock wave, thus larger flow losses generated, and the thrust coefficient decreased. At the *NPR* of 13.88 and *SPR* of 1.0, about a decrease by 0.5% was found when bypass injection varied from 90 ° to 130 °

3.3. The influence of bypass injection position on vector performance of a SVC nozzle with bypass injection

The interactions of bypass injection flow and secondary injection flow have influence on the shock waves and vortices, resulting in the re-distributions of pressure on nozzle walls and the change of vector performance. The strength of jets interactions are the critical factor affecting vector performance, and the distance between the two jets are direct affecting variable. In this section, the influence of bypass injection positions ($X_{j.ad.}$) on the performance of a SVC nozzle with bypass injection was investigated with X_j of 0.79, θ_{ad} of 90 and $A_{s.ad}$ of 0.08. $X_{j.ad.}$ has a varies from 0.69, 0.74, 0.84 to 0.89. The first two bypass injection positions locate upstream of secondary injection slot, and the other two downstream of secondary injection slot.

The flow fields of a SVC nozzle with bypass injection varying with bypass injection positions at *NPR* of 13.88 and *SPR* of 1.0 is shown in Figure 10. It is seen that the change on distance between two jets has no influence on the partten of vortices. There are still two pairs of counter-rotation vortices and a lip vortex. But the size of the vortices changed. With the jets distance change, two aspects of flow fields altered, the separation length of boundary layer and the pressure distributions on wall between jets. Comparing the configurations of $X_{j.ad}$ of 0.69 and $X_{j.ad}$ of 0.74, Figure 10 (a) and (b), it is seen when the two jets get closer, the jet penetration depth was enhanced, and the high pressure zone (boundary layer separation length) increased by about 9%. But the entrainment also increased, which made the pressure on walls between the two jet smaller, seen Figure 10 (e). And it is found the pressure drop has larger effect than the high pressure increase, therefore, the vector angle of $X_{j.ad}$ of 0.74 configuration is a little smaller. Comparing the configurations of $X_{j.ad}$ of 0.74 and $X_{j.ad}$ of 0.84, which are upstream of secondary injection slot and downstream of secondary injection slot respectively, Figure 10 (b) and (c), it is seen that boundary layer separation length of the later one is larger, it is mainly because the moment of secondary flow is larger than that of bypass injection. Meanwhile the pressure distribution between the two jets kept almost unchanged. Thus the vector angle of the configurations of $X_{j.ad}$ of 0.84 is larger. With the bypass injection position moving afterwards continuely, the jets interaction weakened, and the jet penetration and boundary layer separation length decreased. But the pressure on wall between the jet ascended, which finally made the vector angle increase.

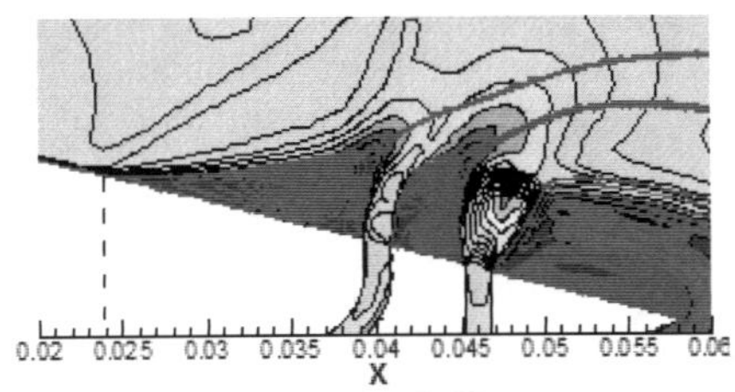

(a) $X_{j.ad}$=0.69

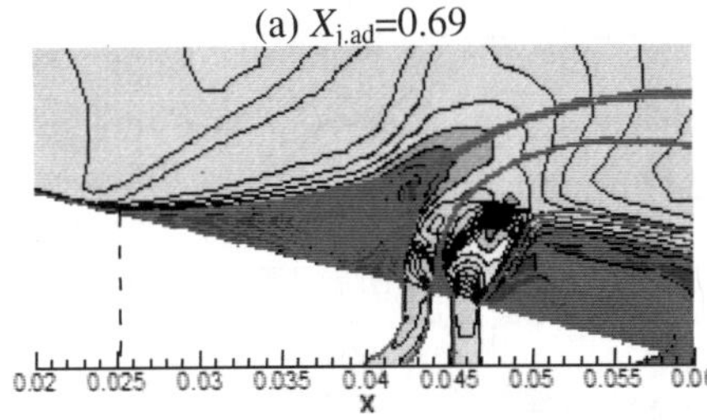

(b) $X_{j.ad}$=0.74

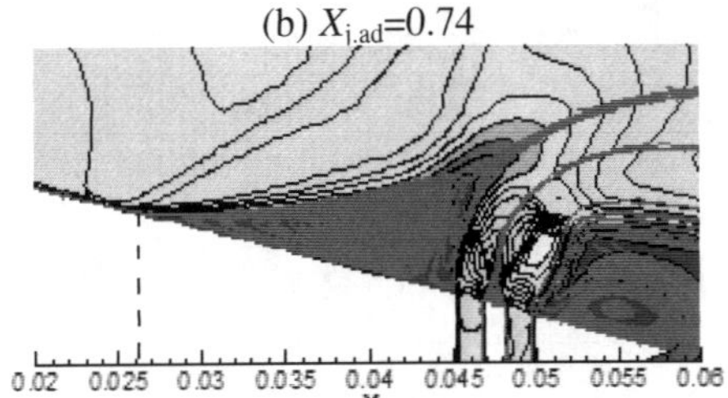

(c) $X_{j.ad}$=0.84

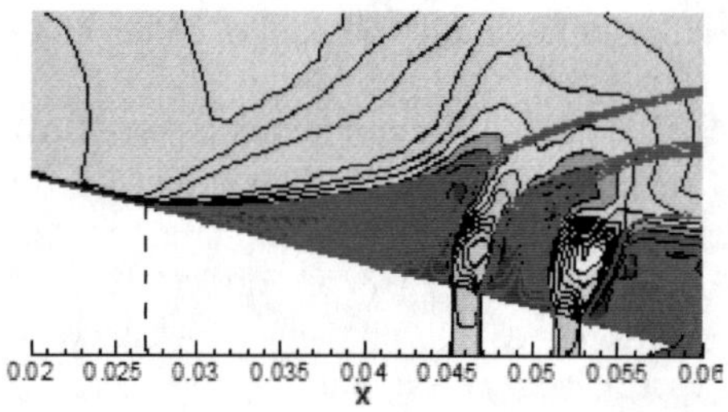

(d) $X_{\mathrm{j.ad}}$=0.89

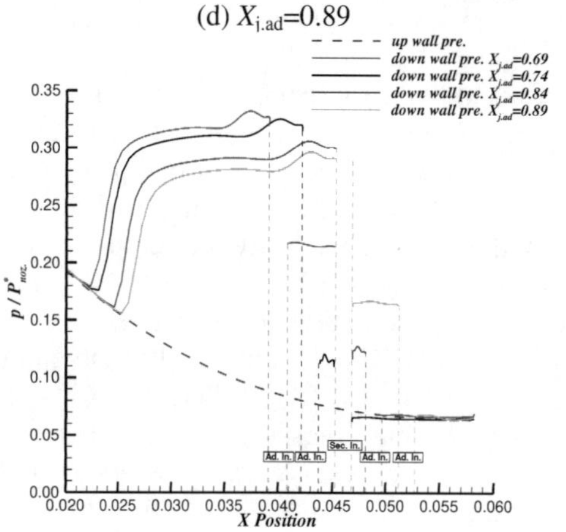

(e) Pressure distributions on nozzle wall (*NPR*=13.88, S*PR*=1.0)

Figure 10 The flow characteristics of a SVC nozzle with bypass injection at different bypass injection positions

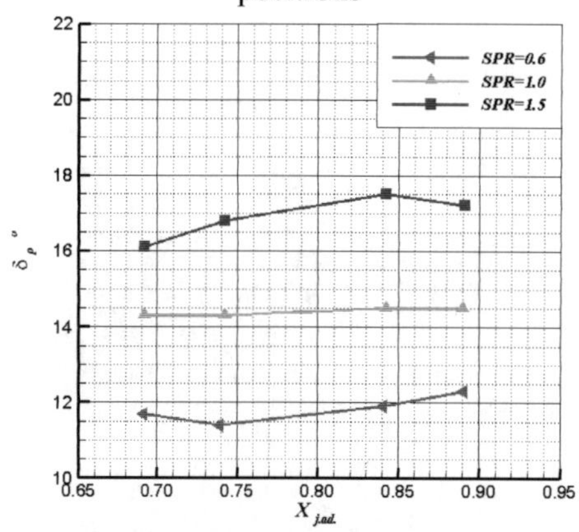

(a) The vector angle

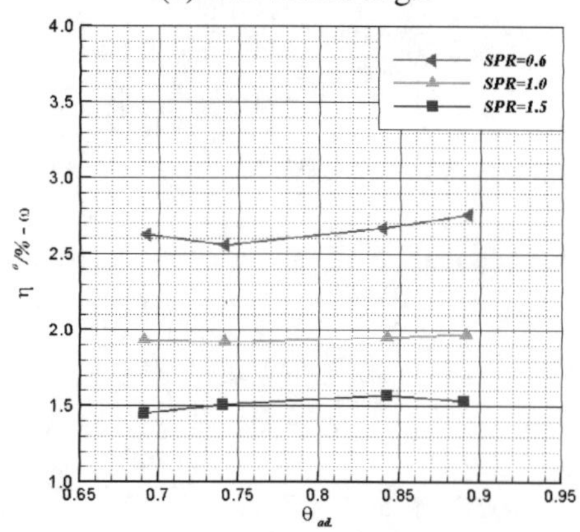

(b) The vector efficiency

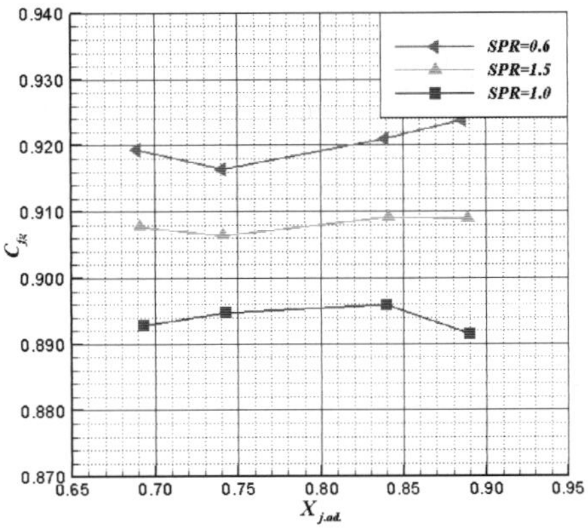

(c) The thrust coefficient

Figure 11 The vector performance of a SVC nozzle with bypass injection at different bypass injection positions

At other *SPR* conditions, the vector angle has the similar change principle varying with bypass injection positons, besides the *SPR* of 1.5 at configurations of $X_{\mathrm{j.ad}}$ of 0.89. In Figure 11 (a), It is seen that the bypass injection is not a critical parameter. At wide working conditions, the vector angle changed within 2° when the bpass injection position moved from of $X_{j.ad.}$ 0.74 to $X_{j.ad.}$ 0.89.

The vector efficiency and thrust coefficient are shown in Figure 11 (b) and (c). It is seen that the vector efficiency is larger than 1.5 °/%-ω and the thrust coefficient is larger than 0.89 at almost simulation working conditions.

4. Conslusions

In the paper, the SVC nozzle with bypass injection configuration was proposed to improve the vector performance of the fluidic thrust vectoring and to reduce the demand on secondary flow extracted from an aero-engine. The flow mechanism of a SVC nozzle with bypass injection and the influence of bypass injection angle and bypass injection position were investigated numerically. Conclusions are as follow:

1) With the assistant of a bypass injection flow, the vector efficiency are improved largely by about 60%, meanwhile the flow fields get more complex. The induced shock wave and vortices brought larger flow losses.

2) By adopting the bypass injection, the vector angle of 16° - 20° and vector efficiency higher than 2.0 °/%-ω were achieved with the corrected secondary flow mass less than 10%, which is desirable.

3) The bypass injection angle is a critical affecting parameter, which has an obvious influence on vector performance when the injection angle increased.

4) The bypass injection position is not a critical affecting parameter, and a vector angle change is within 2 °, when the position moved from $X_{j.ad.}$ 0.74 to $X_{j.ad.}$ 0.89.

References

[1] Arun K. Sehra, Woodrow Whitlow Jr., Propulsion and power for 21st century aviation, *Progress in aerospace sciences*, 2004, 40: 199-235.

[2] D. A. Terrier, F. K. Lu, Aerodynamically controlled expansion nozzle for STOVL aircraft, *AIAA 2003-185*, 2003.

[3] B, Gal-Or., The fundamentals concepts of vectored propulsion, *JPP*, 1990, 6(6):747-757.

[4] Daniel Ikaza, Christian Rausch, Thrust vectoring for Eurofighter – the first steps, *Air and Space Europe*, 2000, 2(1):92-95.

[5] Richard K Scharnhorst. Characteristics of future military aircraft propulsion system, *AIAA 2013-446*, 2013.

[6] Karen A. Deere, Summary of fluidic thrust vectoring research conducted at NASA Langley Research Center, *AIAA 2003-3800*, 2003.

[7] Karen A. Deere, Computatonal investigation of the aerodynamics effects on fluidic thrust vectoring, *AIAA-2000-3598*, 2000.
[8] Yagle P. J., Miller D. N., Ginn K. B.. Demonstration of fluidic throat skewing for thrust vectoring in structurally fixed nozzle, *ASME 2000-GT-0013.*
[9] MyungJun Song, SangHoon Park. Application of Backstep Coanda Flap for Supersonic Coflowing Fluidic Thrust-Vector Control. *AIAA Journal*, 2014 52(10): 2355 -235.
[10] WANG Zhan-xue, WANG Yu-nan, LI Zhi-jie, et al., "Experimental on fluidic thrust-vectoring nozzle based on shock control concept," *JPT*, 2011, 31(6):751-756.
[11] ZHANG, J. D., WANG, Z.X., Numerical research on two types of fluidic thrust vector, *Acta aerodyanmica sinica*, 2012, 30(2): 205-209.
[12] SHI J. W., WANG Z. X., ZHOU L., ZHANG X. B., Investigation on flow-field characteristics of shock vector controlling nozzle based on confined transverse injection, *Journal of Engineering for Gasturbine and Powe*r, 138（10）: 1 - 10.
[13] Chiarelli C., Johnsen R. K., Shieh C. F., Fluidic scale model multi-plane thrust vector control test results, AIAA 93-2433, 1993.
[14] Wing D. J., 1994, Static investigation of two fluidic thrust vectoring concepts on a two-dimensional convergent - divergent nozzle, *NASA TM-4574.*
[15] Giuliano V. J., Wing D. J., Static investigation of a fixed-aperture nozzle employing fluidic injection for multiaxis thrust vector control, *AIAA-97-3149*,1997.
[16] Anderson C. J., Giuliano V. J., Wing D. J., Investigation of hybrid fluidic/mechanical thrust vectoring for fixed-exit exhaust nozzles, *AIAA-97-3148*, 1997.
[17] Spaid F. W., Zukoshi E. E., A study of the interaction of gaseous jets from transverse slot with supersonic external flows, *AIAA Journal*, 1968, 6(2): 205-212.

IOP Conf. Series: Journal of Physics: Conf. Series **1215** (2019) 012039 doi:10.1088/1742-6596/1215/1/012039

RSM Performance Characteristics Improving Due to the Application of the Nanostructured Carbon-containing Coating "Superlattice"

R Tellez [1*], A Zhdanov[1], D Kononov[1], D Gavrilov[1], V Morozov[1] and L Belyaev[1]

[1] Department of the mechanical engineering, Vladimir State University, Gorky str. 87, Vladimir, Russia

rikardo.tezhez.b@mail.ru

Abstract. This article presents the results of the experimental researches on nanostructured carbon-containing PVD coatings based on titanium and chromium (Cr/C; CrN/CNx, Ti/C; TiN/CNx) deposited on a disk Ø 20 x 5 mm made of steel ShKh15 (ШХ15). Coatings were obtained on the UNICOAT 600SL installation (Russia). In the course of the research, such parameters as coatings adhesion to the substrate, their hardness and elastic modulus, as well as friction coefficients were determined by using the indentation, scratch testing and tribological testing methods, respectively, on the CSM equipment (Switzerland). The low values of the friction coefficient (Cr/C – 0,02, Ti/C – 0,07; CrN/CNx - 0,15; TiN/CNx – 0,13) suggest that they are promising to be used in the friction pairs of the drive actuating mechanism, and the high hardness values (from 3,000 to 3,700 units on the HV scale) are promising in highly loaded drives under extreme operating conditions.

1. Introduction

Nanostructured carbon-based superlattice coatings have been of particular interest due to their practical use in friction pairs of actuators and mobile supports [1].

The properties, structure and adhesion [2] of PVD coatings determine a number of factors: degree of ionization; flow velocity and density of the sprayed particles, temperature optimization for coating deposition; application of substrate sputtered cleaning, accelerating stresses, displacements, etc.

The presence per se of a solid chemical element (carbon C) in the coating composition cannot be a guarantee of wear resistance improvement. Different deposition modes, tooling configuration, pre-ion etching or alloy addition, and many other features determine the structure of the coatings themselves and the interface boundary structure "coating-substrate" and the coatings properties [3], in contrast to the coatings obtained by laser ablation [4-8 , 11, 12].

2. Main part

For comparative experiments, nanostructured carbon-containing coatings [13] (obtained on the UNICOAT 600SL system) based on titanium and chromium [14] *(Cr/C; CrN/CNx, Ti/C; TiN/CNx)* deposited on a disk (Ø 20 x 5 mm) made of steel ShKh15 (ШХ15) were used. Experimental researches have been carried out on the CSM equipment (Switzerland) using the indentation, scratch testing and tribological testing methods. The main results are given below.

2.1. Obtained coatings

During the superlattice coatings deposition on the UNICOAT 600 SL system, a scheme was used in which two magnetrons are located in the centre of the chamber. With this magnetrons arrangement, the products pass through the treatment zones opposite each magnetron, and an uniform multi-layer periodic structure is created. In the internal part of the chamber there are magnetrons with the atomized target materials Ti and C, (Cr and C) and a table on which the products are located. The table realize planetary movements during the coating deposition.
The engineering process of the superlattice coating deposition consisted of several stages:

2.1.1. Stage 1. Vacuum chamber pumping to the "base" pressure. The purpose of this operation is to obtain a "base" vacuum in the processing chamber. The requirements for the "base" vacuum are determined by two parameters: the pressure magnitude till the $P_{баз}$ is pumped and the amount of impurity gases and vapours leakage (flow rate) into the vacuum chamber Q_{H}.

2.1.2. Stage 2. Products sputtered cleaning. Products sputtered cleaning is a final cleaning of the products surface in vacuum under the action of surface bombardment with the working gas ions, which are carried out in order to ensure a high coating adhesion.
For the sputtered cleaning execution, the working gas (argon) is supplied into the chamber. When the gas pressure is Pp = 0,20...0,28 Pa, a glow discharge (the argon ions (Ar+) source) is emitted in the chamber by applying a negative voltage U_{CM} = 800 ... 1000 V to the equipment and by turning on the magnetrons at low discharge currents (up to 1...3 A on every one). This method of sputtered cleaning is called " Magnetron Sputtering Glow Discharge"

2.1.3. Stage 3. Metal precoating deposition. In this operation, a metal layer of titanium (or chromium) is applied, at the values of current I= 15...20 A, which ensures a high adhesion between the substrate and the reactive layer

2.1.4. Stage 4. Reactive coating deposition. The vacuum coatings that are formed on the surface of a substrate as a result of a heterogeneous reaction between the deposited metal atoms and the reactive gas molecules (nitrogen, methane, oxygen, etc.) are known as "reactive". In this operation, internal transient and external reactive layers of coatings are applied using a dual magnetron operating in a bipolar pulsed mode. Reactive layers are applied from two unbalanced magnetrons of the dual system, equipped with specific targets from (Ti, Cr and C), depending on the type of coating obtained.
When Cr/C (Ti/C) coatings are obtained, a current I=15...20 A is applied to the magnetrons, and a voltage u=40...80V to the product. Under these conditions, an intense targets bombardment and an effective transfer of metal and carbon atoms occurs to the work surface, forming a multilayer coating.
When CrN/CNx (TiN / CNx) coatings are obtained, a current I=15...20 A, is applied to the magnetrons, and a voltage u=40...80 V, this process occurs under a constant continuous feed of working gas (nitrogen) into the chamber. Under these conditions, an intense targets bombardment, the connection of atoms with a nitrogen gas and the transfer of the obtained nitride compounds to the work surface, forming a multilayer coating.
The cooling products operation is usually carried out at the end of the manufacturing process and is intended to reduce the temperature of the products heated during the coating deposition process prior to the discharge into the atmospheric air chamber. This reduces the influence of some negative factors that can lead to the obtention of defects in the applied coatings.

2.2. Determination of physico-mechanical characteristics
The indentation tests [9] were carried out on the nanoindentometer "CSM Micro scratch tester" using a Vickers diamond indenter. The tests were carried out according to the following method: 1) test specimen fixation on the device; 2) basic setting-up assignment of the indentation mode; 3) indentation execution; 4) photographing of the indentation marking; 5) analysis of the results.
The load aplication was realized under a certain level (50 mN), then the exposure was carried out for a limited period of time at this load level and unloading (Fig. 1). For each specimen, a series of tests was

performed to determine the following parameters: indenter hardness (Hit), Vickers hardness (HV), given elastic modulus (E*).

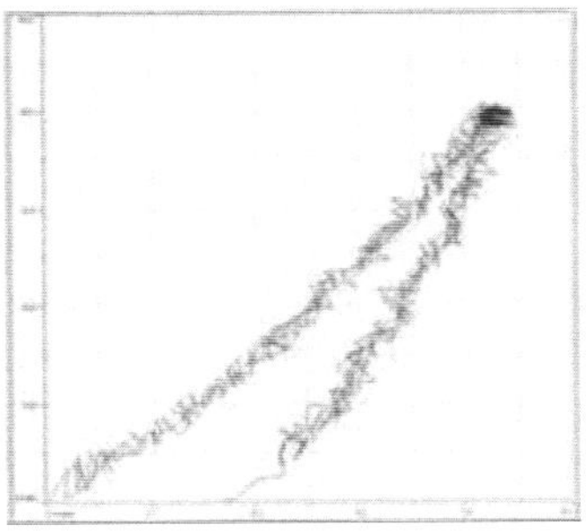

a) Cr/C

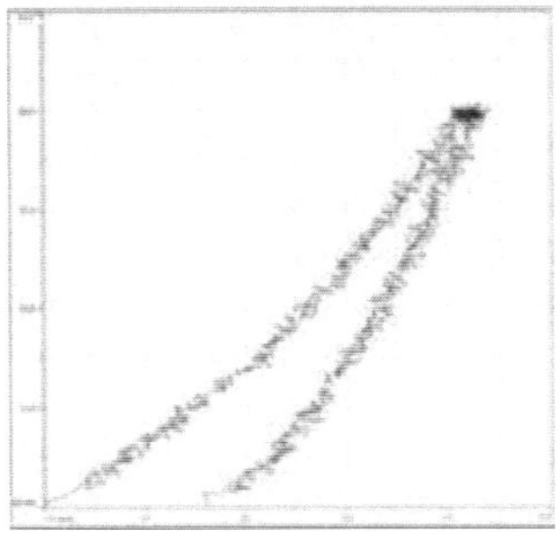

b) CrN/CNx

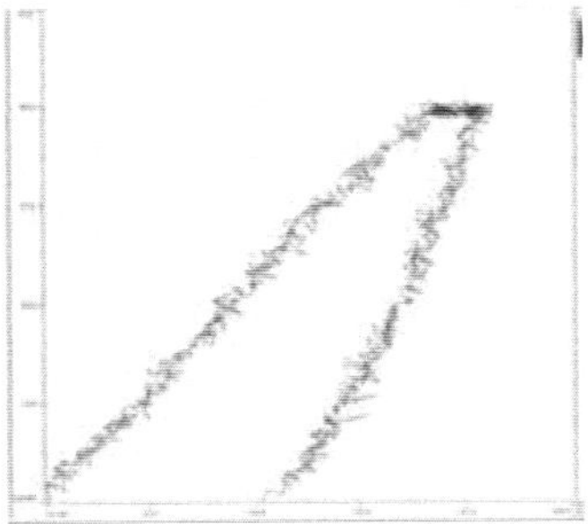

c) *Ti/C*

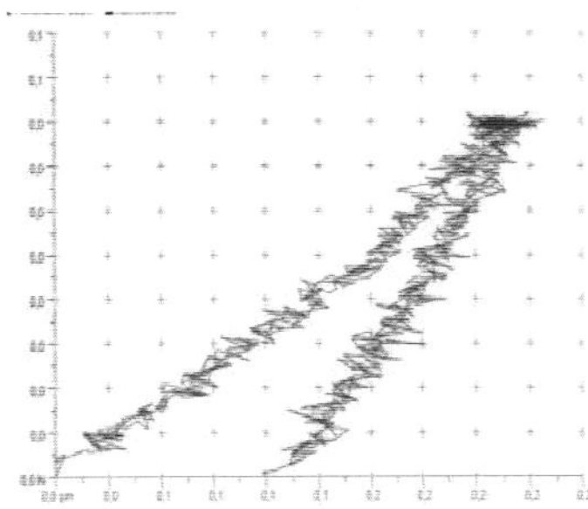

d) *TiN/CNx*

Figure. 1. Indentation curves under at a maximum load 50 mN

The software allowed to automatically process the results and build the "load-penetration depth" diagrams using the Oliver-Pharr method. The quantitative estimation of the examined coatings was carried out on the basis of these curves (Fig. 1) and indentation impress. Fig. 2 shows an example of an impress with a size of 3 x 3 μm; Table 1 shows the average values of the physico-mechanical characteristics for the examined specimens.

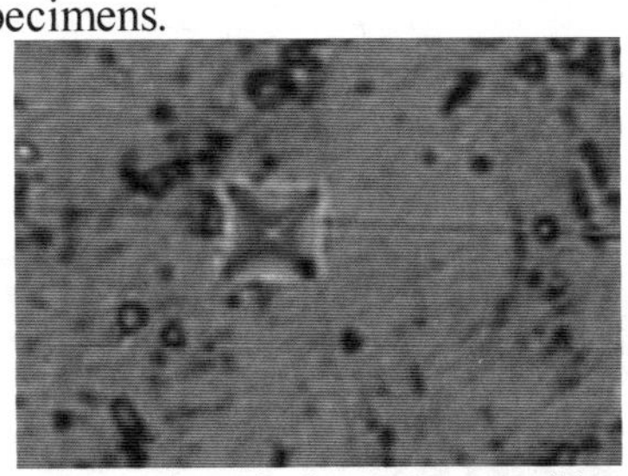

Figure. 2. Impress indentation marking (x50)

Table 1. Indentation measurement results

Coating	Physico-mechanical characteristics		
	Hit, GPa	E*, GPa	HV
TiN/CNx	39	250	3700
Ti/C	35	280	3450
Cr N/CNx	31	310	2970
Cr/C	39	230	3690

The scratch method testing was carried out on the nanoindentometer "CSM Micro scratch tester" using a Rockwell diamond indenter and linear scratching. According to the results, it is concluded the critical load at which the coating begins to split off from the substrate.
The tests were carried out according to the following method: 1) fixing the test sample on the device; 2) preliminary scanning of the surface and determination of its profile; 3) parameters assignment for scratch testing in the settings; 4) scratch testing execution; 5) panoramic image building of the scratch marking; 6) analysis of the results.
The software allowed in automatic mode to process the obtained diagrams "load penetration depth - acoustic emission signal". The scratch testing parameters were the same for all specimens: initial load 0,1 N, final load 15 N; load rate 8,94 N/min; scanning load 0,03 N; scanning load speed 2 mm/min; measurement length 5 mm; indenter radius 100 µm. Fig. 3 presents the results of scratch testing. The numerical values of the scratch tests are shown in Table. 2.

Table 2. Scratch testing and tribometrics results

Test №	Load at which the first shearing distortion is formed, H	Friction coefficient value
Cr/C	4,64	0,025
CrN/CNx	4,85	0,15
Ti/C	5,83	0,072
CrN/CNx	5,69	0,13

Tribological tests were carried out according to the following procedure: 1) fixing the test specimen on the swivel slide; 2) adjustment of the movable rod and fixing on it a rotating ball-indenter; 3) setting testing parameters (material, specimen speed, ball pressing force, ambient temperature, test time, etc.); 4) tests execution; 5) loads graphs obtaintion, abrasion depth and friction coefficient; 6) wear track research and analysis of the obtained results.
The tests of all three coatings were carried out with the following identical parameters: track diameter 11 mm, load on the specimen 10 N, sample speed 20 cm/s, travel 300 m, the counterbody material was steel ShKh15 (ШК15), the counterbody geometry was a ball with a diameter of 6 mm. Fig. 4 presents the friction coefficient changes graphs for the investigated coatings and the average values of the friction coefficient are shown in Table 2.

Scratch Markings

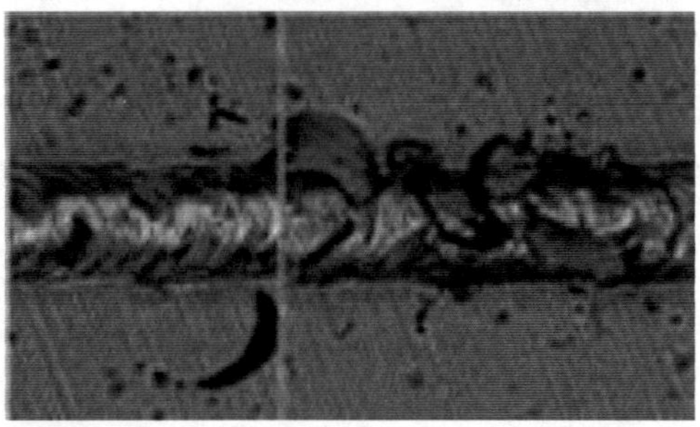

Scratch Testing Load graphs

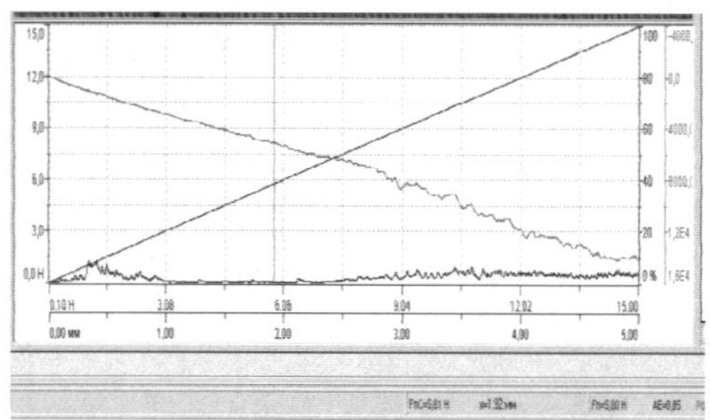

a) Cr/C

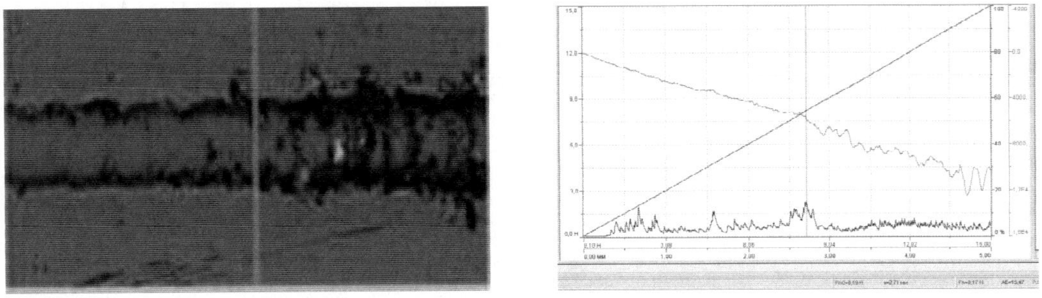

b) CrN/CNx

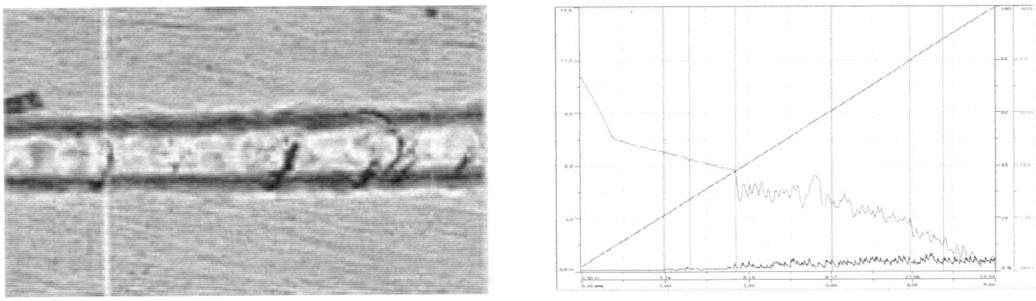

c) *Ti/C*

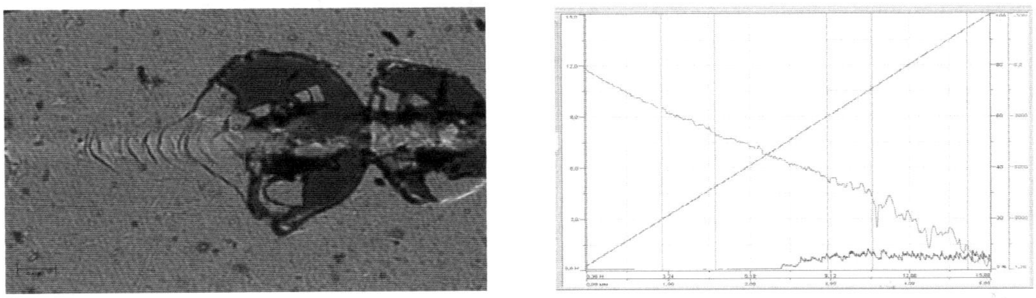

d) *TiN/CNx*

Fig. 3. Scratch testing results

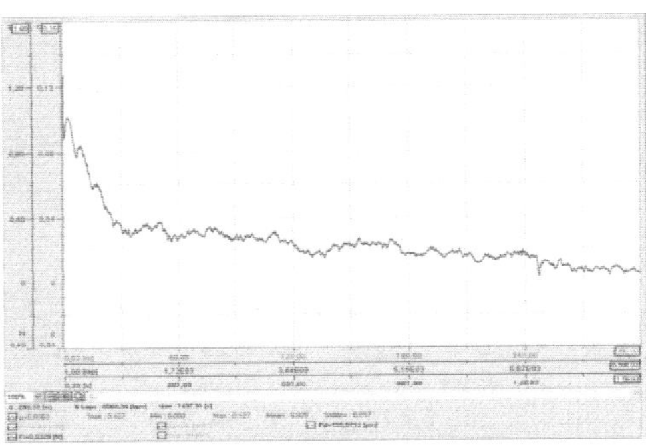

a) Cr/C

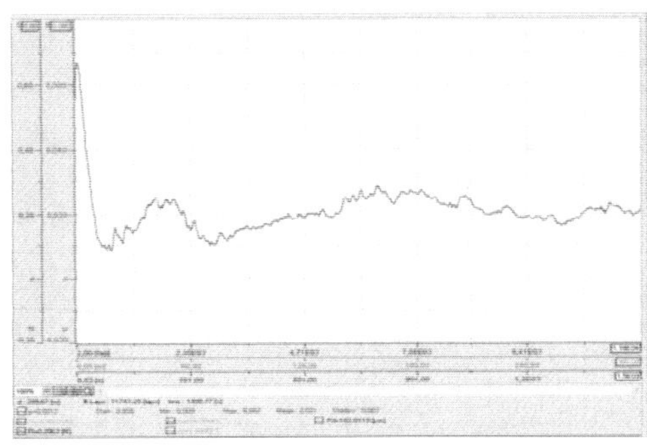

b) CrN/CNx

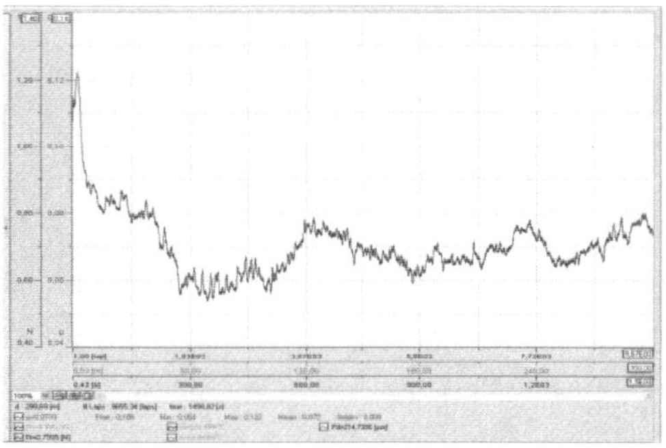

c) *Ti/C*

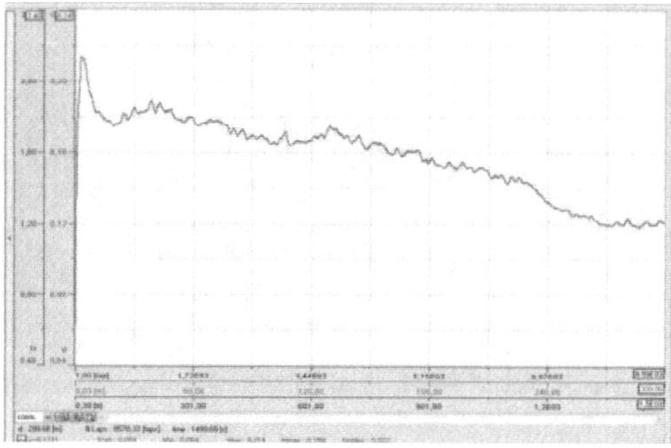

d) *TiN/CNx*

Fig. 4. Friction coefficient variation graphics obtained from the tribological tests

3. Conclusions

The analysis of the experimental results shows that the hardness of the investigated coatings ranges from 3,000 to 3,700 units on the HV scale, and it is not inferior than the known coatings, even than those which do not contain carbon (for example, the traditional coatings TiN 2900 HV, TiAlN 3200 HV).

The modulus of elasticity of the investigated coatings shows low values: 230 - 310 GPa (in comparison: TiN 361 GPa, TiAlN 650 GPa).

The realized tribological tests revealed low friction coefficient values for coatings based on C. For the coating (AlSiCr)C: the coefficient of friction Cr/C is 0,02, *Ti/C is 0,07*, CrN/CNx is 0,15, TiN/CNx is 0,13 (in comparison, the TiN and TiAlN coefficients are much larger).

Low friction coefficient values allow to suggest to use carbon-containing coatings in moving friction pairs and actuating mechanisms. High values of hardness characteristics allow to use these coatings in high-loaded drives and extreme operating conditions.

References

[1] Levashov E. A., Measurements uniformity ensuring of physico-mechanical and tribological properties of nanostructured surfaces // URL: http://www.galvanicworld.com/netcat_files/899/750/h_9e07867a656af04723527438d4d9c9c5.

[2] Lunyov V. M., Nemashkalo O. V., 2010, Adhesive characteristics of coatings and methods for their measurement, *ФІП ФИП PSE*. **8**, № 1. pp. 64-71.

[3] Morozov V. V., Sysoev E. P., 2011, Nanotechnology in ceramics: monograph, *Nanofilms, nanocoatings, nanomembranes, nanotubes, nanorods, nanofibres*, **2**, Vladimir State University Publishing house: Vladimir State University, 167 p.

[4] Li X. F. et al., Formation of 100-nm periodic structures on a titanium surface by exploiting the oxidation and third harmonic generation induced by femtosecond laser pulses, 2014, *Optics Express*, **22**, №23, pp 28086-28099.

[5] Prokhorov A.M. et al., 1990, Laser Heating of Metals, Bristol: A. Hilger, pp 239.

[6] Ashfold M.N., 2004, Pulsed laser ablation and deposition of thin films, *Chem. Soc. Rev.* **33**, pp 23-31.

[7] Khorkov K. S., Abramov D. V., et ali., 2017, Laser femtosecond nanostructuring of the tungsten surface, *Izvestiya Rossiiskoi Akademii Nauk* [Bulletin of the Russian Academy of Sciences], Seriya Fizicheskaya [Physics]. T. **81**, №12, pp. 1619-1623.

[8] Semaltianos N.G., 2010, Nanoparticles by Laser Ablation, *Critic. Rev. Solid State Mater. Sci.* **35**, pp 105-124.

[9] Kononov D. M., Zhdanov A. V. and Korolev A. N., 2011, Study of the properties of nanostructured PVD-coatings based on carbon, *Sovremennye problemy nauki i obrazovaniya* [Modern problems of science and education],№6, pp. 130-130.

[10] Khrushchov M. M., 1950, Modern theories of the antifriction properties of babbitt alloys, *Friction and wear in machines*,№6.

[11] Alifanov E. V., et ali., 2018, Features of the friction mechanism of elastomeric materials of various types. Theoretical and practical aspects (review), *Trudy VIAM [Proceedings VIAM]* №. 1 (61).

[12] Barberoglou M. et al., 2013, The influence of ultra-fast temporal energy regulation on the morphology of Si surfaces through femtosecond double pulse laser irradiation, *Applied Physics A.* **113**, №2, pp 273-283.

[13] Voznesenskaya A. A., Zhdanov A. V. and Morozov V. V., 2015, Research of the surface properties of carbon coatings, *Sovremennye problemy nauki i obrazovaniya* [Modern problems of science and education], №1-1, pp. 199-199.

[14] Tsyganov I. A., Pozdnyakova A. I., Richter E., Maytz M. F., 2007, Preparation of hemocompatible coatings on the basis of titanium using the method of plasma-immersion ion implantation and metal deposition, *Solid-state physics* **1**, 56 Bulletin of N. I. Lobachevsky State University of Nizhny Novgorod.

IOP Conf. Series: Journal of Physics: Conf. Series **1215** (2019) 012040 doi:10.1088/1742-6596/1215/1/012040

Study on Properties of Interlayer Short Fiber Reinforced Carbon Fiber/Epoxy Composite Laminates

Weiwei Liu [1], Weiwei Li [1], Chengyan Ma [1] and Shuai Wang [1]

[1] School of Mechanical Engineering, Northwestern Polytechnical University, Xi'an 710072, People's Republic of China

Abstract. In order to improve the interlaminar delamination resistance of the carbon-fiber/epoxy composite laminates, short fibers were applied between layers. The effects of length, density, type of short fiber on mode I fracture toughness(G_{IC}) and mode II fracture toughness (G_{IIC}) were investigated by three-factor and four-level orthogonal test combined with range analysis and scanning electron microscope observation. Finally, parameter optimization and test verification. The results show that the order of factors influencing G_{IC} and G_{IIC} within the selected parameters are: fiber type, fiber density and fiber length. When Kevlar fibers with a length of 11mm and a density of 20g·m^{-2} were laid between layers, the G_{IC} was 451.4J·m^{-2}, which was 82.2% higher than that of the specimens without short fibers in the layers. Kevlar fibers with a length of 11 mm and a density of 5g·m^{-2} were layered with a G_{IIC} of 2034.3 J·m^{-2}, which was increased by 240.5% compared with the specimens without short fibers in the interlayer.

1. Introduction

With the development in the field of aerospace, the material requirements are increasingly strict, and the traditional metal materials can not fully meet the performance requirements. Therefore, lightweight and high-strength composite materials are increasingly used in the aerospace industry, of which carbon fiber reinforced plastics (CFRP) laminates account for a large proportion[1]. At present, most of CFRP laminates are laid by prepreg, then cured and molded. It is easy to process and design. But the composite laminates have no reinforcing fibers in the thickness direction and are only bonded by the resin, and the interlaminar properties are much lower than the in-plane mechanical, which limits the application range of the laminates.

In view of the shortcomings of poor interlaminar performance, the interlaminar reinforcement technology is adopted to form the microstructure that hinders the interface fracture by introducing the reinforcing material and improve the fracture toughness. Interlaminar reinforcement technology can be divided into two categories: one is the use of fibers throughout the structure, such as "Z-pin"[2], stitching[3]. While the overall fixed technology to introduce defects, reduce damage tolerance, has complex process and narrow scope of application[4-6]. Another type of reinforcement technique involves the addition of a ductile material such as short fiber, particle[7], nanomaterial[8][9], or thermoplastic resin film between layers to form a second phase with a toughened structure between the layers. The technology is low cost, wide applicability. But, a good enhancement effect requires adding a larger amount of particles, films or nanomaterials, which leads to an increase in the quality of the composite laminates and the loss of in-plane mechanical properties. Compared with the addition of particles, nanomaterials and thermoplastic resin film between the layers or physical macroscopic Z-pin and stitching method, short fiber reinforced technology has many advantages such as good enhancement effect, less impact on the structure weight, simple process and low cost[10].

At present, a large number of studies have been made on the technology of interlaminar short fiber reinforcement and significant achievements have been achieved. In 1994, MS Sohn and XZ Hu

[11][12] proposed for the first time that Kevlar chopped fibers were added between carbon fiber/epoxy composite layers to reinforce the composite. During the lamination process, a small amount of kevlar fibers, short cut 5-7 mm in length, were uniformly and non-directional dispersed between the prepreg layers and were press-cured. So that chopped fibers in the resin-rich layer between the layers formed a very thin reinforced interlayer, interwoven with prepreg sheets. The experimental results showed that the G_{IC} and G_{IIC} of the test specimen both increased more than one time. Subsequently, Sohn et al [13] added Kevlar fibers between the laminates layers for impact testing. The results showed that under impact load, Kevlar fiber had poor reinforcing effect and fiber breakage was not obvious. Yasaee et al [14]compared the reinforcing effect of aramid short fiber, carbon short fiber, thermoplastic resin film and other materials. The results showed that aramid short fiber achieved excellent enhancement effect on both G_{IC} and G_{IIC} and resulted in the least structural weight gain. Choi et al [15] added aramid fiber to the surface of carbon fiber resin-based composites to improve the bonding properties. ST Cholake et al [16] modified structural epoxy resin by adding short milled carbon fiber (SMCF) particles. Subsequently, the SMCF modified epoxy resins were used to fabricate unidirectional carbon fibre reinforced laminates and tested under Mode I crack opening. The SMCF reinforcement showed 50% and 64% improvement in the laminates G_{IC} after adding 5% and10% SMCF by wt. Z Mo et al [17] made the interlayer-toughed carbon fiber/epoxy resin (CF/EP) composite laminates with short ramie fiber (SRF) through hot press laminated molding process. The test results indicated that the insert of SRF significantly improved the CF/EP composites G_{IC} and G_{IIC}. The maximum increase of G_{IC} was 34.24%, and the maximum increase of G_{IIC} was 69.54%.

Currently, there is a certain research foundation for interlaminar short fiber reinforced laminates. The research directions mainly focus on the test of short fiber reinforced specimen and the establishment of mechanical model. But there are few researches on the selection of short fiber reinforcement parameters. Therefore, in this paper, in view of the poor interlaminar properties of the CFRP laminates, short fiber reinforcement was carried out. The types of short fibers, short fiber length and short fiber density were studied. Taking G_{IC} and G_{IIC} as targets, the parameters were optimized and the influence of short fiber parameters on the composite laminates was analyzed, which lays the foundation for the extensive application of CFRP laminates.

2. Test preparation

2.1. Specimen preparation

The raw material is T800 carbon fiber reinforced epoxy composite unidirectional prepreg tape. The prepreg tapes are cut by hand, laminated and autoclaved (figure 1). Then the laminates are cut by special diamond grinding wheel to meet the standard requirements.

Figure 1. Solidified laminates (It has not been cut into a standard test specimen).

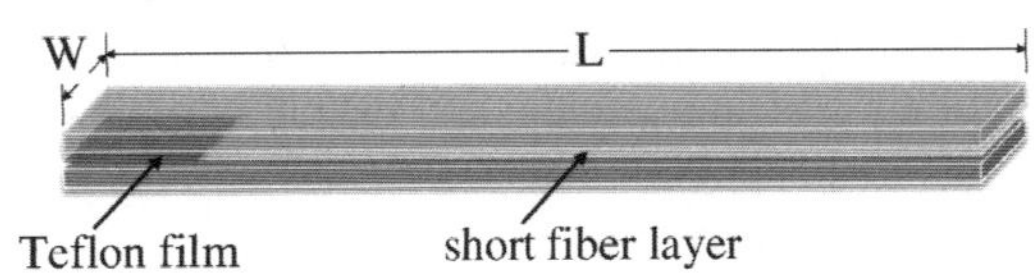

Figure 2. Laying diagram between layers.

As shown in figure 2, the double cantilever beam (DCB)test according to HB7402-1996 standard, the specimen length of 180mm, width 25mm, 24-ply of unidirectional (0° oriented) plies. A Teflon film of thickness 20μm and length 50mm was used as a crack starter, placed on the mid-plane of the laminates. The end notched flexure (ENF)test according to the standard of HB7403-1996, the test specimen is 140mm long and 25mm wide with 24-ply of unidirectional (0° oriented) plies. A 20μm thick, 40mm

IOP Conf. Series: Journal of Physics: Conf. Series **1215** (2019) 012040 doi:10.1088/1742-6596/1215/1/012040

long Teflon film was placed on the mid-plane of the laminates. Hand-cut short fibers were added evenly between the two types of specimens, the use of electronic scales to control the spread of staple fiber surface density.

2.2. Orthogonal design

As shown in table 1, taking three factors of fiber length, fiber density and fiber type of staple fiber, taking four levels for each factor, an orthogonal experiment table of three factors and four levels was established.

Table 1. Orthogonal test table.

Number	Fiber length (mm)	Fiber density ($g{\cdot}m^{-2}$)	Fiber type
1	3	5	Kevlar
2	3	10	carbon fiber
3	3	15	glass fiber
4	3	20	basalt fiber
5	7	5	carbon fiber
6	7	10	Kevlar
7	7	15	basalt fiber
8	7	20	glass fiber
9	11	5	glass fiber
10	11	10	basalt fiber
11	11	15	Kevlar
12	11	20	carbon fiber
13	15	5	basalt fiber
14	15	10	glass fiber
15	15	15	carbon fiber
16	15	20	Kevlar

3. Test and data processing

The DCB tests and the ENF tests all adopt the electronic universal testing machine CSS-88010 manufactured by Changchun Testing Machine Research Institute. The maximum tensile load is 10KN, and it is loaded by controlling the displacement.

3.1. DCB Test

The DCB test requires gluing hinges on both sides of the composite laminates, with the center of the hinge 25 mm from the top of the prefabricated crack, as shown in figure 3.

With the intersection of the two hinge centers and the composite plate as the starting point for effective cracking, the initial crack is 25 mm. To facilitate the observation of cracks, apply a white correction fluid on the side of the specimen and mark lines on the side. Prefabricated crack tip as line 0, line 1 from the 0th line for the 20mm, then draw a line per 10mm, a total of 7 lines. The test loading rate was set at 1 mm/min and the unloading rate was set at 20 mm/min. The experiment used a high-definition digital microscope to observe the crack propagation, using a magnification of 25 times.

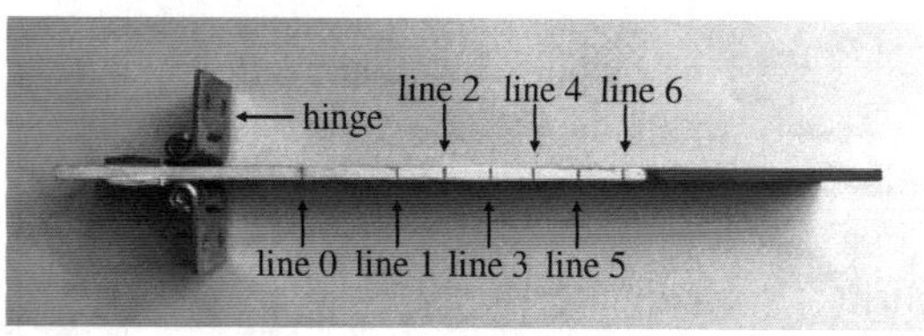

Figure 3. DCB specimen.

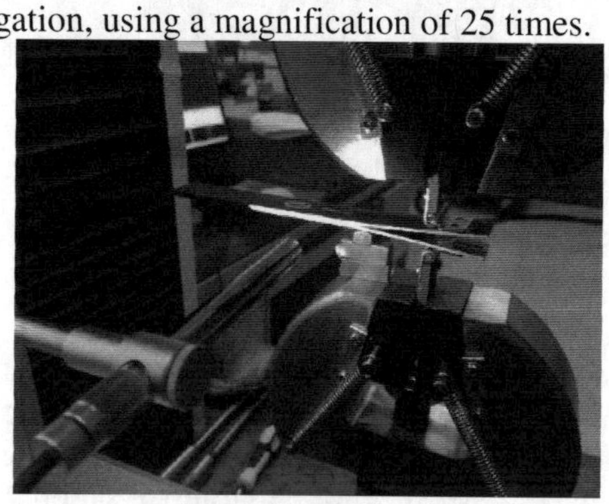

Figure 4. DCB test.

When the effective crack length is 45mm, that is, when the crack propagates to the line 1, unload immediately and record the load and displacement at this moment. When unload to load and displacement is 0, pause. And continue to re-load until the crack growth to the second line (an effective crack length of 55mm), immediately unload. Then when the load and displacement of 0, reload. This operation has been repeated until the effective crack reaches 95mm distance, the line 6, while stopping the test. The test process is shown in figure 4.

3.2. ENF Test

To facilitate the observation of crack propagation, similar to the DCB test, a white correction fluid was applied to the side of the test piece and a marking line was made at the tip of the prefabricated crack, 45 mm away from the edge of the test piece, as shown in figure 5.

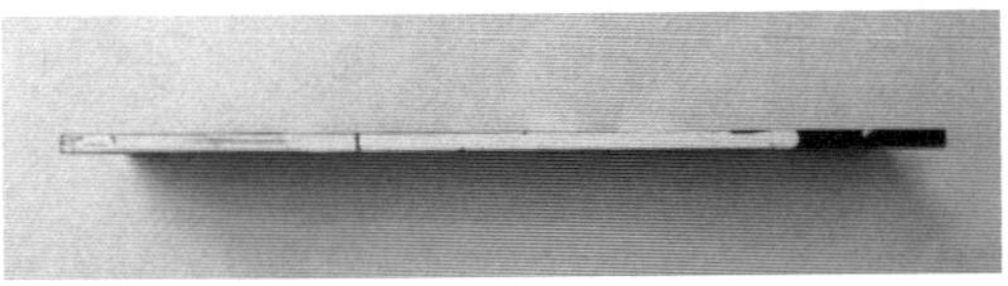

Figure 5. ENF specimen.

Figure 6. ENF test.

First put the specimen into the support. The center of the supporting cylinder on the side of the initial crack is 15 mm from the edge of the specimen. Adjust the span between the supports to 70mm and ensure that the loading head is in the middle of the support. Span between the support refers to the distance between two cylindrical centers of the support.

The effective crack distance is calculated from the center of the bearing cylinder, at which point the effective crack length is 25 mm. Load was started and crack propagation was observed by an electron microscope. When the crack spreads to the mark line, it is unloaded and stops when the load drops to zero. Adjust the span between the support to 100mm, to ensure that the loading head is in the middle of the support, and adjust the position of the test piece so that the center of the support cylinder is 20mm from the edge of the test piece. Due to the crack extension of 5mm, but also adjusted 5mm, then the effective crack length continues to 25mm. Load head reload, when the load drops, stop the test and unloaded, record the maximum load and displacement. The test process is shown in figure 6.

3.3. Experimental data processing

G_{IC} calculated according to equation (1).

$$G_{IC} = \frac{mP\delta}{2Wa} \times 10^3 \tag{1}$$

Where G_{IC} is mode I fracture toughness, and the unit is $J \cdot m^{-2}$. P is the load when reaching the mark line, the unit is N. δ is the displacement when reaching the marking line in mm. W is the width of the specimen in mm; a is the effective crack length in mm. And m is the fitting coefficient of the compliance curve, calculated using equation (2).

$$m = \frac{\sum_{i=1}^{k} \lg(a_i) \cdot \lg(C_i) - \frac{1}{k}(\sum_{i=1}^{k} \lg(a_i))(\sum_{i=1}^{k} \lg(C_i))}{\sum_{i=1}^{k} (\lg(a_i))^2 - \frac{1}{k}(\sum_{i=1}^{k} \lg(a_i))^2} \tag{2}$$

IOP Conf. Series: Journal of Physics: Conf. Series **1215** (2019) 012040 doi:10.1088/1742-6596/1215/1/012040

i represents the i-th line, k is the number of valid data points. a_i represents the effective crack length before the i-th loading, in millimeters. C_i is the compliance, and the calculation formula is δ_i / P_i, the unit is mm/N.
G_{IIC} calculated according to equation (3).

$$G_{IIC} = \frac{9P\delta a^2}{2W(2L^3 + 3a^3)} \times 10^3 \quad (3)$$

Where G_{IIC} is mode II fracture toughness in J·m^{-2}. P is the maximum load in N. δ is the displacement at maximum load in mm. a is the effective crack length in mm. W is the width of the test piece, the unit is mm. L is half the span of the support, the unit is mm.

4. Results Analysis and Discussion

4.1. Test results

The G_{IC} of the laminates without short fibers is 247.7 J·m^{-2}, and the G_{IIC} is 597.5 J·m^{-2}. The results of the laminates with interlaminar short fiber reinforcement are shown in table 2.
According to table 2, get G_{IC} range analysis table and G_{IIC} range analysis table , as shown in table 3 and table 4. The M value is the average of the corresponding experimental values at different levels. For the fiber length, the corresponding levels of M1 to M4 are 3mm, 7mm, 11mm and 15mm respectively, and for the fiber density, the corresponding levels of M1 to M4 are 5 g·m^{-2}, 10 g·m^{-2}, 15 g·m^{-2}, 20 g·m^{-2} respectively. The different levels of fiber types corresponding to M1 to M4 are Kevlar, carbon fiber, glass fiber and basalt fiber.
It can be seen from the R value in table 3 that the fiber type has the greatest influence on the G_{IC}, followed by the fiber density, and the fiber length has the least impact.
It can be seen from the R values in table 4 that the fiber type has the greatest influence on the G_{IIC}, followed by the fiber density, and the fiber length has the least effect.

Table 2. DCB test and ENF test results.

Number	G_{IC} (J·m^{-2})	Increase percentage (%)	G_{IIC} (J·m^{-2})	Increase percentage (%)
1	260.9	5.3	1389.5	132.6
2	242.5	-2.1	1339.3	124.2
3	222.5	-10.2	1131.7	89.4
4	204.7	-17.4	1426	138.7
5	203.8	-17.7	1595	166.9
6	321.6	29.8	1349.9	125.9
7	233.9	-5.6	1278.2	113.9
8	233	-5.9	1283.8	114.9
9	206.7	-16.6	1360.5	127.7
10	210.5	-15.0	1343.3	124.8
11	336.7	35.9	1615.9	170.4
12	272	9.8	1391.6	132.9
13	191.2	-22.8	1389.9	132.6
14	216.6	-12.6	1430.1	139.3
15	224	-9.6	1245.3	108.4
16	345.1	39.3	1555.5	160.3

IOP Conf. Series: Journal of Physics: Conf. Series **1215** (2019) 012040 doi:10.1088/1742-6596/1215/1/012040

Table 3. G_{IC} range analysis.

factor	Fiber length	Fiber density	Fiber type
M1	232.7	215.7	316.1
M2	248.1	247.8	235.6
M3	256.5	254.3	219.7
M4	244.2	263.7	210.1
R	23.8	48.0	106.0

Table 4. G_{IIC} range analysis.

factor	Fiber length	Fiber density	Fiber type
M1	1321.6	1433.7	1477.7
M2	1376.7	1365.7	1392.8
M3	1427.8	1317.8	1301.5
M4	1405.2	1414.2	1359.4
R	106.2	115.9	176.2

The results of range analysis show that the order of influence of various factors on G_{IC} and G_{IIC} is: fiber type, fiber density and fiber length.
According to the range analysis, the optimum level of G_{IC} in the selected parameters range: length 11mm, density 20 $g{\cdot}m^{-2}$, fiber type Kevlar fiber. The optimal combination of G_{IIC}: Length 11mm, density 5 $g{\cdot}m^{-2}$, fiber type Kevlar fiber.

4.2. The effect of short fiber length
Through the results in table 3 and table 4, the influence of fiber length on G_{IC} and G_{IIC} is obtained as shown in figure 7.

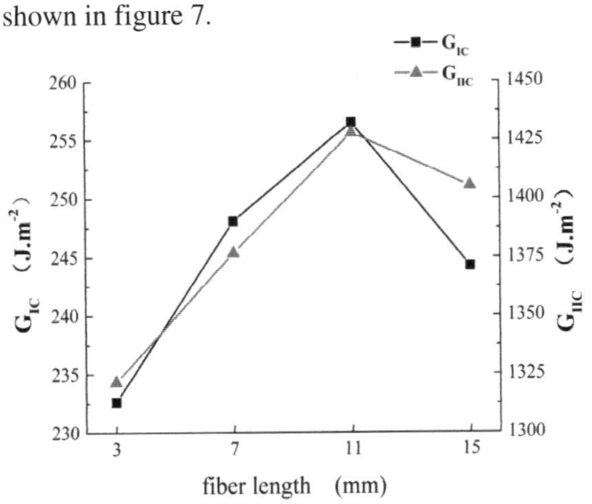

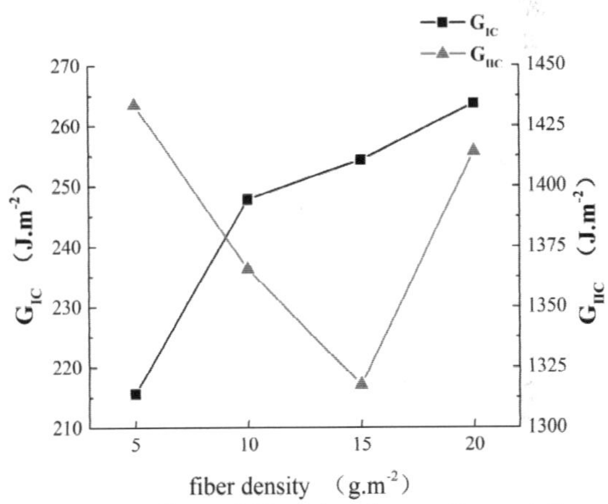

Figure 7. Effect of fiber length on G_{IC} and G_{IIC}. **Figure 8.** Effect of fiber density on G_{IC} and G_{IIC}.

G_{IC} first increases and then decreases with the increase of fiber length. When the fiber length is 11mm, the G_{IC} is most improved. When the fiber length is 3mm, the G_{IC} increase is the least. When the fiber length is 7mm and 15mm, the difference of G_{IC} value is not obvious. When the short fiber length is small, the short fibers can't enter into the resin matrix to form a bridging effect, and the force of fiber pulling out is relatively small, and its G_{IC} is relatively low. When the fiber length is too short, the fibers form an impurity for the composite laminate, reducing the G_{IC}. When the fiber length is longer, it can't completely enter the resin matrix, and can only be simply lapped in the in-plane direction. As the crack propagates, longer short fibers do not provide good bridging and assist in crack propagation. The G_{IC} of 15mm short fiber is not high.
With the increase of fiber length, G_{IIC} firstly increases and then decreases. G_{IIC} reaches its maximum at a fiber length of 11mm. For G_{IIC}, shear stress is applied to the composite laminates. When the fiber length is too short, the contact area between the resin and the short fiber is too small, and the G_{IIC} is not significantly improved.
When the length of the short fiber is too long, the resin can not all wrap the short fiber, the short fiber can not fully infiltrate, and the enhancement effect of the interlaminar performance is not significant.

4.3. The effect of short fiber density

Through the results in table 3 and table 4, the influence of fiber density on G_{IC} and G_{IIC} is obtained as shown in figure 8.

As the fiber density increases, the G_{IC} increases, but the increase rate decreases. For the DCB test, the load direction is perpendicular to the crack plane and the improvement of G_{IC} only depends on the bridge force. Short fibers between layers are disordered. When the short fibers and crack surface are not parallel, short fibers provide bridge force. As the short fiber density increases, the number of short fibers that are not parallel to the crack surface also increases, providing greater bridging forces, impeding the cracking, and thus increasing the G_{IC}.

With the increase of fiber density, G_{IIC} firstly decreases and then increases, and G_{IIC} is larger when fiber density is smaller and larger. At the fiber density of 5 $g \cdot m^{-2}$, G_{IIC} reaches the maximum. With the increase of the density of short fibers, the amount of interlaminar resin is certain, and the amount of short fibers is greater than the amount of resin able to be infiltrated, so interlaminar resin can't combine well with short fibers, resulting in a decrease of G_{IIC}. When the fiber density reaches 20 gm^{-2}, the short fibers form simple laps, because of the large number of interlaminar short fibers. Resin plays the role of glue between short fibers. This increases the friction at crack propagation and increases G_{IIC}.

4.4. The effect of short fiber type

Through the results in table 3 and table 4, the influence of fiber type on G_{IC} and G_{IIC} is obtained as shown in figure 9.

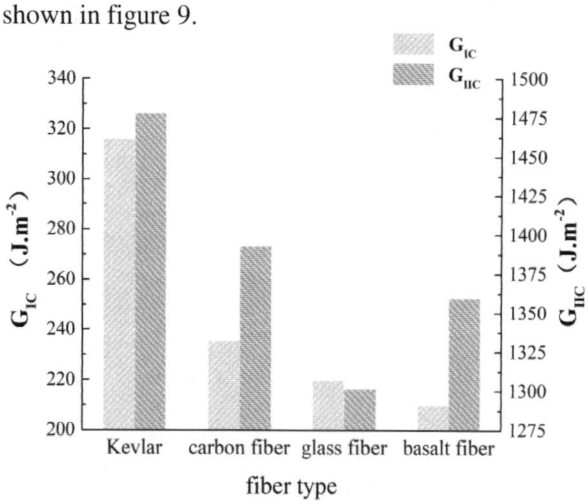

Figure 9. Effect of fiber type on G_{IC} and G_{IIC}.

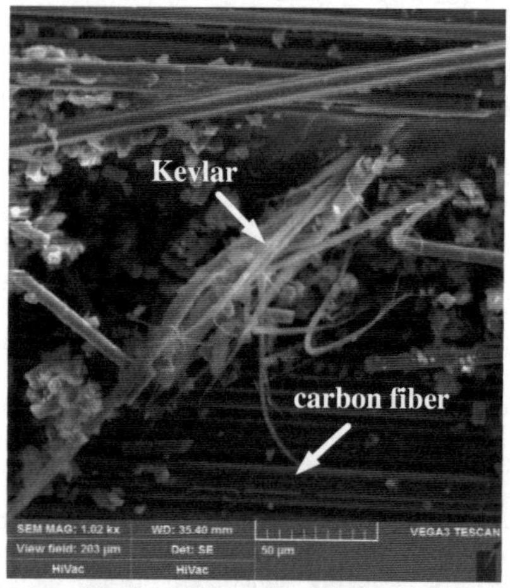

Figure 10. Scanning electron microscope image of Kevlar short fibers (high magnification).

It can be seen from the figure that Kevlar fiber has better enhancement effect on G_{IC} and G_{IIC}. Carbon fiber secondly, the reinforcing effect of glass fiber and basalt fiber is poor.

In order to further study the influence of fiber types, the fracture surface of DCB test specimens were observed by scanning electron microscopy (SEM). Kevlar images at different magnifications are shown in figures 10 and 11; carbon fibers are shown in figure 12; glass fibers are shown in figure 13 and basalt fibers are shown in figure 14.

Figure 10 shows the high-magnification interlaminar short Kevlar image. In the figure, the carbon fibers are neatly arranged, while one end of the Kevlar fiber is embedded in the inside of the substrate, and the end of the other end is severely bifurcated, which is a marked trace of fiber tensile fracture. This shows that the Kevlar fiber plays a good bridging role in the propagation of mode I cracks, and eventually tensile failure, thus increasing the interlaminar fracture toughness.

Figure 11. Scanning electron micrograph of Kevlar short fibers (low magnification).

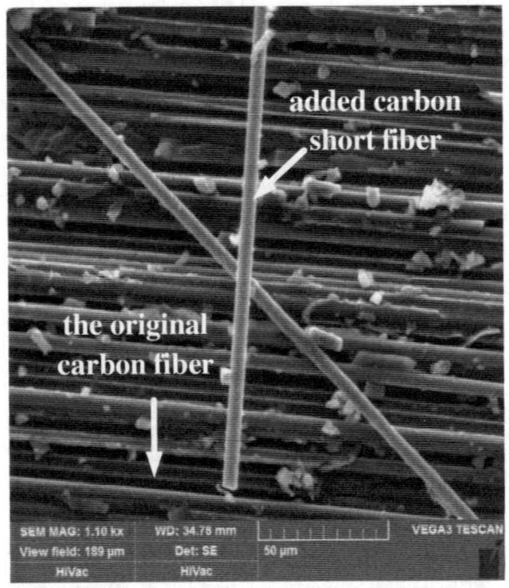

Figure 12. Carbon short fiber SEM image.

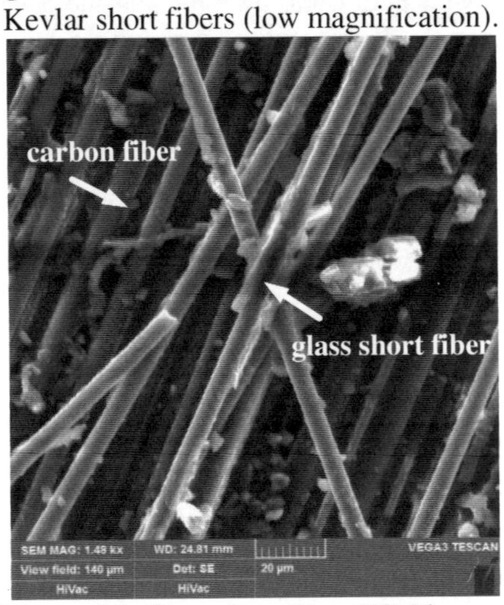

Figure 13. Glass short fiber SEM image.

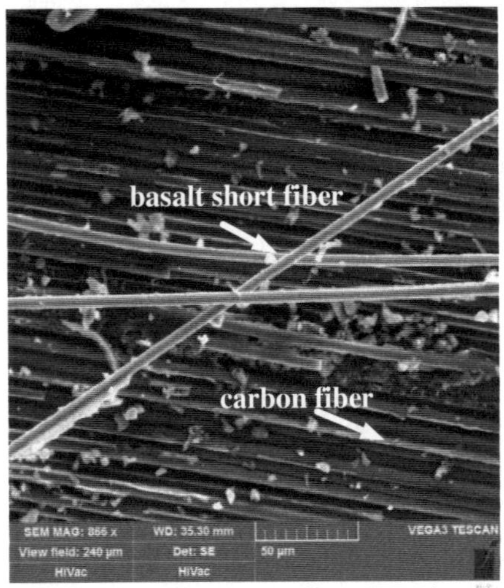

Figure 14. Basalt short fiber SEM image.

Figure 11 is a scanning electron microscope image of Kevlar short fiber at low magnification. It can be seen that there are many traces of fiber pull-out and tensile fracture. And in the layer form a random messy distribution of short fiber toughened structure. Kevlar fiber embeds in the resin matrix, resulting in anchoring. Not only the fracture energy of the resin, but also entanglement between the Kevlar fibers and the energy extracted by the Kevlar fibers from the resin increase the interlaminar fracture toughness.

In figure 12, figure 13 and figure 14, the fiber topography is almost the same. And a large amount of entanglement between the fibers does not occur. Kevlar fiber has better toughness and can produce greater bridging displacement. However, the interlaminar carbon fiber, glass fiber and basalt fiber are fractured under a small bridging displacement, and a large number of effective bridging fibers can not be observed during experiments. Therefore, Kevlar fibers can produce more energy consumption than the other three fibers and significantly improve the interlaminar fracture toughness.

The effects of adding three short fibers on interlaminar performance are different and related to their resin compatibility. Carbon fiber has good compatibility with epoxy resin. Glass fiber is an inorganic surface, which has poor affinity with resin. Basalt fiber contains a lot of calcium oxide components, but too much calcium oxide causes poor fiber toughness, smooth texture and small surface roughness, so it is poor in combination with resin. For the addition of short fiber layer technology, it is recommended to choose Kevlar, and try not to choose carbon fiber, glass fiber and basalt fiber.

4.5. Parameter optimization and verification

When the length of Kevlar fiber was 11mm and the density was 20g·m^{-2}, the G_{IC} increased most. The optimum level combination was tested. The average value of G_{IC} was 451.4 J·m^{-2}. The G_{IC} increased by 82.2% over the composite laminate without the addition of short fibers.

When the length of Kevlar fiber was 11mm and the density was 5 g·m^{-2}, the G_{IIC} was the most improved. The average G_{IIC} value was 2034.3 J·m^{-2}. Compared with the composite without the addition of short fibers, G_{IIC} increased by 240.5%.

5. Conclusions

The results show that the order of influence of various factors on the G_{IC} and the G_{IIC} is: fiber type, fiber density and fiber length.

The addition of Kevlar fibers enhances the G_{IC} most dramatically. And G_{IC} increases with increasing fiber density. With the increase of the length of short fibers, G_{IC} first increases and then decreases. For G_{IIC}, adding Kevlar fibers increases the most. With the increase of fiber density, G_{IIC} first decreases and then increases. As the fiber length increases, G_{IIC} first increases and then decreases.

Within the parameters of the study when Kevlar fibers with a length of 11mm and an areal density of 20 gm^{-2} were added into the interlaminar layers of composite laminates, the G_{IC} was increased by 82.2% compared with the non-reinforced composite laminates. When Kevlar fibers with a length of 11mm and an areal density of 5 g·m^{-2} were added into the interlaminates, the G_{IIC} increased by 240.5% compared with the non-reinforced composite laminates, which significantly improved the interlaminar properties of the composites laminates.

References

[1] Ma L, Zhang J, Yue G, et al 2015 Acta Materiae Compositae Sinica.**32**(2),317-22
[2] Li C, Wu Z, Meng Z and Li M 2014 Appl Compos Mater.**21**(4),615–31
[3] Turki Y, Habak M, Velasco R, et al 2014 Int J Mach Tool Manu.**87**,61-72
[4] Mouritz AP 2007 Compos Part A.**38**(12),2383-2397
[5] Steeves CA and Fleck NA 2006 Int J Solids Struct.**43**(10),3197-3212
[6] Chang P and Mouritz AP 2006 BN Cox. Compos Part A.**37**(10),1501-1513
[7] Borstnar G, Mavrogordato MN, Yang QD, et al 2016 Compos Sci Technol.**133**(14),89-96
[8] Ashrafi B, Guan J, Mirjalili V, et al 2011 Compos Sci Technol.**71**(13),1569-1578
[9] Fan Y, Gu Y, Deng H, et al 2015 Acta Materiae Compositae Sinica.**32**(2),332-340
[10] Sun Z, Jeyaraman J, Shi S, et al 2014 J Adhes Sci Technol.**28**(18),1835-1845
[11] Sohn MS and Hu XZ 1994 Scripta Metallurgica Materialia.**30**(11),1461-1472
[12] Sohn MS and Hu XZ 1994 Compos Sci Technol.**52**(3),439-448
[13] Sohn MS and Hu XZ 1995 Composites.**26**(12),849-858
[14] Yasaee M, Bond IP, Track RS, et al 2012 Compos Part A.**43**(1),121-128
[15] Choi I and Dai GL 2013 Compos Part A.**48**(1),1-8.
[16] Cholake ST, Moran G, Joe B, et al 2016 Constr Build Mater.**111**,399–407
[17] Mo Z, Hu C, Huo J, et al 2017 Acta Materiae Compositae Sinica.**34**(6),1237-1244

IOP Conf. Series: Journal of Physics: Conf. Series **1215** (2019) 012041 doi:10.1088/1742-6596/1215/1/012041

Research on the Application of Supportability Analysis Technology in Ejection Seat

Wensheng Wang[1,3], Liying Jin[2,a] , Jing Liu[3] and Xi Sun[3]

[1] Aviation Key Laboratory Science and Technology on Life-Support Technology, HubeiXiangyang, 441003, China

[2] School of Mechanical Engineering, Xi'an Jiaotong University,Shaanxi Xi'an 710049, China

[3] Aerospace Life-Support Industries, Ltd., HubeiXiangyang, 441003, China

ggnjinliying@126.com

Abstract. Ejection seat is a complex system that can be both one-time-used and continuous-used, in which the supportability analysis technology plays an essential role. Supportability analysis includes Reliability-Centered Maintenance Analysis (RCMA, for short), Corrective Maintenance Analysis, Level of Repair Analysis (LORA, for short), Operation and Maintenance Task Analysis (O&MTA, for short), Support Equipment Requirement Analysis and Spare Parts Requirement Analysis (shortened as "6A"). Firstly, in this paper, the concept of "6A" technology and their relationship are introduced. The general "6A" working process of a certain type of ejection seat is then studied. In the end, the "6A" work is carried out on a certain type of ejection seat according to the analytical steps and requirements to "6A".

1. Introduction

Ejection seat [1] often serves as relatively comfortable seat and provides necessary restraints for pilot during normal flying. While it can automatically initiate ejection rescue program after activating the ejection handle when an irreparable fault happens to the aircraft, and the pilot will be safely catapulted out of the aircraft, remaining survival and waiting for rescue based on the conditions provided by the seat. Therefore, supportability analysis [2] is an important part that affects the design of ejection seat, determines the support resources, and improves the operational efficiency of ejection seat.

The "6A" constitutes the main content of supportability analysis [3-5], and is the main basis that affecting equipment design and support resource [6]. The application of "6A" technology in the ejection seat can provide clear requirements of preventive maintenance and repair maintenance, determine resource requirements, so as to provide reliable data for the formulation of a reasonable and effective maintenance scheme for the ejection seat. With urgent research requirement and reliable technological support, the "6A" technology shall be effectively applied to the development of ejection seat [7, 8].

2. "6A" technology

The "6A" technology is the core of supportability analysis technology. RCMA keeps the inherent reliability and safety of equipment with minimal resource consumption, and determines the preventive maintenance requirements of equipment based on logical decision method [9]. On one hand, it provides necessary input for design modifications, on the other hand and more importantly, it helps to identify work items that require preventive maintenance and to provide input data for O&MTA.

Corrective Maintenance Analysis determines the type of repair maintenance and the possible working frequency, and proposes recommendations of maintenance level. LORA performs non-economic or economic analysis for a product that is expected to fail [10], so as to determine the level of viable repair or end of service. O&MTA guarantees the resource requirements for application and repair tasks (including preventive maintenance tasks, remedial repair tasks, and application tasks), provides support for determining related transport requirements, provides data for evaluating alternative support plan, and provides raw materials for the formulation of various support documents and schemes. Support Equipment Requirement Analysis determines the project and technical requirements of necessary support devices (including tools) during operation and maintenance of equipment, providing a basis for the formulation of supportive plan and equipment development. Spare Parts Requirement Analysis determines the required spare parts items and quantities for the equipment during application and repair, especially for initial application and maintenance.

As an important analytical work in supportability analysis, “6A” shows inseparable relationships between each item. First of all, the work of “6A” should be carried out on the basis of fault mode and impact analysis in obtaining information generated by Failure Modes and Effects Analysis (FMEA, for short), such as "failure mode", "failure cause", "failure effect", "severity", "design and improvement measures" and "application and compensation measures". RCMA is mainly used to determine the important function product of ejection seat, and the logical determination is applied to determine the type of preventive maintenance and the maintenance interval. Corrective Maintenance Analysis mainly determines the required repair type of ejection seat. LORA mainly decides the site for ejection seat repair. O&MTA, based on the requirement of preventive maintenance and repairable maintenance, determines the support resource required for ejection seat repair. Support Equipment Requirement Analysis mainly determines the type and quantity of support equipment at each maintenance level. Spare Parts Requirement Analysis mainly decides the items and quantity of spare parts required in application and maintenance. In short, the “6A” items are interdependent and mutually influential, they can optimize the support plan and determine required support resource by analyzing the supportive designs that affect the operation of equipment.

3. Application research of "6A" in a certain type of ejection seat

3.1. The Composition of a certain type of ejection seat

A certain type of ejection seat is mainly composed of seat structure, seat system and seat power device. Seat structure includes comfort component, umbrella box, chair basin, slide rail component and personal survival kit. Seat system includes ejection control system, separation system, shoulder belt control system, lower limb restraint device, arm limit system, belt control system, high-speed airflow protection system, electrical system, remote ignition system, oxygen supply ventilation and anti-load pipeline system, stability system and rescue umbrella system. The seat power unit consists of a twin-barrel ejector, a rocket motor, an attitude rocket and an umbrella mechanism.

3.2. The “6A” working flow of a certain type of ejection seat

The general “6A” working flow of a certain type of ejection seat is shown in Fig. 1, including collecting the functions, structures, and other data of the analyzed objects (ejecting seats, sub-systems or parts, etc.). Carrying out FMEA works at corresponding function or hardware levels, carrying out RCMA based on FMEA work and the characteristics of the certain type of ejection seat, determining the type of preventive maintenance, performing repair maintenance analysis based on FMEA results and carrying out, LORA, carrying out RCMA, repair maintenance analysis, LORA, and other product data (such as inspection, packaging, handling, storage, transportation, etc.) for O&MTA, determining the required support resource for application and repair tasks, determining the support resource requirements at each maintenance level based on the required support resources.

3.3. The “6A”characteristics of a certain type of ejection seat

Support analysis is highly targeted, which means the prioritized logistics support analysis should be carried out in accordance with the supportive characteristics of different types of equipment. The analysis results will be more reasonable, and will provide better service to the supportive design of equipment. A particular type of ejection seat is composed of electro-mechanical and initiating explosive device, and has high reliability requirement, many parts of which are not used in normal flying, but only used in emergency ejection. Therefore, great differences remains between the ejection seat and other machinery, frequently used equipment, forming unique characteristics of the seat in application and maintenance support.

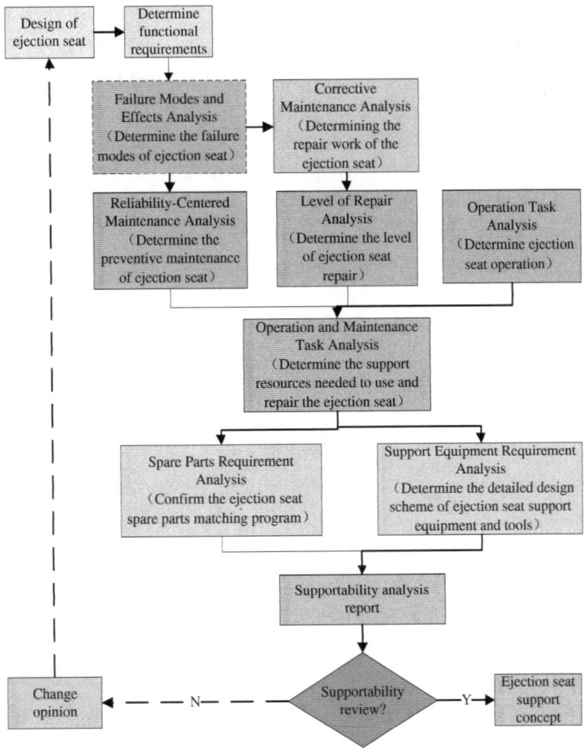

Figure 1. The general "6A" working flow of a certain type of ejection seat

According to the requirements of the host machine, a certain type of ejection seat adopts the second-level maintenance system, so its application and maintenance support shall be accordingly arranged. Under the current condition of planes with high work attendance and high operational efficiency, a rapid detection and maintenance requirements to aircraft equipment is put forward, thus the first-level maintenance of a certain type of ejection seat is mainly undertaken by field security personnel with field support devices and tools on seat inspection, maintenance and units replacement. The second level maintenance is carried out in aircraft back shop, in which all ejection seats of the certain type will be finished. In addition, the storage function of a certain type of ejection seat belongs to the maintenance system, while it turns to be out of the maintenance system in the emergency ejection phase. Some success or failure products of certain type of ejection seat cannot be repaired or tested, such as ejection, rocket engine, etc. These unrepairable parts are determined as life parts. All life parts shall be accurately determined and strictly replaced according to their service life. Therefore, according to the characteristics of a certain type of ejection seat and the actual application and maintenance situation, the "6A" analysis of a certain type of ejection seat focuses on the subsystems or components that are not used in normal flying. In this way, the design and analysis of "6A" can be carried out to highlight its purpose and effectiveness.

3.4. The "6A" working of a certain type of ejection seat

The failure mode, influence and harmfulness of a certain type of ejection seat have been analyzed before supportability analysis. In order to ensure the comprehensiveness of the supportability analysis, the failure mode, influence and hazard analysis of a certain ejection seat are started from its

IOP Conf. Series: Journal of Physics: Conf. Series **1215** (2019) 012041 doi:10.1088/1742-6596/1215/1/012041

components. Due to the complex structure of a certain type of ejection seat and the limited space of this paper, only the supportability analysis of ejection control system is talked about by considering the complicated structure of the ejection control system which contains not only electromechanical parts but also initiating explosive devices.

3.4.1. Reliability-Centered Maintenance Analysis. According to the FMEA analysis results of the ejection control system, the first layer of logic decision map is used to determine whether the impact of the functional failure of the ejection control system is of obvious security, mission, economy or of concealed safety, mission and economy. The cause of each functional failure is considered afterwards by using the second-level logical decision diagram in determining the type of preventive maintenance for the ejection control system. The RCMA record diagram of the ejection control system is obtained through logical decision diagram analysis, as shown in Table 1. The impact of the failure mode of the ejection control system is analyzed as a safety impact, so the "mission impact" and "economic impact" are not listed in Table 1. The types of preventive maintenance work for the ejection control system obtained from Table 1 are functional detection, application inspection, timed scrap, and timed repair.

3.4.2. Corrective Maintenance Analysis. According to the results of the FMEA analysis of the ejection control system, the corrective maintenance analysis was carried out. The result shows that only disassembly and replacement were performed for ejection control system maintenance, seeing Table 2.

3.4.3. Level of Repair Analysis. When determining the repair level of the ejection control system, non-economic analysis should be carried out at first in determining the reasonable repair level. If it cannot be determined, economic analysis will be performed to select a reasonable and feasible repair level or directly scrap. Based on the non-economic impact factors (safety, confidentiality, current maintenance options, mission success, handling, transportation and transportation, support equipment, manpower and personnel, facilities, packaging and storage) as specified in GJB2961, the perform non-economic analysis to the ejection control system was performed. The result shows that the replacement of central gripper, spring, program control box and seal ring were completed at base level and the replacement of detonation bomb is completed at grass-roots level due to the technical requirements of maintenance personnel, seeing Table 3.

3.4.4. Operation and Maintenance Task Analysis. When determining the tasks for an ejection control system, the work profiling in the expected working environment should be clarified. Since a certain type of ejection seat is disposable during emergency ejection, the operational tasks of the ejection control system mainly include storage, transportation, operation inspection and normal pre-flight operation. According to the results of RCMA, repairability maintenance analysis and operation analysis of the ejection control system, the application and maintenance analysis table of the ejection control system is obtained. Due to the limitation of space, only the O&MTA of central pull ring pull-out force inspection is given in Table 4.

3.4.5. Support Equipment Requirement Analysis. The requirements of support equipment is determined according to the O&MTA, while the preliminary plans for the support equipment and tools required are made based on the progress of the product development. The existing or common support equipment and tools of the force are encouraged to be used, only can the new equipment and tools be designed and manufactured when the existing equipment and tools fail to meet the requirements of the application and maintenance of new research equipment. According to the O&MTA of the ejection control system, the special support equipment and tools required for the ejection control system are determined, seeing Table 5.

IOP Conf. Series: Journal of Physics: Conf. Series **1215** (2019) 012041 doi:10.1088/1742-6596/1215/1/012041

Table 1 ejection control system RCMA record sheet.

Product name	Failure mode and reason		Logical decision graph (Y or N)											Failure consequences	Job type	Working time	Maintenance interval	Maintenance level
			Fault impact					Security impact										
			1	2	3	4	5	A	B	C	D	E	F					
	Start the ejection delay	Central pull ring pull-out force excess	N			Y		N	N	Y	N	N	Y	Covert security	Function detection	Regular work in use	10 years	D
		Firing mechanism pin pull force inadequate	N			Y		N	Y	N	N	N	Y	Covert security	Use check	Regular work in use	10 years	D
Ejection control system	Unable to start ejection	Detonation bomb not working	N			Y		N	N	N	N	Y	Y	Covert security	Timed scrap	Regular work in use	2 years	D
		Abnormal connected switch on and off of program control box	N			Y		N	Y	N	N	N	Y	Covert security	Use check	Regular work in use	10 years	D
	Initiating ejection function drops significantly	Leakage of the pipeline components	N			Y		N	Y	Y	N	N	Y	Covert security	Use check Timing disassembly and repair	Regular work in use	10 years	D

IOP Conf. Series: Journal of Physics: Conf. Series **1215** (2019) 012041 doi:10.1088/1742-6596/1215/1/012041

Table 2 Corrective Maintenance Analysis.

Product name	Failure mode and reason		Maintenance work			Technical basis
			Numbering	Work type code	Work explanation	
Ejection control system	Start the ejection delay	Central pull ring pull-out force excess	1	G	Disassembly and replace the central pull ring	Design requirements
	Unable to start ejection	Firing mechanism pin pull force inadequate	1	G	Disassembly and replace the spring	Design requirements
		Detonation bomb not working	1	G	Disassembly and replace detonation bomb	Design requirements
		Abnormal connected switch on and off of program control box	1	G	Disassembly and replace program control box	Design requirements
	Initiating ejection function drops significantly	Leakage of the pipeline components	1	G	Disassembly and replace the seal ring	Design requirements

Note: G - disassembly and replace.

Table 3 ejection control system LORA.

Serial number	Failure mode and reason		Maintenance level (O, D)			Remarks
			Disassembly and replace	repair	scrapped	
1	Start the ejection delay	Central pull ring pull-out force excess	D		D	
2	Unable to start ejection	Firing mechanism pin pull force inadequate	D		D	Subject to technical requirements of maintenance personnel
		Detonation bomb not working	O		O	
		Abnormal connected switch on and off of program control box	D		D	
3	Initiating ejection function drops significantly	Leakage of the pipeline components	D		D	

Note: O – grass-roots level: D – base level.

IOP Conf. Series: Journal of Physics: Conf. Series **1215** (2019) 012041 doi:10.1088/1742-6596/1215/1/012041

Table 4 O&MTA of central pull ring pull-out force inspection.

Operation sequence number	Operation step description	Spare parts, Consumables, fuel, Oil material and Pyrotechnics			Ground support equipment and tools		Precautions
		name/ model	quantity	attributes	name/ model	quantity	
1	Disconnect the connector plug under the program control box	Cotter pin	1	spare parts	Central blasting mechanism insurance pin, needle-nosed pliers, spring balance, detonation craft bomb	One each	When using the spring balance to pull the central pull ring for measure force, two people should cooperate with each other. One person pulls the central pull ring and one person should observes the pull pin. so as to ensure that when the central pull ring pulls the pin, it shall immediately stop pulling the spring scale.
2	Disassemble the detonation bomb according to the disassembly and decomposition steps of the firing mechanism in the "Disassembly and Disassembly Procedure"						
3	Replace the detonation bomb with the detonation craft bomb, and complete the installation of the firing mechanism and the craft bomb according to the assembly and installation steps of the firing mechanism and the detonation bomb in the "Assembly, Installation Procedure"						
4	Pull-out the ejection handle insurance pin						
5	Pull the central pull ring along the direction of the central pull ring with a spring balance to check whether the tension meets the requirements of "135N~225N"						

3.4.6. Spare Parts Requirement Analysis

Spare parts are required for the preventive and repair maintenance of the catapult control system. The corresponding maintenance tasks cannot be completed if the required spare parts are not sufficiently provided. Therefore, in order to ensure the in-time maintenance of the ejection control system, required spare parts should be prepared according to the result of the application and maintenance analysis seeing Table 6.

IOP Conf. Series: Journal of Physics: Conf. Series **1215** (2019) 012041 doi:10.1088/1742-6596/1215/1/012041

Table 5 the special support equipment and tools required for the ejection control system.

Serial number	Name	Quantity	Remarks
1	Electrical system function detector of ejection seat	1	Special support equipment
2	Ejection seat airtight checker	1	
3	Central blasting mechanism insurance pin	1	Special support tool
4	detonation craft bomb	1	
5	firing pin prominent amount template	1	
6	Upper spring top	1	
7	The primer seat	1	
8	Rocket motor craft bomb	1	
9	plug	2	
10	Airtight inspection joint	1	
11	Steel needle	1	

Table 6 Spare Parts Requirement Analysis for the ejection control system.

Serial number	Name	Quantity	Remarks
1	Cotter pin	1	
2	Insurance wire	0.5 m	
3	Seal ring	21	
4	Spring	1	
5	Primer	2	
6	Cut copper wire	1	
7	Gasket	1	
8	Cut pin	1	
9	7012 extreme low temperature grease		
10	734 epoxy adhesive		
11	Thread locking sealant 243		

4. Conclusions

The “6A” work is an important method in synchronizing the development of support resources during the development of a certain type of ejection seat, and is capable of regulating support resource during equipment deployment, which plays an irreplaceable role. Through the 6A work carried out to the certain type of ejection seat, the preventive and repairing maintenance of a certain type of ejection seat was clarified, and the requirements for the equipment, tools, spare parts and other support resource for the application and maintenance were determined. The determined support resource also plays their function during the test process, thus further providing a guarantee for the establishment of a cost-effective guarantee system for the host machine or for the troops. In view of the importance of the supportive analysis technology in a certain type of ejection seat, it is recommended that: ① supportability analysis should be carried out as early as possible, and should be hold throughout the development of the certain type of ejection seat; ② FMEA is an important foundation for 6A work, so it is necessary to clarify the information connection between FMEA and 6A before conducting the supportability analysis; ③ Due to the complex structure of a certain type of ejection seat, emphasis should be clarified during supportability analysis; ④ The process of supportability analysis is cumbersome and complicated. Manual analysis is time-consuming and laborious and has poor results, so computer-aided technology can be considered for analysis.

5. References

[1] Jian Bu, The Ejection Seat System Analyze And Optimization [D], Jilin university, (2017).
[2] Shuyue Yin, Kun Shang, Discussion on Environmental Test of Electronic Equipment Working off-aircraft with Ejection Seat [J], Electronic Product Reliability & Environmental Testing, (2018), 36 (1) : 18-20.
[3] Yanliang Li, Feng Qin, Reliability - centered maintenance analysis software development and application [J], Equipment management and maintenance. (2017), 11: 13-14.
[4] Jiying Li, Sihua Zhu, Zhihong Ji, Yuan Sun, Shipborne aviation equipment repair level analysis [J], Journal of Naval Aeronautical and Astronautical University, (2017), 32 (4): 383-388.
[5] Jianzhong Zhao, Wen Ye. Application and maintenance work analysis in determining resource requirements for missile equipment support [J], Precision manufacturing and automation, (2014), (2), 2-5.
[6] GJB. 3872, General requirements for comprehensive equipment support [S], (1999).
[7] Xiangjun Duan, Xinjun Yang, Failure Analysis and Study on the Fracture of the Connecting Lugs on the Ejection Seat of Ejection Seat [J], Aviation Precision Manufacturing Technology, (2017), 53 (5): 42-47.
[8] Yan Zhang, Tao Li, Xiangjun Duan, Failure Analysis and Research on Support Fracture of Detonation Transmission Mechanism of Certain Rocket Ejection Seat [D], Aviation Precision Manufacturing Technology, (2018), 54 (1): 35-39.
[9] Lin Ma, Analysis and evaluation of supportive design [M], Beijing: National Defense Industry Press, (2012).
[10] GJB. 2961, Level of Repair Analysis [S], (1997).

Acknowledgments

We would like to thank the anonymous reviewers whose thoughtful comments improved the quality of this paper. This work was jointly supported by the Aerospace Life-Support Industries, Ltd. and the Aviation Key Laboratory Science and Technology on Life-Support Technology.

IOP Conf. Series: Journal of Physics: Conf. Series **1215** (2019) 012042 doi:10.1088/1742-6596/1215/1/012042

The Effect of Sealing Ring Gap on Hydrogen Peroxide Electric Pump

He Yuanbo[1], Pan Hongliang[1], Liu Yang[1,3], Qin Fei[1] and He Bin[2]

[1]Northwestern Polytechnical University, Science and Technology on Combustion, Internal Flow and Thermo-Structure Laboratory, Xian, Shaanxi,China
[2]AECC Xian engine Control Technology Co., Ltd., Xian, Shaanxi, China
[3] grants: This work was financially supported by the Fundamental Research Funds for the Central Universities of china (3102016ZB046)

Abstract. Chained decomposition will take place in terms of overheat of high concentration hydrogen peroxide, which will further cause explosion, in the electric pump pressurized liquid propellant rocket engine, which takes hydrogen peroxide of high concentration as oxidizing agent, the design of oxidant pump should consider safety factor for temperature rising of hydrogen peroxide and the influence of pump efficiency on quality control of motor system simultaneously. It's very important to know the effect of sealing ring gap at the bottom of impeller on temperature rising of bearing chamber and pump efficiency. So, this paper firstly establishes mathematical modal about the effect of sealing ring gap on temperature rising of bearing chamber and pump efficiency loss, to gain influencing curve, and then it analyzes parameters of cooling flow and mechanism of temperature rising, and give reasonable gap values, which take both safety and efficiency into account. Finally, 3D simulation calculation is carried out through pumplinx to verify the accuracy of this influence model.

1. Introduction

In May 2017, "Electron" rocket, developed by American rocket laboratory, was launched in New Zealand [1]. The power of this rocket is "Rutherford" liquid oxygen/kerosene engine, in which pump is driven by motor to realize propellant pressurized transportation; in this way, the cost per launch can be reduced to 5 million dollars[2]. So, it has opened up a new development idea for low-cost rocket. And in china, Liu Yang and others [3-6] have carried out some related analysis and research. On the other hand, high concentration hydrogen peroxide is an ideal low-lost and green propellant[7], which was ever applied in "AR" series[8] and "RD-150" serial engines[9]. Besides, experts such as Lin Ge[10], Wang Guangwei[11], and Park[12] also carried out plenty of researches and tests about hydrogen peroxide thrust chamber and turbine pump. Because of its various merits like non-toxic, high density, easy to store at room temperature, and self-inflammable, hydrogen peroxide is very suitable to be served as modern rocket engine oxidizer [13-14], which may further decrease development and operation Difficulty of rocket. However, hydrogen peroxide is unstable, it will decompose in case of overheat and lead to explosion (limiting temperature is 408K in case of 90% concentration) [15-16]. Therefore, for electric pump pressurized liquid propellant rocket engine, which takes battery as energy, and considers hydrogen peroxide as oxidizer, both safety requirement of hydrogen peroxide, and high efficiency demand of electric pump system should be taken into account simultaneously during pump designing.

In hydrogen peroxide pump, lots of friction work for bearing will be converted into heat when running at great speed; and this heat needs to be brought to outside pump through cooling section consisted by hydrogen peroxide within pump, to realize pump heat flow balance. Sealing ring gap, set at the bottom of impeller (figure 1), can be used to restrict cooling flow of bearing. But over large of the gap may lead to excessive loss of pump efficiency, while, if the gap is too small, insufficient cooling flow will be caused for bearing, which may further trigger the risk of overheat of hydrogen peroxide cooling flow. Hence, the gap value is one of the important design parameters for hydrogen peroxide electric pump.

For this reason, the paper explores the effect of sealing ring gap values on temperature rising of bearing cooling flow and pump efficiency loss, and reveals influence law of the gap for hydrogen peroxide pump performance, to obtain rational value for sealing ring gap.

2. Mechanism analysis

2.1. Structure demonstration

Pump structure is shown as figure 1, radial sealing ring is adopted at both inlet and bottom of impeller to limit leakage, which is also a common design structure; however, there are few research related to sealing ring gap at home and abroad [17].Angular contact ball bearing is selected and installed at bottom region of impeller, high pressure hydrogen peroxide liquid at outlet of impeller flow into bearing chamber after being throttled by sealing ring gap δ (at the bottom of impeller) . After bearing cooling (when flowing through bearing ball gap), it flows into low-pressure chamber (in front of chamber) through Return pipeline, to form closed loop.

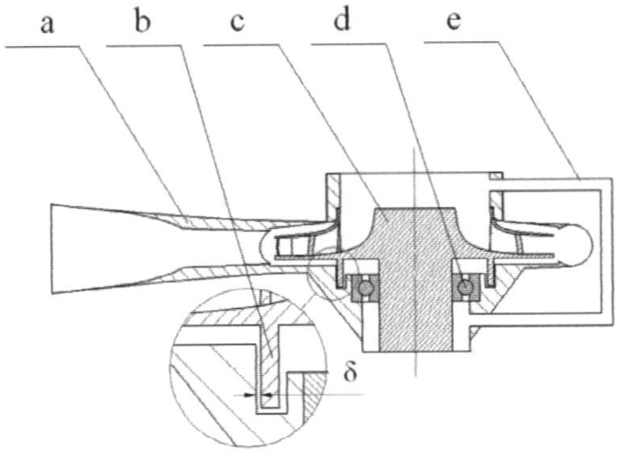

a.Volute casing b.sealing ring c. Impeller d. Bearing e. Return pipeline
Figure 1. Structure diagram of hydrogen peroxide pump.

2.2. Mathematical model

As this structure belongs to multi-disciplinary system, if we want to gain temperature rise of bearing, joint solution needs to be conducted for cooling flow model, impeller axial force balance model and bearing power-heat conversion model etc,.

2.2.1. Cooling flow model. Coolant flow of hydrogen peroxide is subject to mass conservation law, the major throttle site in flow channel are sealing ring gap and return pline. And ring gap flow model is built according to ring-slot flow formula [18]

$$q = \pi d_3 \delta^3 (P_2 - P_3)/12\mu l_1 \quad (1)$$

While, Return pipeline flow model is established based upon equation of pipeline pressure loss [19]

$$q = d_4^4 (P_3 - P_1)/8 v l_2\, k \quad (2)$$

$$Re > 2000,\ k = 6.8 \left(\frac{q}{v d_4}\right)^{4/3} \quad (3)$$

$$Re \leq 2000, k = 1 \tag{4}$$

Where, q is cooling flow, δ is sealing ring gap (at the bottom of impeller), P_1 indicates pump inlet pressure, P_2 means impeller outlet pressure, P_3 is pressure of bearing chamber, d_1 indicates sealing ring diameter at top of impeller, d_2, d_3 and d_4 indicates impeller diameter, sealing ring (the bottom of impeller) diameter, and Return pipeline diameter respectively, l_1 and l_2 are separately represent length of sealing ring gap (the bottom of impeller) and Return pipeline, besides, v is kinematic viscosity of hydrogen peroxide, μ is dynamic viscosity of hydrogen peroxide, and k is correction factor.

2.2.2. One-dimensional steady-state axial force balance model of impeller. According to axial pressure distribution on both sides, axial force balance model of impeller is set up to calculate axial load of bearing.

$$\frac{\pi p_2\ (d_1^2 - d_3^2)}{4} + \frac{\pi d_3^2 p_3}{4} - \frac{\pi d_1^2 p_1}{4} - F_Z + F_0 = 0 \tag{5}$$

Where, Fz is axial load of bearing, and Fo is pre-tightening force.

2.2.3. Bearing power-heat conversion model. As power consumption and heating value are decided by frictional torque of rolling bearing, bearing torque equation can be built with common method[19], and frictional heat of bearing can be calculated accordingly.

$$N = 0.1047Mn \tag{6}$$

$$M_1 = f_1 F_Z d_m \tag{7}$$

$$M_0 = f_0 (vn)^{2/3} d_m^3 \times 10^7 \tag{8}$$

f_0 *equals to 6.6 with reference to* [19]

f_1 *equals to* 7.37×10^{-4} *with reference to* [19]

Where, N is frictional heat of bearing, M is total frictional torque of bearing, M_0 and M_1 indicates lubricating oil torque and external load torque of bearing respectively, while, dm is average diameter of inner and outer ring, f_0 indicates bearing structure and lubrication coefficient, f_1 represents bearing load coefficient , and n is rotating speed.

2.2.4. Transfer model. According to energy conservation law, cooling heat flow within cooling channel (at steady state) is balanced with frictional heat-power of bearing.

$$Cq\Delta T - N = 0 \tag{9}$$

Where, ΔT indicates temperature rising of hydrogen peroxide coolant, C is specific heat of hydrogen peroxide.

2.2.5. Model related to effect of sealing ring gap on pump efficiency loss. This equation example only analyzes the effect of sealing ring gap on pump efficiency loss, which mainly consists of two parts, namely, the influence of gap leakage on pump volume efficiency loss, and frictional efficiency loss caused by bearing load variation.

$$\lambda_1 = q/Q \tag{10}$$

$$\lambda_2 = N/Q\ (P_2 - P_1) \tag{11}$$

$$\lambda = \lambda_1 + \lambda_2 - \lambda_1\lambda_2 \tag{12}$$

Where, Q is total flow at pump outlet, λ is total efficiency loss induced by gap, λ_1 indicates volume efficiency loss, and λ_2 indicates frictional efficiency loss of bearing.

2.3. Analysis of temperature rising and efficiency loss effect

Numerical solution is carried out for δ within 0.01mm to 0.1mm scale by combing above mentioned mathematical model with pump performance and structure parameter in this example, to gain influence curve of δ on bearing cooling flow q, temperature rising ΔT,bearing load Fz and pump efficiency loss λ.

Table 1. pump structure and performance parameter

N (rpm)	Q (kg/s)	P_1 (MPa)	P_2 (MPa)	P_3 (MPa)	d_1 (mm)	d_2 (mm)	d_3 (mm)
35000	4	0.5	7	5	28	52	28
d_4 (mm)	dm (mm)	l_1 (mm)	l_2 (mm)	V (mm^{2}/s)	M (pa·s)	F_0 (N)	T_0 (K)
3	18	12	400	1.14	0.00158	100	300

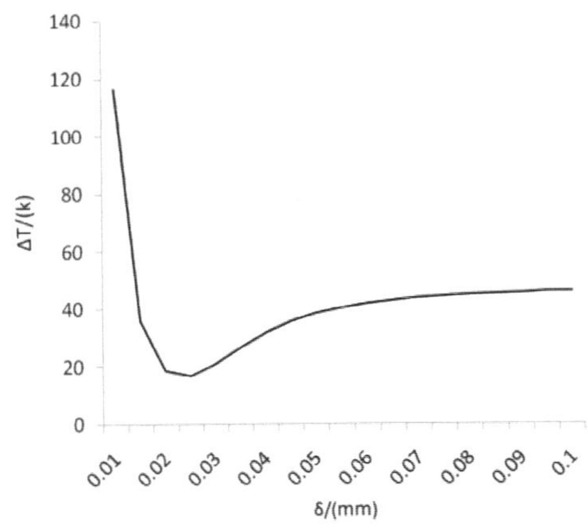

Figure 2. Influence curve of gap on coolant temperature rising.

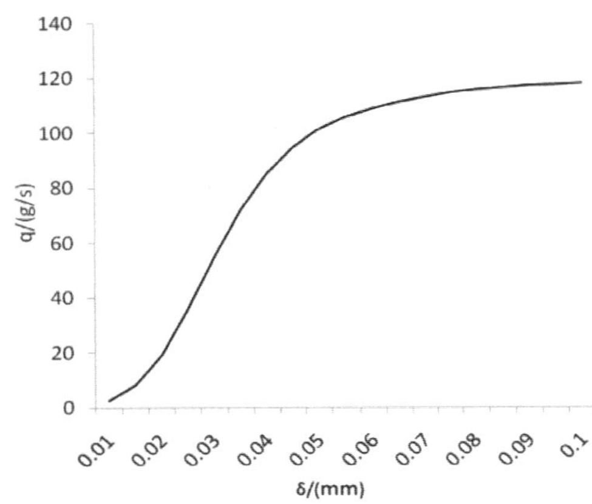

Figure 3. Influence curve of gap on cooling flow.

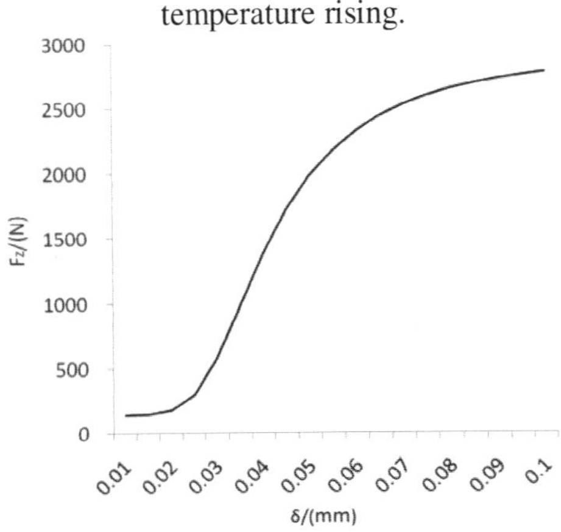

Figure 4. Influence curve of gap on bearing load.

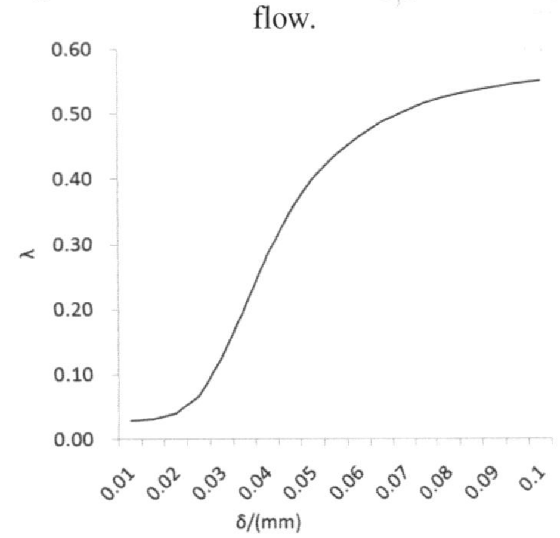

Figure 5. Influence of gap on efficiency loss.

It can be seen from figure 2 that when δ increases from 0.01mm to 0.02mm, temperature rising within cooling channel promptly decreases from 116K to 18K, and keeps at trough zone ΔT≤20K in 0.02 to 0.03mm section, afterwards, it slowly rises to 45K and levels off. Cause of such formation: gap throttling effect is prominent when δ=0.01mm, axial load of bearing is smaller, frictional torque is mainly lubricating oil moment with high rotating speed, but cooling flow is only 2.3g/s (figure 3), which cannot make valid cooling to bearing, and further lead to temperature rising of oil lies in the peak; when gap δ increases to the area of 0.02mm to 0.03mm, bearing load has little change in this zone, however, cooling flow increases sharply to over 20g/s, and coolant temperature rising lies in the trough; afterwards, δ continues to increase, pressure at bearing chamber goes up, which leads to axial load F_Z rises greatly (figure 4), and accelerates bearing frictional heating, so, cooling temperature goes up again till F_Z in steady state, and ΔT keeps at about 45K.

From figure 5, we may know that gap throttling effect is obvious when $\delta \leq 0.025$mm, and growing rate of λ is slow; when $\delta > 0.025$mm, influenced by the increase of bearing load, efficiency loss λ rapidly goes up to 0.5 and tends to be stable after Fz being in steady state.

To sum up, if sealing ring gap is in the range of 0.02 to 0.03mm, coolant temperature rising within bearing can be ensured<20K(lower range), meanwhile, power loss influenced by gap is less than 12%. So that's the reasonable range we get.

3. CFD simulation verification

Is conducted for hydrogen peroxide pump in this example by adopting fluid finite element analysis software pumplinx, to verify the accuracy of numerical analysis result of one-dimensional model.

3.1. Modeling and meshing

In this case, UG is utilized to integrally model centrifugal pump from inlet section to outlet of volute casing, and to simply the bearing portion; besides, binary tree algorithm based cartesian grid is employed to divide impeller, volute casing, inlet, bearing chamber and Return pipeline, and the total meshing quantity is 205,343 . For sealing ring gap (research focus) ,pumplinx gap mesh tool is used to make encryption meshing, and quantity of encrypted mesh is 300,000.

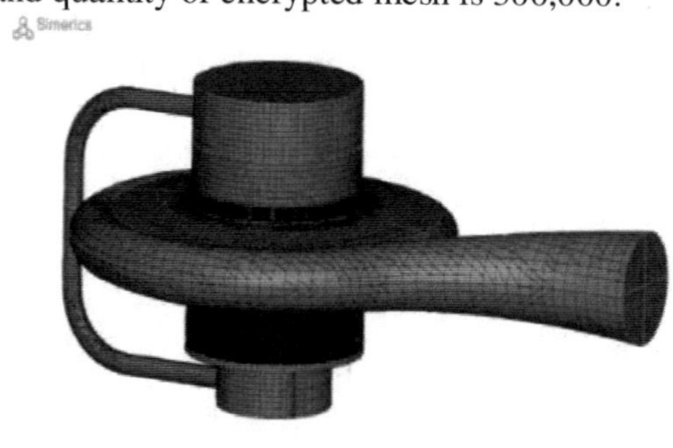

Figure 6. Centrifugal pump meshing.

Unsteady simulation method is adopted in this example, what's more, standard k-ε turbulent model is selected, the mean residual is less than 10^{-3}, upwind scheme is utilized for pressure term in control equation, and second order upwind is employed for speed term. For boundary condition, pump inlet is set as total pressure inlet, and volute outlet is set as outlet of a given flow; in heat transfer model, heat flow values for inside and outside bearing slide are offered according to model computing results, the given temperature is 300K at pump inlet, volute outlet is set as outlet of heat flow, and all of other walls are insulated boundary.

3.2. Simulation results and analysis

Simulation computation is done separately for δ=0.01mm,0.02mm,0.03mm,0.05mm ,and 0.07mm, the result of temperature field for bearing chamber and cooling channel is shown in figure 8-12.

CFD simulation result (in figure 7) indicates that variation trend of temperature in cooling channel is consistent with numerical results, deviation range between ΔT simulation data and numerical calculation is within 3.4%-11.3% , verified that one-dimensional calculation is relatively accurate, and can be used to rapidly design sealing ring gap.

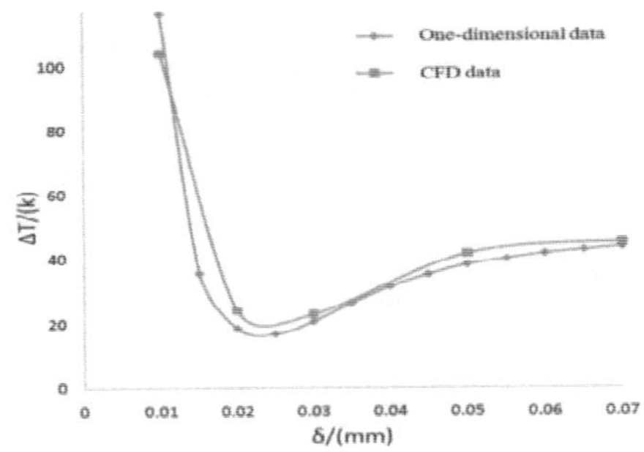

Figure 7. Contrast of ΔT simulation and numeric curve.

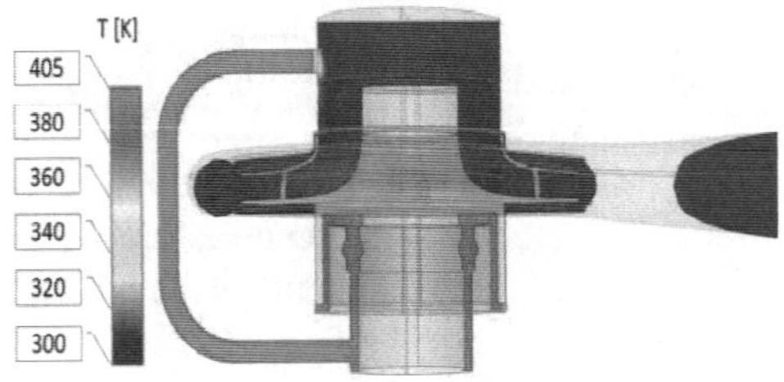

Figure 8. δ=0.01mm temperature rising of pump section.

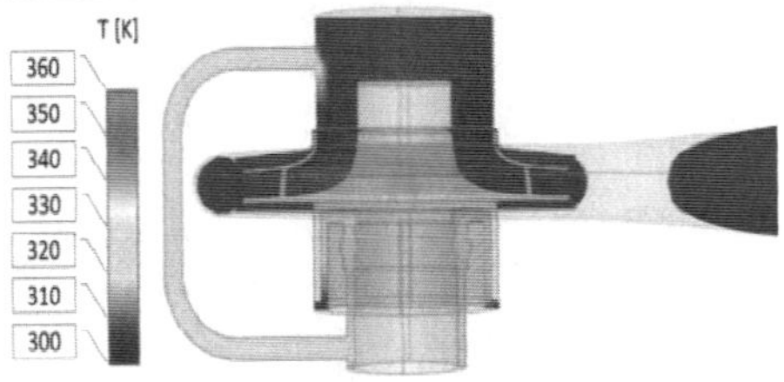

Figure 9. δ=0.02mm temperature rising of pump section.

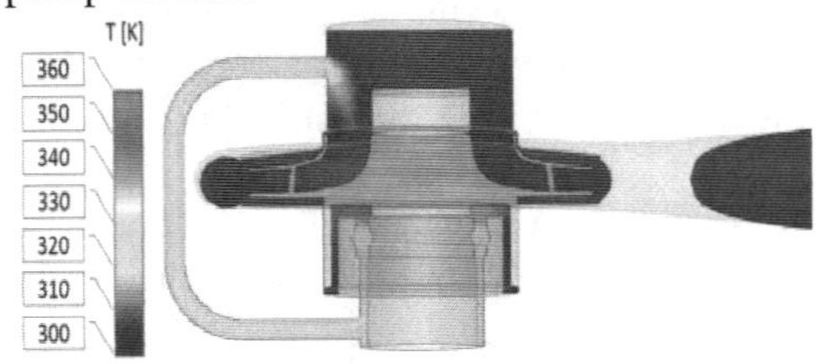

Figure 10. δ=0.03mm temperature rising of pump section.

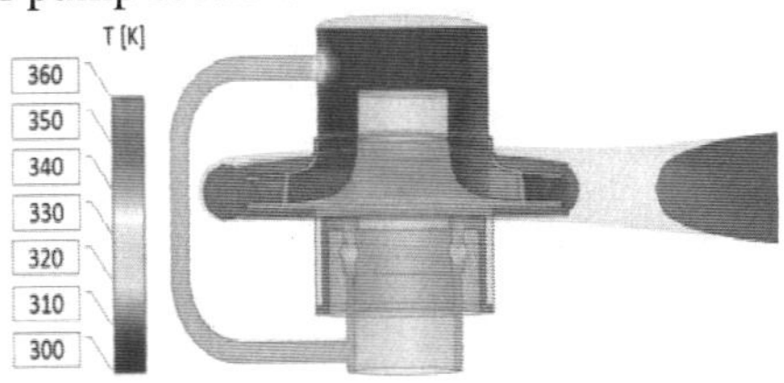

Figure 11. δ=0.05mm temperature rising of pump section.

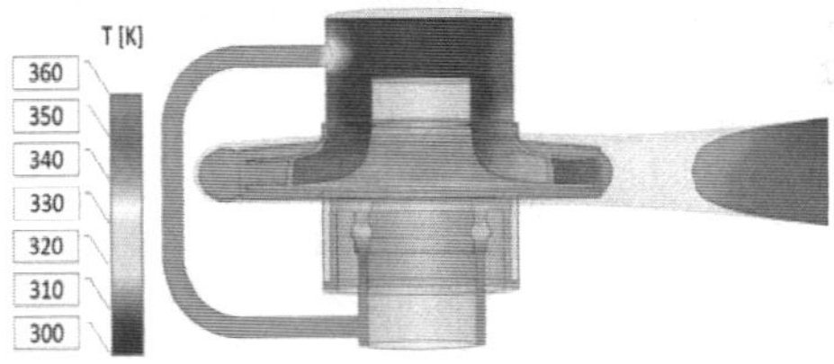

Figure 12. δ=0.07mm temperature rising of pump section.

4. Conclusion

Influence law of sealing ring gap on coolant temperature rising and efficiency loss is obtained by establishing flow model of cooling channel for hydrogen peroxide electric pump, heat transfer model, and bearing power-heat conversion model, and the rational gap size, which both hydrogen peroxide safefy and pump efficiency are taken into account, is confirmed. Accuracy of calculation model is verified through CFD simulation computations. This paper provides a feasible plan for rapidly designing critical gap size of hydrogen peroxide electric pump, and in the future, model parameter can be further optimized through pump test and verification.

References

[1] Rocket Lab's Electron Rocket Reaches Space, Satellite Today, May 29, (2017)
[2] Morring,Frank,Jr., Norris,Guy, Electric rocket.(Rocket Lab), Aviation Week & Space Technology, April 27, (2015)
[3] Yang Liu, Jian-gang Yang, Ben-shuai Fu, Guo-qiang He, Pei-jin LIU,System Desigh of Elcctric Pump Feed Liquid Propellant Rocket Engine,7th CSA/IAA Conference, 2017.11
[4] Yang Liu,Yuan-bo He,Jian-gang Yang,Chun-bo Hu, Guo-qiang He, Pei-jin Liu,Xiao-fei Zhu, Concept analysis of electric pump pressurized liquid rocket engine,The thirty-eighth technology exchange meeting of Aerospace propulsion Technology Information Society ,2017.8, 352-358.
[5] Jian-gang Yang,Yang Liu,Yuan-bo He, et al., Quality sensitivity analysis of electric pump pressurized liquid rocket engine feed system [C]. The thirty-eighth technology exchange meeting of Aerospace propulsion Technology Information Society, 2017.8, 352-358.

[6] Yang Liu, Ben-shuai Fu, Jian-gang Yang, Yuan-bo He, Pei-jin Liu, Analysis of electric pump pressurized liquid propellant rocket engine system, The 5th Manned Space Conference, 2018.10

[7] Musker A,Highly stabilized hydrogen peroxide as a rocket propellant., AIAA034619,(2003)

[8] Butler Kathleen N.AR2-3engine refurbishment and gas generator testing , AIAA992738,(1999)

[9] V. I. Arkhangelsky. History of development of hydrogen peroxide Lpreatnpo energomash, IAC(2005)

[10] G.Lin, Q.Ling, F.Li,A Study of thrust chamber technology using hydrogen peroxide, Journal of Rocket Propulsion, (2005)

[11] W. Wang, H.Chen, W. Chen, Y.Du, Development of a peroxide/kerosene turbopump for afterburning cycle for advanced upper stage engine, Journal of Rocket Propulsion ,4 (2016)

[12] Park, Daejong, Jang, et al. Preliminary Test of Turbopump Using Radial Turbine Driven by Hydrogen Peroxide Gas Generator,AIAA (2013)

[13] J.C.whitehead, M.D.Dittman, A.G.ledebuhr. Progress toward Hydrogen peroxide Micropilsion, AIAA, (1999)

[14] M.V.entura,P.Mullens,TheUse of Hydrogen peroxide for Propulsion and power, AIAA,(1999)

[15] D.Sengupta, S.Mazumder, J.V.Cole, et al. Controlling non-catalytic decomposition of high concentration hydrogen peroxide, ADA-426795. Springfield. NTIS.

[16] Z.J.Wu, G.Zhou,Explosion Causation Analysis of High Concentration Hydrogen Peroxide in Unsealed System, Chinese Journal Of Explosives& Propellants, 4(2006)

[17] B.Gao, Z.Wang, Yang L, et al., Effect of wear-ring clearance on performance and flow characteristics of centrifugal pump, Journal of Drainage and Irrigation Machinery Engineering ,(2017)

[18] Aero Jet engines Automatic control Handbook， National Defense Industry Press,6(2005)
Wang Zhenhua,Bearing Practical Handbook, shanghai scientific and technological literature press

[19] Wang Zhenhua,Bearing Practical Handbook, shanghai scientific and technological literature press，(1996)